P9-BHT-408

REMOTE-SENSING APPLICATIONS FOR MINERAL EXPLORATION

REMOTE-SENSING APPLICATIONS FOR MINERAL EXPLORATION

Edited by

William L. Smith

DH&R Dowden, Hutchinson & Ross, Inc.
STROUDSBURG, PENNSYLVANIA

Library of Congress Catalog Card Number: 76-14805
ISBN: 0-87933-230-1

79 78 77 1 2 3 4 5
Manufactured in the United States of America.

LIBRARY OF CONGRESS CATALOGING IN PUBLICATION DATA
Main entry under title:
Remote-sensing applications for mineral exploration
Includes index.
1. Prospecting—Remote sensing. I. Smith, William L.
TN270.R46 622'.1 76-14805
ISBN 0-87933-230-1

Exclusive distributor: **Halsted Press**
A Division of John Wiley & Sons, Inc.
ISBN: 0-470-15154-4

PREFACE

The purpose of this book is not to provide a textbook or reference handbook on remote-sensing capabilities and applications, but to take a broad look at the early returns from a new technology as they relate to mineral exploration and foreseeable problems in mineral resources management. It is an attempt to synthesize new concepts and capabilities that have been gained largely in the past few years since NASA's Earth Resources Technology Satellite (LANDSAT) was launched into orbit and started delivering high-quality, high-resolution multispectral images of the earth, providing the basis for a vast new source of geological information. The primary benefit is clearly the synoptic scale of LANDSAT data products, which permits the geologist to think on a broader conceptual scale as to the relationship of mineral deposits to a wider structural and lithologic format.

Two diametric points of view regularly confront the writer when dealing with mineral economists and exploration geologists. The first is that current mineral or energy shortages will persist or worsen due to geological, demographic, and other restraints, which cannot be solved by applied science. The second point of view is that new technology in exploration, metallurgy, the materials sciences, and particularly in remote sensing will soon solve our resource problems if not provide the key to the treasure house. Between the warnings of alarmists and the assurance of the program enthusiasts lies room for an appeal to realism.

Let us take a dispassionate look at our current mineral supply picture: critical minerals, fossil fuels, nuclear resources, and geothermal sources. Let us consider foreseeable mineral and energy resource problems. Can we achieve an assured supply of the resources vital to industrial civilization? What are the benefits of accelerated discovery? Let us take a hard look at what actual information can be provided from LANDSAT technology or possibly from future earth resources remote-sensing missions. We shall consider the differing views of some of the foremost qualified investigators of the applications of remote-sensing data to mineral resource and management problems. Finally, we shall consider remote-sensing applications in the developing areas of the world, a possible partial solution to industrial demands on finite ores and fuels, an unknown reserve for industry—while other technologies address the problems of geothermal and solar energy, nuclear breeder reactors, substitute materials, and new methods of extractive metallurgy and benefication of currently submarginal deposits. A successful application of remote sensing to mineral exploration and the resulting accelerated discovery of resources in one sense would borrow from the future. But this new information if properly applied at decision-making levels might aid in locating enough currently economic reserves to significantly shorten the critical interim period in which population requirements meet or exceed predicted resource availability.

Let us not oversell space technology. Satellite remote sensing is not necessarily expected to discover mineral deposits per se, although it can identify promising areas for exploration. Remote sensing may only assist in locating additional indicators of our unrenewable resources, and perhaps even contribute to our rapidly exhausting the supply. However, this, the latest of the technologies that may contribute to keeping resource availability abreast of industrial demands, arrives at a time when human requirements are catching up to reserves and when science has not yet found the technical answers to many future needs. Science will find these answers, and satellite remote sensing can be of great assistance in this interval if the data returned are properly used by geologists, industry, and our planning agencies.

In 1967, the late William T. Pecora, then Director of the U.S. Geological Survey, said in an address to the American Congress on Surveying and Mapping:

> One of our fundamental problems today—one that is basic to many of the seemingly more immediate and pressing problems, is the task of supplying the raw material that our modern society requires. This demand can be met in the future only if we locate and exploit new resources and devise more efficient ways of using those that we already know about. If our ability to find and efficiently utilize resources does not accelerate, and accelerate rapidly, the industrial civilization we now enjoy will crumble within a few decades, for the economic status of any nation is almost always a direct function of the use it makes of available natural resources.*

These words apply not only to geologists and mineral economists but also to the highest levels of government responsibility. Even though space-age technology will possibly provide us with additional resources, the increasing needs of industry and the general rise in the world standard of living preclude our exploitation of such reserves on the dictates of the market alone. Long-term needs, conservation, consideration of environmental quality, and stockpiling against emergency contingencies must be elements in the modeling and interpretation of resource data by those areas of government that formulate policy for legislation and in the functional areas in which are made the decisions and actions required for resource management. Much of the technical capability needed to guide mineral resource management may be derived from satellite experiment data as quantitative input for making models for extracting predictive information.

LANDSAT represents an entirely new technology. It is more than a high-altitude extension of aerial photography. It is the precursor of higher orders of natural resource remote sensing, and with optimum management of the information derived it is the forerunner of a major source of vital data for the next several critical decades. The geological applications of LANDSAT data have only been studied for three years, and the practical and economic applications are only now being developed. To date, the geological studies have been largely observational, and the degree to which new mineral discoveries have been aided by LANDSAT is controversial. However, LANDSAT has been able to relate the position of known deposits to previously unknown structural lineaments, and is able to identify various surficial indicators of the possible loci of obscured mineralization and better define the controls of mineralization patterns. The more important indicators are based on observable symmetry or geometry of structural features, apparent geological anomalies, and regional trends. Other indicators include relevant mineralogical or geochemical patterns, changes in vegetation cover, and other secondary guides. The methodology is just developing for effective use of space imagery for locating such indicators of mineral occurrences

*William T. Pecora, Surveying the Earth's Resources from Space, 27th meeting, American Congress on Surveying and Mapping, Washington, D.C., Mar. 10, 1967.

and for understanding the tectonic framework of mineralized areas.

The mineral picture is one of constant depletion of current ores and fuels and exploration for new ones. Accelerated discovery through remote sensing could ease many foreseeable problems and better enable industry to program extended operations; when user agencies are able to undertake longer-range planning, more confident management will be possible in those areas of the mineral economy where an assured supply is critical.

The 1952 Paley Commission suggested that new technology in exploration might save us from running out of many major mineral resources. New developments in geophysics and geochemistry in part answered that need. Some 20 years later it was again apparent that we would face shortages of fuels and critical minerals, and the Secretary of the Interior warned that we required new technology to locate new supplies. At the March 1974 meeting of the American Society of Photogrammetry, John T. Awald, Denver Mineral Exploration Corporation, replied to Secretary Morton's comment with the observation, "It is with divine timeliness that remote sensing has arrived as a powerful new aid in the search for mineral deposits."*

WILLIAM LEE SMITH

*J. T. Awald, A Technology of Renovate the Search for New Mineral Deposits, 40th annual meeting of the American Society of Photogrammetry, St. Louis, Mo., Mar. 1974.

ACKNOWLEDGMENTS

The editor wishes to acknowledge his gratitude to the several distinguished co-authors of this book, who provide its real substance and who were selected as the foremost investigators of the applications of remote sensing to mineral exploration and resource management. There was unavoidable redundancy among the various contributed chapters as received, due to the limited amount of significant findings available for reference. In a few cases this redundancy has been retained inasmuch as the different writers approach our subject from different standpoints, and a variety of interpretations provide greater insight into the elements of this developing science.

Robert A. Summers of the Office of Laboratory and Field Center Coordination, U.S. Energy Research and Development Administration, Washington, D.C., is preeminently suited to provide the Introduction. He has had extensive experience with earth observations from space since the inception of the LANDSAT program. At System Planning Corporation he was responsible for several comprehensive program analysis and planning efforts for the LANDSAT Earth Observations Program and for economic evaluations of LANDSAT data utilization for the U.S. Department of Interior and the Agency for International Development. The latter work involved LANDSAT data application projects in Africa, Asia, and South America and evolved into a group of pilot project grants in the same areas. Summers was formerly Program Manager, Advanced Programs and Technology, for the Earth Observations Program, NASA Office of Applications, where he organized and directed all advanced studies, integrated program planning, and related technology development for NASA earth-looking missions.

Bernardo F. Grossling is a Research Geophysicist and Advisor to the Director of the U.S. Geological Survey, Reston, Virginia. He is also an authority on the economics of national security. He was a factor in the establishment of Chile's petroleum industry as Chief Geophysicist and Exploration Advisor to the Empresa Nacional del Petroleo. He was a Research Geophysicist with Standard Oil of California, and has been Chief of the Engineering Division and Technical Advisor to the Inter-American Development bank. He is particularly qualified to discuss the economic justification for the gathering of raw remote-sensor data and how to interpret it so that it becomes useful for private and public users (Chapter 3).

Enrico P. Mercanti is Assistant to the Chief, Office of Mission Utilization, NASA Goddard Space Flight Center, (NASA/GSFC). Greenbelt, Maryland. He has the responsibility of coordinating LANDSAT data user affairs. Previously, at NASA headquarters he was coordinator of several hundred LANDSAT and SKYLAB investigations during the initial investigator selection process. He is largely responsible for compiling and editing the *Proceedings* of the second and third Earth Resources Technology Satellite-1 symposia held by NASA/GSFC in March and December 1973. Chapter 4 provides a valuable overview of LANDSAT applications, not restricted to sensing for minerals alone.

Chapter 6 is co-authored by Robert S. Houston and Ronald W. Marrs, Department of Geology, University of Wyoming, Laramie, and by Nicholas M. Short and Paul D. Lowman, Jr., Earth Resources Branch and Geophysical Branch, NASA/GSFC, Greenbelt, Maryland. The investigators at the Geology Department of the University of Wyoming, with Short of NASA as co-investigator, undertook an analysis of the LANDSAT imagery of Wyoming as it is related to natural resources, emphasizing geology (tectonic framework, lineaments, distribution of fold structure), and a LANDSAT study of the geology of an ice-free test site in Antarctica. Chapter 6 is based in part on "Earth Observations from Space: Outlook for the Geological Sciences," N. M. Short and P. D. Lowman, Jr., NASA/GSFC, X-650-73-316, October 1973. These co-authors provided the first comprehensive evaluation of the geological applications of remote sensing from orbital altitudes. Lowman's investigations of geological applications go back to the analysis of terrain photography from early sounding rockets and Gemini spacecraft, and include an interpretation of the Apollo 9 multispectral photography. Short, who is also author of Chapter 7, which is concerned with exploration for fossil and nuclear fuels from space, has been NASA's primary spokesman for geological applications through the LANDSAT Symposia Discipline Summary Reports and through his analysis of the significant results and projected applications obtained through LANDSAT-1 principal investigator interviews.

Chapter 8 is a timely consideration of the use of remote sensing and associated techniques in the area of energy development. John E. Johnston was formerly Liaison Officer for the U.S. Geological Survey's Earth Resources Observation Systems Program (EROS); he coordinated private industry activity with EROS/LANDSAT program requirements and formulated interagency operation plans related to data management and applications studies. He has undertaken extensive research on the geology of fossil fuels and nuclear energy resources as well as on remote-sensing technology applicable to resource and environmental conservation. At present he is Staff Geologist, Branch of Coal Research, U.S. Geological Survey National Center, Reston, Virginia. He is also assistant editor of the *Manual of Remote Sensing,* American Society of Photogrammetry. The chapter co-author, Frank J. Janza, is Executive Director of the Institute for Technology and Society, Sacramento State College, California. His present position follows a significant career in the development of remote-sensing techniques and their applications for addressing societal problems of such accelerating urgency as energy and mineral resources. He is also editor of the first volume of the *Manual of Remote Sensing,* American Society of Photogrammetry.

R. Michael Hord of the Institute for Advanced Computation, Falls Church, Virginia, and formerly of the Earth Satellite Corporation, Inc. (EarthSat), has specialized in computer-oriented techniques for image processing and pattern recognition research. He is responsible for the conceptualization and structuring of a variety of analytical problems and is uniquely qualified in the areas of photogrammetry, statistical analysis, simulation, and system optimization. Chapter 9, which is concerned with digital image enhancement of LANDSAT multispectral scanner data, includes some exceptionally good computer products pertinent to mineral exploration. Material of this quality has contributed to the leadership position of EarthSat in the area of digital image processing of earth resource data. Hord has directed software development specializing in user applications of digital image processing. He is Chairman of the Automatic

Imagery Pattern Recognition Committee of the Electronic Industries Association, where he has directed standardization activities for the digital image processing community. Prior to joining EarthSat, Hord, as a Senior Physicist with Itek Corporation, was involved with the design of state-of-the-art optical apparatus, including mapping cameras and spaceborne, large-aperture imaging systems, and with the development of techniques for automatic photointerpretation.

Chapter 10 is authored by Robert K. Vincent, President of Geospectra Corporation, Ann Arbor, Michigan, where he conducts data processing and image interpretation in the area of applied geological remote sensing with satellite and aircraft data. Vincent was formerly associated with ERIM, where he developed the spectral ratioing technique that for the first time made it possible to image a single class of minerals. He is best known for his ratio images of iron oxides, which may prove to be useful as guides to other mineralization. Vincent discusses geochemical mapping by spectral methods and provides some convincing examples of the potentialities of this new technique.

Chapter 12 is a contribution by Carlos E. Brockmann, Alvaro Fernandez, Raúl Ballón, and Hernán Claure of the Programa del Satelite Technologico de Recursos Naturales, Servicio Geologico de Bolivia (GEOBOL). Under the directorship of Brockmann, the LANDSAT-Bolivia program has extracted the maximum possible geological information from LANDSAT products, including the first thorough natural resource inventory made from a single frame. Other investigations have defined obscured structures in petroleum regions and identified lineaments in mineralized areas. Several structural sections have been revised, and extensive photomapping has been done at scales of 1:1,000,000 and 1:250,000. The writers are uniquely experienced in the use of LANDSAT products for locating guides for the exploration for mineral deposits.

Fernando de Mendonça is General Director of the Instituto de Pesquisas Espacias (INPE), Conselho Nacional de Pesquisas, Sao Jose dos Campos, Sao Paulo, Brazil. INPE is in the forefront of utilizing LANDSAT technology for mapping remote areas, defining geological structure, and providing a data base for LANDSAT surveys pertinent to the development of vast relatively undeveloped regions. Under Mendonça's direction, Brazil has established a LANDSAT receiving station at Cuiaba and processing facilities at Cauchera Paulista. In 1974, INPE hosted a meeting of the Committee on Space Research (COSPAR) and a workshop on space applications of interest to developing countries. Chapter 13, which is co-authored by Aderbal C. Corréa, Fernando de Mendonça and Chan C. Liu, concerns applications of LANDSAT imagery to geological studies in Brazil and is included as uniquely significant to the theme of this book.

Ravi D. Sharma is Head of the Planning and Training Division of the Indian National Remote Sensing Agency, Secunderabad (NRSA). Sharma was previously a consultant to the NASA Office of Manned Space Flight at Bellcomm, Inc., an employee of the Environmental Research Institute of Michigan (ERIM), where he undertook remote-sensing studies for the Infrared and Optics Division, a consultant to the Space Systems and Applications Division of System Planning Corporation (SPC), where he conducted a study of the earth resource survey applications of the Space Shuttle sortie mode for NASA, and Scientific Secretary to the Chairman of the Indian Space Research Organization (ISRO). Chapter 14 demonstrates the applications of geological remote sensing to other disciplinary areas including land-use analysis and drainage mapping. Chapter 14 is co-authored by D. N. Raina, Geology Division, Indian Photointerpretation Institute, Dehra Dun, and by Mukhtar S. Dhanju, Remote Sensing and Meteorology Applications Division, Space Applications Center, ISRO, Ahmedabad.

The concluding chapter is authored by Sid Verner, Office of Monitoring, U.S. Environmental Projection Agency, Washington, D.C. Verner was formerly Senior Scientist at the Washington office of the Illinois Institute of Technology Research Institute, where as a NASA consultant he was responsible for several studies related to remote sensing and applications. As he notes in his chapter, the mineral industries tend to degrade environmental quality except where specific precautions are instituted. If, as we envision, remote sensing will contribute to new mineral ventures, it is the responsibility of the industries involved to limit such despoliation and to control the discharge of pollutants. Where this is done, industry will contribute to our economic and social progress and at the same time preserve the quality of the environment. Where this is not done, either the environment suffers from our enterprise or governments must monitor source emissions for compliance with the law. Remote sensing not only will assist in our search for mineral resources, but will aid efforts to preserve our natural heritage through monitoring its changes.

The editor also wishes to acknowledge the previous work in remote-sensing applications for mineral exploration that has been drawn on repeatedly in the preparation of this book, including the papers and summary reports of the first three LANDSAT symposia under the auspices of NASA/GSFC, the several International Symposia on Remote Sensing of Environment held at the University of Michigan, the Symposium on Management and Utilization of Remote Sensing Data and the various meetings of the American Society of Photogrammetry. "Earth Observations from Space: Outlook for the Geological Sciences," by N. M. Short and P. D. Lowman, Jr., NASA/GSFC, 1973, has been referenced repeatedly inasmuch as it was the initial source on satellite applications to geology.

Much of the editor's contributions have drawn upon his experience and reports to NASA, the Department of State, and the Department of Interior from his work at Battelle Memorial Institute, Bellcomm, Inc., and System Planning Corporation. (The editor is currently with the Environmental Research Institute of Michigan, Arlington, Virginia.) Much of the material on ERTS applications in developing countries is taken freely from his contributions to "An Economic Evaluation of the Utility of ERTS Data for Developing Countries" prepared for the Agency for International Development by the Environmental Research Institute of Michigan, System Planning Corporation, and Mathematica, Inc., 1974. He is indebted to former colleagues in the U.S. Geological Survey, the aerospace industries, and several mining and exploration companies for various assistance.

In the area of economic geology the editor has relied upon data and positions taken in "United States Mineral Resources," by D. A. Brobst and W. P. Pratt, U.S. Geological Survey Professional Paper 820, 1973, and material presented at hearings before the Joint Committee on Defense Production, 92nd Congress. He also has been influenced by positions taken in *Affluence in Jeopardy,* C. F. Park, Jr., 1968; "The Energy Outlook for the 1980's," W. N. Peach, Joint Economic Committee, U.S. Congress, 1973; and "Limits to Power Growth," E. S. Cheney, in the Geological Society of America publication *Geology,* June 1974. Geological terminology, classification of mineralization, and the concept of guides or indicators are those found in *Igneous Rocks and the Depths of the Earth,* by R. A. Daly, 1933; *Economic Mineral Deposits,* by A. M. Bateman, 1950; and *Mining Geology,* by H. E. McKinstry, 1948. Important sources of current information have been publications of NASA's Office of Applications and the reports of the Principal Investigators in the LANDSAT Program as released by the National Technical Information Service, U.S. Department of Commerce.

The following organizations have granted permission to quote specific material from their published works: the American Congress of Surveying and Mapping, the Society of Mining Engineers of the American Institute of Mining, Metallurgical and Petroleum Engineers, the Geological Society of America, the AirForce Association, the American Society for Photogrammetry, W. H. Freeman and Company, Publishers, and the Eastman Kodak Company.

The writer is indebted to the following for expert assistance: Dorsey Clement, Ed Clement, and Judy Kolos for graphics; Lin Gambatese for copy editing; Donna Knoerr for translations; Elaine Allen, Jean Bellany, Joyce Chin, Bettie Magee, Margaret McCormick, and Wendy Shinn for typing and preparation of the manuscript; and to Frances Atkiss for indexing.

CONTENTS

REMOTE-SENSING APPLICATIONS FOR MINERAL EXPLORATION

1 INTRODUCTION

Robert A. Summers

Robert A. Summers is currently on the staff of the Assistant Administrator for Field Operations, U.S. Energy Research and Development Administration, Washington, D.C. He was formerly Vice-President, System Planning Corporation.

In his monumental work *Ascent of Man (1)*, Bronowski states:

> I use the word "ascent" with a precise meaning. Man is distinguished from other animals by his imaginative gifts. He makes plans, inventions, new discoveries by putting different talents together, and his discoveries become more subtle and penetrating as he learns to combine his talents in more complex and intimate ways. So the great discoveries of different ages and different cultures in technique, in science, in the arts express in their progression a richer and more intricate conjunction of human faculties—an ascending trellis of his gifts. . . . In every age there is a turning point, a new way of seeing and asserting the coherence of the world.*

Bronowski attempts to relate the history of science as it is intertwined with the progress and development of mankind and the relationship to social and other issues. In viewing the ascent of man as a slow climb out of the briny waters onto the land, then standing erect, and then proceeding through processes of conceptualization, invention, and development, I would suggest that an important step has been the creation by man of the ability to leave his planet and then to observe his planet from a significant distance, that is, earth orbit or beyond. Certainly, this is an accomplishment by man of perhaps equal significance to almost any other major accomplishment described by Bronowski.

Through the synthesis of chemical knowledge leading to propulsion technology, electronic knowledge leading to control systems, and optical systems technology, it has been possible for man to build orbital vehicles that can view the earth in a wide variety of spacial, spectral, and temporal scales and resolutions. It is possible to keep the earth under nearly continuous examination in terms of its reflectance of solar light at all wavelengths and its emission of heat energy through thermal infrared measurements. It is also possible to examine the earth in terms of its emission of radiation in other parts of the electromagnetic spectrum, including microwave radiation, long wavelength radiation, as well as the reflectance properties of the Earth in the radio and microwave range using man-made emitters. This kind of periodic and/or continuous examination of the earth in its entirety as a

*Reprinted by permission of the publisher, Little, Brown and Company, Boston.

planet is certainly a significant benchmark in the ascent of man.

BACKGROUND

Periodic observation of the earth's surface from near-polar orbiting satellites was initiated on an experimental basis by TIROS 1 in 1960. The first operational system was ESSA-1 some six years later in 1966. Continuous observation of the earth was first demonstrated experimentally from geostationary orbit on ATS-1 in 1966; the first dedicated operational system, SMS-1, was obited in May 1974. These meteorological systems, whose principal purpose is cloud-cover imaging, are of generally low spatial resolution (~1 km) and limited to one spectral band in the visible and one in the thermal infrared region. With wide field-of-view systems in near-polar orbit, it is possible to view any portion of the earth twice a day. By particular selection of the orbital altitude and orbit plane inclination, a sun-synchronous orbit is established, which retains approximately the same local sun time at each observation. In this way, the parameters of viewing angle and solar illumination angle are held largely invariant on a day-to-day basis (but changing slowly through the seasons). The current satellite-based remote sensing of natural resources is, for the most part, an evolution from the meteorological satellite program technology. These evolutionary natural resource related systems are characterized by higher spatial resolution and higher spectral resolution, but at the expense of lower temporal resolution. The latter results from a desire to retain near-orthographic quality in the imagery through the use of narrow view angle optical systems (10 to 11°). As a result, these images can directly meet national mapping standard accuracies up to scales of 1:250,000 or better, and, by utilizing the basic digital data directly, can provide useful, map-like products up to scales of 1:64,000 and 1:30,000.

The first Earth Resources Technology Satellite, now known as LANDSAT, was launched in July 1972; although designed to last only one year, it has had more than three years of useful life. The LANDSAT-2 was launched in January 1975 and has also performed effectively. The primary sensor for these systems is a multispectral scanner producing, for each scene, four "images" of approximately 80-m spatial resolution in four bands of the visible and near-infrared spectrum: 0.5 to 0.6, 0.6 to 0.7, 0.7 to 0.8, and 0.8 to 1.1 μm. These data can be stored in on-board tape recorders or transmitted directly to ground. The basic data are a digital bit stream telemetered to three U.S. receiving stations, Greenbelt, Maryland, Goldstone, California, and Fairbanks, Alaska. This digital bit stream is recorded on high-density magnetic tape, which is processed into computer-compatible tape (CCT) and analog imagery at NASA-Greenbelt. The principal distribution source for these data is the EROS Data Center, Sioux Falls, South Dakota, operated by the U.S. Department of Interior.

Remote sensing from aircraft using cameras and scanners has been a well-established component of natural resource information gathering for many years. The current NASA satellite program does, however, provide some particular innovative features as follows *(2)*:

1. A spacecraft platform from which to obtain synoptic, repetitive viewing of the earth's surface in near-orthographic form and at essentially fixed viewing and solar illumination angles.
2. More precise quantitative multispectral measurements (from either spacecraft or aircraft) of electromagnetic radiation reflected and emitted from earth's surface features.
3. Digital computer-based techniques for automatically extracting information from remote-sensing data.
4. Decision-oriented resource and environmental models that can make effective use of remote-sensing data.

It seems clear, however, that any future operational system related to natural resource management functions will depend on a combination of spacecraft, aircraft, and ground systems. These modes are essentially complementary for both technical and economic reasons.

Although the technological basis for these satellite-based earth-observation systems derives from the imperatives of the space and defense arena, their major impact will be in the economic sectors of life on earth, including mineral exploration, agricultural yield prediction, rangeland management, water resource management, cartography, land-use planning, and environmental monitoring. Many agencies of the U.S. federal government are involved in determining how this new information source may improve the efficiency of their activities in these areas. Similarly, private-sector interests, especially mineral and petroleum exploration companies, are pursuing this course actively on their own. The U.S. federal government, through the National Aeronautics and Space Administration, has made this technology available to all nations of the world, with particular emphasis on United Nations members and developing nations. Many symposia and training sessions have been held before and during the operating periods of these earth-observation satellites. Results to date clearly indicate that applications of this information product of space technology will have a major impact on activities in the developed and developing world over the next few decades.

In all fairness, the world is just learning how to use this new information source and it is difficult to predict just how it will finally have its impact. Evidence to date, however, indicates a number of successful applications or potential applications, including, specifically, mineral exploration. Although the space segment has been developed and fielded by a single nation, there are a number of nations now involved and interested in the reception of data and the processing of it through their own national or regional ground stations. Operational stations already exist in Canada and Brazil, and agreements have been signed for stations in Italy, Iran, and Zaire. The development of ground stations and possibilities of international arrangements for both the ground and space segments of a future international system have been the subject of debate in the United Nations and, in particular, in the Outer Space Affairs Committee as well as its Technical and Legal subcommittees. Thus, international institutions for utilizing what is essentially global information are just emerging.

INFORMATION, DECISION MAKING, AND POTENTIAL ECONOMIC BENEFITS

To yield economic benefits, these remotely sensed data must be transformed into information in a format useful for decision making by resource managers. Whether this decision making relates to the mineral exploration process or crop yield prediction, the information has value insofar as it facilitates "better" decisions. The "better" decision is broadly defined as that for which there is a greater expected present value. To evaluate the consequences of decisions in terms of present values recognizes that time is a major dimension of the problem, and that it is the time distribution of all costs and benefits—using an appropriate discount rate—which will be a major factor in the decision process.

In looking at the potential economic applications of these data, it is clear that modeling is essential for this new source of data to be transformed into interpretations and predictions that can be utilized to effect decision making by resource managers, either on a national, regional, or international basis. A simplified model of an operational remote-sensing system is suggested in Fig. 1 *(4)*. The basic components shown are the observation systems (spacecraft, aircraft-based) that provide data, the earth-science-based models that produce information on the present or future status of a resource phenomenon, and the management decision models whereby management actions are generated (e.g., drill here or plant more corn there). It is suggested that the system is essentially a closed loop in which management actions as well as natural phenomena affect earth conditions. A more detailed view of such an earth-observation-system operational model is given in Fig. 2 *(4)*. It should be noted that meteorological data constitute major, first-order inputs to vegetation-related (e.g., crop yield) or water resource management problems. The necessary earth science and decision models are currently under development.

In looking at the potential of this new information source for both the developed and developing world, at least one study has suggested that the impact will be sooner and relatively greater in the developing

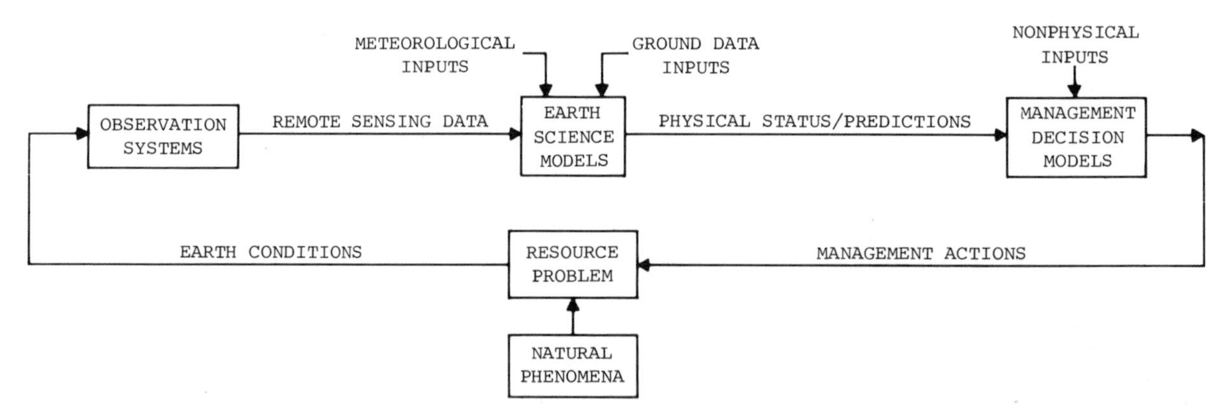

FIG. 1
Model of an operational remote-sensing system.

world, where even one-time coverage provides a new plateau of information beyond that which might have been available even in decades of ground-based activity *(3)*. Aircraft can do some of the job, but the processing of data is more complicated and expensive and such an operation is in many cases beyond the capability of nations in the developing world. In the developed world, on the other hand, the information base is considerable and detailed, so that it will require a great deal of development of sophisticated inferential models before this tool can have a similar impact as in the developing world. This comes about in part since we are just learning how to use these data; with what we have learned, we can take relatively large steps in the developing world, but in the fine tuning required for the developed world, we are just beginning to learn how to use these data. This, of course, impacts the basic issue of domestic program justification.

It is perhaps worthwhile to view the decisions that

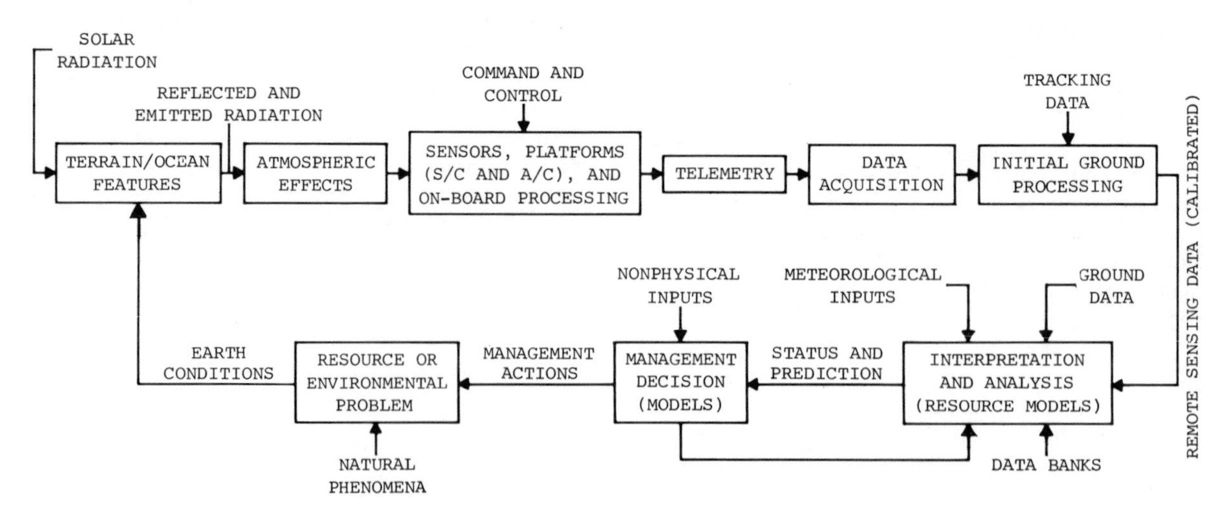

FIG. 2
Operational model of an earth-observation system.

may be impacted as being related to either static or dynamic processes. In this context, geological phenomena and mineral exploration are often viewed as static situations. However, even these phenomena are indeed dynamic, and it is only a question of time scale; the so-called static phenomena of geological and mineral exploration have characteristic time scales of thousands of centuries, whereas the usual dynamic phenomenon (e.g., crop growth, flooding) may be matters of hours, days, or weeks. Typical among the dynamic phenomena are agricultural yield prediction, range management, and water resource management. In the intermediate time scale, although much closer to the traditional dynamic case, is the land-use planning area, which is heavily cultural related and occupies an intermediate temporal position.

The question of the need for repetitive coverage for geology has been addressed by C. W. Mathews, former NASA Associate Administrator for Applications, in the FY 76 Budget Authorization Senate hearings *(5)* as follows:

> There has been some mention in the recent past that geology is quite static and that applications in this area, therefore, do not justify repetitive coverage. The point is that geology may not change, but its appearance does. Repetitive coverage does assist in geologic interpretations because of variations in sun angle effects over the seasons, snow cover enhancement, and changes in vegetation and moisture that are associated with different rock types and/or structural controls. It is estimated that at least 3 to 5 years, perhaps even 10 years, of seasonal imagery will be needed to show a full range of variations in vegetative cover, soil moisture, water in fractures, snow enhancements, and illumination conditions related to sun angle and atmospheric interference. Some dynamic processes, such as volcanism, glaciation, offshore sedimentation, and landslides, will continue to justify repetitive coverage.

Numerous attempts have been made to quantify the benefits and costs of a future operational remote-sensing program. It is possible to partition the impact into two parts: (1) that deriving from a one-time coverage of the world at the spatial resolution of LANDSAT, including perhaps four-time or seasonal coverage, and (2) the monitoring of dynamic situations, which may require observations at intervals of minutes, hours, or days, some of which may ultimately require geostationary platforms for continuous observation. The one-time, or even four-seasonal, coverage of the earth largely affects nonrenewable resources such as are related to geology and mineral exploration. Subsequent periodic monitoring of the earth is considered to be more closely related to the other dynamic applications enumerated above. If there exists any rationale for national sensitivities to the distribution of remote-sensing data, it seems to relate more directly to the one-time coverage. It is, in a sense, enigmatic, in that the one-time coverage leads to obvious statements related to cartography and mineral exploration; however, these are very difficult, if at all possible, to transform into credible potential benefit statements. On the other hand, dynamic phenomena require considerable sophistication in modeling; however, once the predictive and decision-making models are properly formulated, the calculation of benefits will probably be less uncertain than in the one-time coverage case.

SOME NATIONAL AND INTERNATIONAL ISSUES

A key issue, at least for the present decade or so, relates to the fact that the present nationally funded LANDSAT system is dependent on domestic U.S. budget justification. However, in the early stages of application, it appears that it is outside the United States, and particularly in the developing world, where significant tangible early benefits will be readily identifiable. It is therefore important to inquire how the system can be continued unless some allowance is made in the domestic analysis for international and developing world impact. Studies have also been started on possible international institutions or organizational alternatives for a future operational earth resources survey system (*12, 13*). There is indeed a continuum of possibilities, although the major basic ones are (1) a U.N.-affiliated agency, (2) a COMSAT/INTELSAT-type organization based on international treaty, (3) a loose consortium similar to the present NOAA model in meteorology, (4) a formalized consortium with a number of nations making contributions and bound to perform under treaty obligations, and (5) a private company. There are of course many variations of each of these organizational themes. For example, one can separate the financing, development, and operation of the systems. One can also conceive of selective pricing policies in the case of some kind of private or multinational private organization.

In examining these organizational alternatives, one key issue is how to pay for the system; this leads to the issue of pricing policy. Clearly, and particularly in view of the potential impact in developing countries, it is important that the pricing be such as to enable these nations to have ready access not only to the data but to the capabilities of employing it effectively in their national decision making. The latter suggests a need for multilateral and bilateral arrangements that can ensure this equitable utilization capability. Many related activities are now underway under the sponsorship of the U.S. Agency for International Development, the World Bank, FAO, and others. Other countries, such as Canada, Sweden, the United Kingdom, and the Netherlands, also have their own international bilateral technical assistance programs.

In view of the potential applications in the developing world, the issue of equity is a crucial one. It is useful first to define equity, and at least two definitions are significant in the present context. First, and perhaps somewhat more obvious, is the issue of equity as perceived by individual nations in the international arena; that is, as they might be represented in the United States and in the Outer Space Affairs Committee. Each nation, and particularly developing nations, wishes to be assured that it has access to all the available data on its country and the capability for interpreting and utilizing it in national decision making. However, a second and more subtle equity issue is essentially an internalized one in terms of the developing world: what is the impact of this new technology on the incomes and quality of life of the poorer people in the developing world? This has been treated in considerable detail by economists such as I. Adelman, and such work suggests that in some cases the introduction of technology, including perhaps technology leading to improved information on mineral resources or more accurate crop yields or rangeland animal-carrying capacity information, when applied in the present social and economic context of these developing countries, may well, for a protracted time, actually reduce the income of the lower 40 percent of the population *(11)*. It is suggested that such timely activities as land reform and massive educational programs could prevent or mitigate this undesirable effect. Thus, these economists suggest that, although the developing country's GNP may rise, the plight of the poorer segments of the population may indeed grow worse, at least for a certain period of time. On the other hand, the discovery and exportation of mineral resources, for example, may provide resources to the developing country, facilitating significant social and economic progress. This issue, while not central here, does warrant attention in the general context of applications of remote sensing.

Numerous attempts have been made and are now underway to assess the potential value of remote sensing from satellites both quantitatively and qualitatively. One of the earlier studies was conducted by the National Academy of Sciences in 1968 *(8)*. This was before the launch of LANDSAT-1 and undoubtedly had an impact on the final approval of the flight program. NASA, in the summer of 1974, conducted a comprehensive study in Colorado at which user groups, including specifically state, county, municipal, and private-sector users, were able to express their views *(10)*. In 1974, the National Academy of Sciences' CORSPERS Committee conducted another study, which addressed both the technical and institutional implications of the ERTS/LANDSAT program *(9)*. The conclusions were that significant benefits were to be derived and that there was utility in quite a few areas. However, it was suggested that perhaps the next level of technology should be investigated before converting to an operational system. As of this writing, the National Academy of Sciences has convened a Committee on Remote Sensing in Development, which is specifically reviewing the potential of this technology for the developing world. Of particular interest are potential institutional arrangements and data policy. In addition, a number of contractor-generated cost-benefit studies have been conducted, generally with rather favorable quantitative results *(3, 6, 7)*.

Although the benefits may be difficult to estimate—and estimates vary widely, from very conservative numbers that barely pay for a potential operational system to very large numbers, which appear to be possibly excessive—there is little doubt that natural resource management, including mineral exploration, will undergo major changes in future decades.

PRACTICAL UTILIZATION OF REMOTE-SENSING DATA

The technical and analytical tools available for examining and interpreting data from the LANDSAT system are also of interest. The original data archiving was in analog form. The process of converting what is basically a digital bit stream from the satellite directly into analog form, with the dynamic range limitations inherent in photographic emulsions, provides a built-in degradation of the data. This analog format has been useful for a one-time look at the entire world, and particularly the developing world, but there exist now digital interactive tools with which it is possible to look at the basic computer-compatible digital tape (CCT), which retains all 64 signal levels in each of four channels without any loss of information. Although this machinery is expensive, it provides potential for very significant advances in interpretive capability. It provides, for example, an ability to examine the spectral and temporal characteristics on a pixel-by-pixel basis so that map-like representations at the 1:60,000 or even 1:30,000 level are of use. We now have a multispectral system with approximately an 80-m spacial resolution, and the technology is available at the next level, through the NASA-developed thematic mapper, to go to a basic 30-m resolution system. There is no doubt that this improved resolution will significantly enhance the information content of the data. It will also tend to exacerbate the data-handling problem. It is of interest that international exploration companies are already exploiting these digital interactive capabilities.

We are attempting here to provide a comprehensive review of the status of remote-sensing applications for mineral exploration, with emphasis on the contributions of satellite-based data. The multiplicity of authors ensures a broad collection of viewpoints, but also permits a certain amount of duplication of material. In view of the early stage in the exploitation of this new technology, such limited duplication is not without its merits.

The field is covered from a number of different approaches: a review of U.S. programmatic results; foreseeable impacts on our mineral and energy problems; improved efficiencies in mineral exploration activities; U.S. data utilization experience in developing countries; examples of practical experience with satellite data in several developing countries; digital data processing techniques, including enhancement and spectral ratioing; and environmental monitoring of mineral-related activities. This comprehensive status review should be useful both to practicing geologists and serious advanced students.

REFERENCES

1. Bronowski, J., *The Ascent of Man,* Little, Brown, Boston, 1973.
2. Summers, R. A., and L. Jaffe, The Earth Resources Survey Program Jells, *Astronaut. Aeronaut.* Apr. 1971.
3. Lowe, Summers, and Greenblat, An Economic Evaluation of the Utility of ERTS Data for Developing Countries, prepared for the U.S. Agency for International Development, Aug. 1974.
4. Summers, R. A., Training and Education Requirements for the Practical Application of Remote Sensing, presented at the Seventh International Symposium on Remote Sensing of the Environment, University of Michigan, Ann Arbor, Mich., May 1971.
5. NASA Authorization Hearings for FY 76 and the Transition Period, Committee on Aeronautical and Space Sciences, U.S. Senate, Mar. 3 and 11, 1975.
6. Earth Resources Survey Benefit–Cost Study, prepared by Earth Satellite Corporation and Booz–Allen Applied Research Corporation for the U.S. Department of the Interior, U.S. Geological Survey, Nov. 1974.
7. Heiss, K. P., and G. A. Hazelrigg, The Economic Value of Remote Sensing of Earth Resources from Space: An ERTS Overview and the Value of Continuity of Service, Econ, Inc., for NASA, Oct. 1974.

8. Useful Applications of Earth-Oriented Satellites, National Academy of Sciences, Woods Hole, Mass., 1968.
9. Remote Sensing for Resource and Environment Surveys—A Progress Review, National Academy of Sciences, CORSPERS, 1974.
10. Practical Applications of Space Systems, National Academy of Sciences, Snowmass, Colo., 1975.
11. Adelman, I., Development Economics—a Reassessment of Goals, American Economic Association Meeting, Dec. 1974.
12. Beilock, M., Systems for Acquisition, Processing, and Dissemination of Earth Resources Satellite Data, prepared for United Nations Outer Space Affairs Division, Dec. 1973.
13. Summers, R. A., and B. W. MacDonald, Preliminary Assessment of the Requirements for the Space Segment of an International Earth Resources Survey System, prepared for the United Nations Outer Space Affairs Division by System Planning Corporation, Feb. 1975.

2

FORESEEABLE ENERGY AND MINERAL RESOURCE PROBLEMS

William L. Smith

William L. Smith, formerly with System Planning Corporation, Arlington, Virginia, is currently with the Environmental Research Institute of Michigan, Arlington, Virginia.

OUR MINERAL ECONOMY

An assured mineral supply is essential to the economic security if not to the very future of industrial civilization. Although there is continuing discovery and effort to maintain adequate reserves, we are continually faced with problems of raw material shortages. New processing techniques have made coproducts and by-products from mill and smelter tailings economic, and many previously submarginal deposits have been found to be amenable to improved metallurgical processes; yet in the United States the industrial demand for an increasing number of minerals regularly exceeds mine production.

Many mineral economists foresee an impending critical increase of the need for several ores. This will certainly lead to an even greater reliance by industrial nations on imports, and to the possible scarcity and increasing costs of several minerals that today are in adequate supply. To plan against such eventualities, it is necessary to markedly accelerate the discovery and availability of critical minerals. At the present time most mineral-producing countries are exploiting their known resources at such a rate that major exploration efforts and radically new exploration techniques will be required unless there are exceptional discoveries of many ores in the near future. It would surely solve many problems if such discoveries were fairly equitably distributed among nations, but gold or phosphate rock is where you find it.

Many potential solutions have been offered: dredge the sea bottoms, build windmills, extract metals from seawater, leach the pegmatites, mine the moon, find plastic substitutes and new alloys where substitution is possible. Many such solutions may be possible and economic decades from now, perhaps sooner. Other alternatives are rigid government control of resources, radical subsidization of uneconomic extractive industries, reliance on imports through cartels (e.g., Organization of Petroleum Exporting Countries, OPEC) for the major minerals, and war. But these extremes and various other proposed political–economic–restrictive measures are costly and at best temporary, artificial programs that tax the healthier sectors of a competitive economy.

One objective of the NASA Earth Resources Survey Program is the application of space technology to the delineation of regions of potential mineral wealth.

However, remote sensing by itself is only another exploration tool, another advance in exploration technique, the next step beyond aerial photography. It is not the key that opens the treasure house or the ultimate panacea for problems of shortages; not LANDSAT-2 nor LANDSAT-C, not the proposed ESO program nor the space shuttle sorties. On the other hand, knowledge that an important new exploration technique is being developed with potentialities of accelerating mineral discoveries should produce a material influence on resource management, on economic planning levels, and on the international policies that a civilized world must address.

But to return to the current mineral supply picture, the United States is fortunate to be well endowed with ample deposits of most essential metals, fuels, and industrial minerals. Mineral raw materials constitute about 5 percent of the GNP, and the mineral industries contribute some 10 percent of the total national income. The annual value of our mineral production exceeds $30 billion, and both the volume and value increase annually. However, although the net supply of most minerals increases yearly, many important mineral stocks continue to decline.

In 1952 the Paley Commission made an evaluation of America's major mineral resources and estimates as to probable demands and sources of supply in the 1970s *(1)*. The Paley Report was of little comfort to the mineral industry. We were about to run out of many major minerals and the future was bleak in general. The only hope the report offered was that possibly new technology in exploration would reprieve us from running out. A dramatic new technology did not arrive but developing technologies like geochemistry led to many major discoveries. Many mineral areas that looked bad in the 1950s have survived or recovered owing to the new discoveries; others less fortunate have been abandoned for reliance on foreign sources of supply. The time for a new assessment arrived in the early 1970s. Much of the mineral industry picture offered little encouragement to mineral economists; but we can expect geologists to continue to discover new ore deposits, and this time there is a dramatic new technology to assist in the exploration.

In 1971 the Honorable Wright Patman chaired Hearings of the Joint Committee on Defense Production, which again considered potential shortages of ores, metals, minerals, fuels, and energy resources, and estimates of future production, consumption, and dependence on foreign sources *(2)*. It was noted that current assessments of raw material requirements of the United States extended to the year 2000, a time no further into the future than the Paley Commission study covered in the past. Representative Patman cited our facing diminishing high-grade reserves, greater competition in purchasing metals and minerals in foreign markets, the reduction in enrollment in college courses related to the mineral industry, greater per capita consumption of raw materials, and attempted to preserve environmental quality. He also noted the Department of the Interior's assessment that in the last 30 years the United States alone had consumed more minerals than the entire world for all time before, and that, extrapolating to the year 2000, the total constant dollar value of demand for minerals in the United States was expected to increase from three to five times the current level.

At the same hearings, Assistant Secretary of the Interior Hollis Dole discussed how foreign mineral sources are becoming less accessible to us inasmuch as foreign consumption is increasing at a rate even greater than our own, and that many easily accessible high-grade deposits are being rapidly exhausted with their place being taken by lower-grade ores or ores lying at greater depth. In addition, exploration and development costs are rapidly rising; it is foreseeable that obtaining venture capital will be an increasingly major problem facing the mining industry. Director of the Office of Minerals and Solid Fuels H. L. Moffett presented a list of selected metals and minerals showing dependency on foreign sources in terms of U.S. consumption (see Table 1).

Nearly all our aluminum, antimony, chromite, cobalt, columbium, manganese, mica, nickel, platinum, rutile (titanium), tantalum, and tin was imported. In some cases the United States had essentially no reserves; in others it was uneconomic to produce from domestic sources. Many essential minerals, such as beryllium, cadmium, and zinc, were imported in amounts approximately equal to domestic produc-

TABLE 1
List of Selected Minerals and Metals Showing U.S. Dependence on Foreign Sources in 1970 (percent of U.S. consumption from foreign sources) *(2)*

Commodity	Total	By source	Quantity of net imports
Aluminum	91	Jamaica 41, Surinam 16, Australia 11, Canada 6, Dominican Republic 4, Guyana 3, Haiti 3, other 7.	4,382,000 tons.
Antimony	94	Republic of South Africa 32, Mexico 20, United Kingdom 14, Bolivia 11, Guatemala 6, France 4, other 7.	18,100 tons.
Asbestos	83	Canada 78, Republic of South Africa 3, other 2	649,400 tons.
Beryllium (ore)	51	Brazil 37, Uganda 4, Republic of South Africa 3, Argentina 3, Mozambique 2, other 2.	4,940 tons.
Cadmium	53	Canada 17, Mexico 16, Japan 7, other 13	3,460 tons.
Chromite	100	U.S.S.R. 33, Republic of South Africa 29, Turkey 18, Philippines 15, other 5.	1,405,000 tons.
Cobalt	93	Congo (Kinshasa) 52, Belgium-Luxembourg 26, Norway 6, Canada 3, other 6.	6,200 tons.
Columbium	100	Brazil 58, Canada 22, Nigeria 12, Congo (Kinshasa) 2; Angola, Argentina, Belgium-Luxembourg, Burundi-Rwanda, West Germany, Mozambique, Portugal, Singapore, Uganda, United Kingdom 4.	2,860 tons.
Copper	6	Peru 2, Chile 2, Canada 1, other 1	9,900 tons.
Fluorspar	78	Mexico 60, Spain 10, Italy 6, United Kingdom, Brazil, Mozambique, West Germany, Republic of South Africa 2.	1,077,000 tons.
Iron ore	33	Canada 18, Venezuela 10, Australia, Brazil, Chile, Liberia, Peru, Sweden 5.	39,400,000 tons.
Lead	38	Canada 11, Australia 9, Peru 8, Mexico 4, Yugoslavia 2, Honduras 1, other 3.	350,000 tons.
Magnesium	0		(31,200) tons.
Manganese (plus 35% ore).	99	Brazil 35, Gabon 31, Republic of South Africa 8, India 4, Ghana 4, other 17.	836,600 tons.
Mercury	38	Canada 31, Spain 3, other 4	17,000 flasks, of 76 pounds each.
Mica	100	India 86, Brazil 13, other 1	5,309,000 pounds.
Molybdenum	0		(55,737,000) pounds.
Nickel	87	Canada 72, Norway 7, Republic of South Africa 1, other 7	135,000 tons.
Platinum-group metals	98	United Kingdom 45, U.S.S.R. 34, Republic of South Africa 8, Japan 3, Canada 2, Colombia 2, Belgium-Luxembourg 1, Norway 1, other 2.	1,009,300 ounces.
Rhenium	7	West Germany 3, U.S.S.R. 2, France 2	210 pounds.
Rutile	100	Australia 92, Sierra Leone 8	242,200 tons.
Selenium	29	Canada 28, other 1	454,000 pounds.
Silver	27	Canada 16, Peru 5, Mexico 2, Honduras 2, other 2	34,686,000 ounces.
Tantalum	100	Canada 46, Congo (Kinshasa) 21, Brazil 17, Spain 5, Burundi-Rwanda 3, United Kingdom 2, Argentina, Australia, Belgium-Luxembourg, Cameroon, Cyprus, Japan, Nigeria, Portugal, Spain, Western Africa, 6.	523 tons.
Tellurium	20	Peru 11, Canada 9	64,000 pounds.
Tin	100	Malaysia 63, Thailand 30, Indonesia 3, other 4	46,100 long tons.
Tungsten	0		(18,171,000) pounds.[1]
Vanadium	22	Republic of South Africa 13, Chile 5, U.S.S.R. 3, other 1	1,033 tons.
Zinc	59	Canada 32, Mexico 10, Peru 6, Australia 3, Japan 2, other 6	795,900 tons.

[1] Net exports largely due to Government sales.
Note: Figures in parens are net exports.

tion. It is well to consider this list of metals and minerals in terms of the 1973–1974 crisis that ensued when Arabian oil was cut off—Arabian oil accounted for less than 3 percent of U.S. consumption.

The best insurance for a healthy, free mineral economy is, obviously, reliable long-term availability of essential ores, whether this supply be domestic or foreign. Historically, ore supplies are first exploited from nearby deposits, often a reason for the location of the industry. Then, as these deplete, or as transportation costs and the market require, lower grades of ore or ores from distant sources are used. Many industries founded on local resources have eventually and successfully depended on imports or foreign captive mines. Other mineral-dependent industries and mining companies are constantly faced with weighing accelerating exploration costs for new, possibly remote deposits against the costs of developing technologies for exploiting low-grade reserves or of buying imported ores, concentrates, or metals. In 1973 the U.S. Geological Survey released a major work on our mineral resource picture *(3)* in which was compiled a most revealing table concerning the more vital mineral resource commodities. It included appraisals of potential resources and noted that many of these are dependent on future geological exploration (Table 2). The writers concluded that but few of the commodities had long-term availability and that serious planning was in order; not only was American affluence in jeopardy *(4)* but world civilization *(3)*.

Within less than a year's time of the referenced hearings of the Joint Committee on Defense Production, LANDSAT-1 was launched into orbit from NASA's California facility and started to transmit high-resolution multispectral imagery. Although the precise role LANDSAT has played to date in locating ore is controversial, it has been sending back data on previously unknown lineaments, fault and fracture patterns, domes and subsidence, and broader pictures of geological structure. This information has prompted several new exploration efforts all over the world; some of these should locate ore or oil.

It was soon determined that preliminary reconnaissance maps of remote regions and of terrain of diffi-

TABLE 2
Potential U.S. Resources of Some Important Mineral Commodities, in Relation to Minimum Anticipated Cumulative Demand to the Year 2000 *(3)*

ST=short tons. lb=pounds.
LT=long tons. Tr oz=troy ounces.
Identified resources: Includes reserves, and materials other than reserves that are reasonably well known as to location, extent, and grade, that may be exploitable in the future under more favorable economic conditions or with improvements in technology.
Hypothetical resources: Undiscovered but geologically predictable deposits of materials similar to present identified resources.
I. Domestic resources (of the category shown) are greater than 10 times the minimum anticipated cumulative demand 1968–2000.
II. Domestic resources are 2 to 10 times the MACD.
III. Domestic resources are approximately 75 percent to 2 times the MACD.
IV. Domestic resources are approximately 35–75 percent the MACD.
V. Domestic resources are approximately 10–35 percent the MACD.
VI. Domestic resources are less than 10 percent of the MACD.

Commodity	Minimum anticipated cumulative demand, 1968–2000 [1]	Identified resources	Hypothetical resources
Aluminum	290,000,000 ST	II	Not estimated.
Asbestos	32,700,000 ST	V	VI
Barite	25,300,000 ST	II	II
Chromium	20,100,000 ST	VI	VI
Clay	2,813,500,000 ST	III	II
Copper	96,400,000 ST	III	III
Fluorine	37,600,000 ST	V	V
Gold	372,000,000 Tr oz	III	Not estimated.
Gypsum	719,800,000 ST	I	I
Iron	3,280,000,000 ST	II	I
Lead	37,000,000 ST	III	IV
Manganese	47,000,000 ST	III	Not estimated.
Mercury	2,600,000 flasks	V	Not estimated.
Mica, scrap	6,000,000 ST	II	I
Molybdenum	3,100,000,000 lbs	I	I
Nickel	16,200,000,000 lbs	III	Not estimated.
Phosphate	190,000,000 ST	II	I
Sand and gravel	56,800,000,000 ST	III	Not estimated.
Silver	3,700,000,000 Tr oz	III	III
Sulfur	473,000,000 LT	I	I
Thorium	27,500 ST [2]	II	Not estimated.
Titanium (TiO_2)	38,000,000 ST	II	II
Tungsten	1,100,000,000 lbs	IV	IV
Uranium	1,190,000 ST	II	III
Vanadium	420,000 ST	II	Not estimated.
Zinc	57,000,000 ST	II	II

[1] As estimated by U.S. Bureau of Mines, 1970.
[2] For thorium, *maximum* anticipated cumulative demand 1968-2000, which assumes commercial development of economically attractive thorium reactors by 1980.

cult access might be made quickly at sizable cost savings. Some mining districts have been plotted against LANDSAT-identified lineaments, and in some cases deposits that were not associated with known structural controls appeared to fall close to these features. Developing countries, with poor or obsolete geologic maps, have been able to initiate efforts in systematic minerals search. But the primary benefit is the scale of LANDSAT products, which permits the investigator to think in a broader conceptual scope as to the relationship of mineral deposits to a wider geological format. LANDSAT products will also provide data important at the planning levels of resources management to better inventory, model, and assess the worldwide resource picture. It is anticipated that the new mapping and exploration techniques derived from satellite sensing will accelerate conventional exploration and should, in turn, advance the discovery of new ores and energy resources.

In 1968, C. F. Park of Stanford University, former Chief of the Metals Section of the U.S. Geological Survey and one of our foremost economic geologists, provided us with *Affluence in Jeopardy (4).* This thought-provoking study of the relationship of mineral supply to political economy plotted the population explosion against future consumption of various mineral commodities. Park emphasized the limited supply of these minerals, the need to manage domestic ores effectively, and the importance of access to resources in remote and underdeveloped areas. Park believes that even if the world is able to increase food production dramatically, widespread poverty could continue and dangerous international political situations arise due to shortages in the minerals that are important to society. In the case of iron, one of the more abundant resources, should the world population double by the year 2000, and if we assign a U.S. 1968 per capita consumption of iron of 1 ton per year, annual production will need to be increased some 12 times. Similar projections were made for other production using Bureau of Mines data. These showed an increase of 11 times for copper production and 16 times for lead at the end of the century.

For all LANDSAT technology may have to offer, requirements like these also mean a need for the rapid development of technologies to get every possible economic ton of ore out of our mines, to develop the marginal deposits, and to greatly expand research in metallurgy and in new mineral and energy sources. A sometimes heavy hand may also be required on the part of those federal agencies whose responsibility is to pursue our mineral policies or to protect environmental quality. Increasing individual and industrial consumption of mineral raw materials undoubtedly will require more, or at least more effective, control over the allocation of mineral production; but that should not mean federal take-over or other extreme measures. The foremost role of federal agencies should be the encouragement of exploration, maximum development of new technology, and accelerated research into the frontiers of economic geology. Greater federal funding of research at geology and mines departments of universities is important. It is likely that LANDSAT data can play a major role in identifying new locations of ore deposition when the data reach a larger number of qualified investigators.

ENERGY RESOURCES

Our long-term problem is ores, not energy. Energy is our present, interim problem. A long-term energy objective is to have a relatively cheap supply so that we may economically win essential metals from lower-grade deposits *(3).*

Contrary to many purposeful warnings, the long-term energy picture for the United States is not as bad as it has been painted. We will have to solve touchy international problems. There will have to be major exploration, increased utilization of unconventional forms of energy, and legislative action; but the potential energy resource situation is brighter than that for a number of metallic ores and chemical minerals. It is up to private and federal science to investigate the practical utility of wind-generated electric power, microwave energy transmission, solar heating, chemical energy conversion, the oceanic and geothermal gradients, and the various proposed earth- and space-based conversion and delivery systems. These could play significant roles in the future; however, it is also quite certain that fossil fuels will remain the primary source of generated energy through the end of this century and probably well into the next.

Let us look at both sides of the energy supply picture and the role that remote sensing might play. Do we have to cut back our energy and mineral consumption to provide for the future? Will we have to temporarily or permanently lower our standard of living or check its rise by allocating a number of commodities for essential future industrial and agricultural usage? Would such sacrifice only solve the problem for a few decades, postponing the time when it all runs out? On the other hand, is adequate potential petroleum, gas, coal-derived energy, nuclear power, and geothermal energy likely to be available for the increasing demands of industrial civilization?

Figure 1, compiled from data for use in the Assessment of Energy Technologies, submitted to the former Office of Science and Technology, Executive Office of the President, shows the projected U.S. energy demand by resource through the year 2020. The chart suggests a growth from about 64×10^{15} Btu to about 300×10^{15} Btu. The most dramatic factor of increase is seen to be in nuclear energy, although requirements are also seen to triple for oil and to increase six times for coal *(5)*. Energy consumption has been doubling over previous 20-year periods. This rate of energy use has required a faster rate of discovery of energy minerals to meet the need to maintain stocks. A summary study of our energy resources released by the U.S. Geological Survey in 1972 *(6)* based on the classification of mineral reserves proposed by McKelvey *(7)* contains the following data shown in Table 3, not including "undiscovered resources."

Let us take a look at the energy resource situation: Cheney *(8)* has suggested that "the major restrictions to power growth are geological, environmental, and political, and that these restrictions are not especially amenable to technological solutions." He also suggests if various "environmental, military, and political

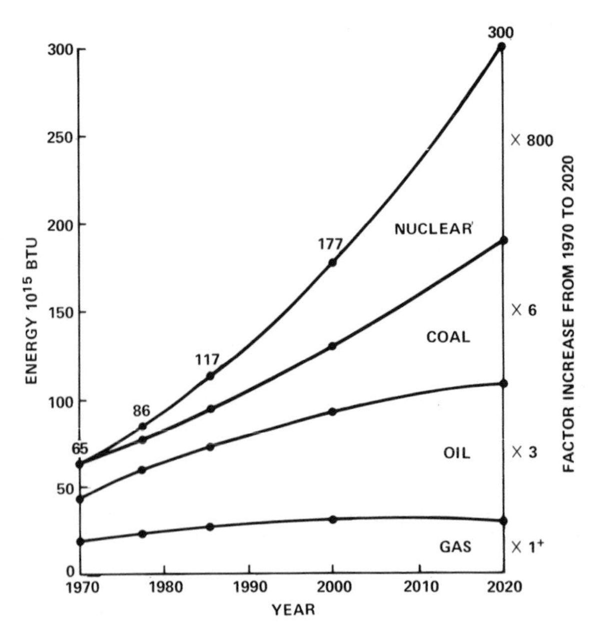

FIG. 1
Projected U.S. energy demand by source *(5)*.

TABLE 3
Identified U.S. Energy Resources

Source	*Recoverable*	*Submarginal*	*Unit*
Coal	390	1200	Billion short tons
Oil and gas liquids	52	290	Billion barrels
Natural gas	290	170	Trillion ft^3
Uranium (conventional)	250	200	Thousand short tons, U_3O_8
Uranium (in phosphate rock)	1000 (paramarginal)	5000	Thousand short tons, U_3O_8
Geothermal energy	2.5	a	10^{18} cal
Oil shale	160–600 (paramarginal)	1600	Billion barrels

[a]"Undiscovered": $> 10{,}000 \times 10^{18}$ cal paramarginal; recoverable at prices as much as 1.5 times those prevailing in 1972.

problems are to be avoided, the price of power must be allowed to rise until it includes most of the indirect societal costs and until the United States is virtually self-sufficient in energy production. Zero per capita power growth might be a resulting consequence." These are the hard facts; however, it is unlikely that the U.S. citizenry and industry would readily adjust to a check in the rise of energy use, let alone holding it to current levels. The political problems that could arise from the proposed solution would seem to be fully as serious as those it might avoid. We would not find such a solution acceptable for more than an emergency interim—not while the United States has enough coal alone to last for centuries, only a small part of our offshore has been tested for oil and gas, and remote sensing is providing encouraging new data on geological structure in several parts of the world. But the price of power must be allowed to rise if we are to consider new sources.

Oil and Gas

Peach *(9)* noted that at one time the world was worried about running out of trees. Our current concern is about nonrenewable resources; however, Peach also noted "a nation which can put men on the moon in the 60's can solve its energy problems in the 70's. But it can't solve them with a 'business as usual' approach." Peach's paper is a long-needed response to testimony before House and Senate hearings to the effect that our resources are again about to run out. Such, he predicts, "is going to make humorous reading by the end of the century." Simply put, he sees the rise in oil and gas prices as making it feasible to develop the alternative energy sources.

Development of these sources may take 10 or more years, but in the meantime there has been the discovery of oil and gas in northern Alaska and Cook Inlet and a potential of further discoveries from the interpretation of LANDSAT data. The United States will certainly continue to maintain current energy levels; the problem is filling demands over the next 10 or 15 years. Although at the moment Arabian oil appears to be our major prospect to fill the gap between what we expect to have and what we will need, the Middle East is not the only answer to our requirement for a foreseeable additional 25 percent of petroleum.

Table 4 shows the projected U.S. petroleum demand versus supply balance extrapolated to 1985, as presented at the 1971 hearings before the Joint Committee on Defense Production *(2)*. At the same hearings it was shown that in 1970 Western Hemisphere production accounted for 96 percent of U.S. oil needs. Office of Oil and Gas data indicated that only 4 percent came from the Eastern Hemisphere *(2)* (Table 5). Estimates of U.S. dependency for 1985, presuming 3,000,000 barrels per day (B/D) from the Alaskan north slope looked as shown in Table 6. Estimates of U.S. dependency for 1985, without Alaskan north slope oil would have required 35 per-

TABLE 4
Projected U.S. Petroleum Demand–Supply Balance, 1970–1985 (thousands of barrels daily)

	1970	1975	1980	1985
Domestic demand [1]	14,860	18,000	22,000	26,800
Exports	240	300	300	300
Total demand	15,100	18,300	22,300	27,100
Less—				
Processing gain	370	500	550	650
Stock reduction	20			
Required petroleum supplies	14,710	17,800	21,750	26,450
U.S. production: [2]				
Crude oil	9,600	11,200	11,900	12,600
Natural gas liquids	1,680	1,900	2,000	2,000
Total	11,280	13,100	13,900	14,600
Required from other sources	3,430	4,700	7,850	11,850
As percent of required supplies	23.3	26.4	36.1	44.8

[1] Projected at a 4-percent annual growth rate. Neither demand nor imports include U.S. military offshore procurements.
[2] Assuming production at approximately present capacity, plus production from the North Slope of Alaska estimated at 500,000 barrels per day in 1975, 2,000,000 barrels per day in 1980, and 3,000,000 barrels per day by 1985.

TABLE 5

		Thousands (barrels/day)
U.S. production	(76%)	11,280
North American imports	(5%)	700
South American imports	(15%)	2,200
Eastern hemisphere imports	(4%)	530
		14,710

TABLE 6

		Thousands (barrels/day)
U.S. production	(44%)	11,600
Alaska North Slope	(12%)	3,000
North American imports	(8%)	2,200
South American imports	(13%)	3,500
Eastern hemisphere imports	(23%)	6,150
		26,450

cent Eastern Hemisphere imports, or 9,150,000 B/D. By 1972, with domestic wells producing near full capacity, nearly 30 percent of the nations petroleum was imported. As of early 1975, the OPEC cartel, those responsible for the oil price crisis, appeared to have the upper hand in negotiations with the petroleum-consuming countries. Although dominated by Saudi Arabia, Kuwait, Iran, and Iraq, OPEC also included six other Mideast producers, plus Indonesia, Libya, Nigeria, Algeria, Venezuela, and Ecuador. Since 1970 the royalties and taxes have increased by an order of magnitude, and Saudi Arabia and other countries have made gestures toward the take-over of producing company properties. Whatever accord we may make with OPEC, our most important actions will be those toward greater self-sufficiency.

If, as presumed, 80 percent of the yet-to-be-discovered petroleum lies offshore, the location of these reserves will be aided very little by studies based on the use of LANDSAT imagery. It is likely that there are sizable oil resources in the offshores of several African developing countries, but even in the case of the United States the Geological Survey estimates a potential resource of 1.0 to 1.5 trillion barrels of oil in place on the outer continental shelf plus 3 to 4 quadrillion ft^3 of natural gas. It has been estimated that of this potential source current methods could recover up to 190 billion barrels of oil and from 800 trillion to 1.1 quadrillion ft^3 of gas. This represents a small part of our 1.8 million square mile continental shelf.

It has been estimated that, in addition to oil, the Alaskan north slope from the Prudhoe Bay area alone also contains 350 trillion ft^3 of gas, or 15 year's worth at current U.S. requirements for gas. It is probable that even greater deposits exist in the Canadian Arctic. John Ricca, Office of Oil and Gas, told the defense production hearings *(2)* that the Potential Gas Committee estimated U.S. gas resources at 260 trillion ft^3 in known fields, with possibly another 335 trillion ft^3 in new fields in producing areas. He considered this resource adequate for all foreseeable demands during the period required to develop supplemental sources from coal. A problem with gas has been FPC interstate price regulations which until recently have not made ventures attractive enough for investment.

The above, however, only represents the optimist's view. In January 1975, the Federal Power Commission indicated domestic natural gas production had peaked, was in decline, and might never again meet our demand. The commission suggested that we "develop plans and policies keyed to the possibility that the nation may indeed be experiencing the early effects of a resource being pushed toward exhaustion." All this was an about face from earlier government statements from the Geological Survey and the Potential Gas Supply Committee, which estimated that we had 568 trillion to 1450 trillion ft^3 of undiscovered reserves in the conterminous United States alone. The FPC cited contrary estimates of 400 trillion ft^3 or less, and we use some 22 trillion ft^3 per year. So, you have your choice for the present, FPC or U.S. Geological Survey. The relatively few new discoveries in recent years admittedly have not been encouraging, but the true picture probably lies above either estimate. The FPC policy is to recommend further leasing of federal lands, reallocation of supplies to high-priority use, and increased incentives for exploration. The latter item essentially means higher producer prices.

In addition to tar sands and secondary recovery sources of petroleum, there is the matter of oil shales, next to coal the most abundant potential source of energy we possess. The current high prices for oil have made economic those shales that are capable of yielding 25 to 40 gal per ton. Deposits believed to contain shales yielding over 10 gal per ton are calculated to contain over 3 trillion barrels, and extensions of these deposits may exceed another 3 trillion barrels. Today some 80 billion barrels are available and consistent with current costs from the Green River shales. Cheney *(8)*, however, notes that "The federal government still owns most of the best oil shale lands; furthermore, only it, not the private energy companies, has nearly instant access to the huge amount of capital (estimated to be at least $12 billion in 1972 dollars) that will be needed to develop the oil shale on a scale that would readily solve the liquid fuel crisis by 1980."

Coal

Peach *(9)* notes "coal accounts for 73 percent of our fossil fuel, enough to last for hundreds or thousands of years." Oil shale accounts for only 17 percent of our fossil fuels. Let us consider our major energy source.

The U.S. Geological Survey estimates remaining U.S. coal resources to total 3224 billion tons *(3)*. Of this, roughly half has been mapped and explored; the remainder is "hypothetical" or extrapolated from data on identified resources into unmapped areas. United States coal fields are considered to be so well known that it may be estimated safely that about half our coal has less than 1000 feet of overburden, and that only 2 percent of this reserve has been mined to date. This is a huge resource, even at 50 percent recoverability. Some 45 billion tons of stripping coal are considered to be recoverable, exclusive of currently submarginal coals.

Underground gasification of coal is feasible technically. Oxidizing agents would be introduced into prepared coal beds and the reaction would produce combustible gases from the coal, which could then be collected underground and brought to the surface *(10)*. Within 3 years a commercial gas plant is scheduled to go on stream using Montana and Wyoming coals. Coal technology includes liquefaction, gasification, or complete oxidation methods. The most likely economic approach is the production of methanol by gasification for synthetic fuels. The economics of this method has to be weighed against the costs of reclaiming surface-mined areas. We also have the opportunity of working with the British or Russians, who have offered to share their knowledge of coal gasification and who have histories of successful operations. It has been estimated that synthetic gas from coal would cost from $0.85 to $1.10 per thousand ft^3. Although this is high compared with a normal $0.25, it is somewhat cheaper than buying it from North Africa at $1.15. Almost any scheme could be knocked out by a lowering of import prices and an assured long-term supply.

Nuclear Energy

In view of the current energy crisis, international political problems related to mineral fuels and foreseeable world demands on conventional power sources, nuclear power plants are taking over an increasingly important fraction of the world's demand for energy. Converter reactors rely upon uranium, and currently known uranium reserves are viewed as adequate for near-future requirements. At present there is a domestic and world uranium surplus. Mine and mill production of U_3O_8 remains high and ore reserves continue to increase. The United States, Canada, and South Africa have the major free-world uranium resources, but new deposits are being developed in Australia, Niger, Southwest Africa, and elsewhere in anticipation of large near-future world requirements. Known uranium reserves are more than adequate to last into the 1980s; however, needs beyond the 1980s may be so great that major efforts will have to be made to discover new large sources. It is probable that much of these resources will be from the developing nations. As for the present time, however, industry is facing the problem of providing incentives for increasing exploration for the reserves that surely will be needed later.

New uranium enrichment technology has progressed as has world enrichment capacity. The Atomic Energy Commission (AEC) made gaseous diffusion technology available to U.S. industry and to some foreign governments where appropriate safeguards are possible. Other advances have been made in nuclear fuel manufacturing and reprocessing. And although the United States is by far the leader in nuclear reactors for electric power and in commercial nuclear power development, several industrial countries have sizable nuclear energy programs, and several of the less-developed countries have announced plans for nuclear power. The U.S. export of nuclear materials is expanding, including enriched uranium, ^{233}U, heavy water, and plutonium. This is indicative of an increasingly high level demand for nuclear fuel for U.S. reactors in other countries.

Converter reactors are considered to be relatively safe, efficient, and produce minimum environmental contamination when considered in terms of the efficient conversion of heat to electricity; disposal problems are not great where plants are properly sited. Nuclear power plants also address the problems of adequacy of conventional mineral fuel sources in view of anticipated population and industrial demands. By the year 2020, nuclear energy is projected as producing 800 times its present energy output *(5)*. According to Peach, uranium should be available at up to $15 per lb. for the next 20 to 30 years, and 1 lb can produce as much electricity as 26 million lb of coal *(9)*. On the other hand, most domestic ores run an average 0.2 percent U_3O_8, and an alternative view is that domestic reserves of this grade could be exhausted as early as 1980. Cheney *(8)* warns that if breeder reactors are not evolved by that time, the United States may experience a U_3O_8 crisis and have to depend on foreign sources, as in our present situation with petroleum.

The U.S. Geological Survey estimates show domestic uranium resources in high-grade deposits as high as 500,000 tons of U_3O_8, about half this amount being known reserves as of 1973 *(3)*. To meet requirements before the breeder reactor is introduced, the other half of the postulated uranium needs to be discovered. There are extensive low-grade U_3O_8 resources, such as the Chattanooga shales, but they are not economic or practical to mine and process by current technology. A major exploration effort in North America and overseas is in order. A possible assist to exploration is the spectral ratioing technique applicable to LANDSAT products.

Geothermal Power

Unlike nuclear energy, geothermal energy production has aroused few fears as to the environment or disposal of waste. The United States has some 100,000 square miles of geothermal sources *(9)*. The technology for utilizing natural steam has been developed and is in use domestically and in other countries. Operating costs are relatively low, and Los Alamos has recently developed a scheme for using geothermal heat per se, which is applicable to wide areas, consisting of dumping water down one hole 15,000 ft deep and collecting steam from an adjacent hole *(9)*. At present, our geothermal power industry is small and restricted to a few thermal areas. The geothermal energy reserves have been estimated both as (1) many times the potential of our reserves of fossil fuels, or (2) as quite limited and depletable. At best, they are poorly known as far as making economic estimates are concerned. Practical considerations require we await new drilling technologies and a means of economically generating electricity from relatively lower temperature sources. The best estimates of world reserves run 2×10^{19} cal of 58,000 mW for 50 years for resources to a depth of 3 km for electrical generation by current techniques. Estimates for U.S. western geopotential alone range from 10^5 to 10^7 mW *(11)*.

Looking at the whole picture, minerals, fuels, and future shortages, we recognize roles for science, engineering, and government. The role for the mineral industry is accelerated exploration and development for quicker availability and an assured supply. There will be a market.

ACCELERATED DISCOVERY AND ASSURED SUPPLY

A report of the Senate Committee on Aeronautical and Space Sciences notes that one LANDSAT user claimed "the value of mineral resources discovered from space over the past year will exceed the total investment in NASA programs over the past 17 years" *(12)*. This, of course, remains to be proved, but the implication that the value of those mineral resources, which may be discovered as the result of the NASA-LANDSAT program, could be of this order of magnitude was enough to enthuse any authorization hearing. Another oversell? Overenthusiastic? No. It could have been stated more cautiously; but the amount of geological content in the multispectral imagery is far beyond that which was expected by most geologists, and the applications of LANDSAT data to mineral exploration and theoretical economic geology are just beginning to take shape. So far they look encouraging.

The Geology Panel of the 1967 Summer Study on Space Applications *(13)* first clarified the possible benefits of accelerated discovery of mineral resources from space. It was suggested that the availability and follow-up of remote-sensor data would result in a stimulation of exploration; new discoveries would ensue plus economic benefits from an acceleration of oil and metal production. The panel, using the introduction of the airborne magnetometer as an example, stated that whenever a new technique is employed an increase in both exploration activity and finding rate occurs. The study noted that, although total reserves would not be increased by the faster finding rate, a significant percentage of the ultimate reserves would be "borrowed" from a future period or brought into earlier production. "The net benefit would then be the difference between the present worth of accelerated production and that of the 'borrowed' production." This concept was not particularly well received in 1968–1969, but it "hit the nail on the head" for 1975–1976.

An accelerated availability of minerals could conceivably eliminate our dependence on some foreign sources and force the costs of some imports downward. A certain result would be drastic revisions in the flow or ores and concentrates in international trade. Accelerated discovery would enable industry to program extended operations, providing a security to many of our mineral-based businesses that is currently lacking. When private and federal agencies are able to undertake longer-range planning, more confident management will be possible in those areas of the economy where assured supply has been essential.

The immediate benefits that could be derived from expanded mining, processing, trade, taxes, etc., are too numerous to list. The location of new sources of ore would have maximum effect in the poorer nations, where inequities in resources and the high cost of imports hamper industrial development. Such smaller and developing countries will require increasingly larger shares of the world's resources as their populations increase and the standard of living of these populations rises. Construction, exploration, and engineering programs related to developing new productive areas, and the associated services and supporting industries, could realize benefits as great as those estimated for the extractive and processing interests.

Assured supply concerns more than fodder for the ore-processing industries. Water is the most essential mineral. Its presence or abundance determines the possibility of human settlement, agriculture, industry, and power generation in many areas. The loss of water supply can return grazing ground to desert and force human populations to move. Many mineral deposits have never been developed because of the absence or high cost of water. Deep wells and desalinization will have an impact on water supply, but remote-sensing data from LANDSAT-1 have already demonstrated applications for monitoring surface water, snowpack, drainage systems, and soil moisture, and the on-board Data Collection System collects and relays abundant information from in situ gauging stations. In arid areas the water supply picture may be indicated by riparian vegetation communities, the distribution of phreatophytes, and seasonal changes in streams and other surface waters. The near-infrared bands of the multispectral scanner are uniquely suited to hydrology, and these water resource problems are at present being investigated.

Salomanson and Rango *(14)* indicate that a large percentage of runoff in the western United States is derived from snowpack in mountain regions and that snow-cover mapping from LANDSAT-1 data will be important to water supply management, inasmuch as some snow-line altitudes have been estimated to the nearest 60 m and the areal extent of cover can be obtained to within 1 percent of drainage basin area *(15)*. Industrial water use is a major management concern in many parts of the country. The value of assured water supply is inestimable, far exceeding a recitation of gallons and dollars.

Close in importance to water requirement is that of productive agricultural land. The petroleum and mineral industries will have an increasing role in providing the essential chemicals such as nitrates (ammonia) derived from hydrocarbons and the phosphate compounds, which depend on phosphate rock and sulfur. These fertilizers are in constant need of replenishment, are of worldwide importance and are not provided by pipeline. Remote sensing for phosphate rock, for example, will require interpretation of multispectral scanner (MSS) imagery of North Africa and the Middle East to better define the regional structure and stratigraphic sequence, with follow-on, low-altitude aerial gamma spectrometric surveys to locate phosphate deposits by emitted radioactivity from their contained uranium *(16)*. In addition, new knowledge as to the origin of marine phosphate deposits makes it possible to more easily prospect for and identify those formations which are favorable or unfavorable for the occurrence of phosphate rock *(17)*. In many parts of the world expanding populations depend on domestic agriculture, and overworked soils are producing insufficient crops, even with improved strains of grains. Imports of adequate fertilizer chemicals are expensive, and long-distance transport of low-value phosphate rock is poor economics. Domestic or regional sources of phosphates are needed in much of Asia, South America, and Africa. In South America alone, the estimated requirement for fertilizer minerals in 1970 was at least five times the consumption in 1965, and the problem is greater in the more populated parts of the world. Phosphorus compounds are absorbed from the soil by vegetation and return to the soil with planet decay. When agricultural crops are harvested, most soils become rapidly depleted of phosphorus and future crops become less abundant. Phosphate fertilizers are in high demand where population pressures require the maximum utilization of agricultural lands. It is estimated that proper fertilization to replace the phosphate removed by crops alone would require double the current production.

It is essential for industry to have an assured supply of clays, salts, and various industrial chemicals. Remote sensing has the capability of providing valuable lithologic information applicable to many industries, but much research in the interpretation of imagery of metamorphic terrains will be required before remote sensing will be of assistance in locating asbestos, talc, and refractories. Vincent *(18)*, however, has indicated that thermal infrared spectral ratioing should have applications in exploration for sand and gravel, surveying of beaches (quartz versus calcite sands and mapping of dunes), and mapping of volcanic ash flows.

National defense requires an assured supply of minerals and metals related to our military economy. Jeffries *(19)* comments that military needs have directly or indirectly been the basis of the development of uranium, beryllium, and titanium ores, and for a significant requirement in copper, steel, aluminum, nickel, and magnesium. The electronics, aerospace, and munitions industries require stockpiles of concentrates, chemicals, alloys, and nuclear products. Efficient management of these industries requires long-term assured supplies of a much longer list of raw materials. Because domestic reserves are clearly inadequate for several such commodities, it behooves the use of space technology for finding favorable areas of exploration both domestically and in the traditionally friendly foreign mineral sources. Assured supplies of such uncommon metals as zirconium, rare earths, cesium, niobium, germanium, and tungsten, although only minor components of their ores, could be important to national security. It is anticipated that LANDSAT data will assist in better defining patterns of mineralization and provide new knowledge about ore genesis related to the occurrence of some of these raw materials.

The initial benefits to be provided to the minerals industry by satellite remote sensing will rarely be the pinpointing of ores, but better knowledge of geology. It is likely that developing techniques for the examination of imagery and tapes will provide information which we may extrapolate to extensions of a mineralized area or to other areas; but the real benefit will be extrapolating the data to concepts and understanding of regional and local ore deposition patterns. Remote sensing provides information on faulting, linears, and lithology, which may serve as indicators to likely areas for follow-on exploration; but in only a few cases so far does it identify known resources. Direct prospecting will not be realized until sensors and computer techniques exceed their current development. Assured supply relies on faster exploration, discovery, and development, and it is becoming increasingly clear that in many areas remote sensing may often be the only means of obtaining the scale of information required for exploration.

To expand the scenario a little further, models of availability and requirements are the basis of forecasting and adapting resource management policies. Conventional surface and aerial methods have provided the data for predictive studies, but it is likely that remote-sensing data can take over much of the role. The employment of this new source of data may permit the identification of problems in their early development, or it may permit a reduction in the time between recognition of supply problems and the implementation of programs to remedy them.

We have a relatively good concept of the world's mineral resource distribution, but our concept of the extent of these resources is hazy, particularly in the remote regions of the earth. What, for example, do we know of structural patterns in much of the Alaskan and Canadian Arctic, the interior of Thailand or Burma, the remote areas of Brazil and Bolivia, or much of central and west Africa. Our knowledge of the mineralization of many such areas is based on very local information, knowledge of the deposit, not the mineralization pattern. A recent LANDSAT-based structural diagram of Thailand showed clearly for the first time two major structural patterns that will be of interest to economic geologists. Recent imagery of a remote Bolivian area showed stream patterns that appear to be controlled by buried folded structure.

It was reported at the Second LANDSAT Symposium that certain petroleum-bearing structures in California appear to fall along LANDSAT-discovered linears, suggesting structural control. Surficial expressions of salt domes have been seen on imagery of the Gulf Coast region, suggesting possible locations for exploration. Short *(20)* noted that "Most geophysical and geochemical surveys take years to 'pay-off'; ERTS should be allowed a similar time-line to prove its worth in geology." However, he also noted, "ERTS images will probably be most valuable to prospecting through their use in improving our understanding of the tectonic framework of the earth's crust. However, ERTS is not likely to reach the status of geophysical or geochemical prospecting or drilling as the prime means of locating new mineral deposits." LANDSAT provides a capability for making better maps and defining structure, but the value of these capabilities as a guide to exploration is limited by the scale of LANDSAT products.

Although it is possible that yet-undetected high-grade ores will be found outcropping in remote parts of the earth, such as the Hamersley Range iron deposits in Australia, such occurrences presumably will be rare. Exhaustive exploration around the major mineral districts has produced relatively few surface deposits in recent years. Most new mineral occurrences have been obscured deposits, many located at considerable depth. The future bonanzas for our mining industry may include the larger outcropping or near-surface, lower-grade ores where commercial exploration and development have been held in abeyance. Geochemical methods have been employed successfully in locating disperse ores such as the bertrandite deposits in Utah. Complex geophysical and geochemical techniques were used to delineate the disperse gold deposit at Carlin, Nevada. It has been forecast that a major activity in minerals in the future will be the exploitation of large, currently marginal deposits by bulk treatment methods.

It is the responsibility of present agency planners to carefully select missions with parameters best suited to the requirements of earth resources data users and to precisely specify systems applicable to user needs, particularly in the case of those narrow-band sensors which apply to geology and geochemistry. LANDSAT has taken the earth's picture; we next require even higher spectral and spatial resolution on products that can be enlarged to scales better suited to exploration.

What Simon Ramo of TRW, Inc., stated in 1967 holds particularly true today *(21)*:

> We can no longer ritualistically recite the intrinsic technological worth to society of the space program. Space must now compete with other national endeavors in a period of tighter funding and increased pressure for alleviating earthbound problems. This calls for a careful reexamination of the priorities and enphases within the entire space enterprise. . . .
>
> We have it within our power of choice to exploit what we have learned . . . so as to produce values for our society in the 1970s substantially greater than what the entire space program will have cost us in the 1960s.

Ramo included mineral resources observation by satellite, among other space applications, as technologically feasible, economically sound, and socially desirable.

RESOURCES AND ENVIRONMENT MANAGEMENT

It would be presumptuous to predict the eventual impact of future remote-sensing data applications on the minerals industry at so early a time in the development of the technology. So far, it has been sufficient to report new developments and to look cautiously into the crystal ball. We can at least conclude that the mineral picture looks somewhat more encouraging because of remote-sensing technology, because a large number of leading geologists are analyzing photographs and imagery to determine their applications, and because the value of the techniques have been accepted in many areas of the developing world where new mineral discoveries are most apt to be made—and upon which the industrial nations may have to depend for adequate raw materials.

In view of the escalating demand for ores and energy, as well as the demand for an increasingly higher standard of living (or consumption) by a dramatically multiplying world population, the mineral and energy resources discovered by remote sensing over the next couple of decades will not solve the ultimate raw material problems. As we suggested at the start of this volume, discoveries will have to be continuous, the initial discoveries hopefully bridging the forecasted period of greater inequities of distribution, further shortages of essential commodities, and attendant domestic and international resource-based crises. Ultimate solutions beyond this possible stopgap are the realms of other branches of science. These responsibilities include developing new energy sources such as thermal and solar energy, making coal gasification and the exploitation of oil shale economic and safe for the environment, and the introduction of the breeder reactor. Political–economic arrangements will undoubtedly have to be made among nations to assure resource availability. We may foresee a possible period of self-sufficiency for energy requirements, but we may never see such a period of self-sufficiency for many essential chemicals and ores. We shall have to increase efforts to find economic methods of using our vast low-grade ores, determine where further substitution of materials is possible, and produce synthetics with a wide range of properties. Science will resolve such of these problems as is possible. When the problems are beyond the realm of technology, it falls to business and government to seek economic and political solutions. It is clearly in the national interest to attain the maximum potential economic self-sufficiency in all possible mineral areas, even if we are not immediately required to rely on the domestic resources. It appears that remote sensing has the capability of providing valuable data for resource inventories and input to predictive models for the use of resource management.

Among the developing countries, shortages and higher prices will have the greatest economic and political impact on the poorer of these nations. Non-oil-producing developing countries are estimated to have had to pay $15 billion more for imports of petroleum products, ferilizers, etc., in 1974 (*22*). The increased prices of these essential imports and foods will result in greater trade deficits. Any reduction of imports would mean a decline in growth rates and would increase the degree of poverty. In this situation, the prosperity of industrial and resource-producing nations produces greater problems for those countries that have little to export and which pay higher prices for imports. The current energy crisis directly impacts the matter of food supply in these areas. Nitrogen fertilizers largely depend on natural gas. With supplies of fuel short and prices high, the situation is directly reflected in fertilizer prices. It is anticipated that combined fertilizer and fuel crises will tax the foreign exchange reserves in several areas and pose problems of adequate farm inputs into the economy *(22)*. In such areas, the location of oil, fertilizer minerals, or minerals for export trade would help solve the imbalance current between affluence (and inflation) in industrial nations and those developing areas with high debts and limited exchange-earning capacity.

As the economic application of remote-sensor products continues to be analyzed, the practical utility of the new source of information becomes clearer. The general approach of most LANDSAT investigations has tended to emphasize technical development and initial research findings rather than the contribution that the data may contain relevant to management decision making; it has been necessary to establish the geological and exploration capabilities of the LANDSAT system. But now that this initial phase is in good hands, it is important to determine how the data may assist mineral resource and environmental planning. The following sections consider evolving resource and environmental issues, some of which are certain to be impacted by remote sensing.

Fuels

Strong societal demands for energy puts increasing pressure on resource agencies and industry to locate additional reserves of fossil and nuclear fuels. We are not about to run out of nuclear minerals or coal or oil, but there are serious problems to face as to permissible increases in energy use, its increasing cost, and where it will come from *(9)*. Peach *(9)* reminds us that we have coal enough to last for centuries and the technology to gasify and liquefy, that there is adequate Alaskan and Canadian oil reserves for three or four additional pipelines, that offshore production of oil and gas accounts for 16 percent of total U.S. production (yet only a small percent of our offshore has been tested), and that we currently have the technology to produce high-quality oil from the Green River shales at competitive prices. In addition, current higher prices make secondary recovery of oil more economic. We have the potential capability of providing 20 to 30 percent of our energy through nuclear plants, and the limited availability of oil and gas warrants the local development of geothermal energy. These developments require the continuance of relatively high prices, however, to receive the necessary congressional support to implement new alternative energy source programs. Peach *(9)* suggests that the U.S. government enter directly into limited oil production and distribution, not through nationalization, but through TVA-type corporations for offshore drilling, oil shale operations, Alaskan oil production, and other areas, including the operation of refineries. Such corporations should not take existing operations from private companies, but would operate some fraction of the increase in petroleum production. As unattractive as such TVA corporations may be to some, alternative suggestions include gasoline taxes, 4-day work weeks, reliance on mass transit, restricted energy use by industry, and rationing of automobile and household fuels.

Under the threat to major U.S. and European oil companies to expropriate concessions, the Organization of Petroleum Exporting Countries (OPEC) has increased its governments' oil revenues radically. Levy *(23)* notes that oil income to the Arab countries has risen from $4 billion in 1970 to an estimated $60 billion in 1974. It is no longer the international oil companies that call the shots in the Persian Gulf area, but the governments of the producing countries. OPEC largely controls not only the price but the availability of a significant portion of the world petroleum. Most proposed solutions for dealing with OPEC are unrealistic stopgap measures requiring austerity by importing countries or payments-in-goods plans. In January 1975, Secretary of State Henry Kissenger discussed the use of force, military action, as a dangerous course to follow to bring down Mideast oil prices; however, such an option was left open if the alternative were strangulation of the industrialized world. Although such action might only be considered in the most extreme emergency, the availability and cost of a mineral commodity had brought the world to this dangerous point. The OPEC cartel not only has dominated the present market; it could adjust the future price of oil to inhibit major investment in developing alternative sources of energy. It was as Park had warned; a dangerous political situation had resulted because of a shortage of a critical mineral *(4)*.

In the final analysis, the solution to the energy crisis appears to be an increase in domestic refining capacity, the initiating of development of alternative sources of energy, and greatly expanded exploration both domestically, offshore, and in those areas where U.S. cooperation may be more attractive than the temporary advantages of alignment with OPEC. Other chapters describe the applications of remote sensing and associated techniques to energy development. It may be generally stated that oil and gas are prime targets for remote sensing from satellite and aircraft, and that spectral ratio techniques apply to the search for both oil and uranium. As for coal, oil shale, and thorium in the United States, large reserves are known; the foreseeable problems are in the areas of extraction, processing, and economics. It is anticipated that the accelerated discovery of energy sources throughout the world will be an important factor in the current international situation, and in the long run will address a foreseeable greater problem than fuel itself: the production of adequate low-cost energy for the economic exploitation of currently subeconomic reserves of nonfuel minerals.

Nonfuel Minerals

Remote sensing is of potential importance in prospecting for minerals such as copper for which undiscovered reserves are probably large and discovery depends largely on improved exploration techniques. On the other hand, there are minerals such as bauxite, which are found in geological settings of relatively limited extent in the United States. Based largely on the minerals reserve picture compiled by the U.S.

Geological Survey *(3, 24),* plus the likelihood of guides to possibly favorable environments being identified, the following list identifies timely important resource subjects for exploration using remote-sensing methods.

Beryllium: deposits usually obscured, but regional geological studies are warranted.
Cobalt: most domestic sources are subeconomic, but Alaska deserves a thorough exploration.
Copper: remote sensing is providing new exploration techniques.
Fluorine: further search for alteration zones and mineralized structures.
Gold: better geologic and geochemical mapping applications may help locate additional dispersed-type deposits.
Molybdenum: adequate reserves, but increasing demand encourages studies relating deposits to plate tectonics.
Nickel: new geologic mapping needed for locating sulfide deposits in the United States and Canada.
Platinum: new structural and lithologic data on Alaska should suggest favorable environments.
Tungsten: deposits usually obscured, better information of broad mineralization patterns needed.
Zinc: probable large reserves await new exploration approach.

Some important minerals are by-products domestically. These include antimony, cadmium, silver, and vanadium. These are tied to copper, lead, and/or zinc production.

Several important minerals are considered unlikely to be found in domestic environments, including asbestos, bauxite, chrome, and economic deposits of manganese. A more thorough search for chrome should be made in Alaska. The search for carbonatites and new environments for columbuim and tantalum might be a rewarding remote-sensing effort. There are possibilities also of finding new geological environments of tin, especially in Alaska. The costs of processing make rutile and mica a matter of imports. Lead is a matter of economics. Industrial minerals and/or chemical minerals are for the most part poor subjects for remote sensing at the present stage of its development. The possibilities of employing remote sensing for the mineral needs of the less-well-endowed or developing nations is a matter of foremost importance to themselves and to the world economy.

CONSERVATION AND ENVIRONMENTAL QUALITY

Industrial society requires an ever-expanding technology for exploiting nature, for converting the resources of our environment into needed products, and for healing the damage such activities inflict on our environment. Population and industrial needs have brought us past the point where we may merely conserve known resources for future use. Economic projections in the minerals field have for some time required estimating the future discovery of as yet unknown reserves in order to predict future allocations of raw materials. Predictions of this nature are realistic, looking at the recent history of geological exploration, but it points up the gap in our resource picture. There is a danger of our not following up with satellite programs at the necessary scale of effort. There is the possibility of current "real-world" problems of inflation, housing, unemployment, peace, etc., cutting into the expenditures needed to prepare against future resource shortages and to monitor environmental problems.

The NAS/NRC Committee on Resources and Man undertook 2 years of inquiry into the basic problems facing mankind, including peace, population, pollution and resources as the critical variables. In a contribution on mineral resources, T. S. Lovering made this observation *(25):*

> The total volume of workable mineral deposits is an insignificant fraction of 1 percent of the earth's crust, and each deposit represents some geological accident in the remote past. Deposits must be mined where they occur—often far from centers of consumption. Each deposit also has its limits; if worked long enough it must sooner or later be exhausted. No second crop will materialize. Rich mineral deposits are a nation's most valuable but ephemeral material possession—its

quick assets. Continued extraction of ore, moreover, leads, eventually, to increasing costs as the material mined comes from greater and greater depths or as grade decreases, although improved technology and economics of scale sometimes allow deposits to be worked, temporarily, at decreased costs. Yet industry requires increasing tonnage and variety of mineral raw materials; and although many substances now deemed essential have understudies that can play their parts adequately, technology has found no satisfactory substitutes for others.*

All of the problems of resource availability cannot be solved by finding substitutes for critical minerals. Park *(4)* notes that there are no known substitutes for structural steel or for mercury in control equipment. Early in the application of aerial methods of remote sensing, Secretary of the Interior Udall commented that the needs of persons now living would include 90 million tons of copper in their lifetime, but that the known reserves were less than half that amount *(26)*. He called upon aerial methods to speed the acquisition of data for resource management use. After 2 years of LANDSAT data, we may conclude that remote sensing has the potential of providing much vitally needed data for problems of such accelerating urgency as mineral and energy shortages and pollution, as well as for food supply and population studies.

As for the increased danger of pollution, Pecora *(27)* noted:

A burgeoning population requires more natural resources each decade. Even if zero population growth is attainable mankind's demand for resources is staggering. Some estimate that cumulative demand will triple or more by the year 2000.

Appalachia is studded with remnants of poorly practiced pit coal mines that operated without regulatory controls. For less than 2 percent of the value of the coal marketed, most of these abandoned sites might have been acceptably restored. Today the cost would reach hundreds of millions of dollars.

A mature nation like the USA so dependent upon its natural resources cannot turn off its economic pattern of development without a major impact on its welfare and way of life. Nor can this nation continue its economic development without more regard for the environmental impact of its industry. . . .

Beside despoilation of our landscape, increased mineral activity, processing, and transportation will bring greater problems of oil spills, the dumping of debris and effluents into streams and into the air, and waste piling on the surface. Inasmuch as remote-sensing technology will afford new opportunities for exploration and hence industrial development, it also provides a new approach to our concern for such alteration of the environment as may accompany accelerated development. The final chapter in this volume specifically addresses environmental monitoring as it applies to the minerals industry. Such monitoring will determine water-pollution sources, concentrations and dispersion patterns, and will provide data for regulatory and enforcement agencies. Signatures will be established for various environmental indexes associated with such problems as tailings disposal, acid mine waters, dump fires, ground subsidence, open-pit mining, spills, etc. A need of environmental management will be for the collection of precise data from in situ stations capable of identifying the various chemical and physical parameters. Environmental management will be concerned not only with the identification of pollution sources but also with the relationship of industrial expansion to the production of such pollution.

*Reprinted from ***Resources and Man: A Study and Recommendations*** by the Committee on Resources and Man of the Division of Earth Sciences–National Research Council with the cooperation of the Division of Biology and Agriculture, by permission of the publishers, W. H. Freeman and Company, San Francisco. Copyright © 1969.

REFERENCES

1. President's Materials Policy Commission, W. S. Paley, Chairman, Resources for Freedom, Washington, D.C., 1952.
2. Hearings Before Joint Committee on Defense Production, 92nd Congress, Aug.–Sept. 1971.
3. Brobst, D. A., and W. P. Pratt, eds., United States Mineral Resources, *U.S. Geol. Surv. Prof. Paper 820,* 1973.
4. Park, C. F., Jr., *Affluence in Jeopardy,* Freeman, Cooper, San Francisco, 1968.
5. An Assessment of Solar Energy as a National Energy Resource, prepared by NSF/NASA Solar Energy Panel, NSF/RA/N-73-001, Dec. 1972.
6. Theobald, P. K., S. P. Schweinfurth, and D. C.

Dunn, Energy Resources of the United States, *U.S. Geol. Surv. Circ. 650,* 1972.
7. McKelvey, V. E., Mineral Resource Estimates and Public Policy, *chapter in* United States Mineral Resources, *U.S. Geol. Surv. Prof. Paper 820,* 1973.
8. Cheney, E. S., Limits to Power Growth, *Geology,* v. 2, no. 6, 1974.
9. Peach, W. N., The Energy Outlook for the 1980's, a study prepared for use of the Subcommittee on Economic Progress, Joint Economic Committee, U.S. Congress, 1973.
10. A Current Appraisal of Underground Coal Gasification, Report to U.S. Bureau of Mines, Arthur D. Little, Inc., Cambridge, Mass., C-73677, Dec. 1971.
11. White, D. E., Geothermal Energy, *U.S. Geol. Surv. Circ. S19,* 1965.
12. NASA Authorization for Fiscal Year 1975, Report of the Committee on Aeronautical and Space Sciences, U.S. Senate, on H.R. 13998, May 6, 1974.
13. Lyon, R. J. P., et al., Geology, Panel 2 Report, Useful Applications of Earth Oriented Satellites, National Academy of Sciences—National Research Council, to NASA, Washington, D.C., 1969.
14. Salomanson, V. V., and A. Rango, ERTS-Applications in Hydrology and Water Resources, NASA-GSFC X-650-73-131, May 1973.
15. Meier, M. F., Evaluation of ERTS Imagery for Mapping and Detection of Changes of Snowcover on Land and on Glaciers, *Proc. Symposium on Significant Results Obtained from ERTS-1,* v. I, New Carrollton, Md., 1973.
16. Altschuler, Z. S., et al., Geology and Geochemistry of the Bone Valley Formation and its Phosphate Deposits, West Central Florida, Annual Meeting Geological Society of America, Miami, Fla., 1964.
17. McKelvey, V. E., *Successful New Techniques in Prospecting for Phosphate Deposits, Natural Resources,* v. II, U.N. Conference on Applications of Science and Technology for Less Developed Areas, 1964.
18. Vincent, R. K., Spectral Ratio Imaging Methods for Geological Remote Sensing from Aircraft and Satellites, *Proc. American Society of Photogrammetry Conference on Management Utilization of Remote Sensing Data,* Sioux Falls, S.D., 1973.
19. Jeffries, Z., Minerals in Man's Future, in *Economics of the Minerals Industries,* American Institute of Mining, Metallurgical, and Petroleum Engineers, New York, 1964.
20. Short, N. M., *Mineral Resources, Geological Structure and Landform Surveys,* Vol. III, Second ERTS Symposium, New Carrollton, Md., 1973.
21. Ramo, S., Space and National Priorities, *Air Force,* v. 50, no. 12, Dec. 1967.
22. The Energy Crunch: What It Will Mean, *War on Hunger,* Agency for International Development, Department of State, Washington, D.C., May 1974.
23. Levy, W. J., World Oil Cooperation or Industrial Chaos, *Foreign Affairs,* July 1974.
24. Pratt, W. P., and D. A. Brobst, Mineral Resources, Potentials and Problems, *U.S. Geol. Surv. Cir. 698,* 1974.
25. *Resources and Man: A Study and Recommendations by the Committee on Resources and Man of the Division of Earth Sciences, National Academy of Sciences—National Research Council with the Cooperation of the Division of Biology and Agriculture,* W. H. Freeman, San Francisco, 1969.
26. Udall, S. L., Resource Understanding—A Challenge to Aerial Methods, *Photogrammetric Eng.,* Jan. 1965.
27. Pecora, W. T., *Nature—an Environmental Yardstick,* U.S. Department of Interior Publication GPC 2401-00214, Albright Conservation Lecture, University of California, Berkeley, Calif., Jan. 1972.

3

GAP BETWEEN RAW REMOTE-SENSOR DATA AND RESOURCES AND ENVIRONMENTAL INFORMATION

Bernardo F. Grossling and John E. Johnston

Bernardo F. Grossling and John E. Johnston are with the U.S. Geological Survey, Reston, Virginia.

We shall discuss here the economic justification for the gathering of raw remote-sensor data, and how to interpret it so that it becomes useful for private and public users. We do not intend to enter into all the ramifications of remote sensing to resource development and environment studies. Briefly stated, remote sensing provides data on the distribution, properties, and characteristics of water, snow and ice, vegetation, soil and rock, and cultural patterns on the face of the earth. Gaps of understanding between the various groups involved are to be expected in such a rapidly evolving field. We shall survey the nature of these gaps; we do not pretend to close them, but the nature of the work to be done to close them should be clearer.

Various users want remote-sensor data formulated into different types of information packages; some want to analyze raw data, while others want only summary information for decision making or for further analysis with other data. It is difficult for the relative few involved in collecting raw remote-sensor data to develop systems for gathering and dispersing earth resources and environmental data to many levels of users.

In the sequel we explore two issues: (1) the bearing of remote-sensing data on economic activity, and (2) how to interpret raw remote-sensor data so that it becomes usable for specific economic activities. Development and conservation of earth resources, on the one hand, and preservation and improvement of environment quality, on the other, are two sides of the same coin. Here we shall emphasize only remote sensing in relation to resource development.

RESOURCE INFORMATION AND ECONOMIC ACTIVITY

First, one has to examine how to decide how much natural resource information, which includes remote-sensor data, is to be gathered in relation to economic activities. For such a study, economic analysis can provide a useful framework but, to be more meaningful, an understanding of the many technical factors involved in the natural resources field and of the decision-making mechanisms of private enterprise is required.

To analyze the subject, one may start by asking some basic questions. What is the influence of natural resource information on the economic development of a nation? When is a lack of publicly available resource information a deterrent to private entrepreneurs, not only in a country like the United States, but also in developing nations? What kinds of information and how much are needed by various decision makers? How are the various items of information used within the fabric of economic decision making? Is it possible to precisely identify such a fabric before development of natural resources? For instance, if a regional geologic survey of an area is contemplated, which projects will be selected for a cost–benefit analysis? To properly answer such questions entails examining a complex branching chain of decisions—sometimes overlapping, interlocking, or apparently unrelated—that may be made by government and private enterprise. These are not academic questions; rather they determine the dimensions of the problem.

Institutional Arrangements for Gathering Resource Information

Governments play a primary role in the resources field. For example, by conducting surveys of natural resources in order to provide a body of public knowledge, which is made available to large and small entrepreneurs. In contrast, resource information obtained by the private sector is usually withheld from public disclosure for competitive reasons. In some cases, government surveys of a regional nature may play a catalytic role in focusing the attention of the private sector on certain areas, which may be undeveloped or economically depressed. Government resource surveys of a regional nature provide a common basis for the various facets of an economy—mining, oil and gas industry, forestry, water utilization, road construction, flood control, irrigation, land utilization, etc. In other instances, government surveys may be required to provide the necessary information for the formulation of natural resource development plans. In this sense, resource surveys are truly instruments of economic policy.

It would seem that such surveys are a precondition to any further planning or actions by government or private enterprise. They provide a minimum threshold of publicly available knowledge. River-flow measurements at a given locality may not seem to be directly related to any specific hydroelectric site. Yet the data may be of potential use to a large but undetermined number of projects—water supplies, flood control, hydroelectricity, irrigation, drainage, etc. To arbitrarily choose one out of all the possible projects would merely make the economic calculations simpler; but it would not add meaning to the subsequent judgment about the extent of the resource information survey.

The motivations and manner of operation of private enterprise are of great relevance to the development of natural resources. In general, private entrepreneurs do not really depend on public information such as that provided by a general resource survey. Integrated mining companies, for instance, already appear to have much information on the mining potential not only of the United States but also of other countries. As an example, in most of Latin America the private sector appears to have enough resource information to proceed with the development of many investment opportunities. It is held back not because of lack of such information but rather by legal and institutional impediments.

In the United States, private enterprise plays a strong role, and the need for resource information is markedly different than in a developing nation. Yet even in Latin America the private sector plays a strong role in gathering information on natural resources. This is particularly true of minerals, including oil and gas, and forest resources.

In developed countries, private enterprise has usually already pulled far ahead of what is achieved in general surveys. Therefore, government surveys of natural resources should aim at opening up new opportunities. One example is provided by Canada, where extensive surveys of almost undeveloped areas are being carried out. In particular, the ready availability of airborne magnetic surveys of extensive areas of Canada has encouraged the private sector to look for mining projects, despite the fact that in most of the areas surveyed there still may be none or only an inadequately developed economic infrastructure.

Nowadays in developing nations the truly effective economic planning, not merely in the resources field, rests in national planning offices usually attached to the office of the president of the nation. Usually in such nations there are strong specialized agencies that, as part of their activities, procure information on various aspects of natural resources. Therefore, a central bureau for getting all the resource information for all the facets of a national economy is not a sound idea. Such a central bureau may be more relevant to countries with a large degree of government control of the economy, but is not too applicable to the highly competitive private enterprise system of advanced Western nations.

Resource Information and the Investment Decision

A simplistic approach to this problem, which has been proposed by some, would be to decide on a specific resource survey only when a specific investment opportunity already has been previously discovered by government planners. They would decide when, how, and how much information is to be obtained. That is, the resource information would be part of a two-step decision process concerning a specific project: (1) whether to gather the resource information, and (2) whether to undertake the feasibility study of the project.

Such a model, by substituting an unknown agenda of investment possibilities for one selected by a government planner, underestimates the potential economic benefits, while charging all the cost of the resource information to one project. To prepare an exhaustive agenda of potential projects is often either impractical or unfeasible. For how can the location and extent of mineral deposits be known ahead of the geologic mapping that is to assist in their discovery? In a highly competitive system such as that of the United States, how is the "present value of the investment opportunities" going to be determined for a regional geologic survey?

How to Decide on the Extent of Natural Resource Surveys

In theory, expenditures on resource information should be considered as ordinary capital investments, making them competitive with other investment opportunities in an economy. As such, they should yield as much as other capital investment opportunities elsewhere in the economy. This objective is correct in principle; the difficulty lies in its application.

First, as we have already noted, for many kinds of resource surveys it is impractical to define in full the agenda of possible investment projects.

A second difficulty is that one cannot assume a simple deterministic link between (1) a decision to invest in resource information and (2) the decision to invest in a specific project. Private entrepreneurs are bound by considerations other than those of government planners, thus breaking a simple cause and effect chain. Market conditions may change unexpectedly after the gathering of the resource information, forcing a postponement or abandonment of a project. Competitive moves might make a project obsolete. For instance, the discovery of gas might make a hydroelectric project unprofitable. Increased cost of capital may also result in the temporary abandonment of a project.

These considerations indicate that the proper model for decision making is not a simple sequential model, but rather one involving multiple alternatives, with uncertainty and risk. To prepare such a model would require a vaster and keener knowledge of the various engineering, technological, and industrial factors and of the investment opportunities than is available to governmental natural resource planning agencies. In fact, a proper judgment as to the extent of natural resource information that is to be gathered by a government survey requires a thorough investigation not only of the investment opportunities that are open for the government itself but also for private enterprise.

FROM RAW REMOTE-SENSOR DATA TO USER INFORMATION

Nature of Remote-Sensing Techniques

Remote-sensing techniques are so meaningful because man's activities take place mostly at the earth's surface. Practically all places of habitation are at the earth's surface; agriculture, perhaps the springboard of civilization, consists in the cultivation of plants at the earth's surface; roads, railroads, and navigation canals extend over the earth's surface; and manufacturing plants are also built at the surface.

Vegetation is vitally dependent on electromagnetic radiation by selectively absorbing or reflecting certain bands of the sun's radiation at the base of the atmosphere. These selective responses to electromagnetic radiation make plants eminently suited for detection and study by remote-sensing techniques. Nature's design, through adaptation and selection, has made them so.

Mining and oil production, on the other hand, entail probing the earth to great depths, much greater than the shallow penetration of the electromagnetic radiation used in most remote-sensing methods. The earth's surface provides a cut across the geology of the earth from which extrapolations can be made to forecast the geology at depth. For remote sensing to achieve its full potential, we should learn the meaning of the various features observed on the skin of the earth.

Let us contrast some characteristics of satellite and aircraft photography of the earth. For a sun-synchronous polar orbit, the satellite height is approximately 920 km, whereas aircraft height for aerial photography usually ranges from 1.5 to 20 km. A suitable scale for satellite mapping is 1:250,000, whereas it is on the order of 1:20,000 for aircraft. The typical coverage of a satellite photograph is 180 × 180 km, whereas that of an aerial photograph is about 20 × 20 km.

Although the earth is large, satellite photography offers the possibility for any individual to recognize on sight which part of the earth a given photograph comes from. About 10,000 satellite photographs are required to cover, with overlaps, the entire land areas of the earth. It seems feasible for an average person to learn to identify them all.

To illustrate some of the basic natural resource and land-use questions that may be studied by satellite imagery, we have chosen a few imageries taken by the NASA Earth Resources Technological Satellite, ERTS-1 (now known as LANDSAT). The imageries were obtained from the Data Center of the U.S. Department of the Interior/Geological Survey's Earth Resources Observational System (EROS) at Sioux Falls, South Dakota. Each image frame corresponds to a vertical view, approximately 100 nautical miles on a side. North is upward in each imagery. These imageries are in false color; the spectral band 4 or "blue-green" is in yellow; the spectral band 5 or "red" is in red; and band 6 or 7, both of which are in the near infrared, is in blue.

An area of Burma, which includes the city of Rangoon, is shown in the imagery of Fig. 1. The center of the picture is at latitude 17°15′N and longitude 96°17′E; the imagery was recorded on October 30, 1972. A portion of the Irrawaddy River delta, known as the Mouths of the Irrawaddy, is in the southwestern portion. On the eastern part is the Sittang River, which flows into the Gulf of Martaban. The city of Rangoon is in the east flank of the estuary at the south-central portion. Much of the lowlands around Rangoon is taken up by rice paddies. Tropical forests extend over a central block north of Rangoon in terrain underlain by Miocene sands and shales. Land use in the flood plains is greatly conditioned by river courses and drainage. The lowlands exhibit a rich pattern of tonal anomalies, which are of geologic interest. A study of silting and marine currents, as for harbor construction, should be facilitated by such imagery.

A portion of the Majunga Basin in the northwest margin of the Island of Madagascar is shown in Fig. 2.* The center of the imagery is at latitude 15°48′S and longitude 46°58′E; it was recorded on October 5, 1972. The delta on the left side is the Betsiboka, and the one on the right is the Mahajamba. An essentially monoclinal sequence of Triassic to Recent sediments, with some Cretaceous interbedded lavas, dips toward the Mozambique Channel in the northwest, away from the crystalline basement on the southeast portion. A significant portion of the Majunga Basin should be offshore. The imagery exhibits a complex interlacing of the outcrop patterns of geologic formations, cuestas, the pattern of watercourses, delta deposition, and agricultural lands. These various patterns are highly interactive, so that their interpretation should be interdisciplinary.

A portion of the state of Jalisco, Mexico, is shown in Fig. 3. The center of the imagery is at latitude

*Figures 2 and 6 will be found in the color insert.

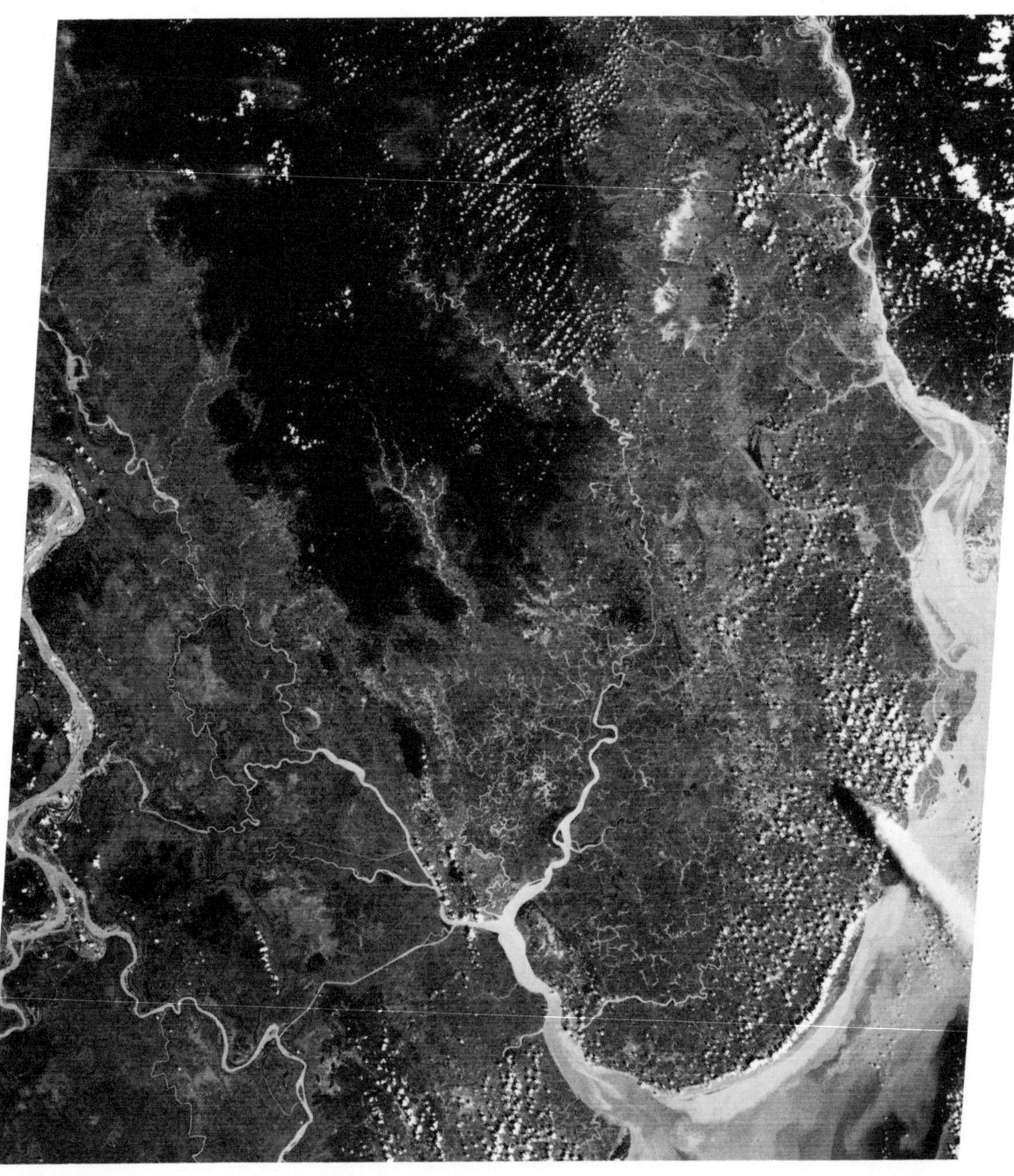

FIG. 1
Rangoon, Burma.

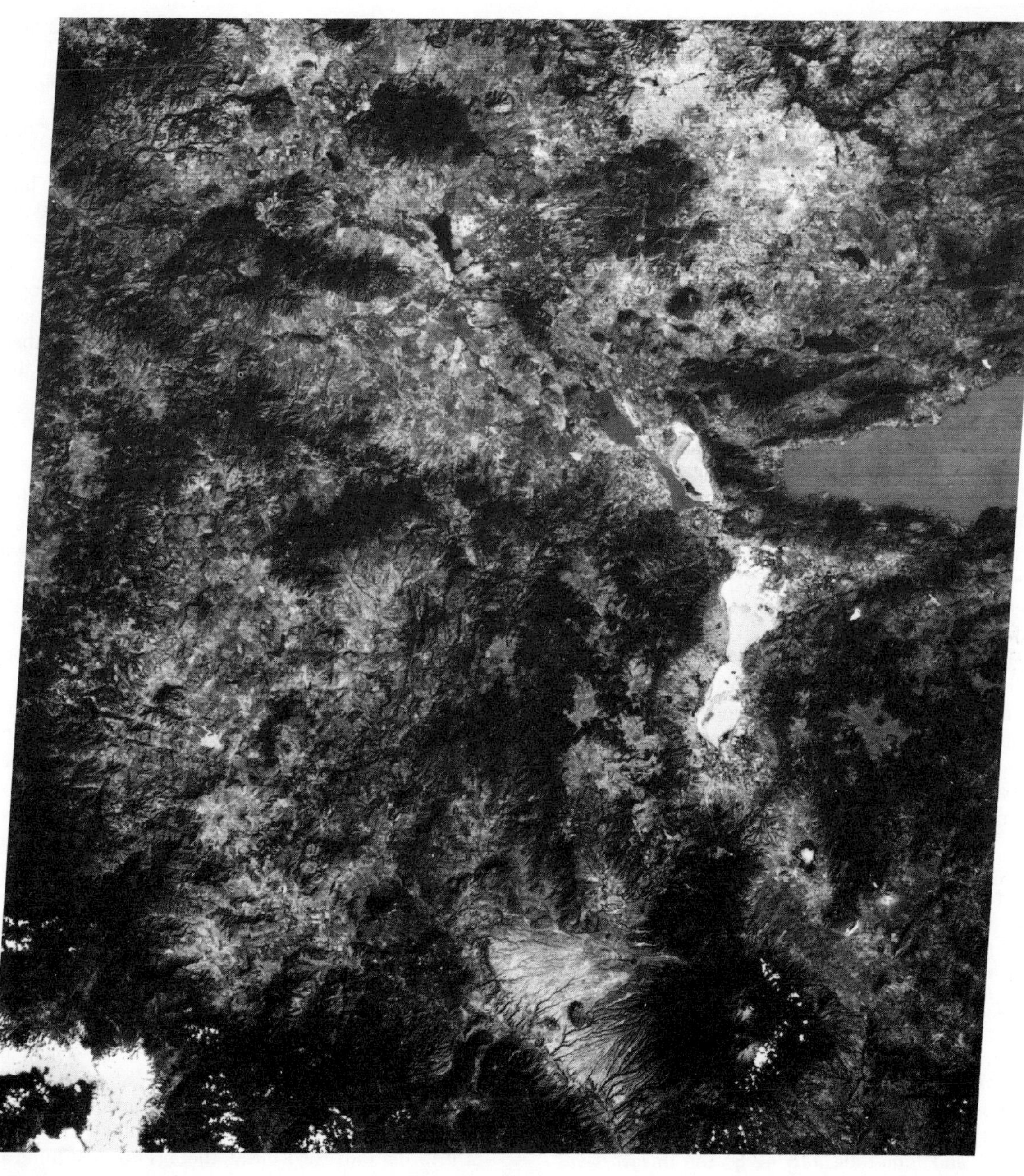

FIG. 3
Guadalajara, Mexico.

$20^\circ 12'$N and longitude $103^\circ 50'$W. The imagery was recorded on December 24, 1973. The city of Guadalajara is just north of the center point, and the west end of Lake Chapala is on the eastern side. Pleistocene and Holocene volcanic rocks underlie most of the area, with a few exposures of underlying Cretaceous rocks. Several iron-ore deposits, a lead deposit, and some gypsum, magnesium kaolin, and clay deposits are known to the southwest of Guadalajara. The extension of agricultural lands is limited.

A portion of international boundary between Montana in the United States and the Canadian provinces of Alberta and Saskatchewan is shown in Fig. 4. The center of the imagery is at lattitude $48^\circ 54'$N and longitude $110^\circ 34'$W. The imagery was recorded on June 23, 1974. The border is clearly shown by the contrast between farming on the south side and grazing on the north side. Repetitive satellite imagery of agricultural areas may permit forecasting crop production.

An area of the United States that includes the cities of Washington, D.C., and Baltimore, Maryland, is shown in Fig. 5. The center of the imagery is at latitude $38^\circ 48'$N and longitude $76^\circ 48'$W; the imagery was recorded on September 23, 1972. The city of Baltimore is shown as the dark-blue area in the north -central part. Dark radial lines indicate roads and other man-made developments extending from the city's core.

Sparrows Point is the dark-blue area just east of Baltimore, and is shown enlarged four times in the inset. The largest steel mill in the world is located at Sparrows Point, which provides an excellent steel-mill site. It consists of a large tract of flat land, a few feet above water, with good foundation ground, which forms a peninsula at the port's entrance. Abundant water for steel manufacture is readily available. The nearby harbor can accommodate deep-draft ships. This location resembles that of Punta do Tubarao, Vitoria, Brazil, where a large steel complex is being developed.

A portion of the Peking Municipality in Northern China is shown in Fig. 6. The center of the image is at $40^\circ 12'$N and longitude $115^\circ 42'$E; the imagery was taken on November 21, 1973. The rectangular outline east of center is the city of Peking, for about 700 years the capital of China, which now contains more than 7 million inhabitants. In the mountains north of Peking, segments of the Great Wall can be recognized. The Wu-T'ai Shan Mountains trend across the picture, and to the southeast is a portion of the North China Coastal Plain.

Peking is an important center with heavy-to-light industries. Coal deposits are exploited on the western side of the Wu-T'ai Shan. The plains produce mainly winter wheat, corn, and kaoliang. The deeply dissected mountains with a rugged and diverse topography make it difficult to lay out routes of roads, railroads, pipelines, and aqueducts through them without having an adequate overview, such as that provided by space imagery.

The Problem of Interpreting Remote-Sensor Data

Basically, remote-sensors measure spectral information. But seldom is spectral information, by itself, sufficient to identify an object. More often, the spatial information is essential, and the spectral information gives corroboration clues or assists in the identification. In other cases, not even the spectral plus the spatial information on a photograph are enough to identify an object.

The desire for simplicity sometimes drives man to engage in wishful thinking. He may run after a concept that, although simple, may be invalid in the real world. About two decades ago some exploration seismologists were seduced by the concept of "formation signature." Each geologic formation would, so to speak, write its signature in the wiggles of reflection seismograms. But the problem is more difficult; one has to consider not only the velocity profile and thickness of the various layers, but also their manner of statistical fluctuation, and even the effect of varying neighboring formations and other physical parameters such as density and seismic attenuation.

In a similar way, now we may be engaging in wishful thinking when speaking of "spectral signature." The recognition of most objects of interest requires mainly spatial information, plus some spectral information. Rigorously, there is no such thing as the spectral signature of an object.

A certain amount of hard work in the interpreta-

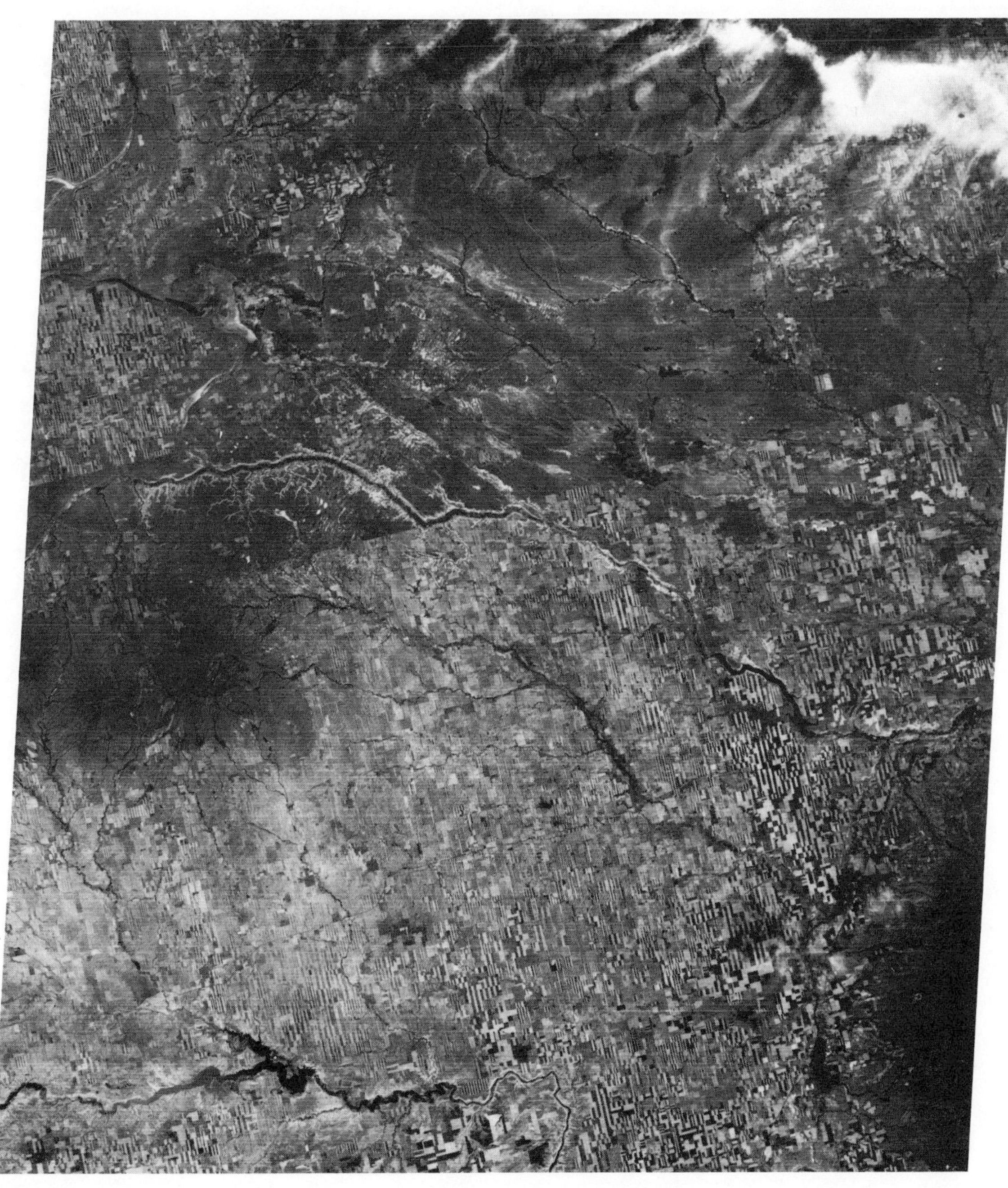

FIG. 4
Montana–Alberta border.

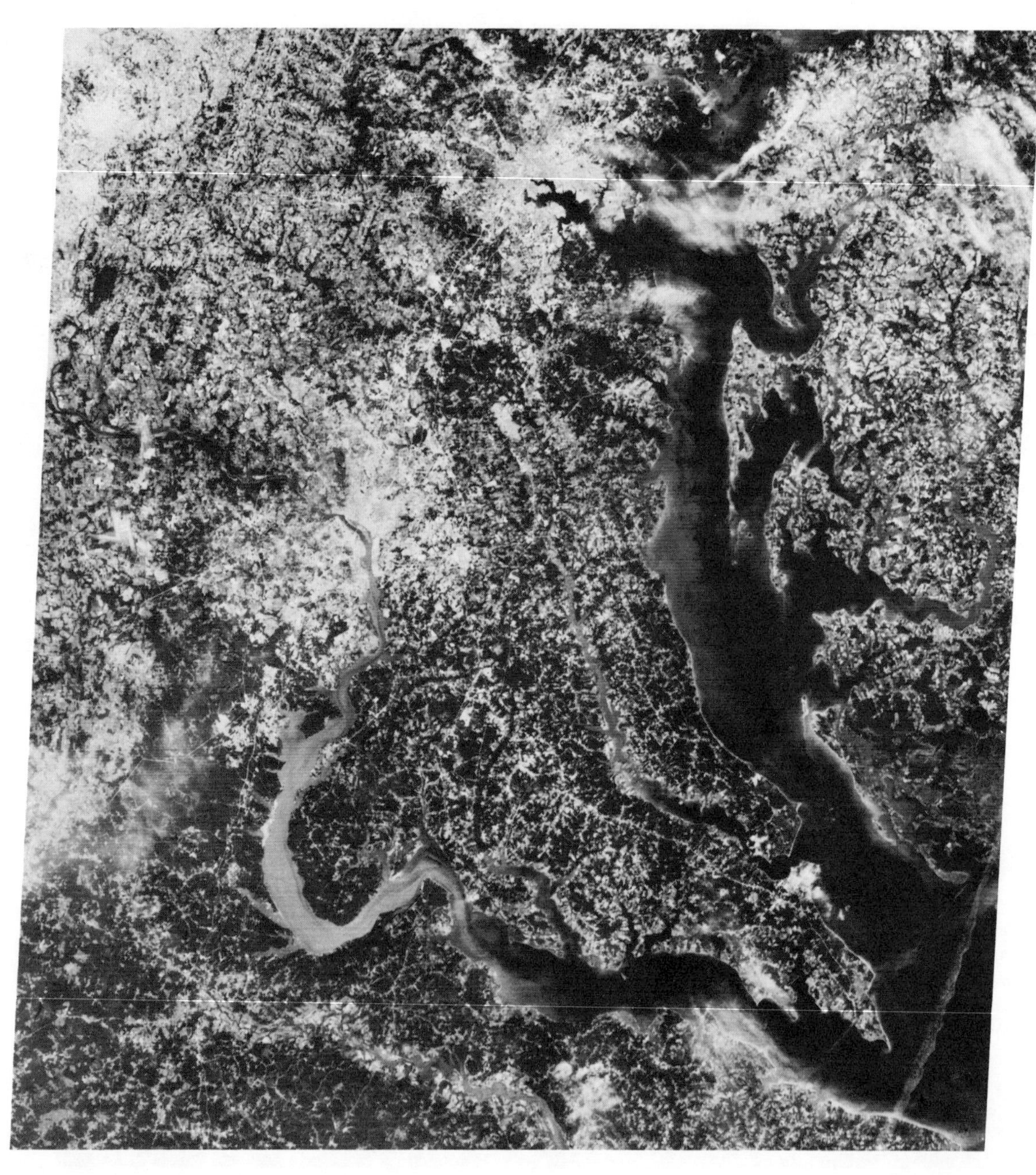

FIG. 5
Washington–Baltimore, USA.

tion of remote-sensor data cannot be avoided. The meaning of the various patterns observed in remote-sensor pictures have to be systematically examined. Each kind of pattern that can be recognized in a picture is like a sieve; we have to establish what kinds of "fish" each sieve catches. Ideally, for each kind of object of interest we want a sieve that separates that and only that kind of object. How good these sieves are depends essentially on the nature of the objects and how they interact with selected bands of electromagnetic radiation. Their success needs to be expressed statistically in terms of probabilities and, in particular, of conditional probabilities, thereby bringing to bear information other than remote sensing.

The basic interpretation problem in remote sensing is the same as the one nature faced in developing the visual system. It took nature over 500 million years to develop the vertebrate eye into man's visual system, probably the most complex natural remote-sensing system on earth. The eye uses the more abundant band of the sun's electromagnetic radiation at the base of the atmosphere, which, as a result, has become our visible spectrum.

From the point of view of object recognition, there are important lessons in nature's experience with the eye. For the recognition of objects the eye does not use primarily spectral information, that is, color, but rather spatial information. Not all the spectral information is used. As a matter of fact, many animals see only in black and white. Even for those few that possess color vision, most of the spectral information is not used. In man, this is because of the drastic simplification implicit in the trichromatic system.

Moreover, the spatial information is not really mingled with the spectral information. That is, the logical abilities for analyzing patterns, determining edges, etc., are wired into the visual system in such a manner that they can be used to analyze any two- or three-dimensional field of data, as mimicked in the trichromatic system.

Remote sensing offers a band of the electromagnetic spectrum broader than the sun's radiation at the base of the atmosphere with which to sense the environment. But object recognition still requires the use of spectral and spatial information. As long as the eye is used in the interpretation, we may resort to the same built-in abilities of the eye to analyze the spatial information, although original radiation may be entirely outside the visible spectrum.

In tackling object recognition, nature's way is often heuristic. Take, for instance, the frog. How does it recognize insects, which constitute its main food? If we were to take the purists' solution, the frog would require a much more complex brain to unfailingly recognize insects, the way a natural scientist recognizes insects. As the now classic experiments of Lettvin, Maturana, and others (1959) have shown, the frog recognizes insects in terms of four decision rules: (1) detection of "sustained contrast," (2) detection of "net convexity," (3) detection of "moving edge," and (4) detection of "net dimming." How did nature hit upon such decision rules? Much cleverness would be required on the part of man, playing the role of a creator of the frog, to find such a drastically simple and obviously effective way of recognizing the frog's main diet.

Interpretation Must be User-Oriented

Remote sensing provides valuable data for the various facets of an economy. But the manner of examining raw remote-sensor data is greatly conditioned to the type of application. It is neither feasible nor practical to interpret all the patterns and textures once and for all for every application. The interpretation process of remote-sensing data is specific to each activity. That is, the proper use of remote-sensor data requires the specific know-how of each activity or industry. Moreover, the needs of each activity vary with the stage of development or progress of each project or operation.

In general, remote-sensor data by itself does not find minerals nor provide the information needed by the various sectors of an economy. The merits of remote sensing should be viewed with proper reference to the context of decision in each activity.

For instance, the conflict between distortion of position and resolution may be resolved differently depending on the application. If we are interested in the search for particular ore deposits, it is not as important to map them very accurately if they can be found on the ground.

That the proper use of remote-sensor data requires the specific know-how of each activity may be understood by looking in some detail at a few examples. Many authors have already shown that remote-sensor data provide valuable information to a wide variety of activities. We do not want here to expand on the checklists of all possible uses; rather we want to inquire into actual specific applications within the range of our experience. The broad generalizations are then not enough. We have chosen geologic mapping, oil exploration, a steel-mill project, and a land-reform project for this in-depth analysis.

Issues in Geologic Mapping

To properly understand the problems of how to utilize remote-sensor data in geology, we must first clarify several concepts in the science of geology. Misconceptions about the types and cost of geologic surveys, about the role of the scale of mapping, and the lack of recognition of the difference between a two- and three-dimensional geology can lead to confusion when examining the economics of resources development.

Ambiguity of geologic interpretation. A basic misconception is to consider that doing the geology of an area is just as simple as taking a photograph. A mere photograph or imagery, by itself, is not enough. Even when an aerial photograph or a remote-sensor imagery is available, abstractions have to be drawn by a human interpreter to outline the geology. In fact, these abstractions are of the same logical nature as those that geologists make when mapping the geology in the field.

The ambiguity of geologic interpretation is especially significant in the making of three-dimensional geologic pictures. Often the geologist finds only a limited number of outcrops, and these are weathered and distorted. Even at outcrops of undisturbed rocks he has to abstract from a great multitude of features, patterns, and textures those to be used in the map representation. From these isolated observations a coherent abstract picture has to be composed. In most cases it is not possible to do this unambiguously; many, or an infinite, number of alternatives may be possible. The geologic picture may have to be modified owing to subsequent observations, whether as the result of geologic fieldwork or of the drilling or mining activities. In other words, doing the geology of an area requires proceeding on a hypothesis, testing it, and then modifying it as required.

All of the above is well known to practitioners of mineral exploration. For instance, oil companies often find themselves baffled in their efforts to find some elusive oil fields, even after an area has been repeatedly mapped by surface geology, and by seismic reflection, gravity, and magnetic methods. It is to be noted that if such ambiguity and uncertainty still persists, even when the amount of information is far beyond what is normally obtained in ordinary regional geologic surveys, then the scanter amount of information of the latter implies a greater degree of ambiguity and uncertainty.

Importance of map scale. First, we note that the choice of scale is a deliberate decision having great impact upon unit costs. It is not an arbitrary decision. The scale basically depends on the minimum size detail that is to be observed and defined in the field and represented in a map. We meet here a situation quite like that of sampling a time function, wherein the sampling interval determines the Nyquist frequency or cutoff of higher frequencies.

Closely related to the choice of scale are two other factors, which we have not chosen to discuss separately: (1) frequency of occurrence of geologic details in terms of size of detail and (2) logistics. As to the first, we want to say here only that the scale should be chosen so as not to lead to sterile mapping efforts, that is, to try to acquire information of a size which is, for all practical purposes, not present in the geology of an area. Logistics, especially means of transportation, is a closely related factor. For instance, helicopters may be useful in reconnaissance work. But as the hopping distance is decreased, eventually it becomes more economical to walk. Thus, the proper choice of scale hinges on a joint consideration of the nature of the geology, the size of detail to be acquired, and logistics.

Maps are ordinarily examined with the naked eye, so that there is a minimum size detail that can be readily comprehended. Such minimum size, as drawn in a map, corresponds to a larger detail in the field, the ratio of the two being the scale (say $1{:}S$). Now, the number of details to be mapped per square mile varies with the square of the linear dimensions; that is, it varies as $1{:}S^2$. This means that a map to the scale of 1:62,500 would contain 64 (i.e., 8×8) times more details per square mile than one to the scale of 1:500,000. As it can be expected that the mapping effort in the field is proportional to the number of details to be acquired, the survey costs per square mile of the two above-mentioned maps should be in the ratio of 64:1. This is a very large factor indeed.

Two- versus three-dimensional geology. An important factor to be considered is the difference between (1) the mere mapping of the surface outcrop pattern and (2) the forecasting of the geology in depth. The first we may call two-dimensional geology. In areas underlain by faulted and folded thick sedimentary sections, two dimensions are not enough. A three-dimensional forecast of the geology is required. Obviously, the unit cost of a three-dimensional geology can be more than an order of magnitude larger than for a two-dimensional geology.

Issues in Oil Exploration

The petroleum industry readily takes advantage of available technologies; remote sensing is one of them. The complexity of the manner in which remote-sensor data have to be stranded into the fabric of decision making may be gauged by considering successive major issues that have to be resolved in oil exploration. To find oil is much more than finding a dome in a space photograph.

A hierarchy of some major issues to be resolved in oil exploration, roughly arranged in their order of appearance during the exploration, is given in Table 1. The left column lists major geologic issues; the center column, major economic issues; and the right column, major management decisions. Many of the issues listed are by themselves complex issues, which in turn may be resolved into secondary issues, and so on. It is to be noted how much more complex these issues are than the raw remote-sensor data.

A satellite photograph may assist at various stages

TABLE 1
Schematic Hierarchy of Issues in Oil Exploration

Geologic Issues		*Economic Issues*	*Major Management Decisions*
To recognize a basin	Identification of specific structures	Assessment of oil potential of area	To undertake reconnaissance
Dimensions of basin	Surface mapping	Marketing considerations	To undertake regional exploration
Sedimentary fill?	Magnetic surveys	Political factors	To undertake detailed surveys
Maximum thickness of sediments	Gravity surveys	Competitive posture	To drill exploration wells
Source beds?	Seismic surveys	Leasing	To abandon or continue
Reservoir rocks?	Wildcat drilling	Profitability	
Geologic history		Growth considerations	
Incidence of metamorphism			
Tectonic deformation			
Differential compaction			
Incidence of groundwater flow			
Identification of possible "plays"			

but, by itself, is meaningless. A fracture pattern may be important or not, depending on many considerations other than the raw remote-sensor data.

Issues in a Steel-Mill Project

Steel mills are examples of large integrated manufacturing plants. We often hear of an iron-ore deposit having been discovered, or rather pinpointed, by some remote-sensor photograph. It is implied that only details are left to be investigated before a steel mill utilizing this ore is established. Rather, the true situation is that there are many iron-ore deposits that are already found, but not yet exploited.

Many issues have to be investigated before a steel mill is established, as shown schematically in Table 2. The realization of a large steel mill project requires several years, usually 5 to 10. The project proceeds in definite stages: feasibility study, preliminary engineering, basic engineering, detailed engineering, and finally construction. The cost of these stages increases almost geometrically with their order. Remote-sensor data can be useful at many stages. For instance, for selecting alternative sites, a port location, and the layout of a railroad to bring the ore. But the specific data required vary with the stage and with the specific industrial organization utilizing it.

TABLE 2
Some Issues in a Steel Mill Project

Mill location selection	Transportation facilities
Alternative sites	Of raw materials to mill
Physical site investigations	Of products to market
Economic considerations	Port development
Site investigation	Labor availability
Extent	Local labor for construction and operation
Foundation conditions	Distribution of skills
Drainage conditions	Training needs
Raw materials (Iron Ore, Limestone, coal, others)	Feasibility study
Location of deposits	Financing plan
Size of reserves	Preliminary engineering design
Grade of reserves	Basic engineering
Metallurgical issues	Detailed engineering design
Utilities (electricity, water, gas, others)	Construction
Location	
Availability	
Characteristics	

TABLE 3
Some Issues in an Agrarian Reform Project

Present	*Foreseen*
Economic and social conditions	Economic and social objectives
Land tenure	Marketing
Lot distribution by sizes	Land tenure
Income distribution	Type of agriculture
Farming practices	Irrigation projects
Hydrology	Land improvement projects
Irrigation systems	Transportation projects
Soils: identification, distribution	Agricultural inputs: fertilizers, seeds, tools and machinery
Condition of land	Legal reforms
Mineral deficiencies	Financing
	Technical assistance
	Administration

Issues in an Agrarian Reform Project

One important use of space photography is for the planning of agricultural development in developing nations. To understand how such data may be useful, we need to look at some of the issues that are involved in, for instance, an agrarian reform project. They are displayed schematically in Table 3.

To properly understand the meaning of the various patterns and features shown by a space photograph in an agricultural area requires much side knowledge. Most of it is not provided by remote sensing itself. For instance, the mosaic of various land partitions cannot be understood without a knowledge of system of land tenure, agricultural practices, climate, and season of the year.

CLOSING REMARKS

Vast amounts of satellite remote-sensor data bearing on earth resources and environment are being collected repetitively. Yet a wide gap exists between (1) raw remote-sensor data (photographs, sensor imageries, line scans, or point measurements) and (2) resource information as needed by various units of a nation's economy. To close this gap, a careful consideration should be given to the overall data-processing system that underlies the gathering and utilization of such information. Such a system should encompass a large variety of governmental and private entities.

REFERENCES

Grossling, B. F., 1969, Color Mimicry in Geology and Geophysics, *Geophysics,* v. 34, no. 2, p. 249–254.

____, 1970, Geology and Investment Planning, *Science,* v. 169, no. 3952, p. 1303–1304.

Lettvin, J. Y., H. R. Maturana, W. S. McCulloch, and W. H. Pitts, 1959, What the Frog's Eye Tells the Frog's Brain, *Proc. Inst. Radio Engrs.,* v. 47, p. 1940–1951.

4

SUMMARY OF LANDSAT APPLICATIONS AND RESULTS

Enrico P. Mercanti

Enrico P. Mercanti is with the NASA Communications and Navigation Division, Goddard Space Flight Center, Greenbelt, Maryland.

On July 23, 1972, LANDSAT-1 was launched from NASA's Western Test Range (Fig. 1). Figures 2, 3, and 4, respectively, identify the sensors on the LANDSAT-1 observatory, indicate the ground coverage pattern, and show a typical daily coverage. Figure 5 describes the multispectral scanner (MSS) orientation. LANDSAT-1 was designed for 1-year operation, but was still providing imagery when LANDSAT-2 went into orbit $2\frac{1}{2}$ years later.

What has been achieved with the millions of LANDSAT data products going to hundreds of scientific investigators and users throughout the world? Conceding the technical and scientific success of the LANDSAT-1 system, what has been the payoff for the time and tax dollars spent on developing and building it? What benefits will be derived from LANDSAT that are of real, everyday, practical use to mankind? To answer these questions, NASA has held three symposia where scientific investigators working with LANDSAT data, including some not funded directly by NASA, were invited to present findings and applications of their investigations. The papers presented at these symposia have been compiled and published as NASA documents available to both the world scientific community and the general public (see Table 1).

Based on the reports of these investigators, LANDSAT has been remarkably successful in achieving its mission of demonstrating the utility of multispectral remote sensing from space in practical earth resources research and management applications.

Many reported results and possible applications are discussed in this chapter. They are grouped by scientific discipline (geology, water resources, etc.), as were the original LANDSAT-1 investigations. These summaries contain specific examples with accompanying LANDSAT imagery or interpretations. For more detailed information on any individual investigation or discipline group of investigations, the reader is referred to the proceedings of the symposia referenced in Table 1. Table 2 summarizes some applications that have been successfully demonstrated by LANDSAT investigators.

FIG. 1
LANDSAT-1 launch from NASA's Western Test Range, Vandenberg, Calif.

AGRICULTURE, FORESTRY, AND RANGE RESOURCES

LANDSAT investigators working within the agriculture, forestry, and range resources discipline have demonstrated the capability to monitor and inventory many resources that can aid in resolving increasingly important food-related problems, both in the United States and throughout the world. Application of their techniques has resulted in crop specie identifications with accuracies as high as 90 percent, highly accurate timber inventories, new and vastly improved soil and range classification maps, and detection of crop and timber diseases and infestations and changes in agricultural land uses.

Figure 6* is an agricultural example of how LANDSAT data are used in making crop classifications. Using digital LANDSAT MSS data of an agricultural area in Holt County, Nebraska, the investigator used a computer to compare and classify crop types by spectral pattern recognition techniques. The result is shown in Fig. 6. Based on ground truth checking, the accuracy of this method of crop classification was 70 to 90 percent using single data sets. Using registered data sets from two or more LANDSAT passes generally improves classification accuracy.

The technique, called "clear-cutting" in the timber industry, means cutting down virtually all usable trees in a given area, while leaving untouched all the timber surrounding the "clear-cut" area. Although it had been known that clear-cut techniques were being used in the heavily forested regions of Oregon, the widespread extent of their use was not fully realized until LANDSAT-1 produced the dramatic scene shown in Fig. 7. In this scene, showing volcanic Mt. Hood and the Columbia River, clear-cut areas are vividly displayed as light patches within the dark red of the forested areas. So numerous are the clear-cut areas that the entire scene assumes a speckled appearance. Imagery such as this can be of immense value to federal and state agencies that have the responsibility of monitoring and protecting forest resources.

*Figures 6, 10, and 18 will be found in the color insert.

FIG. 2
LANDSAT observatory sensor arrangement.

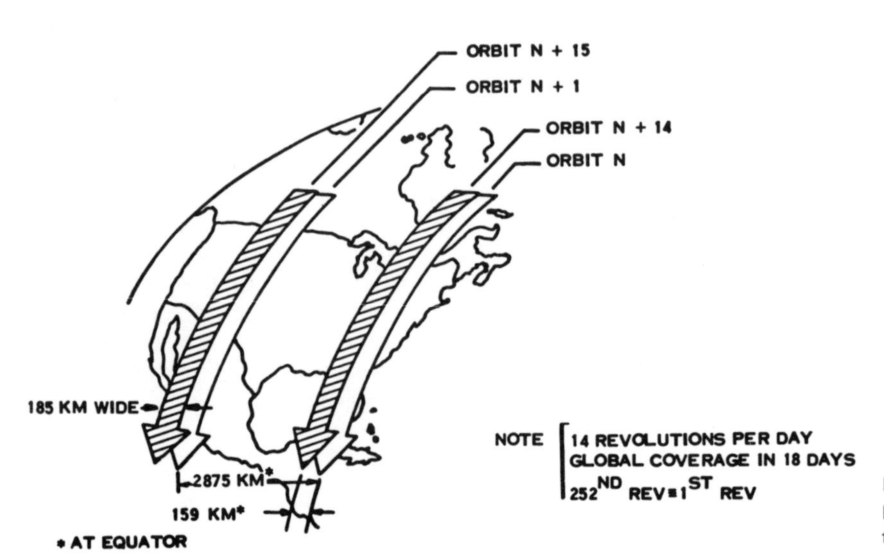

FIG. 3
LANDSAT ground coverage pattern.

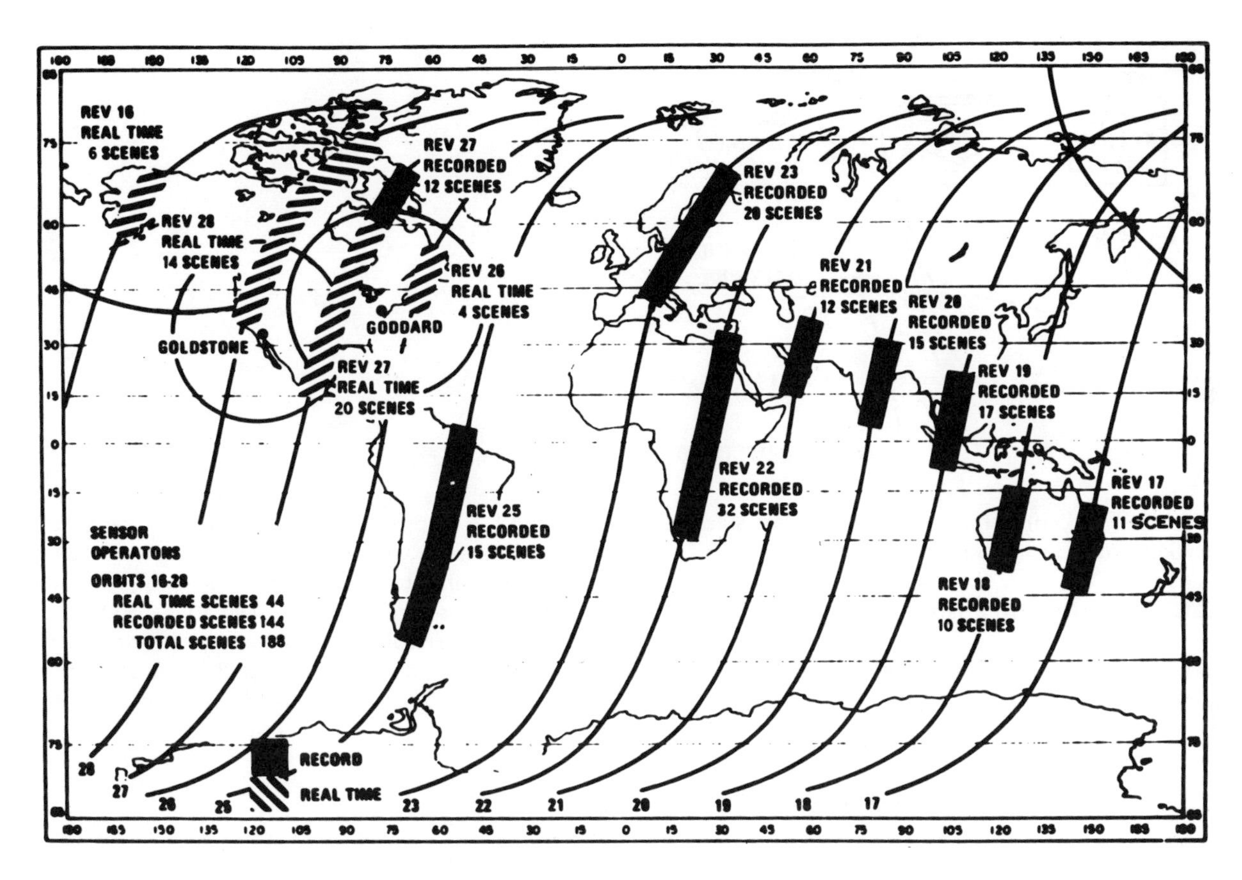

FIG. 4
Typical LANDSAT daily coverage.

Figure 8 illustrates a forestry application of LANDSAT data that can save time and money for resource management agencies. A fire in the Pocket Gulch area of Northern California was mapped by the state's Division of Forestry (photo A in Figure 8) using conventional techniques and by a LANDSAT-1 investigator (B) using LANDSAT imagery (C) taken over the area 10 days after the fire was suppressed. The accuracy of the LANDSAT-derived map was checked by low-altitude aircraft photography (D). Not only was the burned area mapped more accurately [13,340 acres (5401 hectares) on the LANDSAT map versus 10,340 acres (4196 hectares) on the conventional map], but the cost of producing the LANDSAT map was one tenth the cost of producing the state's map.

The Sahel region of Africa has been suffering from drought for the past 5 years. While studying LANDSAT imagery of the region, LANDSAT investigators detected a rare area of vegetation (dark polygonal area in Fig. 9A) within the semidesert region. The unusual linearity of the area's boundaries led to the conclusion that the feature was man-made. On a field trip to the area, investigators discovered that the dark polygonal area was a carefully managed fenced-in ranch, differing noticeably in vegetative cover from the uncontrolled grazing areas outside its fence (Fig. 9B). The ranch demonstrates that controlled grazing

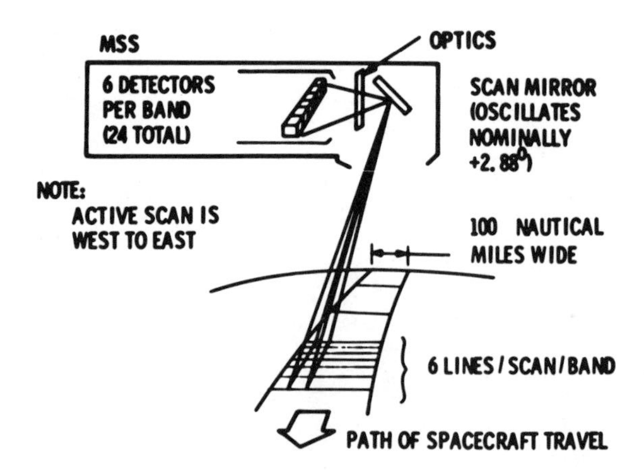

FIG. 5
LANDSAT multispectral scanner (MSS) orientation.

can make it possible for some of the desert areas in the Sahel to be reclaimed for productive use.

LAND USE AND MAPPING

The dynamic nature of land resources requires constant surveillance to ensure their proper use. LANDSAT-1 is providing data especially useful to land-use planners who must have up-to-date inventories of regional land and water spaces to properly plan and propose legislation and regulations, and later to monitor the results of their efforts. Investigators report being able to classify up to 12 separate land-use categories using units as small as 1.2 acres (0.48

TABLE 1

Symposium Date	*NASA Document No.*
Sept. 29, 1972	First ERTS Symposium X-650-73-10
Mar. 5–9, 1973	Symposium on Significant Results Obtained from ERTS-1 SP-327 (Vol. I, Technical Presentations) X-650-73-127 (Vol. II, Summary of Results X-650-73-155 (Vol. III, Discipline Summary Reports)
Dec. 10–14, 1973	Third ERTS-1 Symposium SP-351 (Vol. I, Technical Presentations) SP-356 (Vol. II, Summary of Results) SP-357 (Vol. III, Discipline Summary Reports)

Note: Appendix A compiles and summarizes the findings of the three symposia in the area of geology.

TABLE 2
Some Applications of LANDSAT Data by Discipline

MSS Bands	*Agriculture*	*Forestry and Range*	*Water Resources*	*Geology*	*Land Use*	*Marine Resources*	*Interpretation Techniques*	*Environment and Meteorology*
Band 4 (0.5–0.6 μm)	Crop survey	Vegetation maps	Snow detection Forest snow lines Sediment Turbidity	Sediment Bathymetry Urban areas	Lake eutrophication	Turbidity Shoals Bathymetry Ice Sediment	Roads Wooded areas Shallow water	Water pollution Air pollution Thin cirrus clouds Jet contrails Haze
Band 5 (0.6–0.7 μm)	Crop survey Soils Stress Soil assoc.	Vegetation maps Stress Topography	Water depth Sediment Turbidity	Soil, land surface Urban areas Sediment Fracture detail	Air fields Concrete Roadways Bridges Forests Sediment	Turbidity Bathymetry Sediment Eddies Turbidity plumes	Roads Wooded areas	Water pollution Air pollution Defoliation Thin cirrus clouds Jet contrails
Band 6 (0.7–0.8 μm)	Crop survey Soil assoc	Vegetation maps	Transient snow Lines on glacier Water boundaries Cloud–snow differentiation	Igneous rocks Tectonic features Marshes Lakes Rivers Stream channels	Land–water boundaries Large bridges Geologic features Wetlands	Chlorophyll	Lakes Surface water	Land–water interface
Band 7 (0.8–1.1 μm)	Crop survey Soils Crop differ.	Vegetation maps Land–water boundaries Grass fires	Irrigated fields Wetlands Water boundaries Flood plains Flood mapping	Tectonic features Lakes Rivers Fractures Surface water	Land water boundaries Small lakes Geologic features Urban areas Burned range Land	Shores Ice Chlorophyll Eddies	Lakes Surface water Serpentine outcrop Major soil types Deep water Cloud penetr. Metamorphic rock–alluvium differ.	Land–water interface Defoliation
Band ratios	Irrigation Crop–soil cond. Green biomass Crop differ. Stress Soil assoc.	Veg. conditions Biomass Stress	Water impoundment Vegetation Soil moisture Floodplains	Rock structures Lithologies Geothermal Iron oxide Surface stains Lithologic contacts	Bridges and causeways Urban core Transportation network Commercial areas Residential Vegetated fields	Bathymetry Fronts	Crops Desert veg. Iron oxide Deep water	Veg. vitality Thin cloud–dense cloud differ. Thin cloud–turbid water differ.
Color composites	Crop survey Biomass Crop differ. Soils Soil assoc.	Vegetation maps Biomass Veg. types Timber Range veg. Stress	Transient snow lines on glacier Floodplain Soil moisture	Soil tones Glacial features Pediments Benches Sand bodies Linear features Hazy anomalies Major rock units	Geological texture enhance.		Differ. of geologic units	Vegetation classification Snow–Nonsnow boundary

FIG. 7
ERTS-1 scene of the forests around Mt. Hood, Ore., showing extent of clear-cut areas.

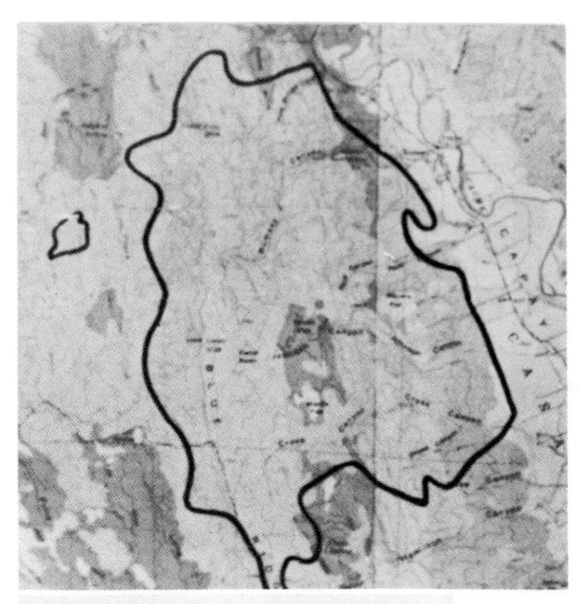

A. (Scale = 1:125,000)

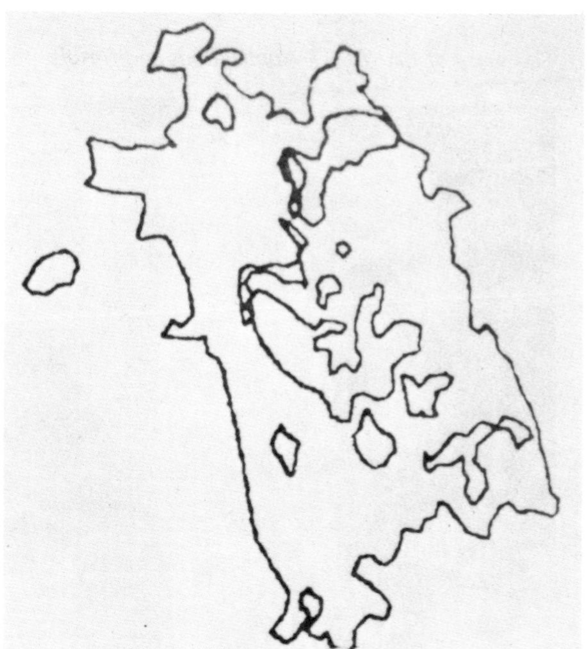

B. (Scale = 1:110,000)

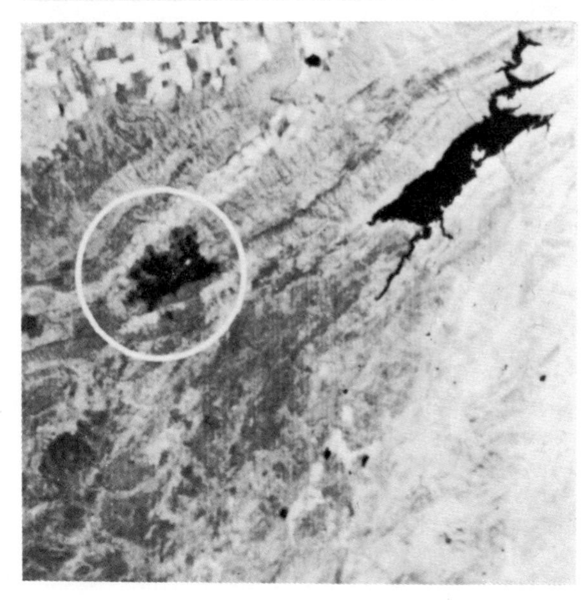

C. (Scale = 1:780,000)

D.

FIG. 8
Pocket Gulch, Calif., fire scene: (A) burned area as mapped by Division of Forestry personnel; (B) burned area as mapped by LANDSAT investigators; (C) LANDSAT image of burn scar used to produce map shown in (B); (D) low-altitude aircraft photo used to check accuracy of LANDSAT map.

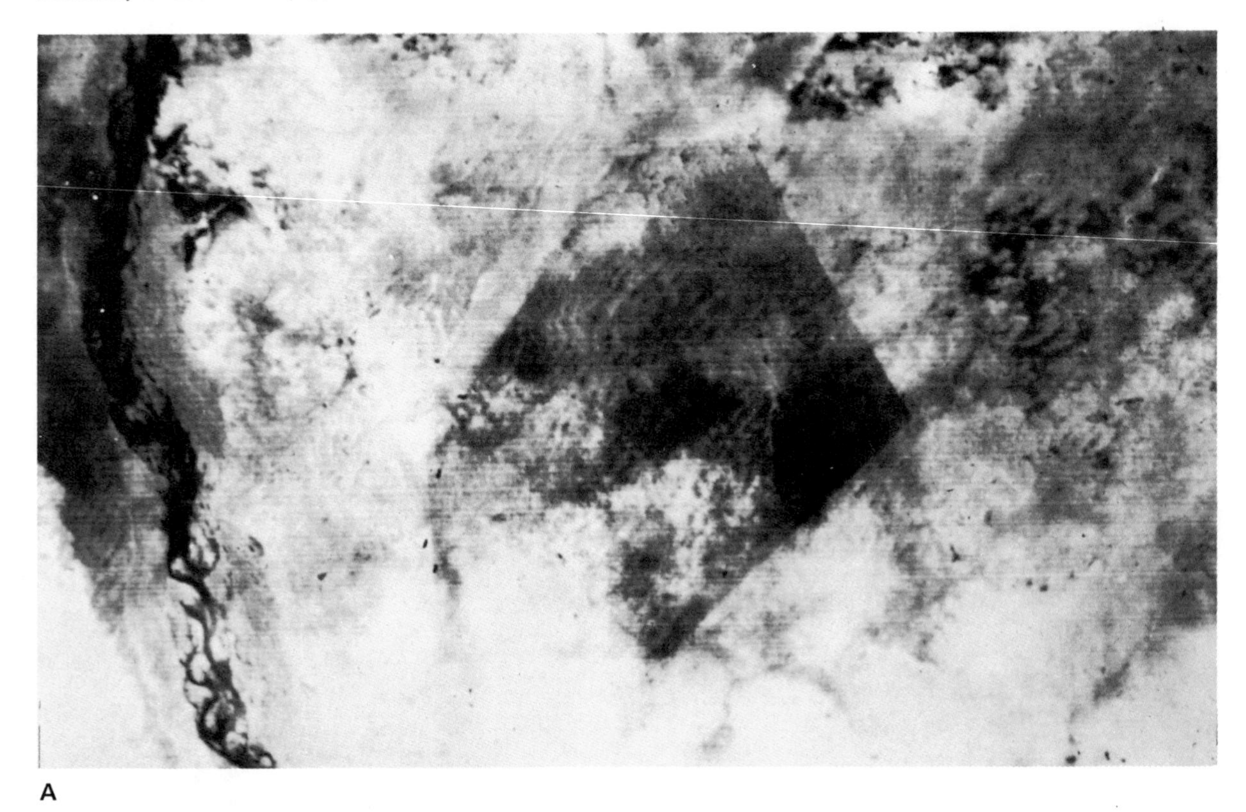

A

FIG. 9 *(above and right)*
Vegetated area in the midst of African desert: (A) portion of LANDSAT scene showing dark polygonal area with linear features; (B) aerial view of ranch fence line showing wide variation in vegetative cover.

hectare). Identifying these categories enables the planner to detect urban–suburban sprawl, conversion of open space to commercial and residential uses, areas of intense construction, and other factors that enter into informed urban and regional planning. In the area of cartography, the U.S. Geological Survey has demonstrated that the LANDSAT MSS bulk image can be used together with appropriate cartographic gridding to make 1:250,000 scale photo maps that meet National Mapping Accuracy Standards.

Figure 10 presents the result of one highly successful LANDSAT-1 land-use investigation. This detailed land-use map of a three-state area (Massachusetts, Connecticut, and Rhode Island) contains 11 separate land-use categories (see legend on Fig. 10). The map was compiled from four LANDSAT-1 images of the New England area taken during clear weather in mid-October 1972. The images were converted into color infrared transparencies, and a single photointerpreter produced the land-use map in draft form within 3 months. It required an additional 2 months to produce the final color-coded map shown in the figure. The investigator estimated savings of approximately 90 percent over what the cost would have been had

B

an equivalent map been prepared by conventional means such as aircraft photography.

GEOLOGY

Geology is the study of the physical structure of the earth. In practical application to present needs, this means the exploration for mineral and petroleum reserves as well as the detection of potential seismic hazards. The earth's crust is made up of a wide variety of materials, each reflecting sensible energy in a unique manner and in a different portion of the spectrum. Investigators have reported that the LANDSAT MSS is capable of retrieving a maximum amount of signature data from geological structures. Furthermore, LANDSAT investigators have found that by manipulating spectral information from the various bands, it is possible to recover surface information (e.g., iron oxide surface stains overlying mineral deposits) unobtainable from a single channel or from a black and white image. Proven data-enhancement techniques, such as color additive viewing, band ratioing, and color photographic compositing can be used with multiband information. The ability to use LANDSAT data for mineral and petroleum exploration and for the detection of geologic hazards enables government and industry to reduce their field exploration and mapping efforts manyfold, with a substantial savings in time and money.

The synoptic coverage of LANDSAT has also proved to be extremely valuable in geologic exploration and mapping. Geologic structures can range in extent from tens to hundreds of kilometers in length, and the detection and mapping of large lineaments, faults, and fractures (critical in a mineral or petroleum exploration program) is aided immensely by using the satellite's synoptic view. LANDSAT imagery allows large areas to be mapped quickly, and regional relationships among lithologic, structural, and geomorphic features can be studied. A study of synoptic data affords a rapid and economic method for isolating key areas for further study. The regional perspective reduces the amount of distracting data to be analyzed, permits large-scale differences to be defined quickly, and at the same time may identify features not visible from the level of the aerial photograph.

Investigators have used LANDSAT imagery for making geological base maps, and have been able to distinguish features such as roads, bridges, lineations, and erosional gullies as narrow as 20 m (66 ft). In many cases, intricate lava flows and regional rock types can be accurately mapped, geothermal springs (as small as 2.5 km^2 in area) identified, and coastal erosion monitored.

Once a spectral signature is correlated to a particular rock or sediment type, the distribution of that surface feature can be accurately mapped at considerable savings. The distribution of surface materials and lineations, of particular importance to construction and civil engineering interests, can be accurately defined on LANDSAT imagery.

One of the first, and subsequently one of the most common, results from LANDSAT investigations in geology dealt with geological structure mapping. There are considerable areas within the United States, and even larger areas outside the United States, that have never been mapped geologically (faults, shears, linears, etc.); LANDSAT data are being used to produce geologic maps of these areas for the first time. LANDSAT can also be used to revise and update existing geological maps. Figure 11 is a geological map of the West Aswan area of Egypt compiled by the Egyptian Geological Survey. Because of the extreme importance of the Aswan Dam and its surrounding area to the economy of Egypt, investigators performed a detailed photogeological investigation of the area using LANDSAT imagery to detect features that could affect seepage from the reservoir (Lake Nasser), drainage patterns, and other features of potential impact to future development. Their investigation produced the new and much more detailed geological map shown in Fig. 12. One highly significant result was the discovery and mapping of several faults intercepting the reservoir that were not shown on previously published maps.

Mineral exploration is an often-cited potential application of LANDSAT data. However, before a new method, system, or instrument is employed by industry, it must show potential use and economic feasibility. With this in mind, a LANDSAT investigator designed an experiment to test the application of photolineament information obtained from LANDSAT imagery to the selection of potential target areas for mineral exploration. Using a test site in the mineral-rich central Colorado area, the investigator studied a LANDSAT image of the area and plotted the photolineaments (Fig. 13A). Next, potential target areas for mineral exploration were selected based on the photolineament data obtained (Fig. 13B). The assumption was made that mineralization is probably structurally related to faults and shear zones that appear as photolinears on LANDSAT imagery. A map of Colorado mineral districts was used to evaluate the target areas. The locations of the major mineral districts is shown superimposed on the LANDSAT image used for the experiment in Fig. 13C. Comparison of Figs. 13B and 13C showed that five of the ten potential target areas coincided with major mineral districts. The results of this experiment suggest that lineaments interpreted from LANDSAT imagery can be used as a guide to the location of metallic mineral deposits. Thus, analysis of LANDSAT imagery can be a very valuable and inexpensive first step in a mineral exploration program, especially if used in conjunction with other sources of geologic information.

Because seasonal variations in vegetation, soil moisture content, and solar angle can dramatically alter the appearance of the earth's surface, repetitive coverage is a valuable, if not essential, tool in geology. LANDSAT data acquired to date have resolved striking differences in geomorphic landform characteris-

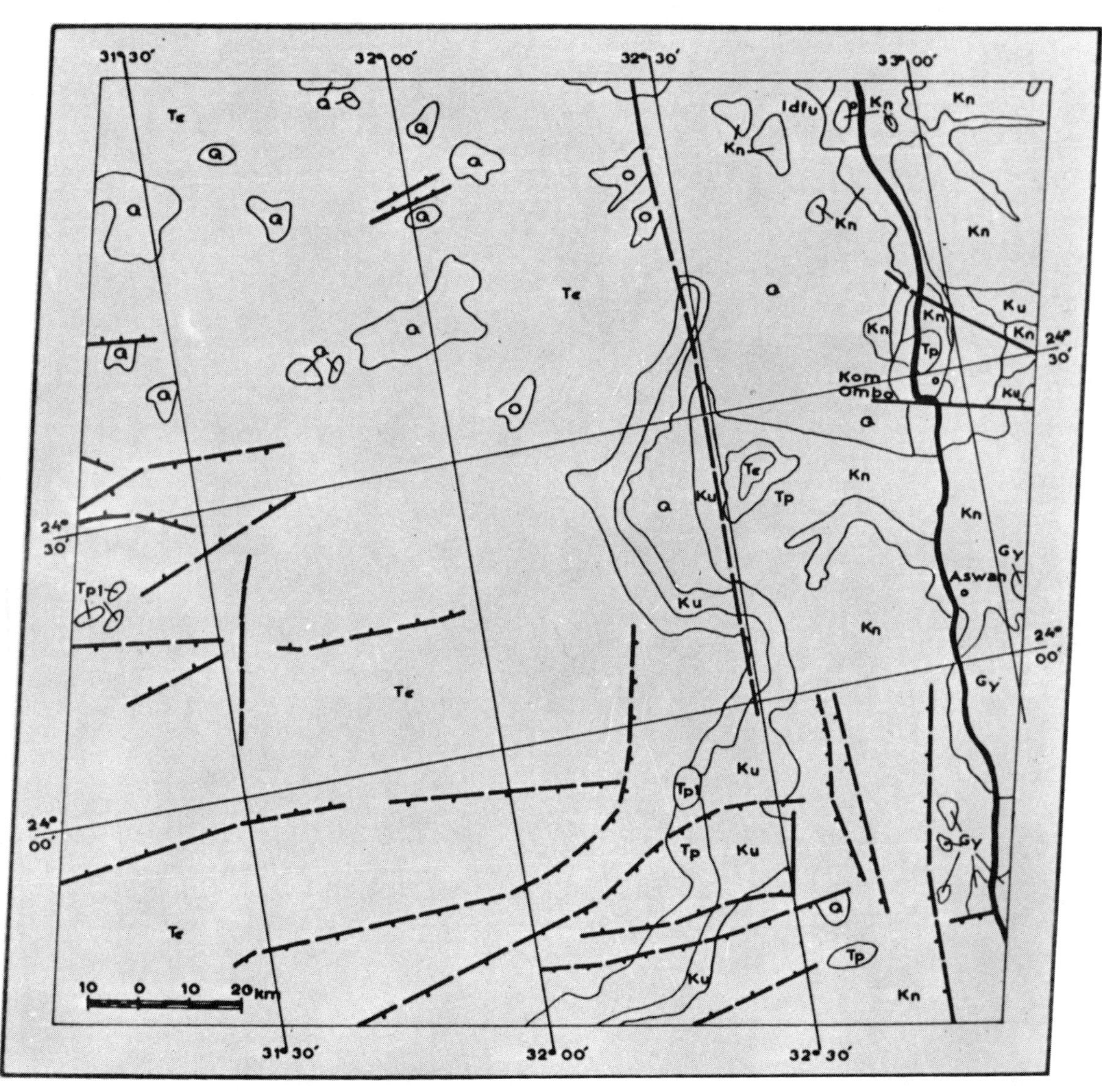

FIG. 11
Geological map of Egypt's West Aswan area compiled by Egyptian Geological Survey.

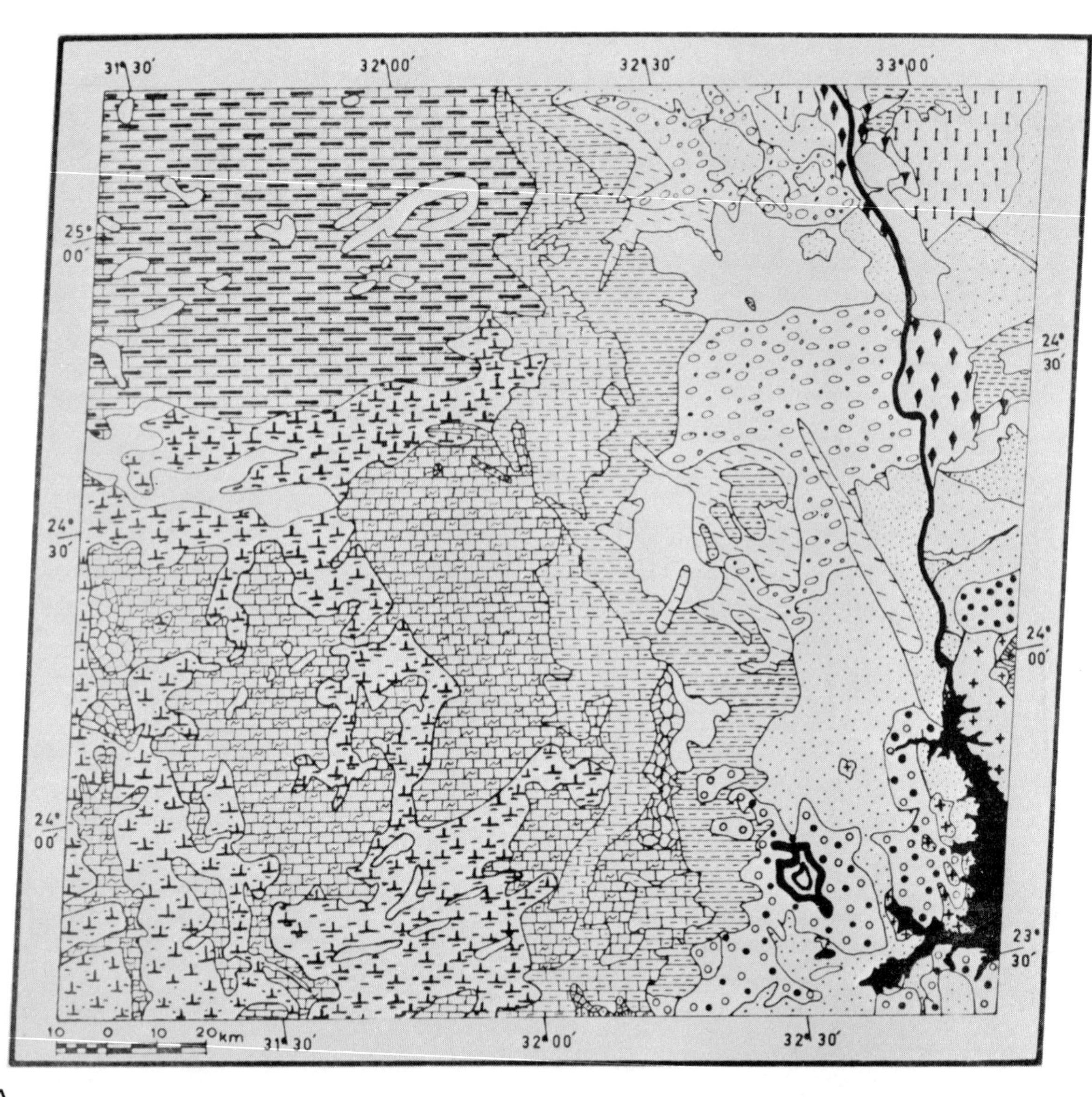

A

FIG. 12 (*above and right*)

Updated and revised geological map of the West Aswan area compiled from analysis of LANDSAT imagery.

tics. In areas where seasonal changes are strong and plants, soils, and rock exposures contribute significantly to image responses, repetitive LANDSAT data can provide valuable information not previously acknowledged. Furthermore, the climatic cycle in an area may be highly variable, and repetitive LANDSAT data, whether studied by cycle or from year to year, may reveal previously undetected geologic features.

The number of reported significant results in geology and the extent to which LANDSAT imagery is

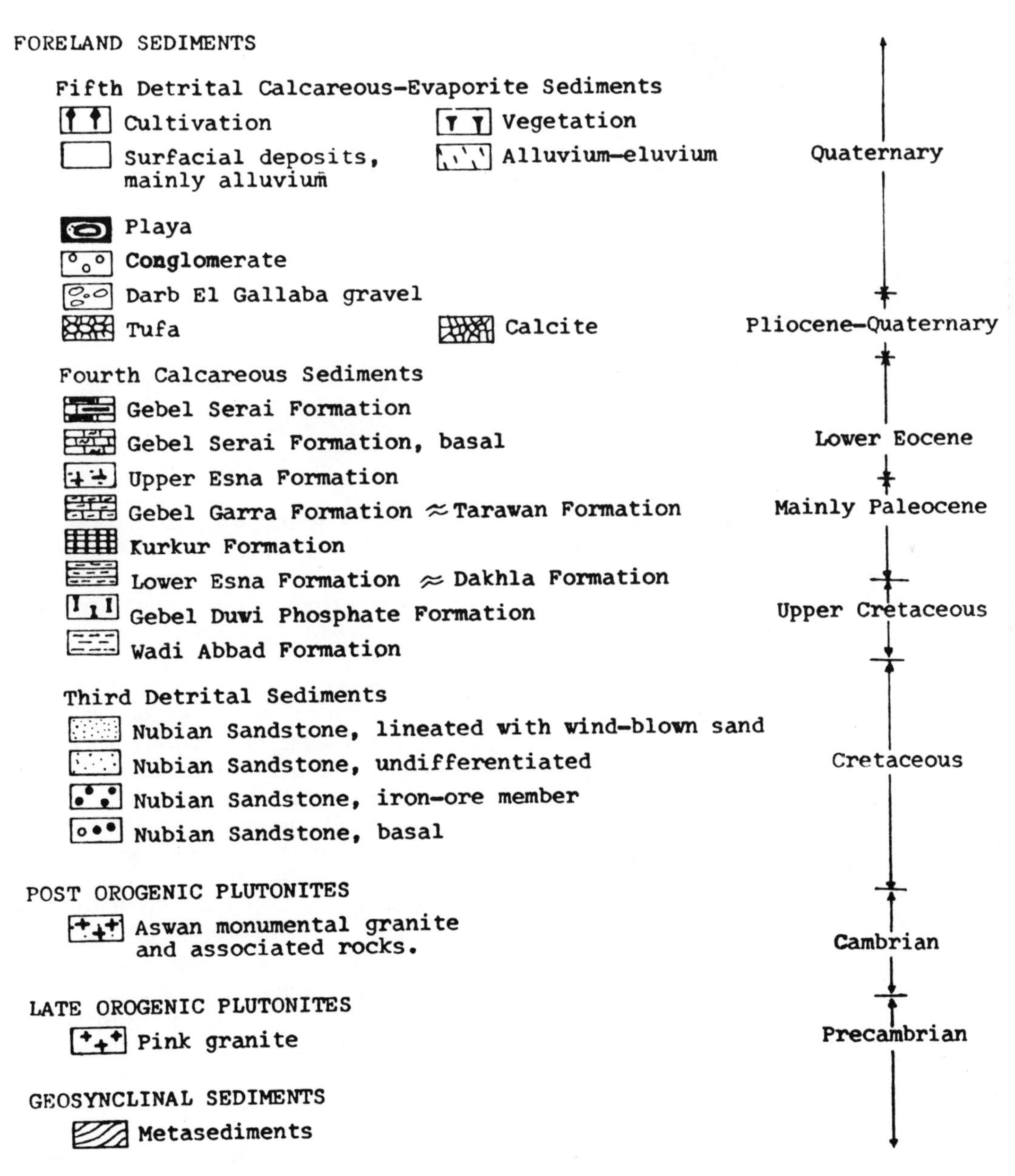
FORELAND SEDIMENTS
Fifth Detrital Calcareous-Evaporite Sediments
Cultivation
Vegetation
Surfacial deposits, mainly alluvium
Alluvium-eluvium
Quaternary
Playa
Conglomerate
Darb El Gallaba gravel
Tufa
Calcite
Pliocene-Quaternary
Fourth Calcareous Sediments
Gebel Serai Formation
Gebel Serai Formation, basal
Lower Eocene
Upper Esna Formation
Gebel Garra Formation ≈ Tarawan Formation
Mainly Paleocene
Kurkur Formation
Lower Esna Formation ≈ Dakhla Formation
Gebel Duwi Phosphate Formation
Upper Cretaceous
Wadi Abbad Formation
Third Detrital Sediments
Nubian Sandstone, lineated with wind-blown sand
Nubian Sandstone, undifferentiated
Cretaceous
Nubian Sandstone, iron-ore member
Nubian Sandstone, basal
POST OROGENIC PLUTONITES
Aswan monumental granite and associated rocks.
Cambrian
LATE OROGENIC PLUTONITES
Pink granite
Precambrian
GEOSYNCLINAL SEDIMENTS
Metasediments

B

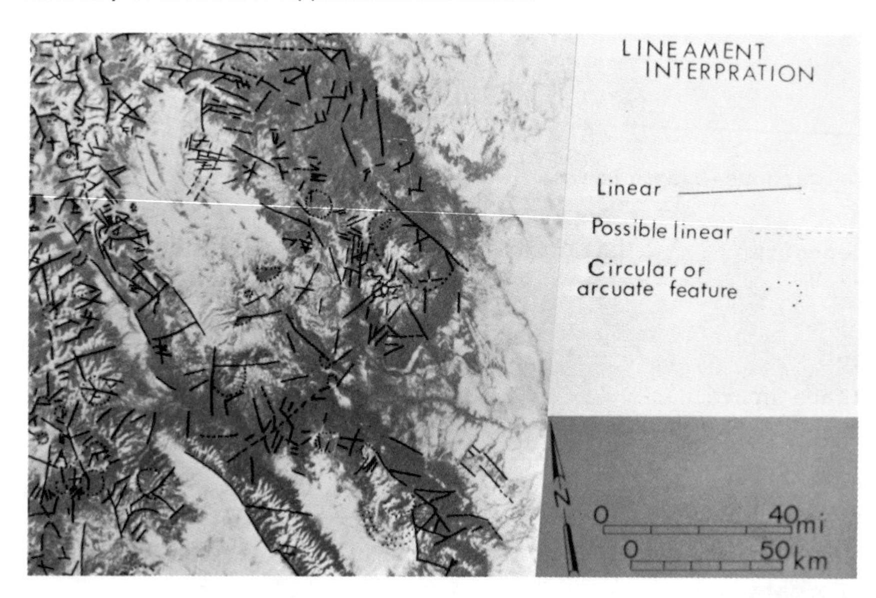

FIG. 13(A)
Lineament interpretation of a mineral-rich test site in central Colorado.

being actively used is graphic proof that the resolution is adequate for a wide variety of geologic applications. One investigator states that geologic features *smaller* than the publicized LANDSAT resolution (gabbro dikes cutting a granite batholith) have been identified under the proper conditions.

WATER RESOURCES

Investigators working with LANDSAT-1 data in the water resources (hydrology) discipline have reported findings that can significantly advance studies in surface hydrology. Application of these experimental results can aid in resolving some of the worldwide shortages of fresh water.

The perspective gained by satellite remote sensing adds areal dimension to conventional hydrologic data collected at point stations. The two data sources combine to provide synoptic information on drainage basin characteristics, hydrologic controls, vegetative indicators of hydrologic conditions, hydrodynamics, lake processes, and distribution of snow and ice.

LANDSAT investigators have reported that snow-covered areas can be measured to within a few percent of drainage basin area and that snow-line altitudes can be estimated to ±60 m. The location of snow lines can be extracted from LANDSAT-1 imagery in more detail than is normally obtained from operational aircraft surveys. Surface water area can be estimated to within a few percent accuracy over large regions, such in the playa lakes regions of the southern high plains of Texas and New Mexico. Lakes as small as 1 hectare can be recognized and mapped accurately at scales of 1:250,000. Other hydrologically significant land-use features, such as impervious area, vegetation, and bare soil, can be mapped. In addition, changes in land use on watersheds resulting from forest fires, clear-cutting, or strip mining can be rapidly noted and this information incorporated into watershed management operations. Relative soil moisture variations such as occur after rainstorms or in irrigated areas can be observed from LANDSAT-1. Another highly significant application that has been demonstrated is the ability of LANDSAT to accurately map flooded areas and to identify and map floodplain features.

The LANDSAT Data Collection System (DCS) has performed a significant role in several water resources experiments. In southern Florida, for example, water-stage information from data-collection platforms installed in the Everglades (Fig. 14) was used in con-

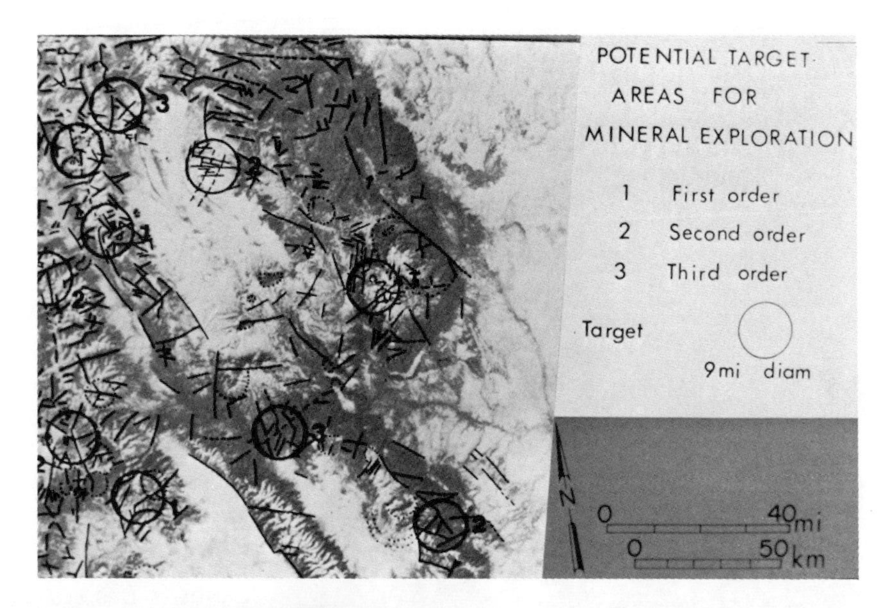

FIG. 13(B)
Potential target areas for mineral exploration interpreted from the lineament analysis shown in Fig. 13(A).

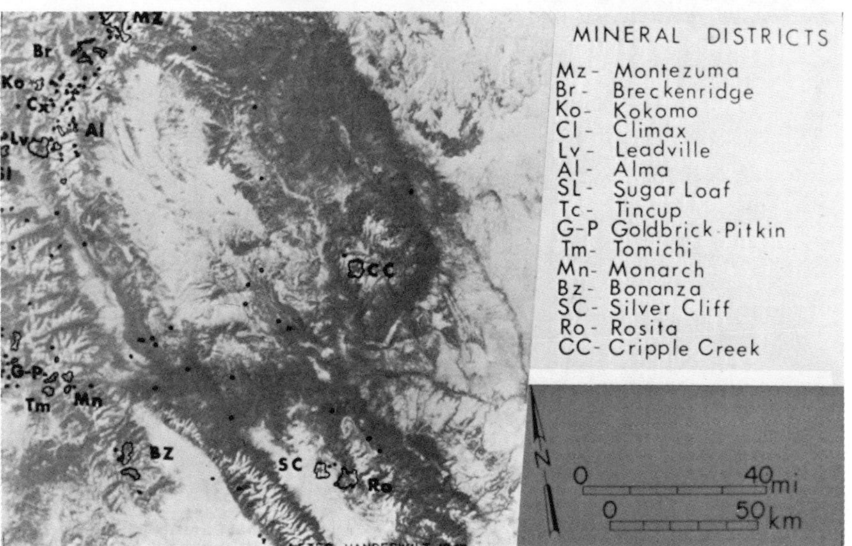

FIG. 13(C)
Known major mineral districts in the test area. Comparison with Fig. 13(B) shows coincidence of target areas with known areas for the following districts: Breckenridge, Leadville–Climax–Alma area, Tomichi, Bonanza, and Cripple Creek.

junction with water-area information from LANDSAT imagery to provide water volume estimates in near real time. This information was used to make management decisions concerning the distribution of water in the area. Figure 15 shows the data relay geometry of the LANDSAT DCS.

In many areas of the world, such as the western United States, snowmelt provides a major part of the annual water runoff that can be used for man's needs. LANDSAT snow-cover images, as shown in Fig. 16, supply data on the speed of melting and the volume of water released. These images show the entire Wind

FIG. 14
LANDSAT DCS data collection platform setup in Florida Everglades.

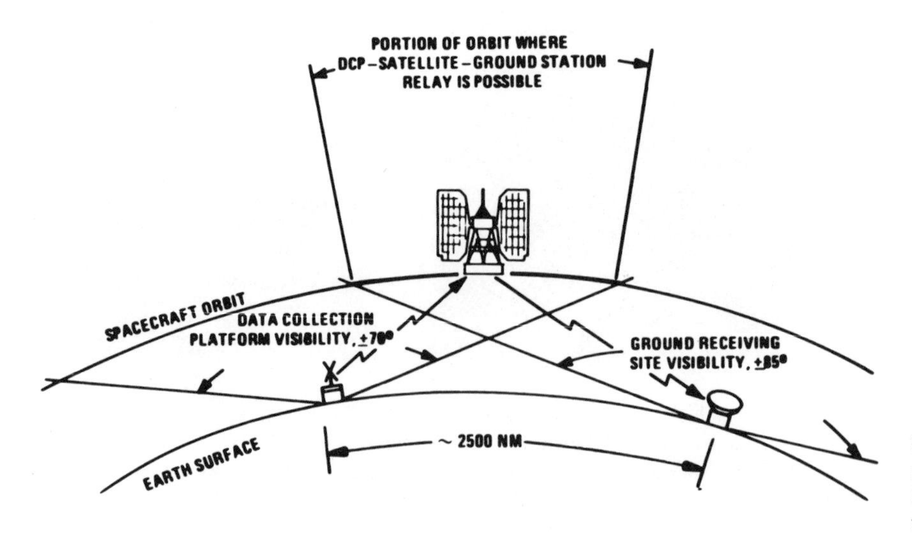

FIG. 15
Data relay geometry of the LANDSAT data collection system.

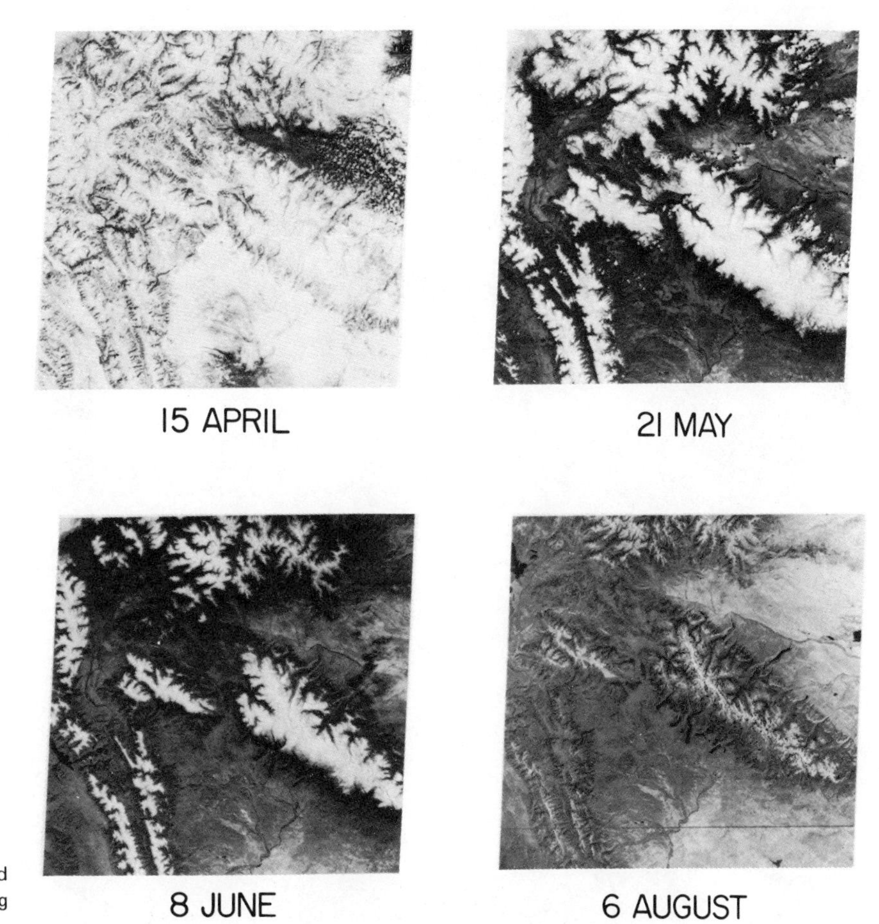

FIG. 16
Four LANDSAT scenes of Wind River Mountains Range showing progress of spring snowmelt.

River Mountain watershed area during the 1973 spring–summer melt. For ungauged, inaccessible, and remote watersheds such as this, or as auxiliary data for conventional snowpack monitoring networks, LANDSAT data can be used profitably for better management of water resources.

LANDSAT-1 imagery has also proved to be effective in identifying surging-type glaciers and monitoring the extent of their areal increase. Figure 17 shows a family of glaciers in the vicinity of Mt. McKinley in central Alaska. It includes uniformly moving types of glaciers, such as Ruth, Kahiltna, and Lacuna, and surging glaciers such as Tokositna and Yentna. Surging glaciers can be distinguished from nonsurging glaciers on LANDSAT imagery by their relatively wiggly-folded moraines. Lacuna glacier has been stagnant for 40 years and its dark surface shows the results of severe melting. Tokositna, however, has just completed a rapid advance that began in 1970, and it exhibits the characteristic wiggly-folded moraines. This LANDSAT image of Yentna showed that its folded moraines have been displaced 6000 ft down the valley from their position on recent maps and 1970 aerial photographs. The other two glaciers

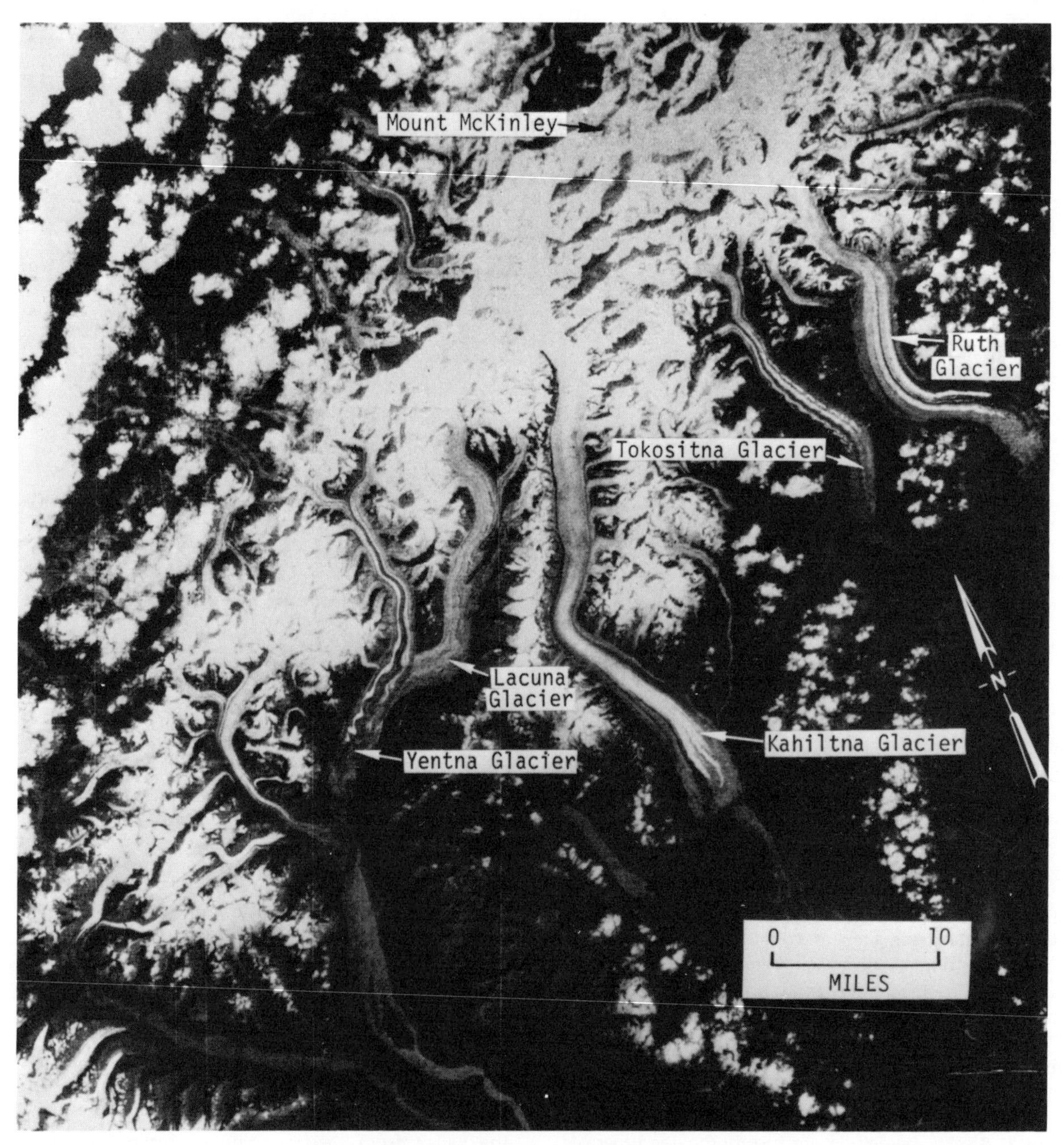

FIG. 17
Surging and nonsurging glaciers in the vicinity of Mt. McKinley, Alaska, viewed from LANDSAT-1.

shown, Ruth and Kahiltna, flow at uniform rates, and their medial moraines (the dark stripes down the ice centers) are straight and uniform. The causes of glacier surges, and why some glaciers surge and others do not, are questions of great scientific concern, but have distinctly practical aspects. Surging glaciers can advance very rapidly over large land areas and can cause devastating floods by blocking and then suddenly releasing large quantities of meltwater. Continuous, repetitive imagery from LANDSAT-1 is aiding glaciologists in observing glacial movement throughout the world.

Figure 18 shows three scenes of Utah Lake, acquired at three different seasons of the year: summer, late summer, and spring. It exemplifies the ability of LANDSAT to monitor lake processes. In the first scene, the lake contains only a small amount of sediment (light blue) and parts of it are clear (dark blue-black). By the late summer, however, huge algal mats (red) cover a significant portion of the lake's surface. In the third scene, large amounts of water have entered the lake from the spring snowmelt and it has become heavily sedimented.

The two LANDSAT images in Figure 19 show the Mississippi River at St. Louis. The image on the left shows the St. Louis area with the river at normal levels; the image at the right shows the same area during the extensive flooding that took place during the spring of 1973. Point A on the figure locates the confluence of the Missouri and Mississippi rivers; point B locates the confluence of the Illinois River and the Mississippi. The two places labeled C on the figure are large areas that were inundated during the flood. These and other images were used by a LANDSAT-1 investigator to map the flood area along almost the entire length of the Mississippi.

ENVIRONMENT

Many types of environmental studies require information gathered on a large scale rather than at single points. A water-pollution plume, for example, can be indicated by field samples taken at scattered points, but its dispersion pattern and flow characteristics can only be determined by larger-scale mapping. The synoptic view provided by LANDSAT has proved to be well suited for environmental studies of this type. The ability of LANDSAT to produce multiband data is also valuable for environmental studies, since it provides a means of discriminating subtle spectral differences in the reflectances from surfaces studied. LANDSAT investigators have reported that the MSS can describe various levels of environmental quality in fields such as water classification, vegetation classification, wildlife habitat monitoring, and strip-mine monitoring. They have found evidence of such occurrences as incipient vegetation disease, algal population increase, and organic versus inorganic content of sediment.

Figure 20 shows an inventory map of strip-mined land in parts of Pike, Gibson, and Warrick counties in southeastern Indiana that was prepared by a LANDSAT-1 investigator. It shows strip-mined areas previously mapped, plus areas newly mapped from the LANDSAT imagery. The utility of LANDSAT for this purpose has been demonstrated by several investigators in coal-mining regions of the Northeast, and the accuracy of the mapping has been verified by ground truth checking. Repetitive coverage by LANDSAT allows state officials to both map newly stripped areas and to periodically check on reclamation progress.

Turbidity swirls in Lake Superior (Fig. 21) have provided dramatic evidence of how LANDSAT data can be of very practical use to city and state governments. The light-toned sediment patterns along the lake's shoreline are obvious in the LANDSAT image. The cross in the bottom-middle portion of the lake marks the spot where a freshwater intake line was placed by a local municipality, at a cost of $8 million. High sedimentation at the intake point made the entire operation unusable. Had LANDSAT imagery been available to guide the placement of the water intake point, this very costly error could have been avoided.

Meteorologists have suspected for years that air pollution causes weather modifications. Figure 22 provides excellent evidence to back up their theories. This late-fall LANDSAT image of the Chicago, Illinois–Gary, Indiana, area shows plumes from seven steel mills and power plants (indicated by arrows) flowing out over Lake Michigan and feeding directly

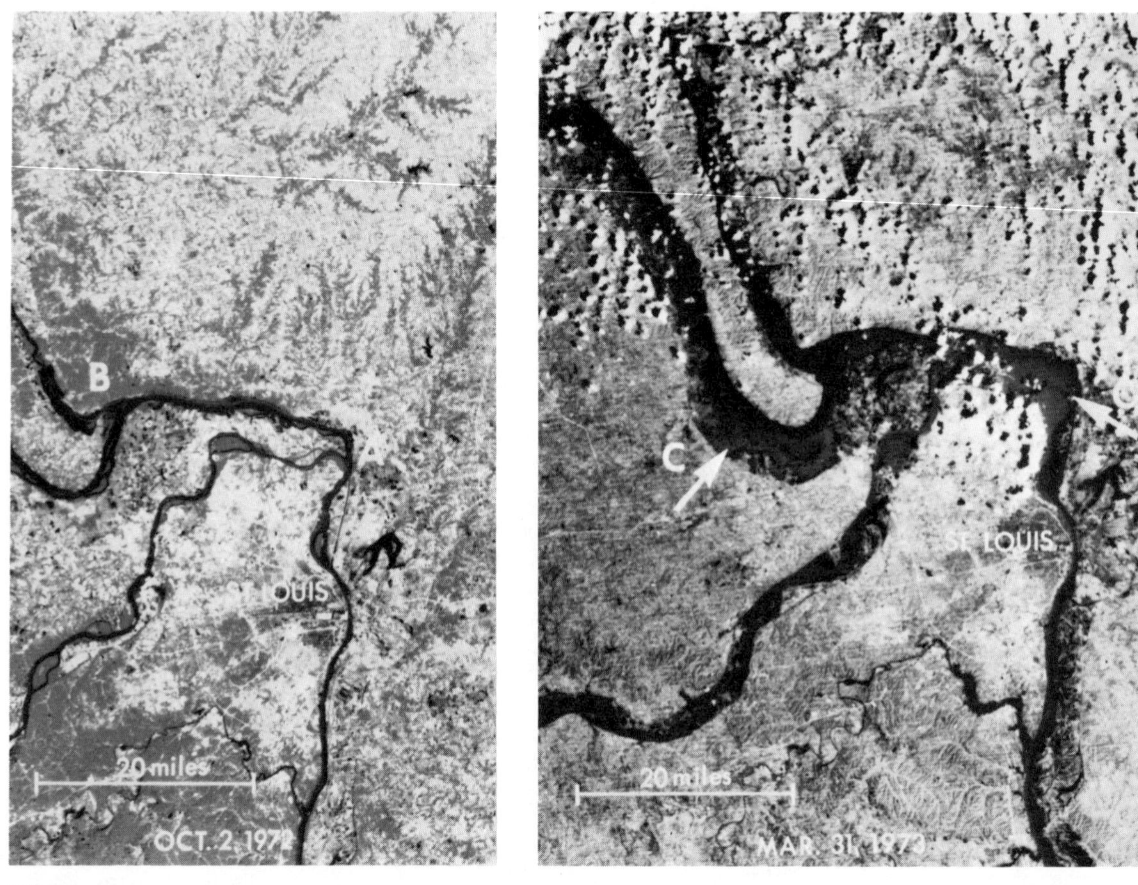

FIG. 19
"Before and after" LANDSAT scenes of the St. Louis area showing extent of the spring 1973 flood.

into the cumulus clouds over the lake. The clouds forming out of the plumes begin noticeably closer to the Indiana–Illinois shore and become denser and more developed as they approach the Michigan shore than do the clouds outside the direct lines of the smoke plumes. Because snow and rain form around the particles contained in the plumes, the clouds influenced by them will more likely produce precipitation than the other clouds. This is believed to be the first clearly observable example of inadvertent weather modification due to man's activities.

MARINE RESOURCES

A majority of the LANDSAT-1 investigations in the marine resources discipline have dealt with coastal processes. Techniques have been developed for defining coastal circulation patterns, using sediment as a natural tracer, and formulating new circulation concepts in some geographical areas. An analytical technique for measuring absolute water depth (bathymetry) based upon the ratios of two MSS channels has also been developed. This technique requires some knowledge of bottom reflectivity, water transparency, and surface characteristics. Initial evaluation of the technique indicates that it is useful in coastal areas of low-to-moderate turbidity for depth measurement up to 9 m (5 fathoms) and for updating the locations of reefs and shoals.

Investigators also report that significant progress has been made in developing techniques for using LANDSAT-1 data to locate, identify, and monitor sea and lake ice. Ice features greater than 70 m in width can be detected, and both Arctic and Antarctic ice-

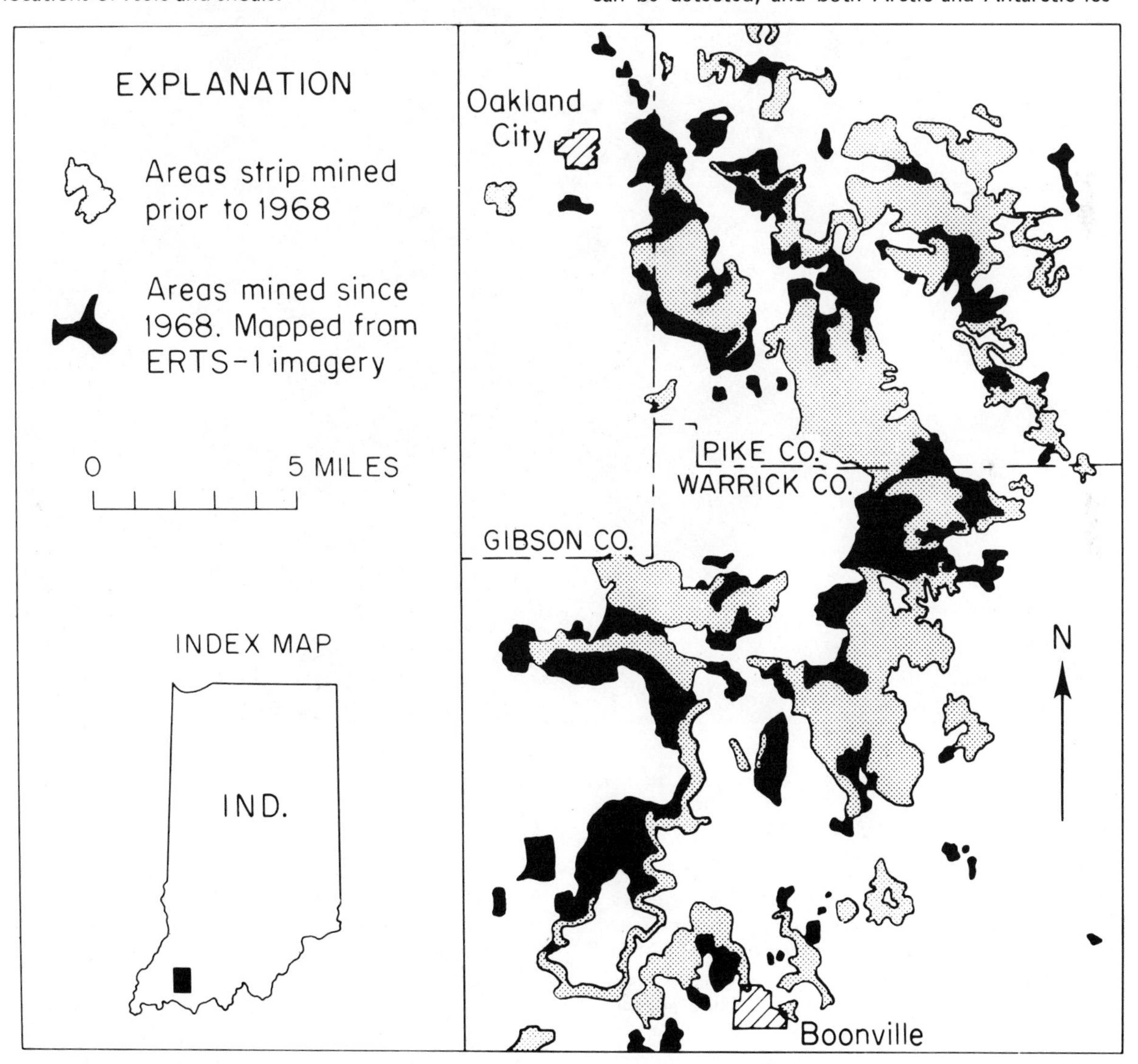

FIG. 20
Strip-mine inventory map of southeastern Indiana compiled from analysis of LANDSAT imagery.

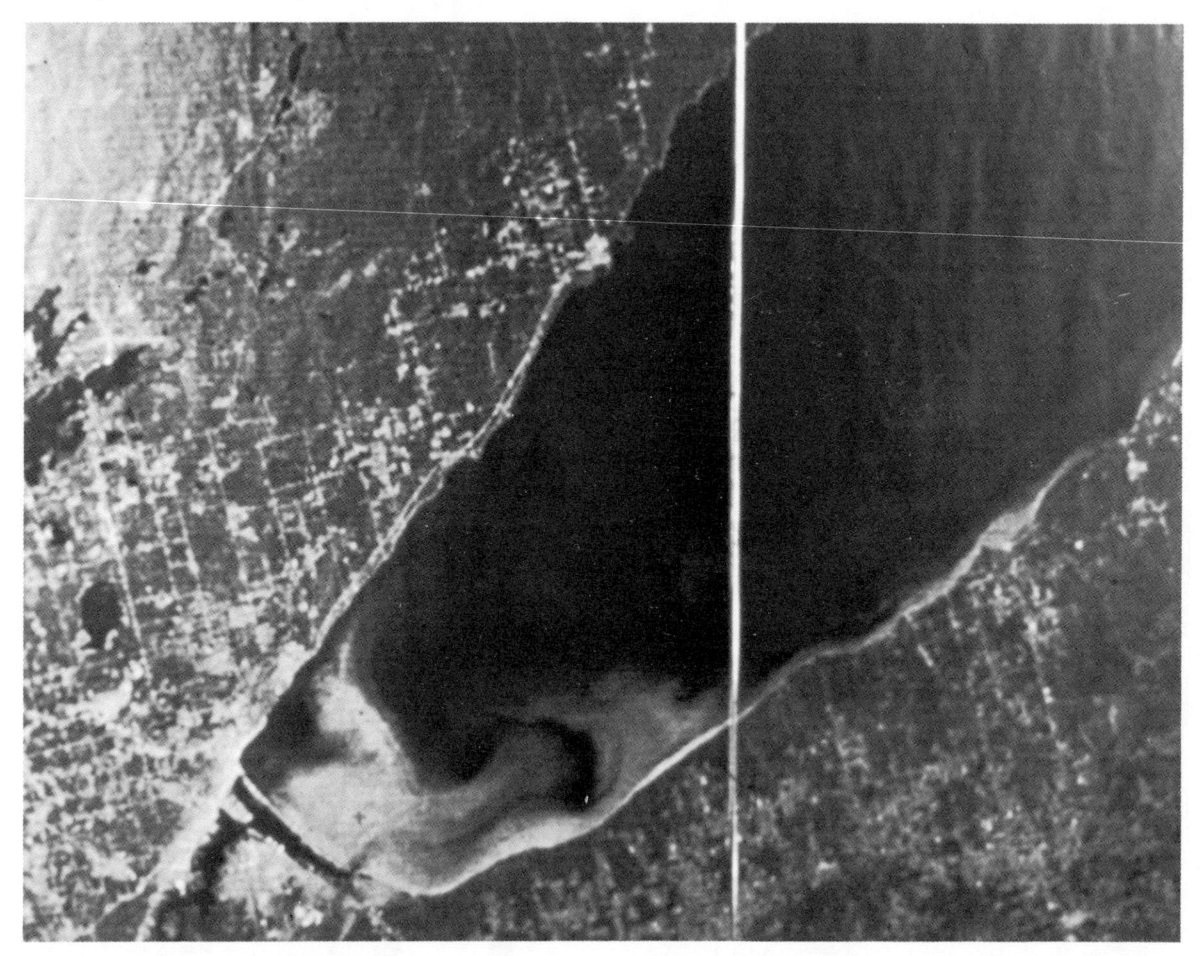

FIG. 21
Sediment plumes in Lake Superior; cross indicates site of poorly located fresh-water intake.

bergs have been identified. Because of the large daily overlap of LANDSAT-1 coverage at high latitudes, some tracking and quantitative measurements of ice movement are possible. The concept of using LANDSAT imagery to replace some present aircraft surveys of Arctic sea ice on a cost-effective basis appears valid.

In the application area of living marine resources, the use of LANDSAT-1 image-density patterns as a potential indicator of fish school location has been demonstrated for one coastal commercial resource, menhaden. Further development of these satellite techniques may yield the only practical method of monitoring and assessing living marine resources on a large scale.

Ocean dynamics is another area for which large-scale synoptic coverage is required. LANDSAT-1 data have been used to locate ocean current boundaries using image-density enhancement, and some techniques are under development for measurement of

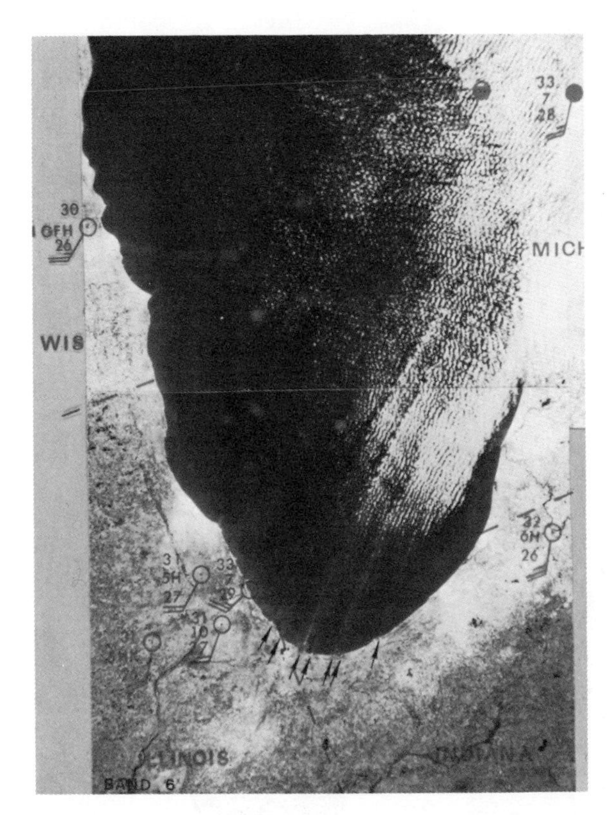

FIG. 22
Weather modification over Lake Michigan caused by air pollution.

suspended particle concentration and chlorophyll concentration.

Figure 23 shows an example from a LANDSAT investigation in this discipline, a study of upwelling and plankton patterns off the northwest coast of Africa, using both LANDSAT imagery and research ship cruises to collect ground truth. The investigator reported detecting the real distribution structure of plankton better with LANDSAT imagery than could be detected using multiple ship cruises.

An excellent example of the ability of LANDSAT to portray sediment is shown in Fig. 24, which is an MSS band 5 scene of Delaware Bay. Investigators used sediment as a natural tracer to delineate currents and aquatic interfaces, or frontal features between areas of water with differing turbidity.

The bathymetry investigation previously referred to has reported a technique for measuring water depth from LANDSAT data that could be used to quickly and inexpensively update shipping charts throughout much of the world's coastal regions. Figure 25 shows all four MSS bands of a LANDSAT-1 frame imaged over the Little Bahama Bank. In the figure, band 4 clearly shows underwater features in the bank north of Grand Bahama Island. Most of these were identified on the International Hydrographic Office depth chart, but there are a number of underwater features in the center of the frame that are not on the published chart. Band 5 shows some of the shallower areas of the bank, but bands 6 and 7 show no underwater features. Bands 6 and 7 do, however, accurately delineate land–water boundaries. The investigator demonstrated that depths can be calculated by ratio processing for those depths that return a signal in both bands 4 and 5. If additional control points are known, and if the reflectances of bottom materials are taken into account, an absolute depth map can be made for depths up to 9 m.

Accurate identification of lake ice and classification of ice types would provide invaluable information for shipping in the Great Lakes, allow more efficient routing of shipping through ice-covered waters, and possibly permit extension of the shipping season. As shown in Fig. 26, LANDSAT-1 provides relatively highly detailed synoptic views of the ice cover on the Great Lakes. Although its 18-day cycle prevents acquisition of imagery often enough to be of operational use in a lake-ice surveillance system, LANDSAT-1 has demonstrated the feasibility of using satellite imagery for this purpose.

APPENDIX A: Summary of findings from geology presentations at the three ERTS symposia

Mineral Exploration

Vincent (Environmental Research Institute of Michigan). "Some geological features smaller than the publicized spatial resolution of LANDSAT, such as the diabasic gabbro dikes in the Louis Lake batholith,

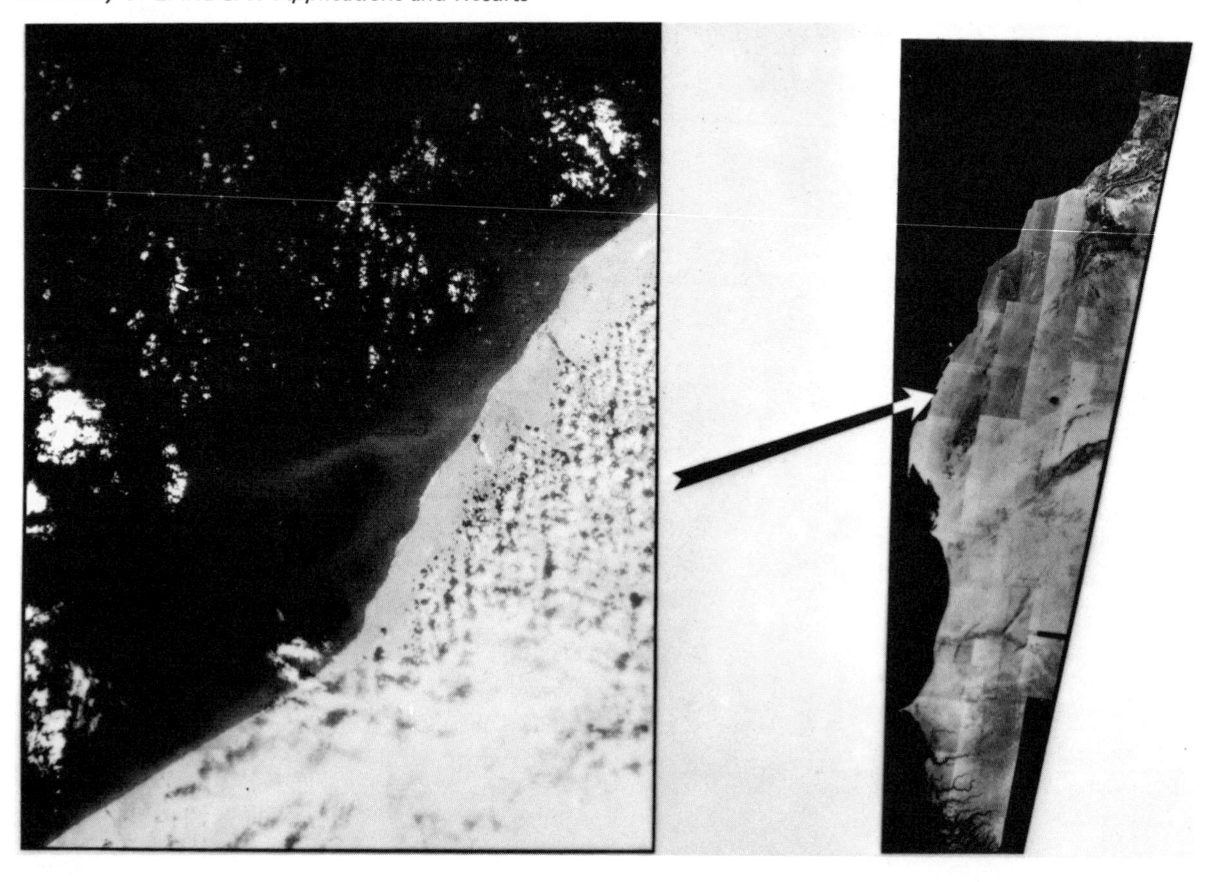

ERTS-1 GREEN BAND 4 NOVEMBER 1972

FIG. 23
Upwelling and plankton patterns off the northwest coast of Africa.

are seen well in LANDSAT images and LANDSAT ratio maps."

Jensen (University of Utah). "Those images that provide areas or features of greatest initial interest are further enhanced when printed in black and white but enlarged four times to a scale of about 1:250,000. Interestingly enough, little detail is lost with the enlargement and numerous geological maps of the area are available at the same scale. A prime example is Humbolt County, Nevada, where regional structural trends, basin and range normal faults, and specific lithologic types can be recognized on the imagery while some are not noted on the geological map. The enlarged LANDSAT-1 image is perfectly adequate as a base map and could have served readily as a field base map for mapping."

Bechtold and Liggett (Argus Exploration Co.) were able to distinguish small-scale faults cutting alluvium and surface tonal anomalies indicative of mineralized areas. "The resolution of the LANDSAT MSS imagery is sufficient for surveys of relatively small faults which cut alluvium."

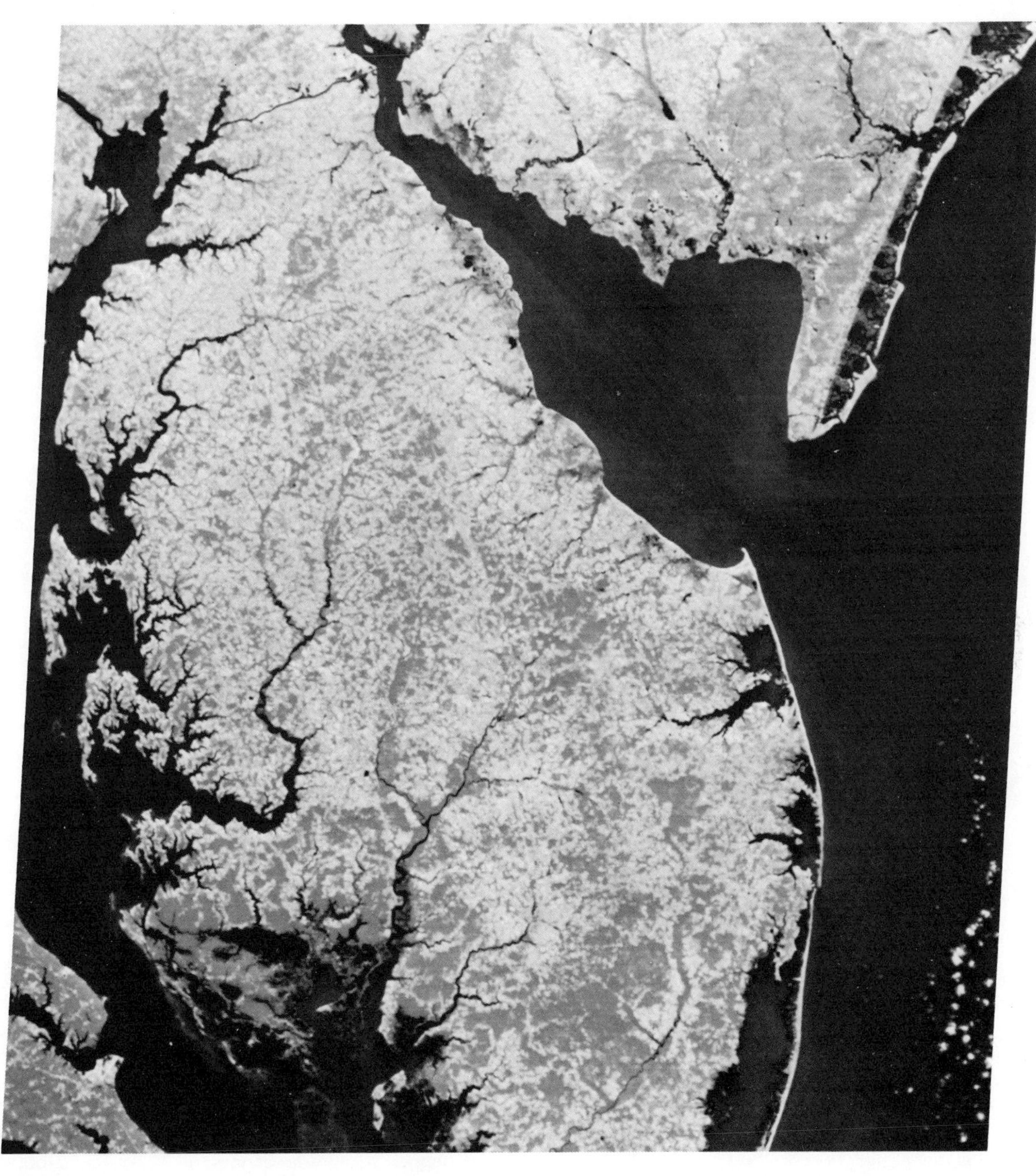

FIG. 24
Sediment distribution in Delaware Bay.

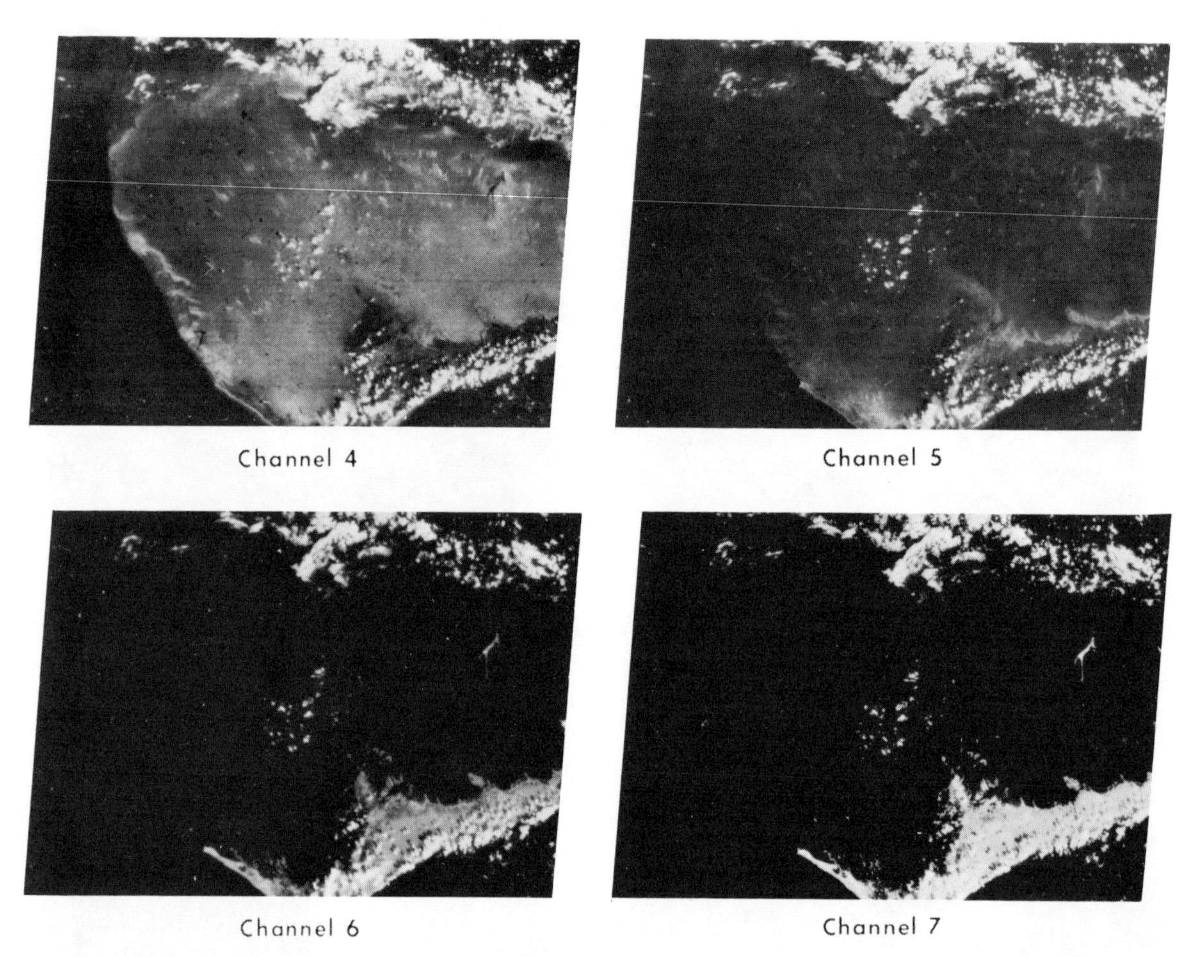

FIG. 25
Four-band LANDSAT MSS data of Little Bahama Bank.

Saunders (Texas Instruments). "The discernibility of geologic–geomorphic data on LANDSAT photo in west Texas that allow the surface reconnaissance mapping of a buried feature with a reasonable degree of accuracy indicates the commercial value of LANDSAT imagery. It can only be concluded from our results that LANDSAT-1 imagery will be of tremendous value in commercial exploration programs."

Lathram (U.S. Geological Survey). "LANDSAT-1 images provide both a new perspective of geologic conditions and new geologic information of great benefit in determining the geologic framework and mineral resource potential of Alaska. They also provide new and excellent operational tools for representing the geology of areas already mapped and for extrapolating geologic knowledge into unmapped areas."

Knepper (Colorado School of Mines) used brand ratios (5/7) and color additive viewing to enhance surface alteration, lineaments, and rock types.

Vincent (Environmental Research Institute of Michigan) used ratio images to enhance surface alteration characteristics of aluminum- and iron-ore deposits.

Bechtold and Liggett (Argus Exploration Co.) used band ratioing and false color composites to enhance surface alterations indicative of mineral emplacement and geothermal energy sources.

Rowan (U.S. Geological Survey) used band ratioing and color compositing to distinguish hydrothermal alteration zones indicative of mineral deposits.

Houston (University of Wyoming) discovered a metasedimentary iron formation using bands 5 and 7 simultaneously.

Saunders (Texas Instruments) found band 5, 7 composites to be useful for mineral exploration in Colorado.

Jensen (University of Utah) found MSS imagery valuable for determining age of strata and structural relationships for petroleum exploration.

Drahozval (Alabama Geological Survey) used the MSS data to plot regional lineaments. A newly recognized pair of lineaments, exceeding 322 km, were plotted and correlated with barite deposits. Indications are that the lineaments are responsible in part for localizing these deposits.

Hoppin (University of Iowa) used the synoptic view to extrapolate the location of mineral deposits, a task which Hoppin indicates would have been very difficult without the data.

Fisher (U.S. Geological Survey) used LANDSAT data to confirm a hypothesis of the distribution of metallic ore deposits by analyzing the synoptic view of a key structural area.

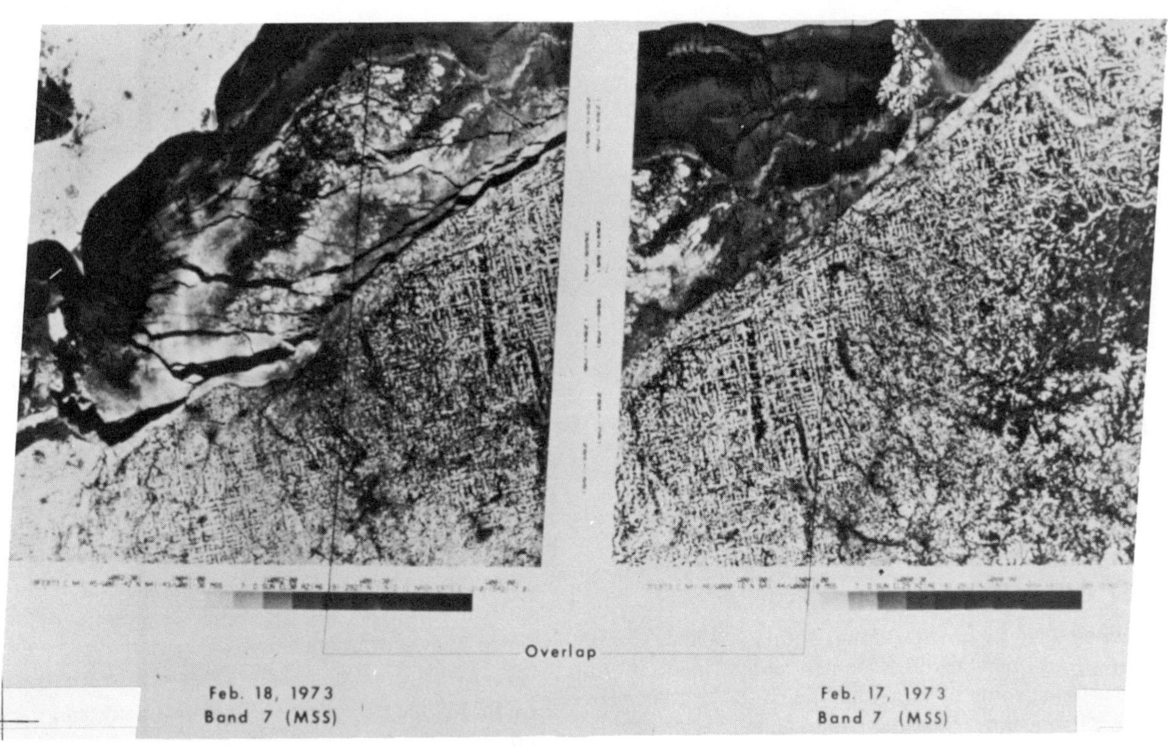

FIG. 26
Twenty-four-hour breakup of ice on Lake Erie as imaged by LANDSAT-1.

Petroleum Exploration

Collins (Eason Oil Co.) reported that LANDSAT resolution is adequate for detecting surface anomalies and lineations indicative of petroleum reserves. "Study of LANDSAT-1 data covering the extensively studied Anadarko Basin of Oklahoma and Texas has shown the LANDSAT system to be an excellent tool for the purposes of petroleum exploration. Types of information derived from LANDSAT data that are useful for petroleum exploration include: a vast quantity of information on linear features; general lithologic distribution; identification of several types of anomalous features of petroleum exploration interest; details of structures controlling hydrocarbon accumulation in some areas; and the regional context of the exploration province as a whole. Preliminary analysis indicates that the use of LANDSAT imagery can substantially reduce the cost of regional petroleum exploration in relatively unexplored areas."

Lathram (U.S. Geological Survey) found LANDSAT resolution useful for distinguishing lithologic differences and structural trends indicative of potential petroleum accumulations in the Alaskan coastal plain. "The comprehensive and orthophotographic portrayal of the structure in this area will provide maps that are significantly more useful to petroleum exploration than using conventional planimetric or topographic bases."

Jensen (University of Utah) found LANDSAT resolution useful in interpreting geologic structures that may be associated with petroleum reserves. "The diversity of mineral and oil deposits, ages of formations and igneous rocks, and the variation in structural features provide a vast potential for the practical application of multispectral imagery to the study of this area."

Earthquake-Zone Investigations

Gedney (University of Alaska) found LANDSAT resolution adequate to discover new active faults in central Alaska, some of potential hazard to the trans-Alaska oil pipeline. Seismic risk maps based on LANDSAT data will aid planners at all levels.

Abdel-Gawad (Rockwell International) discovered evidence of active faulting in regions of southern California thought to be seismically safe. LANDSAT resolution was able to provide the investigator with evidence of transverse stresses on the San Andreas complex.

Wobber (EARTHSAT) recognized complex fracture trends using snow-enhanced LANDSAT MSS imagery. This information is valuable for determining zones of crustal weakness.

Carter (U.S. Geological Survey) found LANDSAT resolution useful for determining and mapping the geologic features and processes that contributed to the Managua, Nicaragua, earthquake.

Merifield (California Earth Science Corp.) analyzed the appearance of important faults using LANDSAT imagery. The San Andreas, Banning, and Mission Creek faults were traced for extensive distances, even though buried by alluvium in many areas.

Geothermal Surveys

Vincent (Environmental Research Institute of Michigan) stated that LANDSAT resolution was found to be adequate for detecting surface anomalies due to hydrothermal alteration, features commonly associated with geothermal energy sources.

Bechtold and Liggett (Argus Exploration Co.) used the resolution of LANDSAT imagery to detect hydrothermally altered zones, tufa chemical sedimentation, and desert oases, all indicative of geothermal energy sources. They also detected surface alterations associated with geothermal activity from multiband imagery and ratioing techniques.

Williams (U.S. Geological Survey) detected geothermal springs as small as 2.5 km^2 in area by noting anomalous snowmelt patterns in LANDSAT.

Rowan (U.S. Geological Survey) found that LANDSAT resolution and computer enhancement techniques (band ratioing and others) enabled him to identify hydrothermally altered areas, commonly associated with geothermal energy sources.

Erosion, Mining Safety, Disaster Assessment, and Sand Migration

Wier (EARTHSAT) detected lineaments from LANDSAT directly related to zones of roof-fall hazard in the King Station Mine, southern Indiana. An accurate roof-fall hazard map was constructed to provide guidance for future mining operations.

McKee (U.S. Geological Survey) found LANDSAT resolution satisfactory for distinguishing sand dune characteristics, including relatively small scale sand patterns. Dune migration and those factors affecting sea sand movement have been measured from LANDSAT.

Morrison (U.S. Geological Survey) found LANDSAT resolution "surprisingly good," and was able to map erosional gullies as narrow as 20 m in width. Morrison used LANDSAT imagery to inventory and monitor rapidly eroding flatlands in southern Arizona. With LANDSAT imagery he was also able to map flood effects with accuracy.

Lithologic and Structural Surveys

Knepper (Colorado School of Mines) used LANDSAT imagery to identify rock structures, enhance lineaments and surface alterations, and map regional rock types. Resolution was reported to be adequate for these uses.

Vincent (Environmental Research Institute of Michigan) found LANDSAT resolution to be adequate for identifying and mapping zones of surface alteration, and identifying small geologic structures such as dikes and other small intrusive structures.

Bechtold and Liggett (Argus Exploration Co.) mapped the distribution of major rock units, and detected surface anomalies caused by metamorphism and hydrothermal alteration.

Martin (Missouri Geologic Survey) was able to distinguish sediment patterns, shoreline configurations, vegetation, and soil and rock types with LANDSAT MSS imagery and computer enhancement techniques.

Goetz (Jet Propulsion Lab) and Billingsley (U.S. Geological Survey) found LANDSAT resolution adequate to map tonal anomalies, major rock units, and tailings from a local copper mining operation. The investigators' ability to map a characteristic rock formation from LANDSAT imagery enabled them to discover a new source of groundwater in an arid portion of the Colorado Plateau.

Houston (University of Wyoming) identified and mapped igneous stocks, plugs, granites, and gneisses, and also soil distribution and an iron-ore (taconite) deposit.

Breckenridge (University of Wyoming) mapped glaciers and glacial deposits from LANDSAT. Glacial deposits represent important aquifers in many parts of the country, and also give clues to ancient climates.

Jensen (University of Utah) found LANDSAT imagery adequate to map structural trends, lineaments, regional-rock-type distribution, and volcanic centers, all of which are useful for mineral and petroleum exploration.

Williams (U.S. Geologic Survey) mapped recent and ancient lava flows, and monitored the growth of the new volcanic island of Surtsey using LANDSAT.

Knepper (Colorado School of Mines) used color additive viewing to distinguish surface alterations and rock types.

Vincent (Environmental Research Institute of Michigan) used band ratioing techniques to detect surface stains associated with mineral deposits.

Bechtold and Liggett (Argus Exploration Co.) used LANDSAT color composites to discover new faults in alluvium that appear to be associated with the fundamental basement structures of the basin and range province.

Grootenboer (Spectral Africa Ltd.) used seasonal data to study structures in Africa. A comparison of winter and summer imagery revealed striking differences in the amount of recognizable detail.

Knepper (Colorado School of Mines) used winter LANDSAT data to produce a lineament map, taking advantage of a snow cover and the low-solar-angle illumination to produce the most detailed map possible. Many new linears were plotted.

Hoppin (University of Iowa) used a winter scene to map geologic structures, finding that the topographic features and drainage patterns are markedly enhanced.

Saunders (Texas Instruments) used repetitive data to perform an exhaustive study of the value of temporal data for geologic studies. Findings indicated that seasonal data are necessary to fully study an area.

Wobber (EARTHSAT) indicated that repetitive

winter data can provide unique fracture data not available during other seasons.

Collins (Eason Oil Co.) found that geologists need data from all seasons to evaluate fully LANDSAT-1 imagery. In his test site, Collins found that spring and fall are both useful, depending on the moisture conditions within the test site.

Gedney (U.S. Geological Survey) used a nine-frame mosaic to study the structures of Alaska's mountains. He reported several new large linears, previously unidentified, trending hundreds of miles. These new linears provide new knowledge of the strip tectonics and seismicity.

Gotham (U.S. Geological Survey) used synoptic MSS data to study the structures of Alaska. In the Alatma Hills area, the regional view revealed a previously unknown series of closely spaced fractures that support current theories.

Isachsen (New York State) reveals that LANDSAT provides good data on structural provinces not available from other sources.

Lathram (U.S. Geological Survey) used the synoptic view to study the complex structure of the northern Alaska petroleum provinces. For the first time, a detailed analysis of the Polony Mountains was possible. He also discovered new large linear structures in Alaska, which can be correlated to known magnetic surveys.

Viljoen (Johannesburg Consolidated Investment Co.) found that LANDSAT data provide a more accurate definition than exists at present of the structure of complex geologic terrain.

Collins (Eason Oil Co.) used the synoptic view of ERTS to map lithologies, finding LANDSAT valuable because (1) major structural and lithologic features are discernible, (2) regional relationships can be studied, (3) reflection and pattern indicate anomalous and possibly important areas, and (4) the resolution and regional perspective permit large-scale differences to be defined and studied.

Cavelier (Bureau of Geological and Mineral Research, Paris) noted a good correspondence between the units seen on the imagery and the concentric strata of Jurassic–Cretaceous formations. He was able to extend their boundaries from the imagery to an area outside the Paris Basin.

Saunders (Texas Instruments) used LANDSAT data to map structures not previously identified from aerial photographs.

Allen (Missouri Geological Survey) used the synoptic imagery to map certain structural linear elements across geologic boundaries.

Geomorphic Surveys

Williams (U.S. Geological Survey) used wintertime imagery and found that the low-angle solar illumination provided considerable detail of volcanic features not previously recognized.

McKee (U.S. Geological Survey) has used repetitive ERTS data to map the dynamic and static nature of shifting sand seas.

Gregory (Gregory Geoscience Ltd.) used winter imagery to map an extensive hummocky morained deposit and summer imagery to map vegetation patterns that are thought to be related to the distribution of lacustrine and marine clay deposits.

Morrison (U.S. Geological Survey) used MSS data to map the anomalies that may signal buried moraines. Morrison indicates that this task would have been very difficult without the synoptic view.

Short (National Aeronautics and Space Administration) reported that Oregon geologists had discovered a feature in Oregon that was totally unknown before LANDSAT. The synoptic view revealed this structure very clearly. He also reported that geologists had discovered from LANDSAT imagery an impact crater in Brazil that was twice as large as previously thought.

ACKNOWLEDGMENT

The author wishes to thank Bruce N. Rogers of the General Electric Co., Beltsville, Maryland, for his assistance in researching and writing.

5

REMOTE-SENSING APPLICATIONS FOR MINERAL RESOURCES

William L. Smith

William L. Smith, formerly with System Planning Corporation, is currently with the Environmental Research Institute of Michigan, Arlington, Virginia.

NASA's stated objectives in mineral resources remote sensing include the development of methods for improving the effectiveness of mineral, hydrocarbon, geothermal and groundwater exploration *(1)*. The review of the 1974 program by the Office of Applications notes that many ores are associated with fracture or fault systems, anomalies in topography, or coloration of surface rock, and that such guides are often distinguishable by the current generation of sensors. Emphasis has been placed on assisting the search for minerals and the inventory of essential resources by the development of methods for rapidly obtaining knowledge of geological relationships and identifying surficial indicators. The status of this research is shown in Table 1.

GEOLOGICAL GUIDES TO MINERAL DEPOSITS

Several types of mineral deposits have various surficial indicators or guides that can often be identified by such remote sensors as cameras and spectral imagers. First, we shall consider the observable geological characteristics of such indicators.

Most mineral deposits are located through the extensive study of topographic and geologic maps in conjunction with conventional aerial and field methods of exploration and laboratory analysis of samples. The initial contribution that remote sensing can offer is better maps derived from data of a synoptic nature. In its present early phases, sensing for geological targets will rely largely on photography and spectral imagers.

Ore deposits are rarely haphazard in occurrence, although they may appear to be randomly located before the mineralization pattern has been determined. Many guides exist that aid in locating mineral deposits by conventional exploration practices. These guides, although not necessarily directly detectable by remote sensors, will be the basis of interpreting return data. The more important guides are characterized by the geometry or symmetry of geological features, apparent geological anomalies, and regional trends. The coincidental occurrence of two or more favorable factors, such as faulting or intrusion meeting receptive lithology or structure, often indicates a favorable location for exploration. Other guides may be relevant mineralogical or geochemical patterns, or such secondary clues to geological phenomena as changes in vegetation, soils, and stream patterns.

Ore deposits themselves are often relatively small

TABLE 1
Status of NASA Research in Mineral Resources, Geologic Structure and Landform Surveys, 1974 *(1)*.

OBJECTIVES*	CURRENT STATUS USING LANDSAT DATA	RESEARCH TECHNOLOGY AREAS	NEEDED
RESOURCES EXPLORATION • MINERAL RESOURCES • METAL ORES • BUILDING STONE, AGGREGATE	• LINEAMENT INTERSECTIONS CORRELATED TO KNOWN DEPOSITS. SOME NEWLY FOUND STRUCTURES HOLD PROMISE OF NEW DEPOSITS-AWAIT GROUND EXPLOR-ATION FOR CONFIRMATION	• IMAGE ENHANCEMENT TECHNIQUES - TO ENHANCE STRUCTURAL FEATURES & SURFACE ALTERATION ZONES ASSOCIATED WITH MINERALIZED AREAS, AREAS OF PETROLEUM CONCENTRATIONS, GROUND WATER ZONES, FEATURES IMPORTANT IN CAUSING LANDSLIDES & EROSION, AND IN IDENTIFYING STRUCTURES, LANDFORMS, AND ROCK-TYPES	• IMPROVED DIGITAL ENHANCEMENT TECHNIQUES & IDENTIFICATION OF OPTIMUM BANDS & RATIOS FOR ENHANCING VARIOUS FEATURES SUCH AS SURFACE STAINING, "TONAL ANOMALIES," ROCK TYPES ASSOCIATED WITH MINERALIZATION • EVALUATION OF USEFULNESS OF SPATIAL FILTER-ING TECHNIQUES TO ENHANCE LINEAR STRUCTURES
• HYDROCARBON RESOURCES • PETROLEUM EXPLORATION • COAL EXPLORATION	• KNOWN OIL-BEARING STRUCTURES IN CAL. CORRE-LATED WITH NEWLY DISCOVERED LINEAMENTS. SURFACE EXPRESSIONS OF SALT DOMES NOTED IN GULF COAST REGION. ANOMALOUS FEATURE IDEN-TIFIED IN ALASKA COULD CONTAIN OIL-BEARING SEDIMENTS-GROUND EXPLORATION NEEDED FOR CONFIRMATION	• MULTISPECTRAL IMAGING FEATURE IDENTIFICATION (NEAR IR & THERMAL CHANNELS) - TO UNDERSTAND THE GEOLOGICAL INFORMATION CONTENT AND EVALUATE THE USEFULNESS OF THE "NON-VISIBLE" MSS CHANNELS FOR GEOLOGICAL APPLICATIONS	• EVALUATION OF GEOLOGICAL INFORMATION CONTENT CONTAINED IN "NON-VISIBLE" MSS CHANNELS
• GEOTHERMAL RESOURCES	• INDIRECT INDICATIONS OF GEOTHERMAL ACTIVITY NOTED BY MAPPING OF SNOW COVER OF FEW SELECTED VOLCANOS		
• GROUND WATER RESERVES	• SUCCESSFUL WELL DRILLED IN LOCAL PERCHED AGUIFER IDENTIFIED IN LANDSAT; FRACTURES CORRE-LATED TO SPRINGS		
GEOLOGIC FEATURES & PROCESSES • STRUCTURAL SURVEYS	• FRACTURES, FAULTS, FOLDS, INTRUSIVES, AND IMPACT STRUCTURES DETECTED WITH VARYING SUCCESS, DEPENDENT ON SIZE, STRUCTURAL SETTING, CONTRAST, EXTENT OF GROUND COVER, SUN ANGLE, ETC. DETECTION OF REGIONAL LINEAMENTS BIGGEST PAYOFF TO GEOLOGY FOR LANDSAT.	• ACTIVE MW IMAGING FEATURE IDENTIFICATION - TO UNDERSTAND THE GEOLOGICAL INFORMATION CONTENT AND EVALUATE THE USEFULNESS OF ACTIVE MW IMAGERY FOR GEOLOGICAL APPLI-CATIONS • AUTOMATED COMPUTER RECOGNITION/CLASSIFI-CATION OF TERRAIN - TO IDENTIFY ROCK TYPES AND MAP ROCKS & SOILS	• EVALUATION OF GEOLOGICAL INFORMATION CONTENT IN ACTIVE MW IMAGERY • DETAILED FEASIBILITY STUDY TO EVALUATE PRACTICABILITY OF AUTOMATED METHODS TO IDENTIFY ROCK TYPES & PREPARE SURFACE MAPS UNDER VARIOUS FIELD CONDITIONS & USING SCANNER SUCH AS MSDS (24 CHANNELS)
• GEOMORPHOLOGIC (LANDFORM) SURVEYS	• MOST HIGHER-ORDER GEOMORPHIC UNITS (MOUN-TAINS, BASINS, LAKES, GLACIAL FEATURES, SAND DUNE FIELDS, VOLCANIC FEATURES, OCEANS, COASTAL FEATURES, ETC.) WELL DISPLAYED - REGIONAL VIEW HIGHLY SIGNIFICANT.		
• LITHOLOGIC (ROCK-TYPE) SURVEYS	• FEW RELIABLE IDENTIFICATIONS OF ROCK TYPES MADE DUE TO RESOLUTION LIMITS (MOST STRATI-GRAPHIC UNITS THINNER THAN SPATIAL RESOLU-TION LIMITS) AND "NON-UNIQUENESS" OF SPECTRAL SIGNATURES FOR INDIVIDUAL ROCK TYPES		

targets located at considerable depth. The following five categories of surficial indicators of possible mineralization or aids to reconnaissance mapping lend themselves to identification by remote sensing:

1. Topography.
2. Igneous and volcanic features.
3. Lineaments and geological structure.
4. Mineralogical–lithological association.
5. Stratigraphic sequence.

The various indicators discussed are not to be understood as showing where to drill but as guides to possible areas of mineralization or accumulations of petroleum. These are the potential areas for follow-on exploration by conventional aerial or surface methods. The IBM study for the NASA Orbiting Research Laboratories *(2)* reviewed the major mineral discoveries of the previous 20 years, noting many rich discoveries had been made by aerial surveys focused on areas where there were promising local indicators. The restriction to these areas was largely due to the high expense of exploration. However, because valuable discoveries also occur in areas where conventional indicators are absent, it was suggested that a likely benefit of resource surveys from space could be the identification of regional guides, which would otherwise remain long unknown.

Topography

Physiographic configurations that are surface expressions of structure, stratigraphy, igneous processes, or lithology can provide much geological and geomorphological information pertinent to reconnaissance mapping and defining locations for prospecting, particularly when their occurrence may be correlated with regional mineralization. These features range in scale from broad regional expressions to discrete topographic units. The following examples of landforms or evidence of surface processes are often clearly discernible on spectral imagery or photographs from orbital altitudes.

Among the broad structural expressions that may be identified are depressions, subsidence, domes, and folding. The limbs of anticlines and the ridges of resistant members of a folded terrain are often clearly defined. Structural features are sometimes indicated by the alignment or parallelism of lakes or the location of swamplands. Radial drainage may indicate doming. Some drainage may be seen to be controlled by joints. Distinctive stream channels may attest to current or preexisting confinement by sandstone ridges. Regional geomorphology may be interpreted or classified, such as in arid areas, by the identification of buttes, mesas, arroyos, yardangs, bajadas, oases, dry lakes, playas, inland deltas, etc. These features are often clearly identifiable in unvegetated areas and may relate to stratigraphy or obscured structure as they influence surface processes.

Discrete terrain features that may aid in identifying hidden geology include evidence of salt domes, cuestas, escarpments, ridges, and a variety of circular features which show on imagery. Diatremes may not be distinguishable from astroblemes in most climatic areas, but cones, craters, and calderas may be distinguished from erosional hills, peaks, ranges, and intermontane basins. Various erosional features such as slides, sinks, and boulders of decomposition assist in defining or delimiting lithologic units. Other erosional evidence such as canyons, cliffs, badlands, valley types, fans, and outwash plains have applications to engineering and surface exploration planning.

Coastal features that may be seen include fiords, deltas, barrier islands, reefs, and lagoons. Terraces, cheniers, and alluvial deposits related to possible placer minerals may be better seen from aircraft altitudes; but estuary details, wetland patterns relevant to dredging, and the location of nearshore bars and shallows may often be seen clearly on LANDSAT products. Certain surface characteristics, such as vegetation density, ash limits, and changes in soils, may be seen. Lava flows and scabland patterns are distinctive, as are some boulder trains, drift cover, and some moraine details. The patterns of engineering and extractive operations, such as tailings, dumps, excavations, construction, and roadways, show up well against a vegetated background, and those related to mining aid in placing known mineral occurrences in the broader picture.

Such a list could go on much further since topography is generally the expression of the other categories of guides to be considered. It is easy to see how such landform information in poorly explored regions can help define prime locations for the search for placer deposits, mineralized structures, traps for oil, or the loci of igneous contacts. Since the topographic features are often clues to subjacent geology, they can be of unique value to exploration planning by narrowing a wide area to a smaller ground of greater potential interest.

There is a tendency to use LANDSAT imagery only as aerial photographs, thus losing the great value of its multispectral properties. Interpreting the physiographic evidence of structure requires more than conventional photogrammetric techniques and the identification of topography and obvious linears. Depressions and elevations on LANDSAT scale imagery may be quite deceptive for mapping applications. However, LANDSAT products are foremost graphic, and the first data applicable to any planned exploration or mapping will be pictures; the initial step in the surface exploration sequence will generally be for verification of the interpretation assigned to the topographic features seen on the imagery.

Figures 1 and 2 are black and white positive images in bands 4 and 7, respectively, of the Arquipelago Dos Bijagos and coastal areas of Guinea-Bissau and the Republic of Guinea. The area is of possible interest for petroleum, low-grade iron ore, and bauxite exploration. This could involve both offshore operations and dredging or drilling of the onshore areas, plus extensive mapping. The images of the west African coast (LANDSAT frame 1104–10522) extend from the Ilha de Jeta, Guinea-Bissau (left), to the Iles Tristao, Guinea (right). The four major rivers are (left to right) Rio Mansoa, Canal do Geba, Rio Grande, and Rio Cacine. Much of the archipelago and

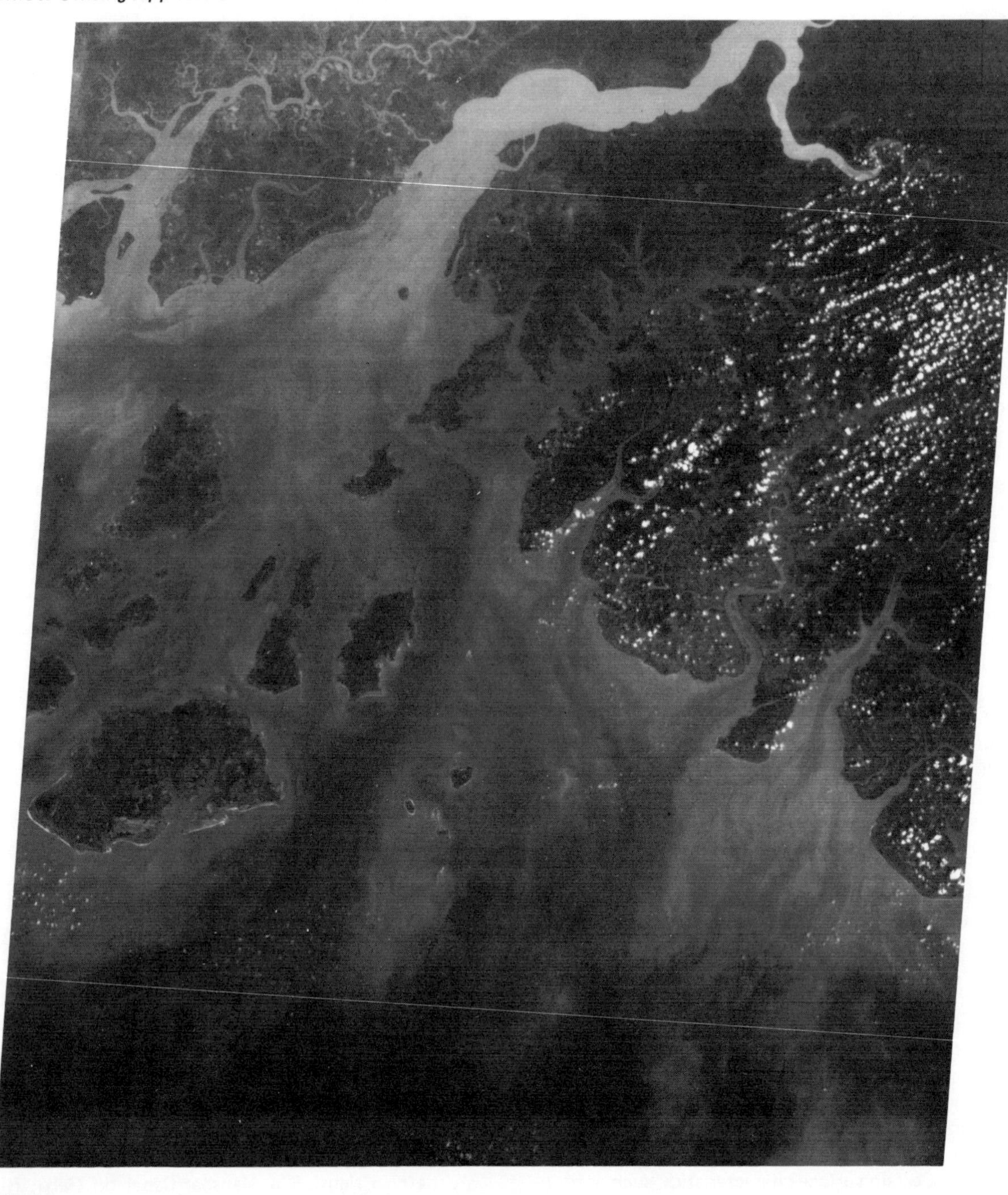

FIG. 1
LANDSAT image of Guinea coast of Africa, band 4 (LANDSAT frame 1104-10522).

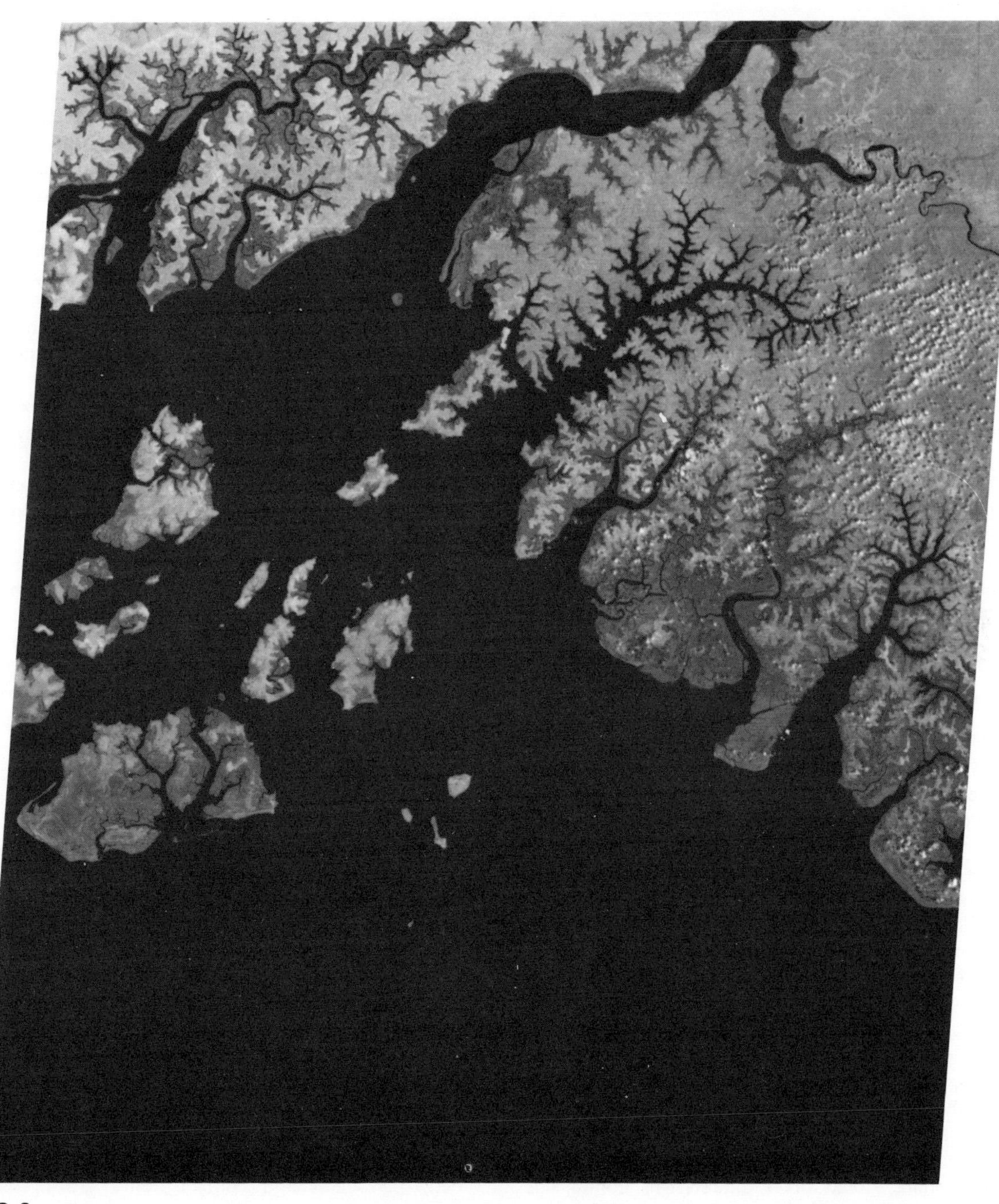

FIG. 2
LANDSAT image of Guinea coast of Africa, band 7 (LANDSAT frame 1104-10522).

coast is mangrove swamp and tidal wetlands. The dendritic patterns are related to different vegetation communities along the rivers, former river outlets, and wetlands. The towns of Bolama and Bissau can be distinguished in the central and north-central parts of the images.

LANDSAT band 4 provides some evidence as to the nearshore bottom configuration and the sediment load carried by the rivers, since clear water is transparent in this spectral range. The information content is of value to coastal engineering and offshore operations. Some cultural features, cleared land, roads, and excavations show against the vegetation. Band 7 shows water as black and clearly defines several gray scales related to vegetation types and soil conditions. The wetlands, sandy coastal areas, mangrove, and swamp, clearly definable on the imagery, show little resemblance to similar patterns on the conventional map (Fig. 3). Note that the Ilha de Orango–Ilha de Orangozinho on band 7 has an entirely different configuration than shown on the map, as do the channelways of Ilha de Maio. This band is pertinent to dredging operations, wetland studies, and hydrology, as well as to forestry and agriculture.

Igneous and Volcanic Features

Many major deposits are formed by differentiation from magmas, simple igneous contact, pegmatite injection, and by ascending fluids that form veins. Economic deposits may form by the contact metasomatic replacement of minerals by reactions with introduced materials, and by the pneumatolytic production of ore minerals by magmatic fluids either within an igneous body or in fissures in the wall rock. Guides to ores formed by igneous processes include the characteristic forms of the intrusive bodies, the nature of the contact of igneous and country rock, and the surficial characteristics of associated mineralization or alteration of country rock. The following are some of the surficial expressions of igneous forms and processes that may serve as guides to locations favorable for mineralization.

The variety of discernible forms include batholithic margins, stocks, dikes, dike swarms, domes, sills (rarely), and some pegmatites. The possible location of veins may only be inferred by the investigator from his familiarity with ore genesis. Similarly, locating any mineralization per se, the probable position of cupolas, or the classic loci of contact skarns may only be alluded to from other geological evidence, although, due to the spectral properties of LANDSAT imagery, information may sometimes be extracted that cannot be acquired by conventional methods. It needs to be emphasized that this information is not the same information obtained conventionally, and it may require a modified approach to the initial sequence of exploration. However, spectral ratio methods and other techniques, when combined with what is already known of an area, will probably in many cases significantly increase the efficiency of exploration. It needs to be further emphasized that any such interpretation of LANDSAT products is highly subjective, and that familiarity with a region plus expert capability in interpretation is of the greatest importance.

Volcanic and thermal features are readily identifiable. Some igneous rocks may be defined by characteristic structures, color, or weathering characteristics. These may be seen as textural or tonal differences on multispectral products or may lend themselves to specific techniques of enhancement. Needless to note, large discordant bodies are more apt to be defined than are sills or other flat-lying concordant bodies. Guides to obscured igneous-derived deposits may be present as characteristic patterns around the ore, such as halos or aureoles, radial or concentric fracturing or dikes, coloration, chemical alteration, domes, or depressions. Only in optimum circumstances will any but the larger of these features be definable on the current generation of remote-sensing products from space.

Most geological studies using LANDSAT products have relied either on the photographic aspect of the imagery or on the interpretation of structure, but more sophisticated spectral techniques are certain to augment this practice, and will provide geochemical information and permit a degree of geochemical mapping practical for assisting exploration. Where topographic expression of igneous bodies can be identified, peripheral and contact zones are preferred locations for geochemical investigations, inasmuch as

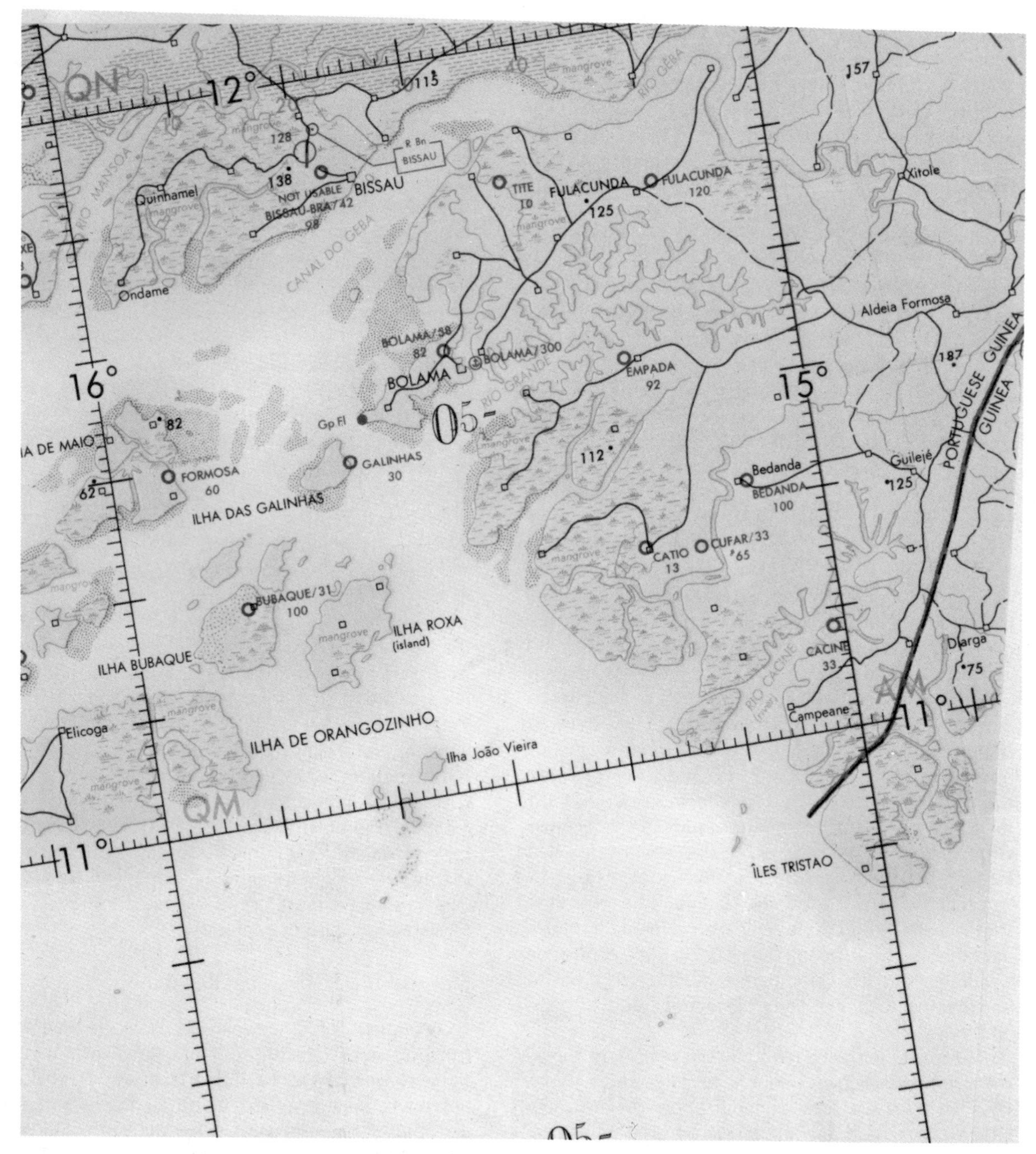

FIG. 3

Equivalent area of LANDSAT frame 1104-10522 on 1:1,000,000 scale, operational navigation chart ONC K-1.

such areas are typified by structural weakness and are often the locations of veins or other accessions from the igneous body. Contact skarns may contain major metallic deposits where reactive rocks are intruded. These areas in particular are pertinent to study by the techniques in development for mapping the iron oxides important to prospecting for various metals and for iron ores themselves.

The nature of the intrusive rock is a significant indicator for locating mineralization. Granites and intermediate igneous rocks are more apt to form contact deposits or produce veins and pegmatites; basic rocks are more apt to contain dispersed ores, such as nickel, chrome, cobalt, or mixed sulfide concentrations. At an igneous contact the country rock is likely to be altered, fractured, or otherwise made amenable to identification. The igneous rock itself may also vary in color, texture, and composition, and in the amount and size of country rock inclusions. Border phases may be identified by shell-like or schistose structures. Pegmatites and late-phase differentiates often associated with larger intrusive complexes may contain deposits of rarer minerals not found as ore in other types of deposits or in the parent igneous body.

Because unique associations exist between mineralization, lithology, and position within an igneous unit, it is important to keep these in mind when analyzing sensor products. When a complex igneous history is involved, it is also important to remember that mineralization derived from metasomatism is apt to be restricted to specific magmas and associated with specific lithologies. It is also pertinent when interpreting the relationship of the igneous body to the regional structure of the host rock to remember that mineralization preferably ascends along fractures and up-dip. Interpreting imagery to locate possible mineralization sites requires all the analysis given to deciphering such problems on reconnaissance maps, and more.

Mineralization is often the result of fluid movement from the igneous source to a receptive lithology. In areas of such mineralization the probable channelways need to be considered. The identification of observable features that might be apophyses should be followed up. Shear zones, where they might be inferred, are possible target areas. The lithology of the intruded rock as related to the possibility of the presence of fissure veining is also worth considering where such formations can be traced.

Several studies related to porphyry copper have been undertaken, but the likelihood of obtaining any detailed information of the deposits from LANDSAT images is poor. The porphyritic copper at Bingham, Utah, for example, is a disseminated hydrothermal replacement consisting of sulfides occupying much of a reactive granite–granite porphyry stock, plus high-grade copper occurring in altered limestones several hundreds of feet away from the central pyrite–copper pyrite mass. These higher-grade deposits are in turn succeeded by lead and zinc deposits at even greater distances from the central ore *(3)*. For all the hopes of LANDSAT capabilities for information of this resolution, little of this pattern may be suggested even with detailed knowledge of the deposit. It suffices that LANDSAT may aid in defining likely locations for exploration for porphyry copper.

Sensor products in no way lessen the role of conventional exploration methods. They are useful tools to assist in defining regional patterns, but much more sophisticated techniques need to be devised before they are applicable to the details of local mineralization. With present methods the chance of locating mineralization sites from igneous or lithologic guides alone is poor on LANDSAT-scale products. This, however, is anticipated to improve with data from postulated mission capabilities such as the Heat Capacity Mapping Mission (1977), which will fly a mapping radiometer capable of discriminating some rock types and structures based on measurements of thermal inertia obtained by day and night. Such products should be most useful for identifying locations for aerial and surface exploration.

Lineaments and Geological Structure

Although some mineral deposits are related to fractures, most fracturing is unrelated to mineralization. Nevertheless, the identification of previously unknown lineaments and fracture systems is the biggest plus for exploration to be derived so far from LANDSAT imagery. Levandowski et al. *(4)*, in a report on the applications of LANDSAT imagery for mapping lineaments favorable to the location of ore deposits in Nevada, demonstrated the value of such data as a supplement to mineral reconnaissance techniques. Fieldwork had established that the major known mineralization in north-central Nevada occurred along northwest trending alignments, and it was believed to be exposed at windows through the Roberts Mountain thrust plate. LANDSAT products showed these domal windows also to be aligned with previously unknown lineaments, which have been interpreted as indications of zones of structural weakness along which ore-bearing injections and fluids have found channels. Lineament maps prepared by the Laboratory for Application of Remote Sensing (LARS) from LANDSAT products showed the mineral belts to coincide with the lineament intersections, and a general correlation of these intersections corresponded with domal windows. LARS suggests on the basis of this relationship that the major intersection concentration areas, which in turn might be buried domal areas, could be potential targets for exploration. The LARS study, employing automatic data-processing techniques of LANDSAT imagery, clearly illustrates remote-sensor applications for locating indicators of mineralization. The ultimate test of the value of all such techniques, of course, will be the discovery of previously unknown ores.

Folded and fractured rock offer a number of surficial indicators helpful for locating areas for mineral exploration. The following include many structural expressions that could initially be seen as lineaments: faults, fracture systems, rifts, troughs, fold axes, stratigraphic ridges, and surface expressions of deep-seated major structure. Other definable phenomena include horst and graben systems, changes in structural or stratigraphic orientation, and various circular structures possibly related to mineralization.

An application of LANDSAT imagery to structural geology studies in Iran was reported by Ebtehadj *(5);* an 800-km-long previously unknown fault was identified and another, 1880 km long, was traced with significant differences in the direction and pattern than are shown on structural maps of the area. Other information derived from LANDSAT studies in this oil-producing country include new details of structure in the Zagros area and the location of a large number of apparent salt domes. Figure 4 is a part of a Gemini 5 photograph (S65-45617) of the Zagros Mountains of Iran, taken from an altitude of 150 nautical miles, which defines faults, the plunge of fold axes, and the dip and strike of the strata *(6).* This photograph prompted one of the earlier structural interpretation studies using a satellite product. The analysis was based on visual familiarity of geological spatial relationships, tonal and textural differences, and the interpretation of such topographic features as cuestas.

The role of faulting and folding in the localizing of ore emplacement is derived, not graphic, information. Most ores associated with folding were emplaced simultaneously with the folding or subsequently and may have little relationship to the location of the flexures. The location of postfolding deposits is largely controlled by associated fracturing or by the role of the folds in influencing the movement of the mineralizing fluids. Extensive fracturing in brittle rock produces breccias, which themselves provide avenues for ore fluids along the plunge of the structure. Understanding the origin of a fracture pattern and the establishment of its geometry aids in predicting the location of possible ores. Inasmuch as mineralization is often at the intersection of fractures, tectonic information derived from study of LANDSAT-identified linears can be of particular value. Many mineral deposits, including petroleum, are located along fault and fracture systems that are themselves visible or have observable relationships to topography. Fractures containing water have been clearly defined on LANDSAT near-infrared imagery as have zanjones, enlarged joint-controlled solution trenches. Exploration technique employing LANDSAT imagery or Apollo photography requires relating observable linears and topographic expressions to concepts of classical geology, the same as with conventional exploration methods. It is also to be considered that the various structural indicators may be more pronounced in areas peripheral to the ore than at its location, many surficial clues being obliterated by subsequent erosion in areas of intensive alteration of country rock.

Mineralogical–Lithological Association

Mineralogical–lithological association is an area that has been less valuable than had been hoped insofar as direct visual applications are concerned;

FIG. 4
Part of a Gemini 5 photograph of the Zagros Mountains, Iran *(6)*.

however, it is of unique importance insofar as the developing techniques of computer enhancement and spectral ratioing apply. Because these methods are topics of other chapters, our discussion here will be brief.

Lowe et al. *(7)*, noted that there are chemically receptive host rocks which have characteristics favoring or receptive to mineralization. Such rocks are usually permeable, porous, or highly fractured, such as breccias or limestones; there are also impervious unreceptive rocks such as shales and schists. The classical economic geologists long ago established the association of ores and minerals with various igneous rock types, such as tin, tungsten, and uranium with acid rock; porphyry copper and iron oxides with intermediate rock; cobalt, ilmenite, and titanium with basic rock; chrome, nickel, and diamonds with ultrabasic rock; and gold with greenstone. In areas where there is a minimum of ground truth, the broader divisions of rock types may be identified if there is adequate exposure and knowledge of weathering characteristics. In highly vegetated areas or where structural evidence or grain is absent, any such assessments are chancy. Generally, a number of rocks may be differentiated if not specifically identified.

Direct identification of gossans has been less successful than was expected. Many alteration zones have failed to show up on LANDSAT products even though there were considerable color contrasts. On the other hand, only the larger blooms or gossans are within the resolution limits of the imagery. In some instances spectral ratio imagery has been shown to be capable of providing information pertinent to major mineralization, variations in mass composition of rocks, specific iron oxides, and other compounds. Topographic expressions of mineralization or alteration represent rare geological conditions. Ridges of cap rock may be the surface expressions of weathered sulfide bodies in one region, whereas in a different climate they might be removed by weathering. In regions of tropical weathering, residual ore deposits are often related to topography, as in the case of some manganese, nickel, and lateritic iron ores. These minerals may concentrate on erosion surfaces from which oxidized gangue components have been decomposed and eroded away.

The major mining districts are generally located in mineralized regions which are part of a pattern that has been determined by the flow of ore-bearing fluids. Alteration of the surface in such areas is often widespread, yet indicators based on coloration alone have been of little direct value in defining the known location of the ores. Until geochemical techniques are better perfected, structural and topographic indicators will provide the major clues in the majority of investigations. However, geochemical techniques in the long run should take remote sensing out of the realm of aerial photography and make optimum use of the multispectral capabilities. Future narrow-band imagers should permit a major step toward that end as such instrumentation is put into use from aircraft and space missions.

Stratigraphic Sequence

Stratigraphy is probably the most disappointing area of LANDSAT applications to geology and mineral exploration. Even where the stratigraphy is well exposed, individual formations are seldom thick enough to be identified, let alone by rock type. Often distinctly separate formations will appear as a single unit on LANDSAT imagery, owing to their similar spectral properties. For the present, at least, identification of stratigraphy will depend largely upon the definition of units by their expressions as ridge makers of other known characteristics, or by vegetational guides. Stratigraphy is a case for which an aircraft subsystem is important to provide essential detail. In some instances of adequate exposure, evaporites, limestones, detrital deposits, shales, sandstones, and other types may be inferred from known properties. The primary interest in stratigraphy is in relation to petroleum exploration and geologic mapping. LANDSAT provides abundant data on major structure and to some degree on internal structural detail, but aerial photography is far ahead of the LANDSAT multispectral imager for stratigraphic studies, particularly when the rock is faulted, or where there are repeated sequences, relatively flat lying formations, or a lack of correlative field data. On LANDSAT imagery, vegetation density often is the most obvious indicator of a particular formation. In areas of uniformly heavy vegetation, however, sometimes little else may be seen on color composites.

Airborne radar or aerial photography is far superior to LANDSAT imagery for the definition of rock units in a stratigraphic sequence. Simonet of the Center for Research in Engineering Science *(8)* used the example of a radar strip obtained by the U.S. Army Electronics Command and Westinghouse Aerospace Corporation to emphasize the importance of look direction as a function in detecting lineaments (Fig. 5). The aerial imagery is like-polarized, HH (top), and cross-polarized, HV, radar of the Potato Hills, east of the Tuskahoma syncline, Ouachita Mountains. The structure has been described both as a window through one of the Ouachita thrust sheets and also as an anticlinorium of closely spaced folds and reverse faulting. Several rock units may be discriminated on the outcropping surface of the major feature. Such detail as this is beyond the resolution capabilities of current LANDSAT technology. Radar imagery will often prove to be an essential companion source of data.

CHARACTERISTICS OF OBSCURED DEPOSITS

Just before LANDSAT-1 was launched, NASA geologist Paul Lowman correctly warned that the mission would not directly detect minerals: "What you will find from ERTS imagery are the areas which look promising. Then you can go in there and further investigate these areas from the ground or through aerial photography." Lowman explained, "You'll be able to pick the big structural features and that is the big thing geologically. If you know an area and you know what kind of structure is associated with ore deposits, this will enable you to pick the most prom-

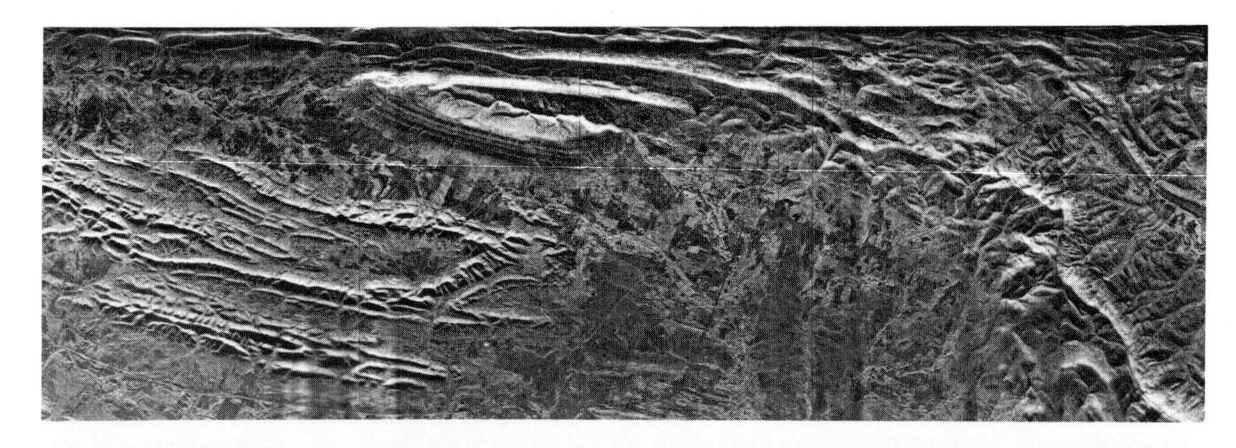

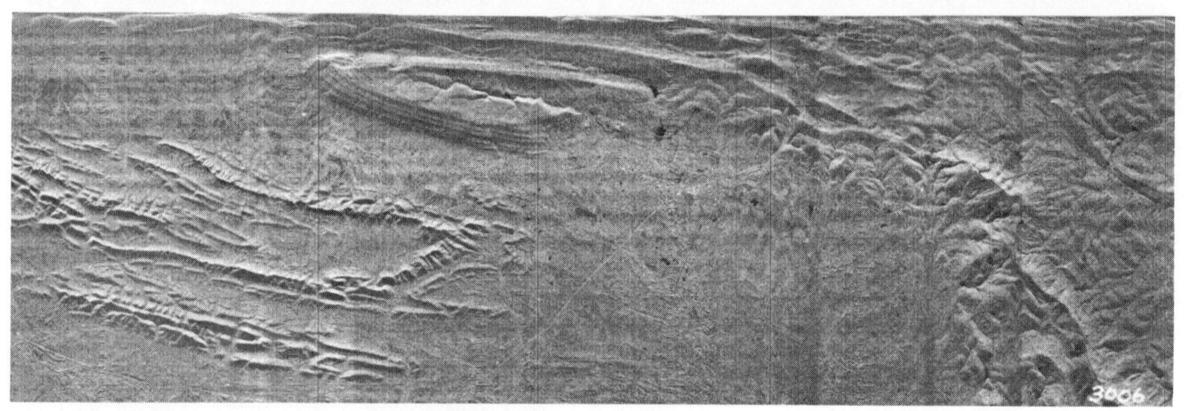

FIG. 5
Radar imagery of Potato Hills, east of Tuskahoma syncline, Ouachita Mountains *(8)*. (Radar images obtained by U.S. Army Electronics Command and Westinghouse Airspace Corporation for NASA.)

ising areas for follow-up aerial photography or for drilling" *(9)*.

Outcropping ores and ores located in obvious near-surface positions have probably already been discovered to a large extent. In recent years the use of conventional airborne instruments has proved to be valuable in regions where a complex geological history has obscured the clues to their location. The role of airborne and satellite-borne multispectral sensors is to identify yet other clues for locating ores in remote areas and in regions where younger formations or structural movements have concealed the deposits or where the orebodies occur at depth.

In a summary study of the economics, geology, and the search for hidden mineral bodies by conventional techniques, Peters *(10)* presented a diagram (Fig. 6) illustrating representative concealed ores. For general agreement with the illustration, Peters's classification is followed below. It may be noted from the diagram that leached and zoned orebodies may have surface expressions of genetically related minerals, which could serve as conventional mineralogical or

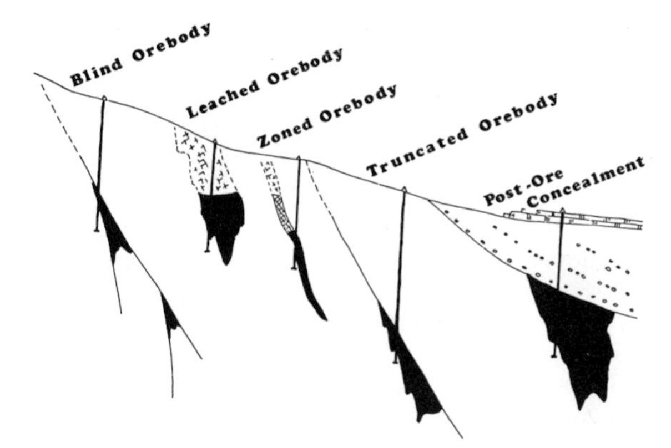

FIG. 6
Representative concealed orebodies *(10)*.

geochemical indicators. Blind and truncated orebodies may in some instances occur along geological structures with surface expressions, although these expressions would not generally serve as indicators of mineralization. In the case of post-ore concealment, younger material completely masks any former surface expression of mineralization. Only where eroded mineralized material has been incorporated into the overburden is there a likelihood of finding indicators by conventional prospecting. Peters noted an observation by McKinstry to the effect that many major mineral deposits may be obscured under lava flows, drift, basalt plateaus, or desert sand, and that probably no ore would have been exposed at Goldfield, Nevada, had the erosion stopped 100 ft higher *(11)*.

Blind Orebodies

Blind orebodies include those whose positions are determined by subsurface structure and lithology, and which may provide little or no surface evidence of their location. Examples include Coeur d'Alene, Idaho, Bisbee, Arizona, and Bernic Lake, Manitoba. Various mineralogical and geochemical indicators are sought in the search for such ores. Both surface and aerial geophysical techniques may be employed. In the type of deposits noted, extensive subsurface exploration and underground geophysical studies may be required. The pegmatite at Bernic Lake is perhaps the world's major occurrence of cesium minerals, yet the deposit has no surface expression and it was discovered by chance when associated dikes were drilled for tin. Such deposits have eluded discovery by surface and aircraft techniques. The football-shaped pegmatite is reported to be oriented with the trend of a major granitic unit, but little is known of its origin. It is hoped that remote-sensing studies of similar mineral problems will turn up broader relationships as to their origins and to possible similar locations for exploration. One anticipated benefit of the new technology is the identification and interpretation of new environments of ore deposition. We can assume that airborne geophysical instruments, when used along the subtle lineaments often discernible on the synoptic images obtained from LANDSAT, will suggest favorable areas for exploring for other types of blind orebodies.

Leached Orebodies

Leached orebodies include those whose surface exposures have been weathered (oxidized) into limonitic gossans, sometimes bearing low-grade surface deposits. The subjacent ore may be unaltered by the leaching, or may exhibit zones of secondary enrichment. Examples include Andean copper bodies and porphyry copper in the southwestern United States.

The surface caps of leached orebodies are often readily identified on aerial color photographs, and some have been discernible on satellite images. The prospects of evaluating the ore beneath leached zones by remote sensors is beyond foreseeable technology and beyond the visualized role of sensing for favorable areas for exploration. There has been much research on techniques for the evaluation of gossans, and the area is still highly experimental and controversial. Induced polarization is a valuable geophysical technique used in delineating ores obscured by oxidized zones *(10)*.

Zoned Orebodies

From the standpoint of sensing for indicators of mineralization, zoned orebodies are similar to leached orebodies inasmuch as the surface mineralogy may be atypical of the economic deposit. Zoned orebodies include such deposits as Walton, Nova Scotia, which has a surface expression of barite, but which grades downward into base metal sulfides, or the deposit at Kutna Hora, Czechoslovakia, which grades laterally from silver to zinc, as determined by hydrothermal conditions at the time of emplacement *(10)*. The characterization of zoned deposits requires extensive mineralogical and geochemical studies. As in the case of leached ores, remote sensing is not foreseen in the role of deciphering problems in complex ore genesis, but of designating environments where mineralization may be likely.

Truncated Orebodies

Truncated ores include those which have been cut off by faulting and part of which may have been moved to another location, often determinable by conventional studies. There is the possibility of considerable geological evidence being exposed at the intersection of the fault with the surface, but in many cases this degree of detail may be below the resolution of satellite-borne sensors. Peters *(10)* gives as an example the Kalamazoo body, 2500 ft deep on the San Manuel fault. It was located by combined geophysical and geochemical techniques, plus structural studies.

Ores Obscured by Post-Ore Concealment

Ores that have been obscured by post-ore sediments or lavas present some of the most difficult problems of exploration geology. Examples include Guanajuato, Mexico; Kirkland Lake, Canada; and Wiluna, Western Australia. Geophysical techniques and geochemical studies on subsurface samples may aid in locating such concealed bodies, but most such orebodies probably lie unsuspected. It is presumed that higher-altitude geophysical instruments may be able to provide deeper structural and anomaly data than are currently possible by eliminating the interference from near-surface anomalies. These are less likely to be located by guides determined from remote-sensor data from space.

What can remote-sensor data provide that relates to mineral exploration? Without recapitulating all the above, we may conclude that the study of linears will identify fracture patterns. Where these are related to known mineralization, their extensions are prime targets for conventional search. Techniques of computer enhancement and spectral ratioing which are being developed indicate that surface alteration and other lithological–geochemical indicators may be identified in some cases. Topographic detail, which shows up like a physiographic diagram on many LANDSAT images, provides most of the information required for the basis of a reconnaissance map. In areas where there has been a minimum of mapping or exploration, this is a lot of geological information; compared to obtaining such information by conventional means, it can be done rapidly and at considerable savings. Remote sensing by no means eliminates the role of conventional exploration, but it provides a large amount of data pertinent to the initial stages of such exploration. To date, the foreseeable major contributions of remote sensing would be defining clues to new areas of interest for exploration and providing a wider format for considering the patterns of mineralization.

In a report to NASA, investigators Lee et al. *(12)* summarized what they felt to be significant findings as to obtaining geological information from satellite imagery. Using both LANDSAT and Skylab/EREP data, they defined three types of regional scale information: rock and soil, geological structure, and landforms. They considered discrimination between soils and rocks and between thick sedimentary sequences to be best because of vegetational differences and topographic expressions. The best-depicted geological features were structural. Differentiating igneous lithology was found to be unsatisfactory. Secondary clues such as drainage patterns and selective vegetation were easily defined. They also concluded that extracting such information is best done by geologists experienced in photointerpretation techniques.

Using remote-sensor data obtained from space in mineral exploration will have several advantages over the use of aircraft-derived data alone in a great many cases. Primarily, there is the opportunity to detect broad patterns, which may not be revealed on aerial photography laydowns. A single image from space may show subtle details that are lost on a mosaic of the equivalent large number of aerial photographs. Whole mineralized districts may be defined on a single image, which makes it possible for the geologist to better consider the bigger tectonic or depositional picture.

In October and November 1973, the various domestic principal investigator users of LANDSAT data were interviewed by NASA to identify significant results that had been achieved *(13).* N. M. Short, chairman of the Geology Working Group, noted,

> The major contribution of ERTS to Geology, first identified within a few months after launch, continues to be its exceptional ability to detect new linear elements (many geological; others of human origin) exposed on the Earth's surface. Those linears which prove to be of structural origin (mainly crustal fractures) will, as they are mapped and verified, provide a new level in understanding the tectonic framework for continental evolution, mineral emplacement, petroleum localization, and potential hazards to man's activities.

In the area of resources exploration, the reports included, in part, the following:

1. Nickel-bearing gabbros in Southwest Africa were better located.
2. Known porphyry copper deposits in Arizona were correlated with structural controls.
3. Metallic deposits in central Colorado were shown associated with intersecting arcuate and linear fractures.
4. Newly mapped linears pass through mercury deposits in California, lead ores in Missouri, and uranium deposits in Colorado.
5. ERTS spectral band ratioing distinguishes iron ore in Wyoming, and hydrated iron oxides (gossans) are identified around the Goldfield, Nevada, mining district.

No new bonanza, but great steps forward in the applications of remote sensing for mineral exploration. Our search is for still-hidden information on subsurface layered rock bearing petroleum or for sulfide ores emplaced along obscured structure, but we have some new ideas of where and how to look. New techniques are developing from LANDSAT technology that will permit the geologist to observe previously undefined phenomena, apply what he learned from Bateman, Behre, or McKinstry, and see a little deeper into the earth.

USES OF REMOTE-SENSOR DATA IN EXPLORATION

Analysis of Surface Characteristics

Exploration for mineral resources, unlike most disciplinary applications of LANDSAT technology, will serve the needs of the investigator and management directly and immediately. Also, whereas most other applications (agriculture, hydrology, oceanography) will require frequent or periodic surveys, remote sensing for geological information may require fewer cloud-free images per representative season per generation of sensor to fulfill the user needs. Repeat coverage is highly desirable, however, as experience shows that the geological content is enhanced differently during repeated imagery of an area, owing to changes in surface conditions and illumination at the time of imaging.

Because mineral exploration depends heavily on the use of maps, it is encouraging to learn that those produced from LANDSAT products can essentially meet the horizontal accuracy standard for a 1:250,000 scale map. The EROS program of the U.S. Geological Survey has undertaken extensive research on preparing photomaps directly from MSS products. Colvocoresses and McEwen *(14, 15)* describe a mapping procedure of making a best fit of a (e.g., UTM) grid to the bulk, system-corrected imagery by means of ground control points. Tests of this procedure on specific frames show positional errors from UTM grid lines in the range of 50 to 100 m (rms*) as compared to a standard accuracy of about 80 rms for a 1:250,000 scale map. At the Third ERTS Symposium, Colvocoresses noted that "applying cartographic requirements is not for the sole benefit of the map maker. All who will use ERTS for quantitative analyses, change detection, or other applications which involve the spatial domain and the figure of the earth,

*Root mean square, error of position.

will find the cartographic requirements to be of value if not essential" *(16)*.

LANDSAT products are particularly suited for mapping topography and structure and for correcting existing geologic maps. Terrain is presented dramatically and lineaments are more clearly defined on MSS spectral images than on any other medium except by computerized reprocessing of LANDSAT data with specialized optical equipment. SLAR can provide as good or better structural information in many cases; however, it provides little information as to lithology and of course lacks the synoptic quality of space imagery. Although simple observation of surface features may often provide the initially-sought-for information, it is probable that more valuable geological content can be found through the more sophisticated analysis of imagery, such as through digital image enhancement, various information extraction techniques, and modeling. However, simple observation of LANDSAT images using light tables, manual plotting methods, and simple diazochrome techniques may permit a preliminary interpretation of structural patterns and allow the observer to visualize the responsible forces or possible structural patterns below the surface.

The correlating of structural linear patterns to tectonic models can produce interesting resource-related information. Saunders et al. *(17)* undertook studies of the commercial utility of using LANDSAT imagery in reconnaissance for mineralization and petroleum, addressing the concept that many mineral deposits and hydrocarbon accumulations are coincident with fracture systems related to major structural events. Two systems of northeast and northwest trending lineaments have been defined in North America and have been interpreted as related to zones of weakness generated early in the Precambrian. Several western U.S. mineral areas are coincident with these belts, and it is postulated that tectonic forces caused movement along these structures, producing shears, which in turn provided avenues for mineral deposition. LANDSAT images of several areas were mosaicked. Identifiable fracture patterns were correlated with tonal linears, indicating lineaments hundreds of miles in length, possibly of continental scale. In the areas studied, many major mining districts were coincident with the LANDSAT-identified lineaments at locations where there is evidence of intersections with transverse linear systems. Structural analysis was undertaken by applying simple strain–shear mechanics concepts to the identified linears, various fracture traces, and tonal alignments.

Rich and Steele *(18)* made an intensive study by plotting the LANDSAT-identified linears of northern California. The linears were compiled on a 1:500,000 scale working map from projections of 70-mm, 1:3,400,000 scale, positive transparencies onto a frosted screen. Eight distinct linear systems were identified, four within the coastal ranges, with only minor overlap between the systems. The authors cited that the repetitive coverage by LANDSAT permitted study of the linear features at the most advantageous times with regard to sun position and ground cover. The linear systems proved to be so well delineated that it has been possible to propose a relative sequence and a tentative structural history of the region.

Abdel-Gawad and Silverstein *(19, 20)* in a study of interpreting tectonic features from LANDSAT imagery suggest that transverse faulting in the central California Coast ranges represents remnants of shears predating the San Andreas system and perhaps formed in Mesozoic time when the Pacific Plate was pushing under the North American Plate. A relationship was identified between transverse faults in the coast ranges and mercury deposits; previously, little had been known as to any regional structural relationship to the mineralization. When the locations of these deposits were plotted against LANDSAT imagery, a clear correlation with the west–northwest shears oblique to the San Andreas system could be defined (Fig. 7). Further interpretation suggests that, although the transverse shear structure appears to be pre-Tertiary, the identification of lineaments across younger formations in the San Joaquin Valley indicates that the shears have persisted through Quaternary time and could still be active.

LANDSAT color composites provide geological information beyond that possible by color photography because of the near-infrared bands (6 and 7). However, false color is sometimes confusing; for situations where "true color" is desirable, a technique has been

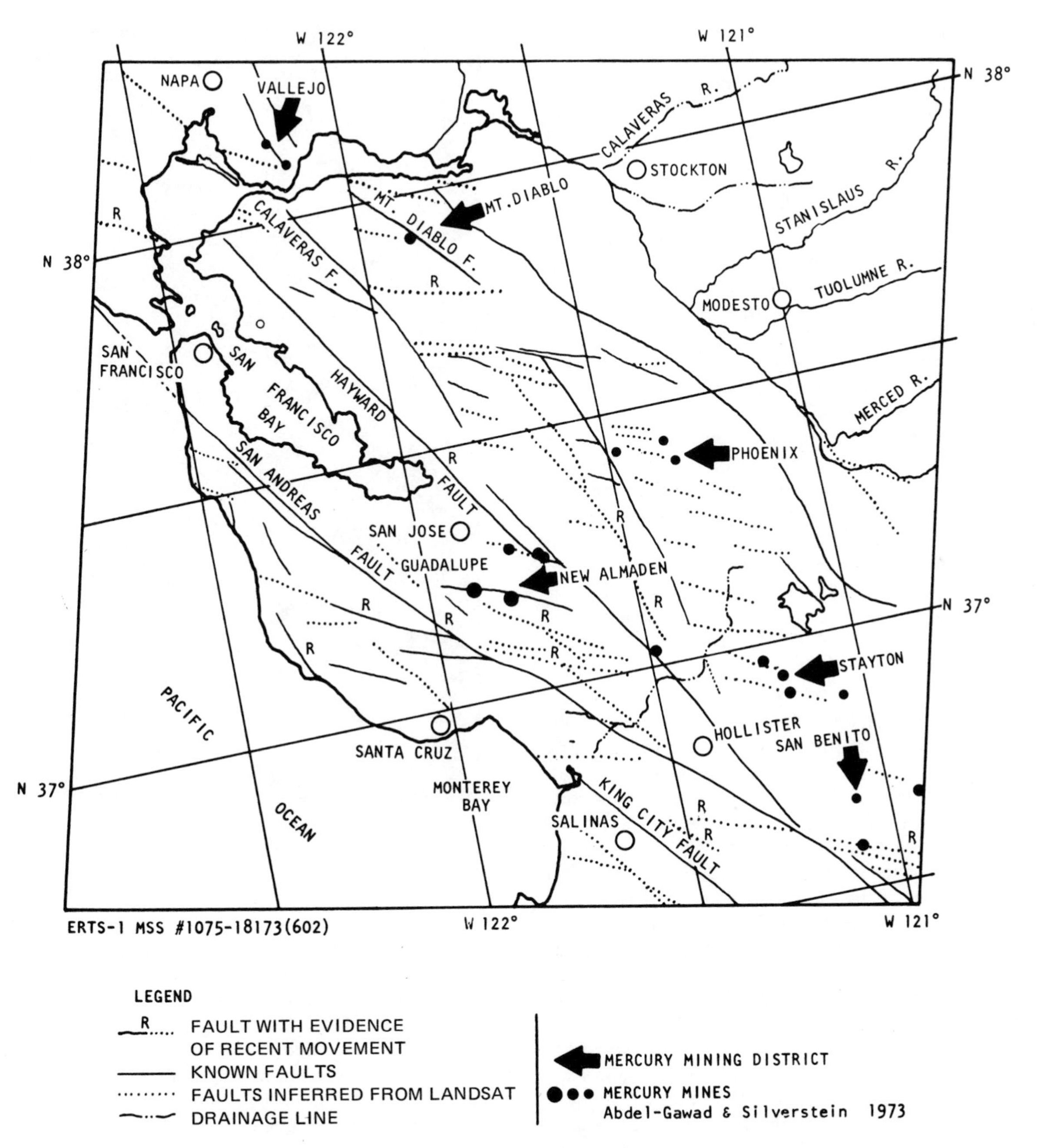

FIG. 7
Relation of mercury deposits to transverse faults, coast ranges of California *(20).*

devised by Eliason et al. *(21)* for producing enhanced "true color" images without black water or red vegetation. The product is more comprehensible visually and will be useful in conjunction with standard color composites for purposes of geologic and topographic mapping.

LANDSAT does not replace the field geologist any more than it takes the place of aerial radar or aerial photography. Estrin et al. *(22)* relate how LANDSAT data, side-looking airborne radar, and color cartographic and infrared aerial photography are used for better directing the efforts of the field geologist in prospecting areas of complex lithology and mineralization patterns. Using the example of fluorite mineralization in the state of San Luis Potosí, Mexico, remote sensing is employed for identifying the surface characteristics of the larger area prior to detailed exploration. Fluorite in the region formed as replacement bodies along fissures, faults, and contacts by hydrothermal solutions replacing limestone country rock and impounding against insoluble rhyolite, as shown in Fig. 8. The sequence of exploration includes the use of LANDSAT data to identify such regional signatures as contacts between rhyolite and limestone, faults, and fissures. Follow-on aerial exploration is undertaken from low-altitude aircraft using color cartographic and infrared aerial photography. Exploration of the narrowed area of interest defines outcrops and soil variations, and provides further structural details. Further studies of the specific exploration zones includes SLAR or photographic methods from aricraft for mapping and precise location of target sites that have been identified. At this point the field geologist undertakes in situ exploration, spared much of the otherwise required search of the entire area. This may be considered a better use of his talents, and his study of the LANDSAT and aerial-sensor products has provided him with valuable insight into the regional patterns of lithology and structure.

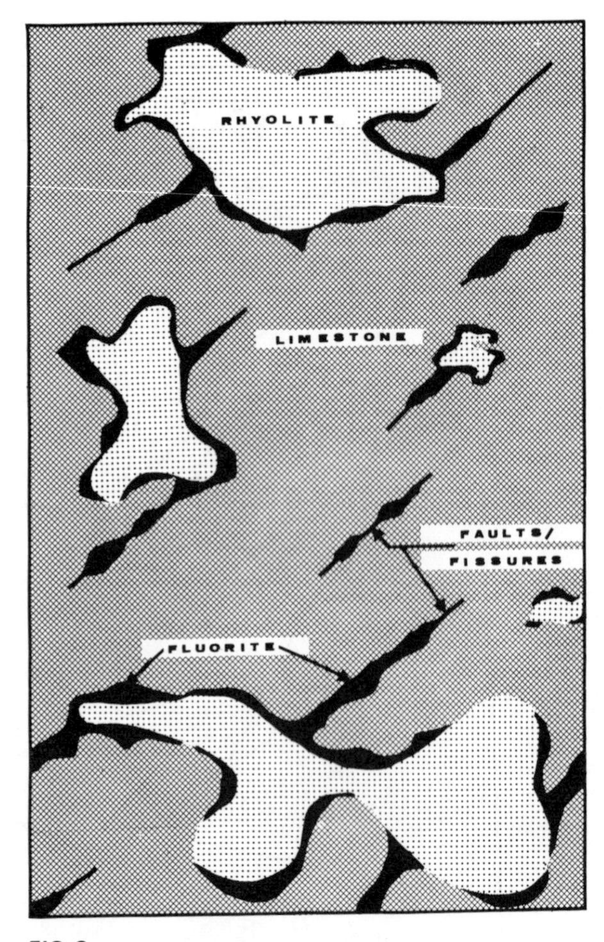

FIG. 8
Fluorite mineralization, state of San Luis Potosi, Mexico *(22)*.

Probably the most significant structural studies to result from the interpretation of space imagery are those undertaken by Lathram et al. *(23, 24)* in investigations of LANDSAT applications to Alaskan geology. Prior to satellite remote-sensor findings, the mineralization pattern in Alaska was believed to be controlled by north-convex arcuate lithologic belts related to regional orogenic features. Nimbus imagery had identified previously unknown northwest and northeast trending lineaments. LANDSAT imagery corroborated these finds, and the lineaments were interpreted as expressions of deep-seated crustal structures. It was hypothesized that favorable areas for exploration would be along belts paralel to these lineaments and at their intersections in particular (Fig. 9). This concept was substantiated by the results of an independent exploration in 1970–1971 based

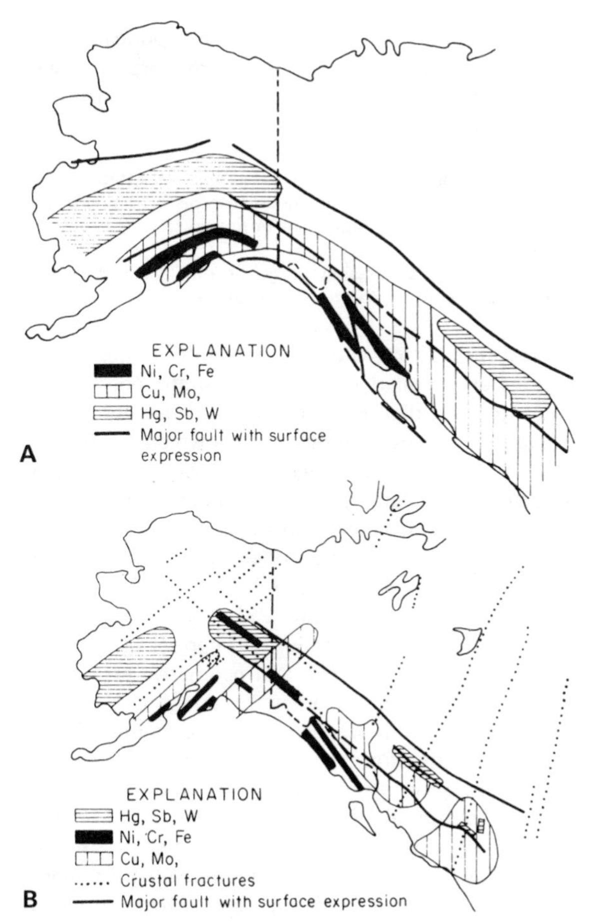

FIG. 9
Areas of Alaska and western Canada considered favorable for location of deposits of selected metals based on extrapolation of geologic conditions at known occurrences: (A) conventional concept guided by north-convex arcuate distribution of lithologic belts; (B) postulated alternative assuming that linears seen on Nimbus IV and LANDSAT images are crustal fractures and have influenced mineralization *(24)*.

on using northeast and northwest fracture systems as a guide, which led to the discovery of several copper–molybdenum deposits. These were not found by the use of LANDSAT imagery; however, they support the hypothesis that was derived from the study of space imagery. The new hypothesis should prove to be a valuable rationale for further exploration in Alaska. Baker *(25)* notes that, although the new copper deposits were not directly discovered by LANDSAT, it is the consensus that such a discovery is soon forthcoming, and that only confirmation in the field is required before a discovery may be specificaly attributed to LANDSAT.

Analysis of Data Products

In general, the use of light tables, projections, and diazochrome sandwiches, and the comparison of standard LANDSAT products to aerial photographs, maps, and radar imagery should be adequate for initial mineral exploration efforts. In many parts of the world the best available maps are the U.S. Defense Mapping Agency modified polyconic *Maps of the World* series, and the U.S. Department of Commerce World Aeronautical and Operational Navigation Charts, Lambert conformal conic projections. These are at 1:1,000,000 scale, the same as the standard 9 X 9 in. format LANDSAT images. In much of the world it is advantageous to undertake initial geological studies on these 1:1,000,000 scale products. Enlargements may be made or purchased for other scales, within practical limits. The aeronautical charts are also available at 1:500,000, 1:2,000,000, and other scales.

Computer enhancement of tapes and ratio image techniques hold more promise for the now-developing applications of remote sensing to mineral exploration from aircraft and satellites. It has been suggested that color-enhanced band ratio techniques may in time prove to be the most valuable computer process for the study of surface phenomena *(13)*. There are examples of the use of stereoscopic techniques, optical Fourier transform analysis for study of fault patterns and regional topographic alignment employing laser illumination of imagery, and mechanical or digital processes of "edge enhancement" by intentional misregistriation of negative transparencies which reduces tonal differences but enhances boundaries of tonal units. Additive color viewing is theoretically very promising, although there have been relatively few geological studies reported so far involving its use. A commercially available four-channel multispectral viewer, for example, permits compositing MSS imagery in a wide range of color balance, using a variety of filter–band combinations with standard NASA 70-mm film chips. Aside from being able to select colors for optimum viewing of surface characteristics, the investigator may select the best color balance for producing higher-resolution color composite photographs for more detailed analysis *(26)*. A variety of optical, computer, and photographic techniques with special applications are now under development, the details of which are beyond the intent of this chapter.

MSS imagery is taken in four separate bands:

1. LANDSAT band 4, green, 0.5 to 0.6 μm (MSS-1).
2. LANDSAT band 5, red, 0.6 to 0.7 μ (MSS-2).
3. LANDSAT band 6, near-infrared, 0.7 to 0.8 μm (MSS-3).
4. LANDSAT band 7, near-infrared, 0.8 to 1.1 (MSS-4).

LANDSAT-2 (1975) is a duplicate of LANDSAT-1. A fifth MSS channel in the thermal infrared is scheduled for LANDSAT-3 (1978). Generally speaking, bands 4 and 6 have demonstrated less value in geological studies; however, they should not be discounted. Band 4, owing to the transparency of water in this spectral region, provides information on sedimentation, shallows, and coastal bottom features, and has applications to some coastal and offshore operations. Band 5 is best for geological structure and for identifying linears, best presents cultural and extractive industry patterns against vegetation and soils, is particularly useful in arid regions, and has applications for monitoring plant effluents and pollution from smelters. Band 6 emphasizes vegetation and has use in locating obscured structure through changes in vegetation type or density. On band 7, as in 6, water is opaque and shows black. Band 7 has applications related to moisture retention and is valuable for locating structure that contains water. Bands 4 and 5 have been ratioed for discriminating iron compounds. Bands 6 and 7 have applications for defining glacial features. Bands 4 and 7 may be ratioed to differentiate sands and to identify surface waters. The experimental studies of the various spectral areas are still largely Edisonian, and new, unexpected applications will continue to be found.

Sabins *(27)*, in a discussion of the applications of image processing for petroleum exploration, suggested that LANDSAT may increase such applications inasmuch as

1. Multispectral imagery is available in digital format.
2. The small scale, minimum geometric distortion, and uniform oblique illumination are optimum for regional interpretation.
3. Digital image processing programs now exist that may have petroleum exploration applications.

He also noted that, although some photometric corrections are applied to airborne remote-sensor imagery, automatic digital image processing is seldom used because of the large volume of data which is required, and also because advantages of image processing of airborne-sensor data have not been convincingly demonstrated for petroleum exploration.

A use of digital-image-processing techniques for scene correction of LANDSAT sensor data has been reported by Bernstein *(28, 29)*. It was stated that precision processing methods can correct MSS imagery to remove radiometric and geometric errors and provide a mapping accuracy of about one picture element (rms) and two picture elements (maximum) through the use of nine ground control points. Photographic imagery produced by this method has an excellent potential for extracting positional information and providing clearer pictures; however, this refinement is beyond the usual requirement for mineral exploration. The processing introduces no radiometric degradation, and there is no loss of resolution in the generation of tapes. The digital method eliminates the manual operations necessary in multiple film-generation stages. There are attendant costs for such a product, but where it is required, the digitally corrected images may have rms figures of less than the standard for line maps.

The use of pseudocolor transformations of LANDSAT images for identifying geological structure has been reported by Lamar *(30)* of RAND Corporation. Based on the idea that the human eye distinguishes in three visual coordinates (hue, saturation, and luminance), whereas only luminance is distinguished on a black and white image, pseudocolor enhancements were produced that have a particular chromaticity–luminance combination for each gray shade of the original image. This permits the use of all three visual coordinates in the enhancement technique and presents each gray scale as a different color. Proposed computerization will permit more sophisticated transformations, using the hue and saturation dimensions.

The Multispectral Photography System, which has been developed by the Spectral Data Corporation, has been used to define soil and vegetation anomalies related to mineralization and is also applicable to geomorphological studies. The aerial technique employs a four-lens multispectral camera and viewer. The viewer permits the investigator to alter the hue, saturation, and luminance, and to produce false color renditions for specific enhancement of surface features. The enhancement method has been used to differentiate among some sedimentary formations, alluvial deposits, and volcanic ejecta, and often clearly defines fault contacts. The method has also been used in in situ studies to detect mineralized trees, where significant spectral reflectance differences were found to exist between conifers that grew on or near a copper–molybdenum anomaly. A correlation was established between soil geochemistry and percent directional reflectance. Yost *(31)* undertook a study of correlating in situ spectral reflectance measurements with LANDSAT-1 MSS imagery. Additive color analysis of LANDSAT spectral band imagery obtained over successive passes showed that the bulk products have the necessary spectral and spatial fidelity to indicate through color differences the presence of soil moisture and temporal changes in dry lakes.

Raines and Lee *(32)* undertook a detailed evaluation of the use of multiband photography for rock discrimination. The study considered the use of multiband photography to differentiate sedimentary rocks; their conclusion was that it is "not statistically possible to select a set of best bands in a practical manner from in situ rock reflectance measurements." Over 8600 in situ measurements of band reflectance of Front Range, Colorado, sedimentary rocks were obtained. Statistical analysis showed that measurements from one site could be used at another 100 miles away, that there is basically only one spectral reflectance curve for the rocks with constant amplitude differences between curves, and that natural variation is so large that some 150 measurements per formation may be required to select "best" filters. Subjective tests supported these conclusions, seriously questioning the value of the aerial multiband photography approach for discriminating sedimentary rock by tonal differences between formations, at least for that area. The investigators determined that equally good tonal discrimination of sedimentary rocks could be obtained from any band, and that the significant difference in reflectances is a relative difference that is essentially uniform throughout the photographic spectrum. It was suggested that crystalline lithologies might be better subjects, that narrower bandwidths might provide better results, or that topography might best delineate sedimentary rocks. Whatever the solution, aerial multiband photography is still in its early stages of development.

A LOOK AHEAD

Various information-extraction techniques and automatic data-processing methods fail as yet to meet information requirements. Malila and Nalepka *(33, 34)* suggest still more advanced techniques are required to better address problems of location, mensuration, and classification accuracies due to geometric distortions of data. ERTS resolution is noted as relatively coarse, and there are a number of problems related to the condition of the atmosphere. A basic problem is the resolution element, which may contain a mixture of materials. Cousin et al. *(35)* describe investigations at Johnson Space Center that include preprocessing, classification, and correlation of data with ground features and analysis of spectral clustered results, including corrections of MSS data for atmospheric scattering and absorption. A two-channel linear discriminant function has been developed for detecting and identifying surface water using LANDSAT-1 data through ratioing bands 4 and 7 *(36)*. Grouping various water classes and edge pixels with a high water content show impressive detection rates. The applicability to geology is the capability of producing the information on a computer map product that is transferable to standard 1:24,000 scale $7\frac{1}{2}$-min maps. The potential applications of this technique are obvious for other ratio data. The computer-generated information may be as current as LANDSAT coverage and may be quickly available for use.

Brewer et al. *(37)* of the Earth Satellite Corporation interpreted the regional structure of a 15,000-square-mile area of Arizona. Several fault systems were identified by trend, including two that are believed to be related to porphyry copper mineralization. The boundaries of several tectonic units that were identified were found to correlate with theoretical shear directions related to the San Andreas stress system. The investigators, using Fourier analysis, identified a fundamental spacing between energy maxima that appears to be related to the distances between copper deposits *(37)*. Overby et al. *(38)*, in studies related to siting and planning an underground coal gasification project, wished to establish the directional characteristics of the rock strata in order to determine the direction of long horizontal injection and production wells. To select a proper site, LANDSAT data and other imagery of deep-coal regions of West Virginia were studied. The locations of gas venting zones were estimated from linears and fractures, and the lineaments seen on remote-sensor products were verified in the field. Such applications are pertinent to studies required for controlling the advance of combustion zones and the movement of gas.

Vincent and Pillars *(39)* note that for Skylab there are 66 spectral ratios that may be constructed from the twelve visible-reflective infrared channels of the S-192 multispectral scanner. Because LANDSAT- and aircraft-originated spectral ratio images have proved to be of use for enhancing various geological phenomena, all 66 of the S-192 spectral ratios were calculated for 235 laboratory spectra of soils, rocks, and minerals, and were converted into ratio codes. This provides a capability to select the optimum ratio images for enhancement of specific targets. In time, such ratio codes may be used in logic design and memory storage in automatic data processing on board future aircraft missions or on the Space Shuttle. Any technological forecast based on future multispectral scanner capabilities should surely indicate opportunities for better geochemical prospecting and mapping employing such methods.

Various studies by industry, NASA, and academic institutions have concluded in advance that the cost of a remote-sensing program to search for minerals would be outweighed by the economic benefits and savings in exploration costs which would accrue. In 1968, Waltz *(40)* correlated the various existing estimates of benefits in terms of annual savings for the 1975 time period. Estimates for geology and mineral resources showed a range of potential annual savings of $15 million to $150 million for the United States alone, and $100 million to $600 million for the world. There has been some slippage in the program, and it will be a few years before it may be determined if the estimates were modest. In many foreign countries a significant gap exists between remote-sensing technology and knowledge of its applications by potential users. In many developing areas much inertia must be overcome before there will be decisions to fund natural resource information systems. Many of the larger or more progressive mining and petroleum companies have realized the potential value of LANDSAT products, and some are taking full advantage. However, many are waiting for more dramatic results before they get involved. South Africa, Japan, Australia, Iran, Canada, Mexico, Brazil, Bolivia, and the major industrial countries of Europe are quite significantly involved in the use of LANDSAT products. It will not be too far in the future when the value of remote-sensor data impacts mineral interests worldwide.

The LANDSAT-1, 2, C program will provide a continuing source of reconnaissance level data, although the addition of the thermal infrared channel on LANDSAT-C is not anticipated to make any major contribution to mineral exploration. The topographic and structural data already obtained will provide for years of study and follow-on airborne and surface exploration. Future programs employing higher-resolution sensors with narrower spectral bands and broader spectral coverage will provide further data to correlate with known geology and to interpret in terms of indicators of mineralization. Many of the sensors required for the next generation of remote sensing have been proved feasible on aircraft flight tests. The problem by the end of the decade will no longer be how to sense for geological data, but how to model it for optimum use by resource management.

A practical use of LANDSAT data is proposed by Sabels et al. *(41)*, who suggest that services similar to those processing geophysical data could be available for earth survey data analysis. The requirements for LANDSAT data analysis are quite similar to those used in geophysics. Both are digitally recorded in similar products. An array processor is proposed as a route for using the mathematical steps most often employed in data correlation. These could include densitometry for color determinations of rocks and soils or contouring, and display of algorithmic relationships in different spectral areas. Such a procedure, patterned after current seismic data processing routes, could provide geologists and management with a service that might prove to be of real value to exploration planning and other applications requiring timely use of a range of earth-observation data.

Computerized multispectral processing and display system services (e.g., General Electric Image-100 and Bendix Multispectral-Data Analysis System) for the present represent the ultimate in commercially available processing of LANDSAT computer-compatible tapes. These are highly sophisticated computer systems; however, manned operation and decision making are kept in the loop. The operator does not become lost in the process of analysis, but controls the analytical functions through a joystick, control panel, and display system. Operational functions variously include the assignment of display colors, color slicing, data normalization, ratioing, cluster analysis, and controlled movement of the displayed image through the field of data. Geological applications include thematic displays, theme area measurements, various enhancement techniques, and numerical and graphic presentations of histograms. Such systems have been used in lithologic and structural surveys, exploration studies for minerals and petroleum, soil typing, well and mine location, and construction siting. The use of the GE Image-100, for example, employing band ratioing techniques, correlated the "hazy" anomalies in the Anandarko Basin with geochemically altered soils and vegetation thought to be caused by hydrocarbon seepage from petroleum traps. The ultimate proof of such studies will be the successful extrapolation of local data to other areas and the location of mineral deposits or petroleum.

THE SHUTTLE ERA

The Space Shuttle Program of the 1980s will carry crew, cargo, and passengers to earth orbit and back. In addition to logistics missions, such as bringing supplies to orbiting vehicles, replacing space station crews, and launching satellites, it will also carry personnel on short-duration sortie missions. Earth applications studies are well suited to 7-day sortie missions for purposes of instrument development and for sensing selected targets. Payloads on such missions will be both developmental and operational. The former will include those of an engineering research nature. The shuttle is ideal for the initial space flight of instrumentation, where it would be accessible to technicians and scientists for calibration and adjustment. These missions will assist in bringing complex multispectral instrumentation into optimum performance. The operational sorties would permit geologists, for example, to participate in target selection and sensor selection to assure reliability of instrument performance. The capacity for a large sensor payload will make it possible to cover the electromagnetic spectrum and permit the intercomparison of several sensors. A geologist could participate directly in the determination of which geologic phenomena may be the better sensed by which techniques or systems. The primary purpose of the geologist on board would not be switching operations, but the development of optimum mixes of sensors and the matching of proper instrumentation with the targets to be studied *(42)*.

It has been shown by the Skylab experience that a trained observer may detect phenomena which might otherwise escape the untrained eye. Of particular relevance to terrain observation are the programmable viewing angles and sun-angle variations. Orbital parameters may be varied for inclination or altitude. The manned capability will permit acquisition and tracking of targets, editing of data, and rapid modification of viewing programs. Payloads are yet to be selected, but it is obvious that a large variety of instruments could be included. Although the large space telescope with 10-ft diameter optics is intended for astronomical use, it is altogether possible that it could be adapted for earth-observation use, particularly from geosynchronous orbit *(43)*. Sortie missions should permit better studies of many surface properties that might serve as mineral indicators, which so far are not practically feasible from orbital altitudes, such as most lithologic varieties, specific major mineralization, subordinate elements, jointing, chemical anomalies, moisture variations, sulfide oxidation, minor color variations, metamorphic features, typical alteration zones, geothermal sources, and various textural, fabric, and aggregate properties.

Only time will ascertain the eventual technical advantages of geologists on shuttle sortie missions. Geology does not require real-time observations; however, the potentialities of real-time experimentation with prototype instruments makes the manned mode attractive. The primary job in geological remote sensing is obtaining photomaps and identifying structure. This can be done by automated satellites; but mapping is not the only requirement for sensing for mineral resource guides. Because man may perform unlimited observational and experimental functions beyond the capacity of automation, the role of a geologist on sortie missions would appear to be desirable. He may prove of limited value for the routine aspects of geological remote sensing, but he might provide a major service in the development of optimum sensing techniques. Howard and Orrok *(44)* cite the use of man in space as an optimization of the man–machine system. They also note Harry Hess's comment that the right man was the critical factor, some being a million times more useful than others.

REFERENCES

1. The Space Applications Program 1974, NASA Office of Applications, Washington, D.C., May 1974.
2. ORL Experiment Program, Vol. B, Pt. 2, Geology/Hydrology, International Business Machines Corp., Rockville, Md., for NASA, 1966.
3. Ridge, J. D., *Annotated Bibliographies of Mineral Deposits in the Western Hemisphere,* Geological Society of America Memoir 131, Boulder, Colo., 1972.
4. Levandowski, D. W., T. V. Jennings, and W. T. Lehman, Applications of ERTS-1 Imagery to Mapping of Lineaments Favorable to the Location of Ore Deposits in North Central Nevada, Laboratory for Applications of Remote Sensing, Information Note 101073, Purdue University, W. Lafayette, Ind., 1973.
5. Ebtehadj, K., et al., Tectonic Analysis of East and Southeast Iran Using ERTS-1 Imagery, Paper G13, Third ERTS Symposium, NASA/GSFC, Washington, D.C., 1973.
6. Ecological Surveys from Space, Office of Space Science and Applications, SP-230, NASA, Washington, D.C., 1970.
7. Lowe, D. S., et al., *Peaceful Uses of Earth-Observation Spacecraft,* v. II, *Survey of Applications and Benefits,* for NASA, University of Michigan, Ann Arbor, Mich., 1966.
8. Simonet, D. S., Center for Research in Engineering Science, University of Kansas, Lecture on Radar as a Remote Sensor for Geoscientists, Stanford University Short Course in Geologic Remote Sensing, Stanford, Calif., Dec. 1967.
9. Lowman, P. D., quoted in NASA Press Release 72-137, Washington, D.C., July 20, 1972.
10. Peters, W. C., Economics, Geology, and the Search for Concealed Ore Deposits, Preprint 69-1-8, Society of Mining Engineers of AIME, New York, 1969.
11. McKinstry, H. E., *Mining Geology,* Prentice-Hall, Englewood Cliffs, N.J., 1948.
12. Lee, K., D. H. Knepper, and D. L. Sawatzky, Geologic Information from Satellite Images, NASA-CR-138230, Colorado School of Mines, Golden, Colo., NTIS-E74-10507, Mar. 1974.
13. Smith, O. G., and H. Granger, eds., *Earth Resources Program Results and Projected Applications,* ERTS-1 Applications Program, v. I, Results and Applications, Johnson Space Center, Houston, Tex., May 1974.
14. Colvocoresses, A. P., and R. B. McEwen, Progress in Cartography, EROS Program, NASA Symposium on Significant Results Obtained from ERTS-1, NASA/GSFC, New Carrollton, Md., Mar. 1973.
15. Colvorcoresses, A. P., and R. B. McEwen, ERTS

Cartographic Progress, *Photogrammetric Eng..,* v. 34, no. 12, Dec. 1973.

16. Colvocoresses, A. P., Toward an Operational ERTS, Paper L15, Third ERTS Symposium, NASA/GSFC, Washington, D.C., Dec. 1973.
17. Saunders, D. F., and G. E. Thomas, Evaluation of Commercial Utility of ERTS-1 Imagery in Structural Reconnaissance for Minerals and Petroleum, Paper G30, Second ERTS Symposium, New Carrollton, Md., 1973.
18. Rich, E. I., and W. C. Steele, Speculations on Geological Structure in Northern California as Detected from ERTS-1 Satellite Imagery, *Geology,* v. 2, no. 4, Apr. 1974.
19. Abdel-Gawad, M., Identification and Interpretation of Tectonic Features from ERTS-1 Imagery, NASA CR-138269, NTIS E-74-10520, June 1974.
20. Abdel-Gawad, M., and J. Silverstein, ERTS Applications in Earthquake Research and Mineral Exploration in California, Paper G22, Second ERTS Symposium, New Carrollton, Md., Mar. 1973.
21. Eliason, E. M., P. S. Chavez, and L. A. Soderblom, Simulated "True Color" Images from ERTS Data, *Geology,* v. 2, no. 5, May 1974.
22. Estrin, S. A., G. V. Wolstenholme, and J. M. Pilner, Remote Sensing and Photogeology Equals Mineral Exploration, Paper 74-130, 40th Annual Meeting American Society of Photogrammetry, St. Louis, Mo., Mar. 1974.
23. Lathram, E. H., I. L. Tailleur, and W. W. Patton, Jr., Preliminary Geologic Application of ERTS-1 Imagery in Alaska, Paper G4, Second ERTS Symposium, New Carrollton, Md., 1973.
24. Lathram, E. H., Geologic Application of ERTS Imagery in Alaska, Paper G3.1, Third ERTS Symposium, Washington, D.C., 1973.
25. Baker, R. N., ERTS Updates Geology, *Geotimes,* Aug. 1974.
26. Leggett, M. A., and J. F. Childs, A Reconnaissance Space Sensing Investigation of Crustal Structure for a Strip from the Eastern Sierra Nevada to the Colorado Plateau, Argus Exploration Company, for NASA/GSFC, CR-133141, NTIS-E73 10774, 1973.
27. Sabins, F. F., Potential Applications of Image Processing for Petroleum Exploration, Paper 74-144, 40th Annual Meeting American Society of Photogrammetry, St. Louis, Mo., Mar. 1974.
28. Bernstein, R., Scene Correction (Precision Processing) of ERTS Sensor Data Using Digital Image Processing Techniques, Third ERTS Symposium, Washington, D.C., Dec. 1973.
29. Bernstein, R., Results of Precision Processing (Scene Correction) of ERTS-1 Images Using Digital Image Processing Techniques, Paper I7, Second ERTS Symposium, NASA/GSFC, New Carrollton, Md., Mar. 1973.
30. Lamar, J., Pseudocolor Transformations of ERTS Imagery, Paper I12, Second ERTS Symposium, NASA/GSFC, New Carrollton, Md., Mar. 1973.
31. Yost, E. F., In Situ Spectroradiometric Quantification of ERTS Data, Paper I19, Second ERTS Symposium, NASA/GSFC, New Carrollton, Md., 1973.
32. Raines, G. L., and K. Lee, An Evaluation of Multiband Photography for Rock Discrimination, Colorado School of Mines, Golden, Colo., for NASA and U.S. Army Research Office—Durham, Remote Sensing Report 74-4, Third Annual Remote Sensing of Earth Resources Conference, University of Tennessee Space Institute, Mar. 1974 (NTIS E74-138233).
33. Malila, W., and R. Nalepka, Atmospheric Effects in ERTS-1 Data and Advanced Information Extraction Techniques, Paper I1, Second ERTS Symposium, NASA/GSFC, New Carrollton, Md., Mar. 1973.
34. Malila, W., and R. Nalepka, Advanced Processing and Information Extraction Techniques Applied to ERTS-1 Data, Paper I5, Third ERTS Symposium, Washington, D.C., Dec. 1973.
35. Cousin, S. B., et al., Significant Techniques for Processing and Interpretation of ERTS-1 Data, Paper I8, Second ERTS Symposium, NASA/GSFC, New Carrollton, Md., Mar. 1973.
36. Procedures Manual for Detection and Location of Surface Water Using ERTS-1 Multispectral

Scanner Data, Science and Applications Directorate, NASA/JSC, Houston, Tex., Dec. 1973.

37. Brewer, W. A., et al., Earth Satellite Corporation, Mineral Exploration Potential of ERTS-1 Data, NASA Cr-138717, NTIS E74-10608, July 1974.
38. Overby, W. K., Jr., C. A. Komar, and J. Pasini, Geologic Investigations for Siting and Planning an Underground Coal Gasification Project, Eastern Section, American Association of Petroleum Geologists, Third Annual Meeting, Pittsburgh, Apr. 1974.
39. Vincent, R. K., and W. Pillars, Skylab S-192 Ratio Codes of Soil, Mineral and Rock Spectra for Ratio Image Selection and Interpretation, Ninth International Symposium on Remote Sensing of Environment, Environmental Research Institute of Michigan, Ann Arbor, Mich., Apr. 1974 (NAS9-13317).
40. Waltz, D. M., Technological Base for Planning of Spaceflight Missions to Obtain Data on the Earth's Resources, Paper No. 68-1074, Fifth Annual Meeting, American Institute of Aeronautics and Astronautics, Philadelphia, Pa., Oct. 1968.
41. Sabels, B. E., et al., Digital Interactive Image Analysis by Array Processing, Paper I13, Second ERTS Symposium, NASA/GSFC, New Carrollton, Md., Mar. 1973.
42. Sharma, R. D., W. L. Smith, and F. J. Thomson, Earth Resources Survey Applications of the Space Shuttle Sortie Mode, System Planning Corporation for NASA Office of Applications, May 1973.
43. Colvocoresses, A. P., Remote Sensing Platforms, *U.S. Geol. Surv. Circ.* 693, 1974.
44. Howard, B., and G. T. Orrok, Use of Man In Space, Bellcomm, Inc., for NASA/OMSF, NASW-417, Sixth Annual Meeting American Institute of Aeronautics and Astronautics, Anaheim, Calif., Oct. 1969.

6

EARTH OBSERVATIONS FROM REMOTE-SENSING PLATFORMS: OUTLOOK

Robert S. Houston
Ronald W. Marrs
Nicholas M. Short
Paul D. Lowman, Jr.

Robert S. Houston and Ronald W. Marrs are in the Department of Geology, University of Wyoming, Laramie, Wyoming. Nicholas M. Short and Paul D. Lowman, Jr., are with the Earth Resources Branch, and Geophysics Branch NASA, Goddard Space Flight Center, Greenbelt, Maryland.

Remote sensing has developed rapidly in recent years in response to its recognition as a useful tool for earth resources studies. But geologists who wish to utilize data generated by remote-sensing platforms must keep in mind that remote sensing is not a panacea. Rather, it is an additional tool that supplements present methodology and can often reduce costs in specific applications. Geologists who maintain that remote sensing has been oversold may not have been realistic in their expectations; a careful review of the applications suggested by the late William Pecora (1967) for remote sensing from space platforms will show that most of his suggested applications have indeed been realized.

Today's sensor platforms include the Earth Resources Technology Satellite (LANDSAT), Skylab, and aircraft, which generate imagery of the earth from distances of 900 km, 440 km, and ~ 700 to 25,000 m, respectively. In addition, photographs taken on Gemini and Apollo missions and images generated by spacecraft such as Nimbus offer views from other altitudes and perspectives. Aircraft images can be obtained from any altitude less than 25,000 m, depending on cost, the intended application, and political factors. The multilevel viewing capacity offered by the various systems and the general availability of images from the National Aeronautics and Space Administration (NASA) earth resources programs give the geologist an opportunity to view the earth at different scales and in different perspectives. An apt analogy for geologists is one made by Short and Lowman (1973, p. 37) between remote-sensing images and the petrographic microscope. The space images give the geologist a synoptic view or regional perspective much like a low-power objective on the petrographic microscope. Using a space image, the geologist may study regional relationships among geologic features in much the same way as the mineralogist studies the relationships among minerals. Aircraft imagery allows more detailed study of individual features, just as the high-power objective is used by the mineralogist to study individual minerals. In many applications, the synoptic view is, perhaps, the greatest single advantage of the space photograph or image. Using satellite imagery, geologists may recognize structures or other geologic features of regional extent usually overlooked in field study or in interpretations of aircraft images. Regional features can be isolated by use of the satellite imagery, and, if necessary, selected areas may be studied in more detail by use of aerial photography and in field studies.

SENSOR PLATFORMS

Much of the remote sensor data currently available to users in the United States have been obtained through three NASA programs: the Earth Resources Aircraft Program (ERAP), the Earth Resources Technology Satellite (ERTS) program, and the Earth Resources Experiments Package (EREP) used during the three Skylab missions. The aircraft program has been actively gathering photographs and other forms of remotely sensed data since 1964. NASA has used both high- and intermediate-altitude aircraft (U-2, RB-57, C-130, P3-A, and Convair 990) in this program, and has employed a variety of sensors. Standard 10-in. aerial cameras have been the mainstay of the aircraft program, serving both as a tool and as a means of gathering reference data for correlation with other sensor data. Other sensors employed in the ERAP program include multiband camera systems, multispectral scanners, thermal infrared scanners, radar, microwave and thermal radiometers, spectrometers, and scatterometers. Data from many of these instruments are available for only a few test areas where they were flown in support of ERAP investigations. Consequently, the general utility of the data from the more exotic sensors is limited, but the growing library of aerial photography provided by the NASA aircraft covers a considerable portion of the United States and represents a significant contribution to the photographic data bank (Fig. 1).

About the same time that the NASA aircraft sensing program was initiated, geologists began to realize the potential of orbital photography for geologic studies. Merifield (1964) was among the first to recognize the value of the synoptic view provided by photography from orbital altitudes. Merifield's work, done chiefly with Viking and Aerobee rocket photography, stimulated interest in orbital photography, and a terrain photography experiment was quickly planned for the last two manned Mercury flights in 1962 and 1963. A number of excellent pictures of remote areas such as Tibet were returned. These demonstrated the potential of orbital photography in geologic mapping and other fields, such as forestry, hydrology, and oceanography. The most important result of this early terrain photography was the interest it generated in imagery of the earth's surface as opposed to meteorological orbital photography. A follow-on experiment was carried out on several Gemini flights. Gemini 4, flown by J. A. McDivitt and E. H. White, returned about a hundred pictures of geologically useful quality. Many of these pictures showed unmapped structures, dune fields, and even a volcanic field in northern Mexico (Fig. 2) that had never been mapped (Lowman and Tiedemann, 1971). The scope of terrain photography experiments was rapidly expanded to cover areas such as hydrology and oceanography, and by the end of the Gemini Program, some 1300 useful color photographs of the earth's surface had been obtained. Hundreds more were taken during the early Apollo missions.

While the Gemini Program was still in progress, planning began for systematic earth resources photography from Apollo missions. The need for multispectral coverage was recognized, and a four-camera experiment was planned for the Apollo Applications Program missions involving use of spent rocket stages as space stations. This concept eventually led to the development of the now successful Skylab program. Color and multispectral photographs were obtained from the SO65 experiment on Apollo 9. This photography was successfully applied in geologic work and other experiments.

Nimbus 1, launched in 1964, provided the first automated satellite system that yielded visual imagery useful to geologists. Because of a more elliptical orbit (lower perigee) than anticipated, many of the TV pictures from Nimbus 1 show more detail than expected (Fig. 3). Subsequent Nimbus satellites, each with improved and more diverse sensor packages, did not produce images of this quality, although many proved suitable for interpretation of major structural features present in the large area (hundreds of miles on a side) covered in the images.

One experiment on Nimbus 5 has direct application to geology. The surface composition mapping radiometer (SCMR) measures emitted radiation in the 8.3- to 9.3-μm range and the 10.2- to 11.2-μm range;

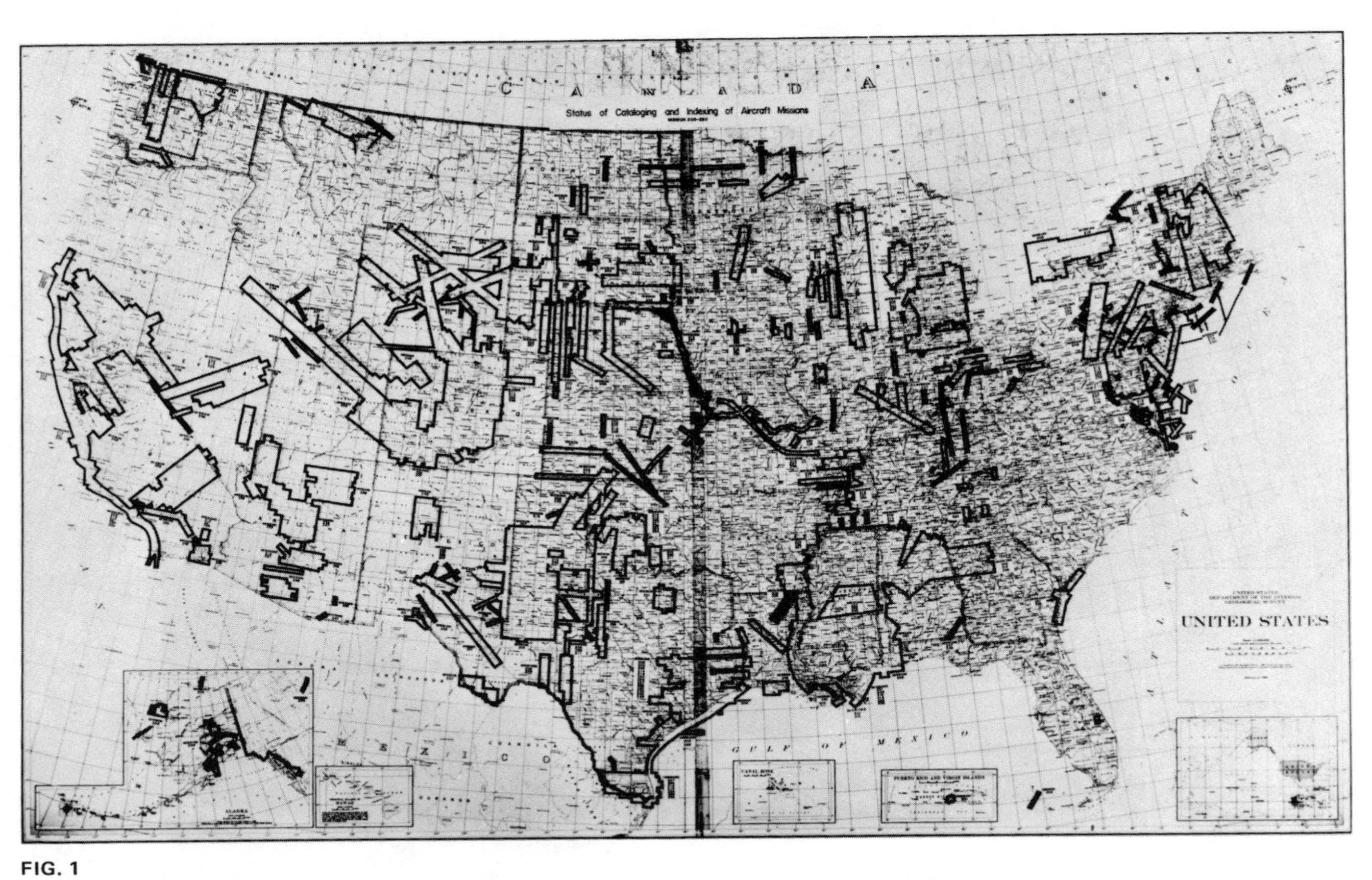

FIG. 1
Index map showing NASA aircraft coverage of the continental United States from missions 200–250.

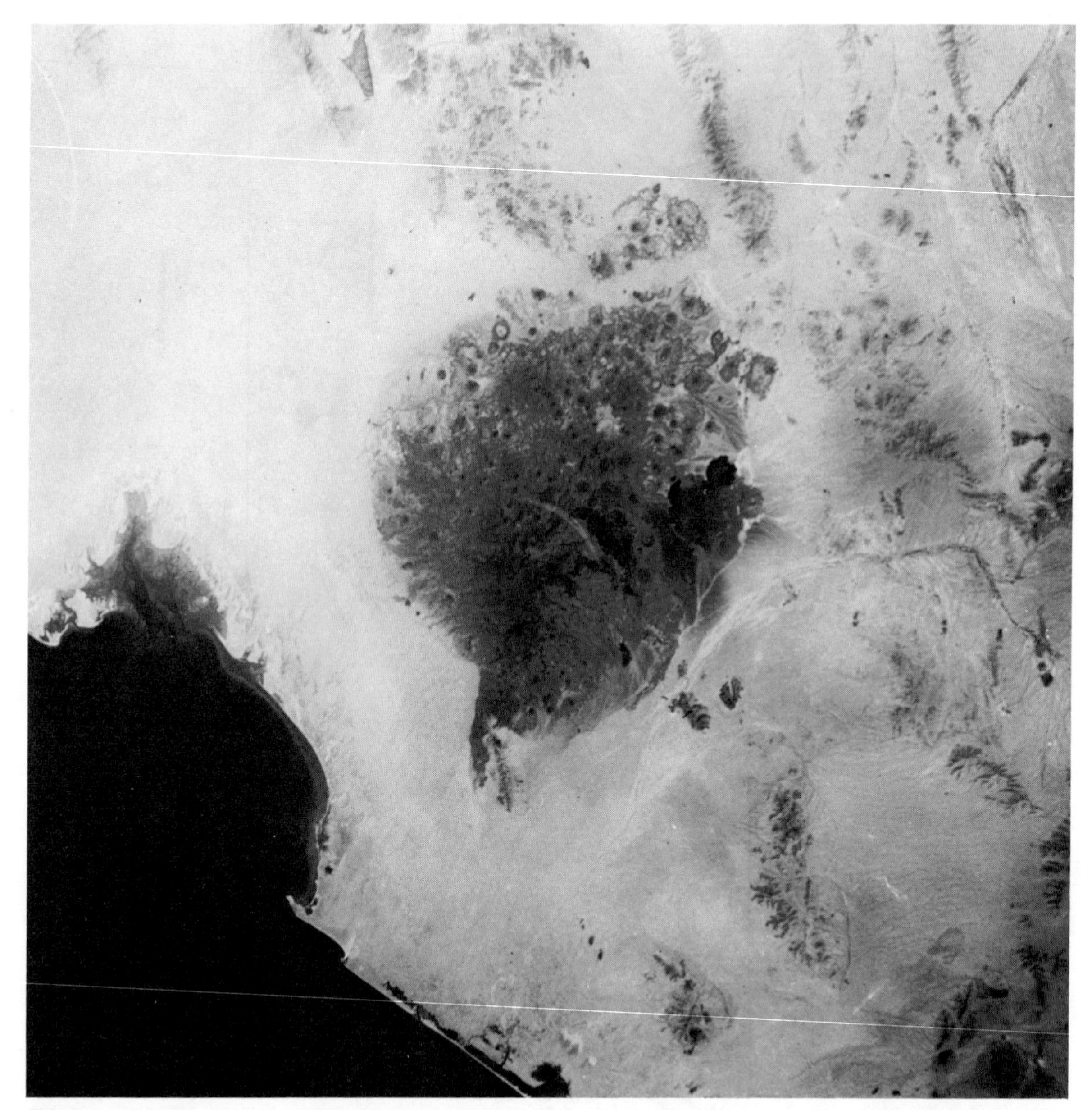

FIG. 2
View showing the entire Pinacate volcanic field. The volcanoes in the field are currently inactive but have erupted in the past few thousand years. Lava flows show up as dark areas near the right edge of the field. This area is part of the Yuman Desert.

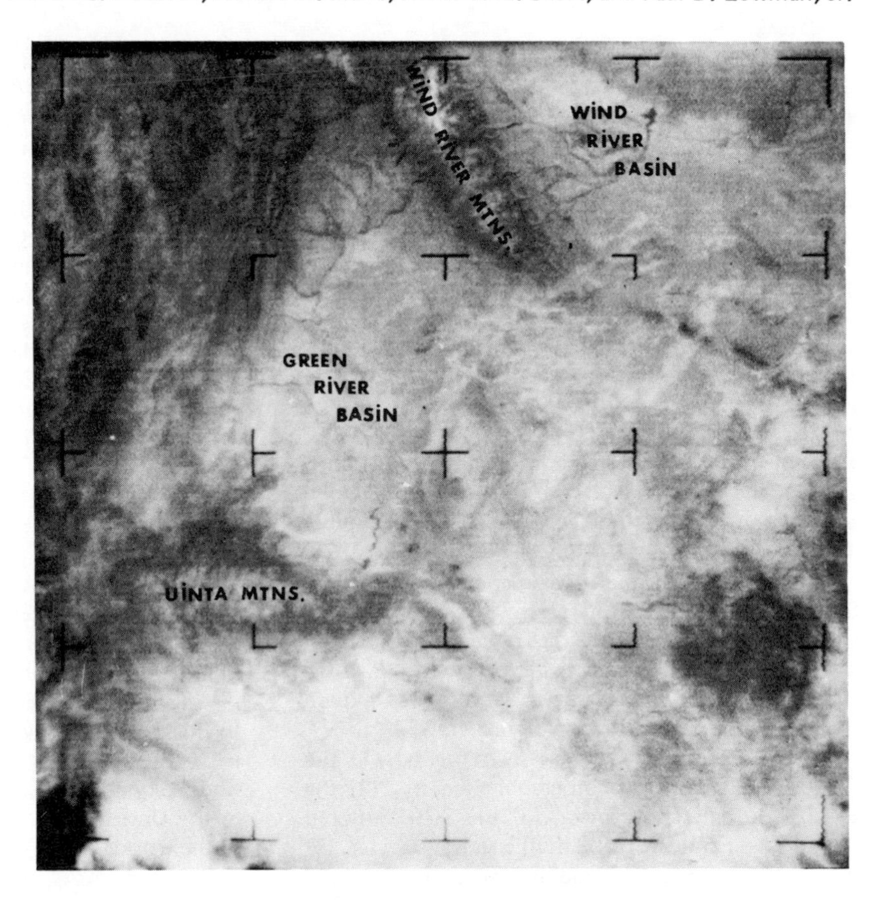

FIG. 3
Nimbus 1 AVCS photograph of southwestern Wyoming, northwestern Colorado, and northeast Utah, taken September 4, 1964.

a third channel operates in the 0.8- to 1.1-μm interval (equivalent to the LANDSAT-1 MSS-7 band). Data from any of these channels can be used to produce photo-images. When used alone, either thermal channel will give information on equivalent blackbody temperatures.

The prime purpose of the SCMR was to determine the silica content of the uppermost layers of terrestrial surfaces. Emissivities of silicates vary with wavelengths as a function of silica content. Temperature decreases are most pronounced near 8.8 μm for silica-rich granites and near 10.7 μm for silica-low dunites (Short and Lowman, 1973, p. 3). Temperature differences calculated from the radiances measured by the two thermal channels are converted to "estimates" of silica content using calibration curves derived from a variety of rock types.

The NASA earth resources program began in 1963 in close cooperation with the U.S. Geological Survey, the Department of Agriculture, and the Navy Oceanographic Office. Under this program, an extensive series of airborne remote-sensing experiments was carried out using sensors that might have eventual space application. This work, performed by NASA and other government agencies, universities, and private industry, greatly stimulated the development of remote sensing. Further stimulus to remote sensing was provided by growing public awareness of environmental problems. These developments culminated in the decision by NASA to develop a dedicated earth resources satellite, originally named the Earth Resources Technology Satellite (ERTS), and a set of earth resources experiments for the first U.S. space station (Skylab).

LANDSAT-1 was the first space platform designed specifically to observe the earth repetitively with sensors that produce high-resolution, multiband imagery and digital data. It was launched on July 23, 1972, from NASA's Western Test Facility in California. LANDSAT-1 carried two earth-looking sensors, a three-band, return-beam vidicon (RBV), and a four-channel, multispectral scanner (MSS). A switching circuit failure caused the RBV to be shut down in August 1972, but it can be reactivated when needed to back up or take over the image acquisition functions of the MSS. The MSS has operated well beyond expectations and is still transmitting data (July 1976). LANDSAT-2 was launched January 22, 1975.

The objectives of the first American space station, Skylab, were to study the earth, the sun, man's response to prolonged space flight, and space technology. The Earth Resources Experiments Package (EREP) was only one of several groups of experiments to be performed by the Skylab crew.

EREP comprised six main instruments: (1) the S-190A multispectral photographic facility, (2) the S-190B earth terrain camera, (3) the S-191 infrared spectrometer, (4) the S-192 multispectral scanner, (5) the S-193 microwave radiometer–scatterometer–altimeter, and (6) the S-194 L-band radiometer. The S-190A multispectral facility obtained 1:2,800,000 scale photography of a 100-mile-wide swath on each EREP data pass. The six S-190A cameras obtained color, color infrared, and four bands of black and white multispectral photography (500 to 600 nm, 600 to 700 nm, 700 to 800 nm, and 800 to 900 nm). The S-190B camera provided high-resolution photography (color, color infrared, or black and white panchromatic) of a 67-mile-wide ground track. On many EREP passes, both the S-190A and S-190B cameras were scheduled to give 60 percent overlap for stereoscopic viewing. The S-191 spectrometer was a pointable instrument (sighted by the crew) that measured infrared spectral radiance values for selected targets (3900 to 15990 nm). The spectrometer has a 1400-ft field of view (1 milliradian, mrad), a radiance accuracy of 6 to 8 percent, and a spectral resolution of 2 or 3 percent (NASA, 1973, pp. 5–12). The S-192 multispectral scanner provides imagery for 13 bands in the 410- to 12,500-nm range (from visible into thermal infrared). The scanner ground track is 42 miles wide and its instantaneous field of view is 260 ft^2 (0.182 mrad). The noise equivalent temperature difference (or noise equivalent reflectivity) ranges from 0.8 to 2.0 percent. The S-193 microwave system (13.9 GHz) serves as a microwave radiometer, a scatterometer, or a radar altimeter by operating simultaneously in several modes. The radiometer responds to a temperature range of 50 to 330°K. The S-194 L-band radiometer measures brightness temperatures at 1.43 GHz with a resolution of ± 1.0°K. Spatial resolution of the S-193 data is such that they are of limited value in most geologic applications.

This payload is, by far, the most complex earth-oriented remote-sensing package ever flown on a civilian space mission. The spectral coverage provided by this array of sensors affords utility in a great number of different applications. The Skylab orbit does not permit global coverage, and EREP data acquisition is further restricted by limits of crew time and recording materials. Even with these rather severe restrictions, Skylab successfully obtained coverage of most of the United States (Fig. 4) and of many other parts of the world.

TECHNIQUES FOR GEOLOGIC ANALYSIS

Automatic Mapping

Images and photographs can be interpreted directly (visual examination of photographs) or after processing (magnetic tapes and enhanced imagery). Reduction of scanner data to a useful image or map format usually requires elaborate redisplay systems or computer processing, but the scanners provide information in both the photographic and nonphotographic portions of the electromagnetic spectrum. Multispectral imagery produced by NASA from the data generated by aircraft, the LANDSAT MSS scanner, and the Skylab S-192 scanner can be interpreted or processed in the same manner as multiband photographs. Few centers are equipped to generate images

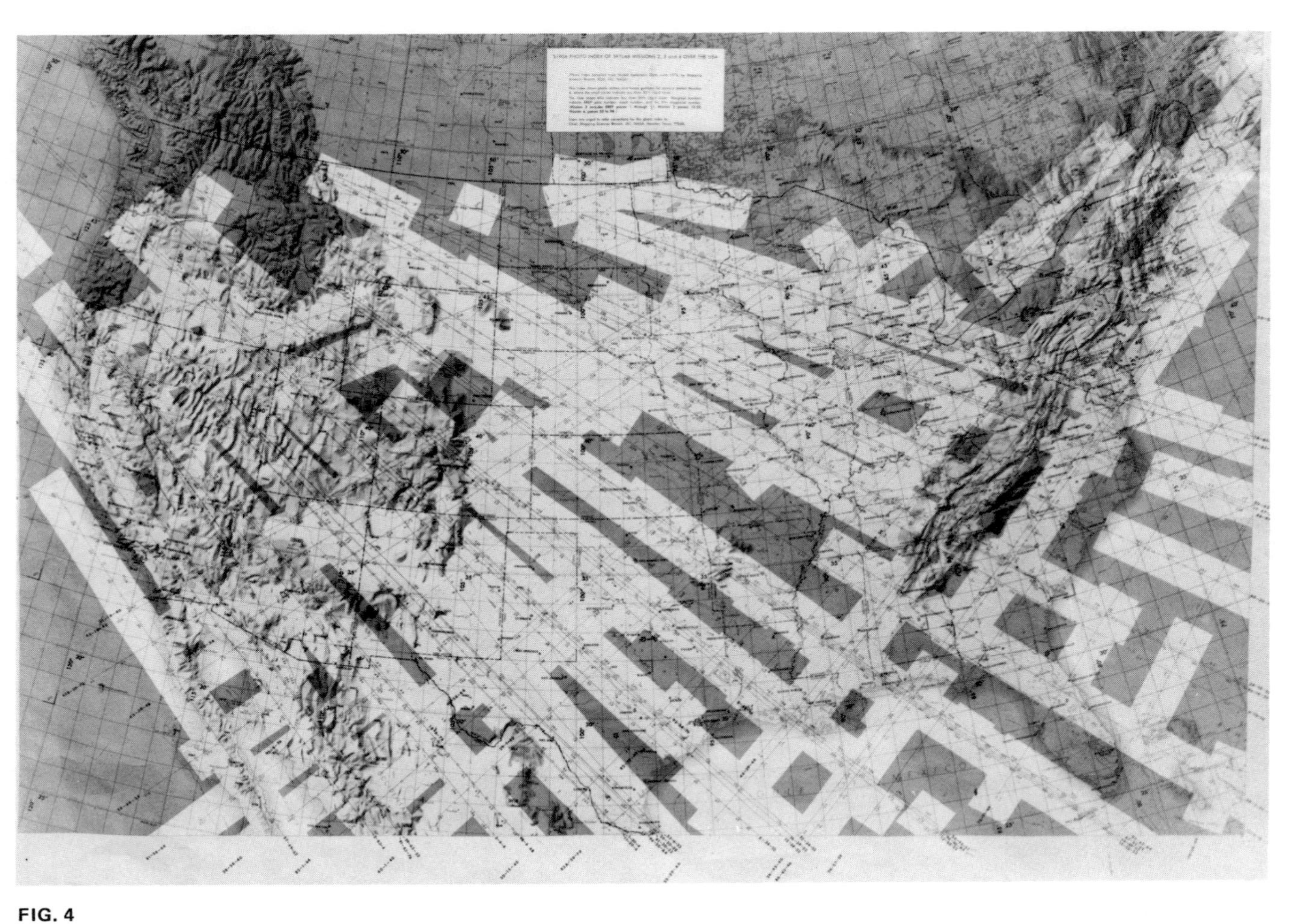

FIG. 4
S-190A photo index of coverage over the United States obtained on Skylab Missions 2, 3, and 4. (Prepared for NASA by the Defense Mapping Agency, Aerospace Center, St. Louis, Mo.)

of this type from magnetic tapes so these scanner images will probably continue to be supplied to users as processed by NASA or regional data centers.

Various techniques have been developed that produce printouts in map form directly from tapes. For example, multispectral tape output may be processed to produce a printout in map form of the scene brightness values for a particular band of the spectrum. In other applications, spectral bands may be combined by direct-additive techniques or statistical methods. The resulting computer-generated maps are often large and difficult to handle, but they have several major advantages over images and photographs. The resolution of the original tape is retained in the computer map, whereas as much as 40 percent may be lost in the reproduction of a standard 9 X 9 in. image. Computer techniques allow band ratioing and multidimensional band combinations that are difficult to make with optical equipment. Computers can be programmed to selectively map densities and band-to-band density combinations that may represent a specific class of objects. In addition, the computer can readily calculate the areal distribution of each class. In summary, the computer offers rapid analysis and much greater flexibility in data processing than visual analyses, but the costs of the analyses are correspondingly higher. The critical question is whether or not the computer can be used advantageously in geologic applications.

Where geologic information is present in the nonphotographic spectrum, computer processing is essential in reducing scanner data to a useful map or image format. In most applications, the geologist will find that the photographic film positive or positive image is not only the most suitable product to work with, but also the only product that he can use effectively. The reason is that the basic product of most geologic studies is the geologic and/or tectonic map. Preparation of such maps requires some method of rock identification and some method of determining the three-dimensional distribution of rocks cropping out at the surface. Rock identification is ordinarily determined by field studies or by using field studies in combination with stereoscopic image interpretation. No computer program has yet been devised to calculate the three-dimensional distribution of rocks exposed at the surface.

Unlike spectra for some surface resources (such as vegetation), rock spectra do not exhibit readily identifiable characteristics. Moreover, the rocks are often covered by vegetation or soils, and their reflectances are influenced by weathering and moisture. Consequently, it is much more difficult to identify them by computer techniques. Iron-rich minerals show strong absorption bands in the near infrared (Rowan, 1972) and a shift in wavelength for peak emissivities in the thermal infrared with changing silica content (Lyon, 1969). These emissivity variations have been used with limited success in identifying rock types under ideal conditions (Lyon, 1970, p. 539), but it is extremely difficult to apply such techniques on a general basis because rock emissivity is affected by a variety of factors (grain size, grain orientation, degree of weathering, vegetation cover, sun angle, sun azimuth, viewing geometry, slope angle, temperature, water saturation, atmospheric conditions, depth of the atmospheric column, and shadowing). Some of these factors can be, and have been, taken into account in computer programs intended to map rocks (Watson and Rowan, 1971; Anuta et al., 1971); but some factors, such as water saturation, may change from day to day and are not easily estimated. Rock, vegetation, and soil mix will change from one area to another, and factors such as grain size and grain orientation may be impossible to measure without petrologic examination. The limitations of present techniques leave us far short of a general capability for automated geologic mapping. Furthermore, the philosophical question arises as to whether automatic procedures are desirable for producing any product as complex and subjective as a geologic map. Perhaps it is more efficient and more desirable to preserve the role of the interpreter and to make geologic maps by a combination of photogeologic and field studies. But research must continue in this area if we are to make geologic maps of inaccessible areas or to map for applications that require the utmost objectivity.

The implication of this review is that automatic mapping does not have the immediate potential in geologic studies that it has in other earth resource studies. This is not entirely true, because, in many areas, geology and geology-related features can be mapped through automated techniques (e.g., mapping of exposed rocks or surface indicator plants). Certainly, computer processing offers potential as a tool for helping the geologist make decisions or distinctions that could not otherwise be made. For example, computer ratioing techniques may prove applicable to mineral exploration when used under controlled conditions with procedures tailored to specific areas and tasks (Vincent, 1973a, 1973b; Goetz and Billingsley, 1973; Rowan et al., 1973).

Visual Interpretation Techniques

Visual interpretation of photographs and images can be done by standard photointerpretive methods (Avery, 1968), which have been in use for over four decades. But new techniques must be applied to take full advantage of the information available from multiband photography and multispectral imagery. Images and photographs that record radiation in spectral bands beyond the visible provide a wealth of new information, and the multiband approach lends additional flexibility to procedures for analysis.

Band Selection

Multiband photography records reflected radiation representing several discrete spectral intervals rather than the broad-band (400 to 700 nm) radiation recorded by panchromatic film. Multiband photographs have three major advantages over the single broad-band photograph: (1) the capacity to present information in wavelength intervals that are particularly suitable for observing given surface phenomena, (2) the capacity for combining bands to enhance given surface resources, and (3) an increase in the dynamic range of the film system. Band selection, to enhance objects with either minimum or maximum reflectance as compared with other objects, has been successful for some geologic applications. For example, near-infrared images or photographs show bodies of water and water-saturated soils distinctly (dark-colored to black) because of the high attenuation of the near-infrared energy by water (Specht et al., 1973, p. 360). Conversely, blue- and green-band images are of great value to sedimentologists who are interested in mapping suspended sediment in water, because water does not attenuate these wavelengths; the sedimentologist can "see through" the water to discern patterns in the suspended particles (Specht et al., 1973, p. 359) or see shallow-water deposits.

Glaciologists and geomorphologists studying active glaciers find that ablating glaciers can be distinguished best in the near-infrared range because of a film of water on the surface of the melting ice. The physiography and structure of glaciers, even with light snow cover, is better defined in the near infrared than in other photographic bands (Benson, 1973, p. 5194). Hydrologists have determined that snow lines are best defined by green-band images because of the lower attenuation of the green light in the meltwater (Barnes and Bowley, 1973, p. 857).

The iron absorption band in the near-infrared spectrum results in dark tones on infrared images of iron-rich rocks. The effectiveness of the imaging system for recording this distinction depends upon limiting the recorded radiation to a narrow wavelength range that includes the actual absorption wavelength (Rowan, 1972). If this is done successfully, distinctions can be made between iron-rich and iron-poor rocks, because the iron-poor rocks retain a relatively high reflectance in the infrared region (Rowan, 1972, pp. 60–69). Therefore, layered mafic complexes and features such as granite–basalt contacts may be mapped using near-infrared bands (Fig. 5).

Red-band images show heavily vegetated areas in dark tones because of absorption by chlorophyll in this wavelength region. The amount of red light reflected or absorbed depends upon the type, quantity, and vigor of the vegetation. Thus, rocks marked by characteristic vegetation may be mapped by use of the red-band image.

In the nonphotographic portion of the spectrum, the information from some aircraft sensors, LANDSAT MSS-7, and the Skylab S-191 and S-192 sensors may be useful in geologic studies. The most obvious applications are in the thermal infrared, where determinations of near-surface temperature gradients and delineation of thermal anomalies can be useful to geologists. Flight timing is usually more critical in these studies, owing to rapid changes in thermal contrast with diurnal variations and weather conditions. Some examples of the use of the thermal bands include studies of thermal pollution in bodies of water (Ruggles, 1969), spring location along shore-

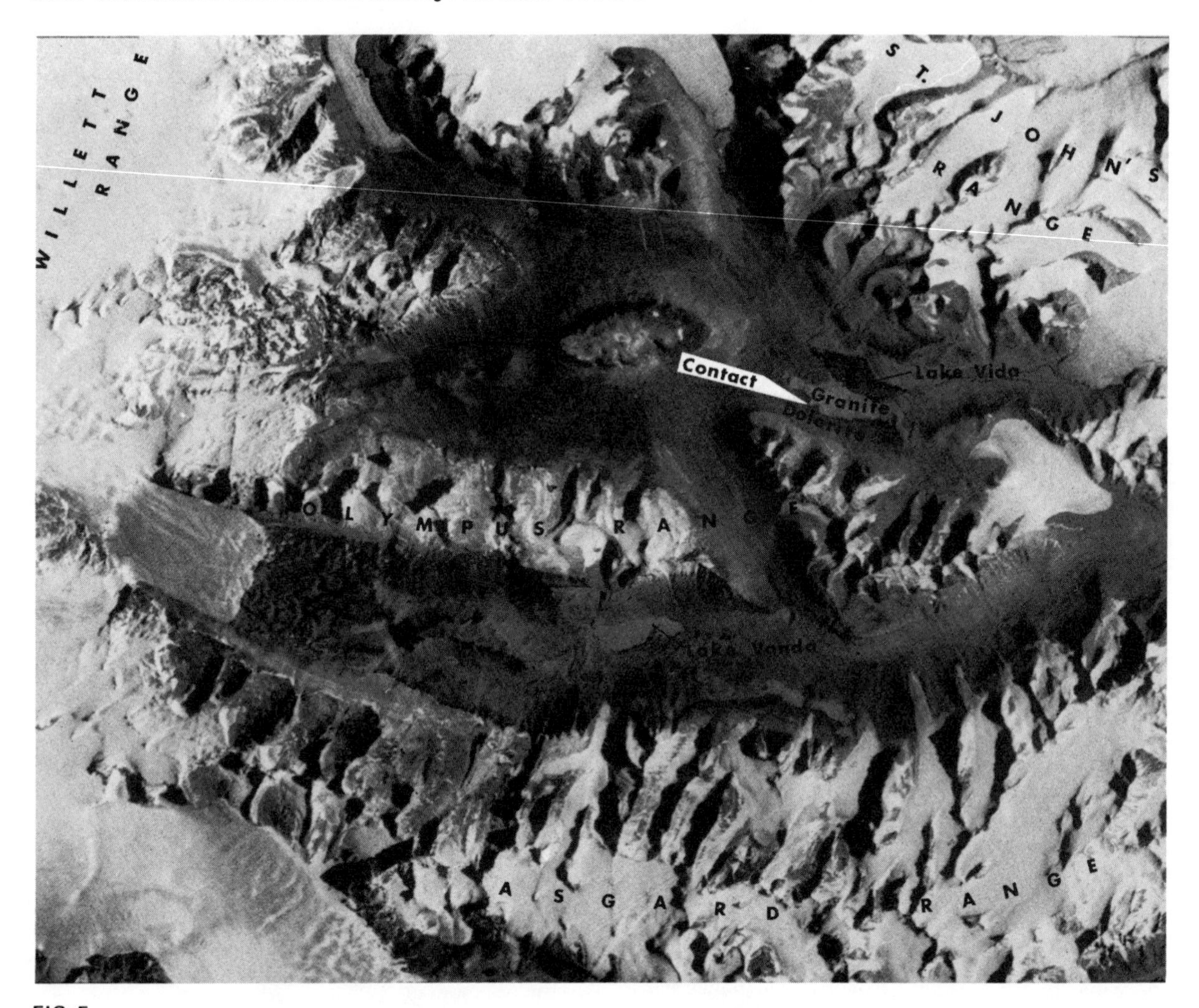

FIG. 5

Enlarged portion of LANDSAT-1 image 1174-19433-7 of the Dry Valleys, Antarctica, showing distinct contact between the Irizar Granite (Late Cambrian and Ordovician) and a mafic sill, the Ferrar Dolerite (Late Triassic or Early Jurassic). Compare with Figure 9B.

lines (Lee, 1969), fire detection (Bjornsen, 1968; Greene et al., 1969), detection of fault-controlled near-surface groundwater movement (Walker, 1972), and mapping of geothermal source areas (Moxham and Alcaraz, 1966).

It is premature to decide whether geothermal areas can be satisfactorily delineated by use of aircraft-mounted thermal sensors or those on Skylab, but studies of Yellowstone Park (McLerran and Morgan, 1964; Miller, 1966) and other known geothermal areas indicate a definite potential for geothermal mapping. The chief problem appears to be one of distinguishing anomalous geothermal areas from other anomalies that are related to differential solar heating or cultural activity. Repetitive coverage with thermal sensors offers some hope for accomplishing the necessary discrimination, but such coverage has, so far, proved impractical.

Radar

The geologic value of side-looking airborne radar (SLAR) is well proved by the fact that contract companies now carry out SLAR surveys on a commercial basis. An example of the high-quality products, containing exceptional amounts of geologic information, appears in Fig. 6. Such surveys can be carried out rapidly, because of the all-weather capability of microwave frequencies and the wide swath of SLAR. Radar is a valuable tool, but power requirements, lack of application in some disciplines, and economic considerations have limited the demand for orbital imaging radar systems.

Under the best viewing conditions (low sun angles) LANDSAT and EREP can produce image mosaics comparable in information content to mosaics of radar images. However, one advantage of radar over photography is its all-weather, day–night capability. Geological features do not change rapidly; so, for geological coverage, one could simply wait for favorable conditions. However, there are some parts of the world, such as the Isthmus of Panama, that are essentially cloud covered all the time.

Radar is a valuable tool for mapping in such areas. A second advantage of radar is control of illumination azimuth. Unlike systems dependent on sunlight, one can obtain almost any azimuth with a self-illuminating system. Also, lower frequencies, on the order of 400 to 1600 MHz (L-band), can provide some penetration of dry alluvium and light vegetation (Dellwig, 1969). This is not possible in the visual and infrared wavelengths. In general, radar offers the user complete control of the incident illumination, whereas with photography he must take what the sun provides.

Because of the obvious advantages of repetitive coverage and platform stability, orbital radar systems may prove useful for certain applications in geology and other fields. Development of such systems is therefore now being studied by NASA for use on future satellite systems. It is impossible to specify, in detail, just what sort of instruments are needed. Present thinking favors X-band (5200 to 10,900 NHz), cross-polarized radar. The synthetic aperture method is a necessity for orbital systems if high resolution is desired, because the antenna size would be prohibitive for a real-aperture space system.

Band Combination

Useful band combinations may be made by superposing positive or negative images (normally using two, three, or four bands) and placing color filters over selected bands so that band-to-band differences in gray tones are displayed in contrasting colors (Yost and Wenderoth, 1971). If an object has a high reflectivity in one band and a low reflectivity in another, and is unique in this respect as compared to other objects, the object is enhanced by placing different color filters on each band and combining them. Vegetation is the classic example, because chlorophyll reflects strongly in the near infrared and absorbs strongly in the red. Therefore, when a red filter is used to project the near-infrared band and a blue for the red band, vegetation will be displayed in varying shades of red (depending on type, amount, and vigor of the vegetation). This color enhancement shows the object of interest in a mode that is easier for the observer to see and interpret, just as the color geologic map shows rock types better than a black and white map. Unfortunately, rocks do not have distinctly different spectra (Fig. 7), so it is generally not possible to use these techniques for rock discrimination (Lyon and Patterson 1969; Lyon, 1970; Raines and Lee, 1974).

A different approach to this problem has been that of Lyon, who has used an aircraft-mounted spectrometer and radiometer to map rocks (Lyon and Patterson, 1966). The spectrometer records the spectral distribution of electromagnetic radiation reflected or emitted from the target. It is then possible to com-

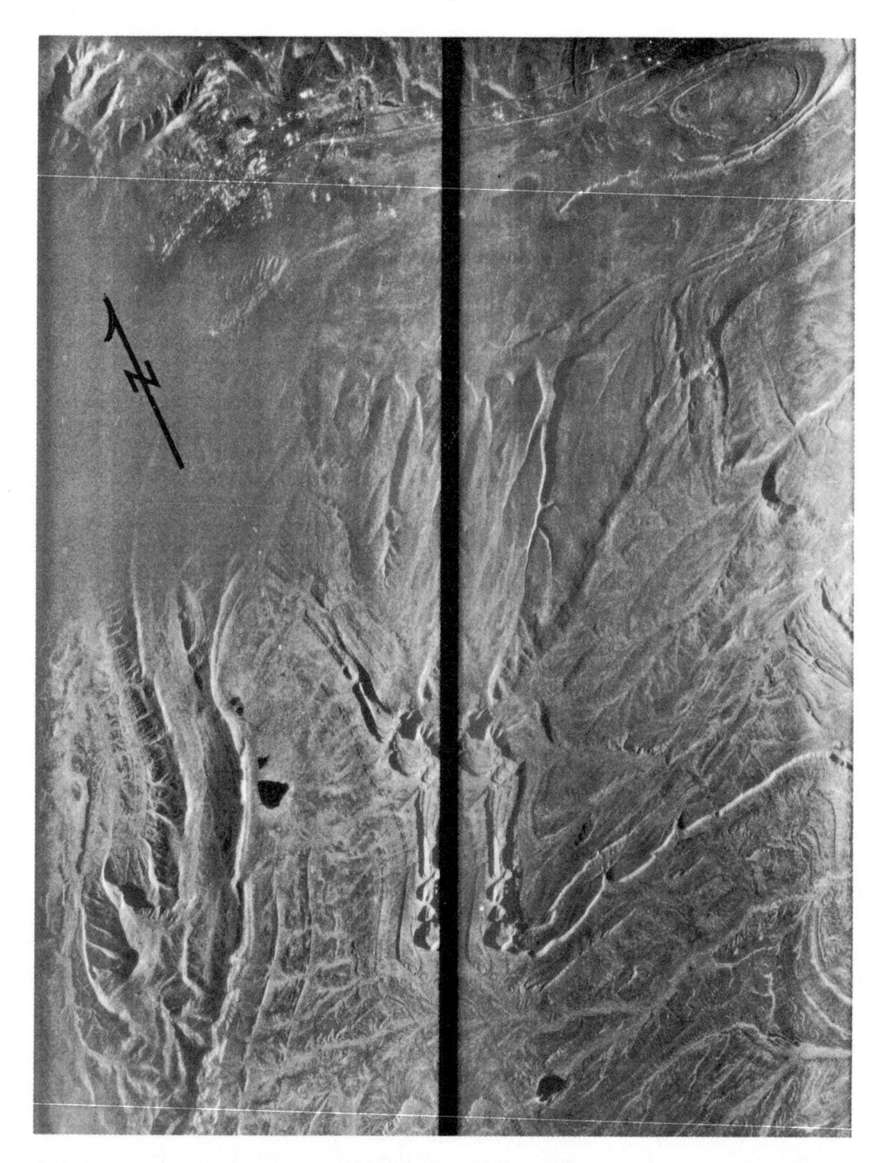

FIG. 6
Side-looking radar imagery of the Rawlins–Sinclair area of central Wyoming. Topographic patterns stand out sharply detailing geologic structures such as that of Grenville Dome, which appears in the northeast corner of the image. (Imagery provided courtesy of Strategic Air Command.)

pare characteristic spectra of common rocks with those of a reference and to define, statistically, the similarities and differences of the spectra. The major limitations of this approach are that the information from the spectrometer represents rocks and other materials associated with the rocks and is confined to a narrow ground path. Consequently, the spectral information cannot be readily extrapolated for mapping purposes. Some success in rock discrimination can be achieved under ideal conditions (Lyon, 1970).

Band-combination techniques seem to work best where rocks are entirely barren or show some vegetation control, so that the color enhancements for rock colors and vegetation can also be used for rock dis-

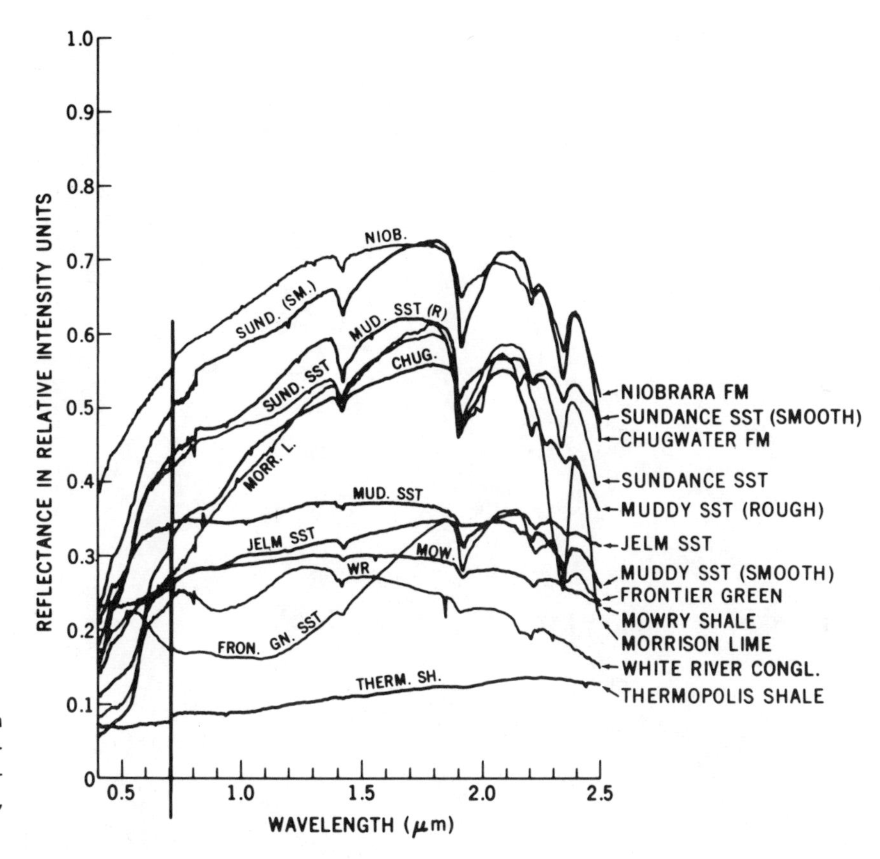

FIG. 7
Spectral response curves of a group of rocks from Wyoming determined on a Cary 90 Reflectance Spectrometer (N. M. Short, unpublished.)

crimination. In situations where there is a variation in water saturation with rock type, the interpreter may also take advantage of the strong water absorption in the near infrared. Under these conditions, seasonal variations in vegetative cover may be significant and strong contrasts may show during wet seasons or after rainfall (Fig. 8*). The LANDSAT repetitive system has been used to advantage where vegetation or rock moisture play a role in rock discrimination, because it is possible to select images on a seasonal basis that show the best contrasts between rock units.

*Figures 8, 13, 17, 30(A), 43, 44, and 45 will be found in the color insert.

Enhancement

Photographs and images usually contain more information than can be resolved by the naked eye. In some applications geologists will find that enhancement of this detail is of great value in problem solving. One very important consideration in applications that require interpretation of minute detail is that the film positives and negatives contain much more information than is normally recorded on prints. Furthermore, there is a systematic loss in resolution with each duplication. It is necessary to limit reproduction to as few generations as possible.

After the initial choice of bands or band combinations to solve a particular problem, the geologist has a

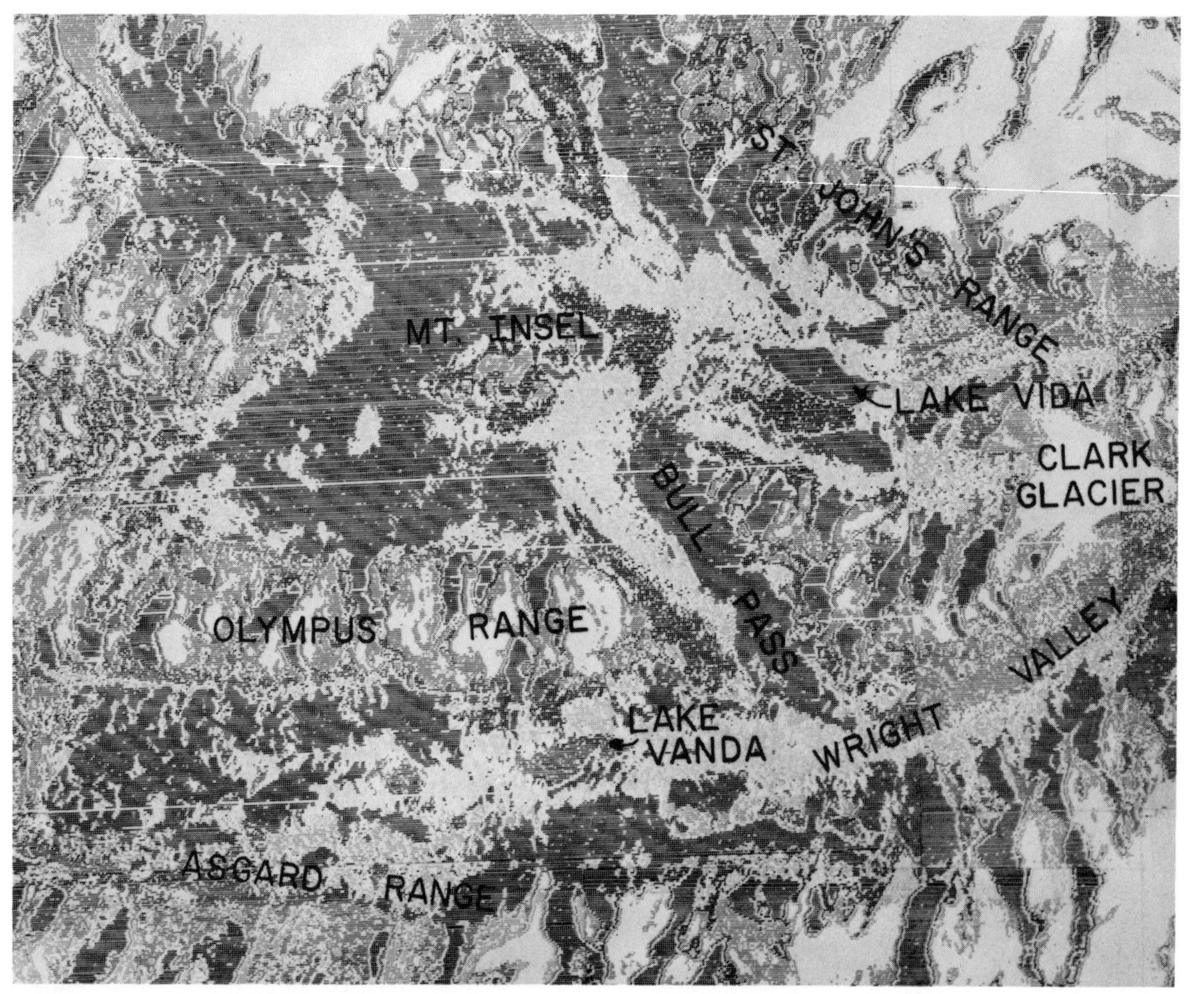

FIG. 9A
Density contour map of the Dry Valleys, Antarctica, produced from a density analysis of LANDSAT-1 image 1174-19433-7 (Fig. 9B). A Joyce Loebl/Tech Ops isodensitracer was used to construct this density contour map. Compare density contour map with Fig. 5.

number of enhancement techniques at his disposal (density analysis, edge enhancement, contrast stretching, and image masking).

Density analysis and contrast stretching. Density analysis refers to techniques by which film densities (a logarithmic expression of film opacity) are accurately measured or subdivided in order to detect and delineate subtle variations in the image tone. Density analysis can be accomplished through photographic, photometric, video, or digital processing. A typical density analysis procedure results in a density contour map (Fig. 9), or density slice. Density slicing can

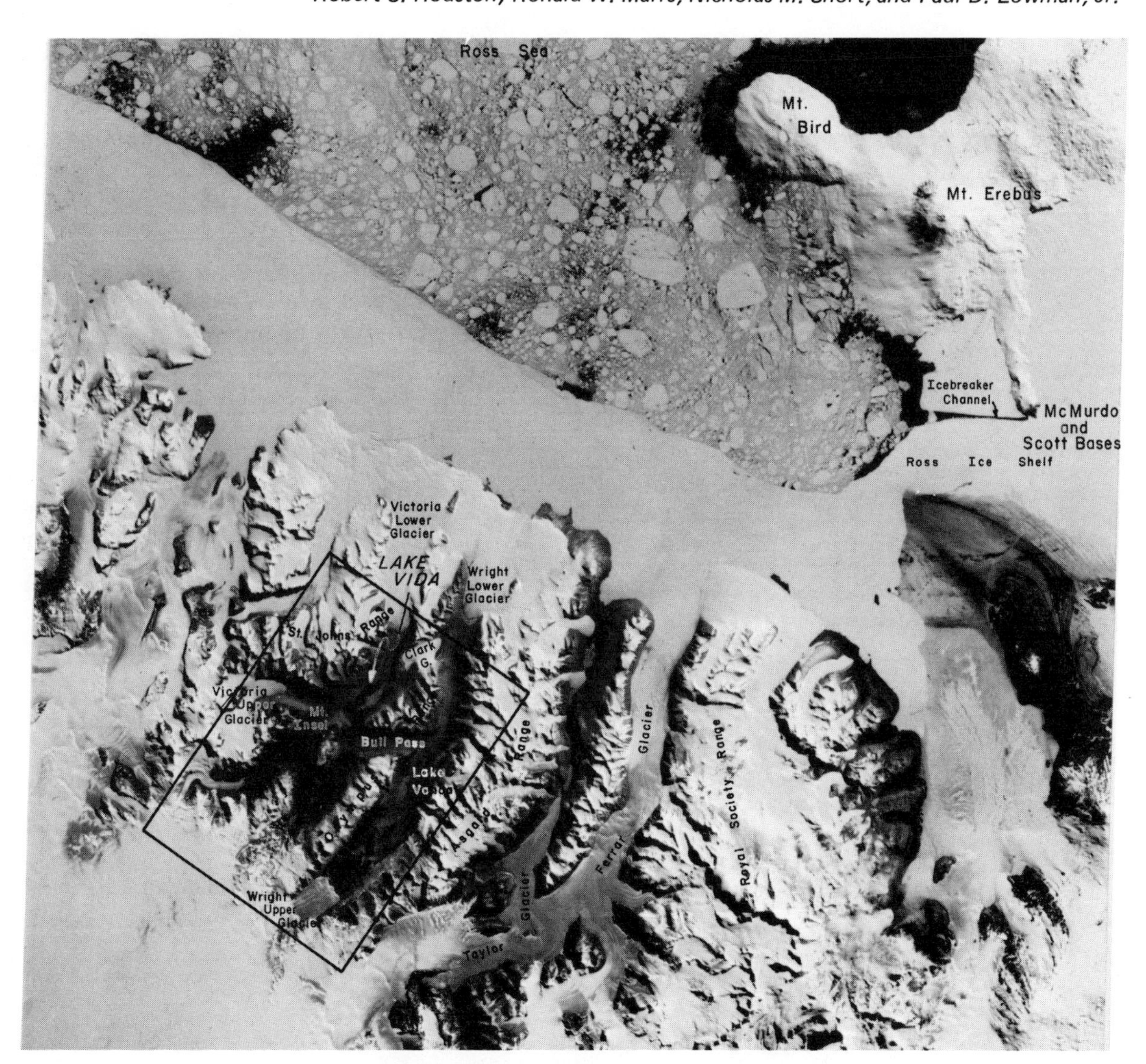

FIG. 9B
LANDSAT-1 image 1174-19433-7 of the Dry Valleys, Antarctica. The Dry Valleys area was a test site for photogeologic interpretation of LANDSAT-1 imagery (Houston et al., 1974).

be accomplished photographically with a set of prints made with various exposures on photosensitive paper (Fig. 10). The resulting prints each represent a distinct density band (with lighter areas washed out and darker areas underexposed). Each image will be a *contrast-stretched* version of the original. The tonal values represented within the reproduced density slice will be stretched across the entire density range of the photographic paper, and the relative contrast between subtly different features should be more apparent.

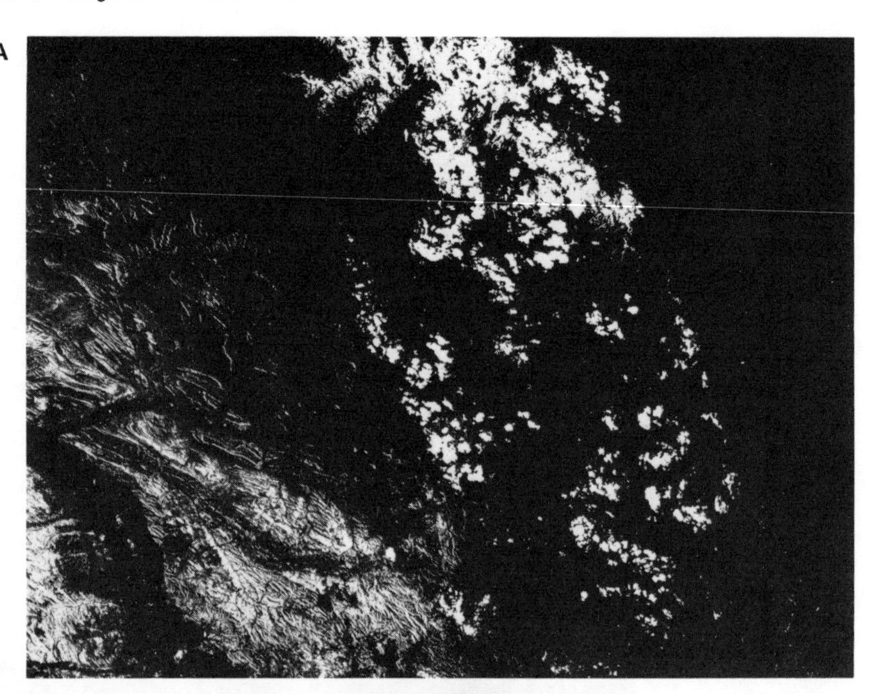

FIG. 10
A series of high-contrast prints from a Skylab S-190B image of the Bighorn Mountains and Bighorn Basin in north-central Wyoming. Each print emphasizes a different range of film densities. Print A shows only the clouds, light-colored barren areas, and a few cultivated fields. With increased exposure, more and more of the terrain included is expressed as light areas on the photograph, until in print D, only shadows, green deciduous vegetation, and forested areas remain dark.

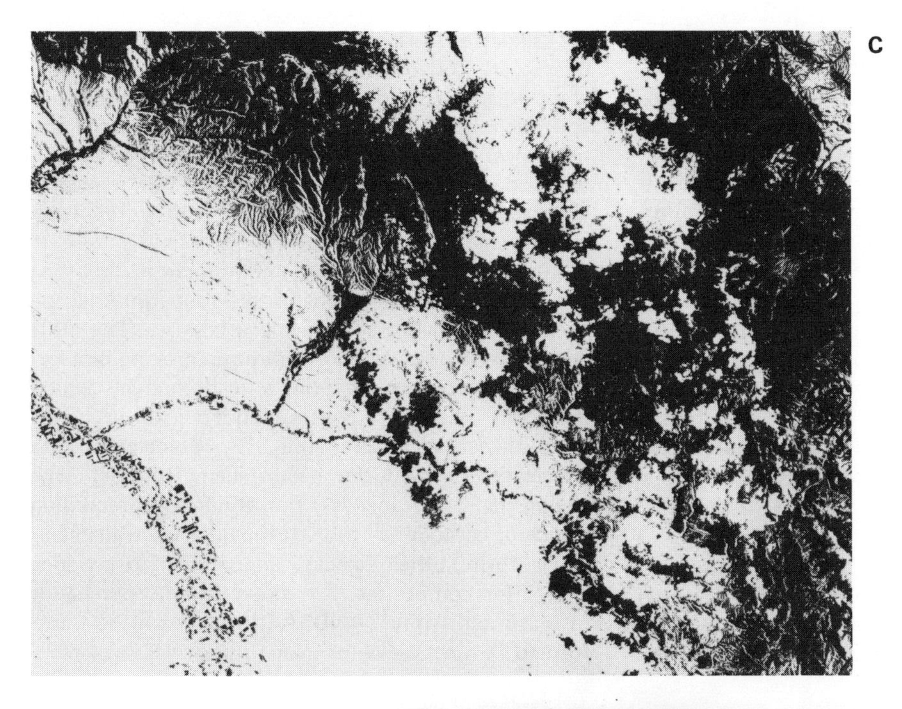

C

D

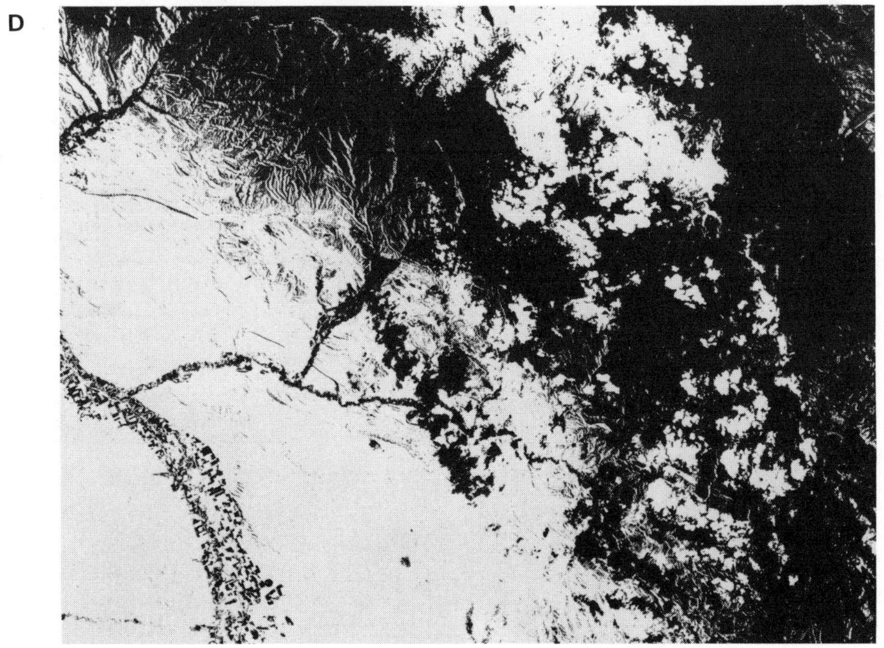

A similar effect can be achieved using a digitizer and computer or a scanning microdensitometer (Fig. 11). These instruments allow the image densities to be measured and then coded or grouped according to some predetermined program. The coded data can then be redisplayed in the form of a density contour map. The density contour map displays subtle film density variations in strongly contrasting shades or colors so that the interpreter can detect and map features that are ordinarily obscure (Buckley, 1971).

Density analysis by computer is readily applied to data that are available in digital form (such as LANDSAT) and provides the least loss in brightness resolution for such data. Photographic or image data must be digitized before being input to a computer, so photographic processing or a scanning microdensitometer may be a more efficient means of density analysis.

Video image-analysis systems (Fig. 12) offer a fourth alternative for density determinations (Schlosser, 1974). A typical video system scans an image by a television camera and subdivides the resulting analog signal into intensity groups. These individual groups are coded (assigned a specific tone or color) and redisplayed on a video monitor as a density contour map. Video systems may lack the spatial and brightness resolution of sophisticated computer systems or scanning microdensitometers, but they provide real-time analysis capability for a relatively modest cost.

Image masking and image combination. Masking is a procedure in which one image is superimposed on another to enhance certain image features. The most commonly employed form of image masking is color-additive image composition, a technique by which false-color images are constructed from various bands of multispectral imagery (Fig. 13). Color composite images can be produced electronically from digital or analog tapes by means of a flying-spot recording device or cathode-ray tube. If the data are available in image form, other devices can be used to produce false-color composites. For example, the four bands of black and white LANDSAT imagery can be transformed into a false-color image using the diazo tech-

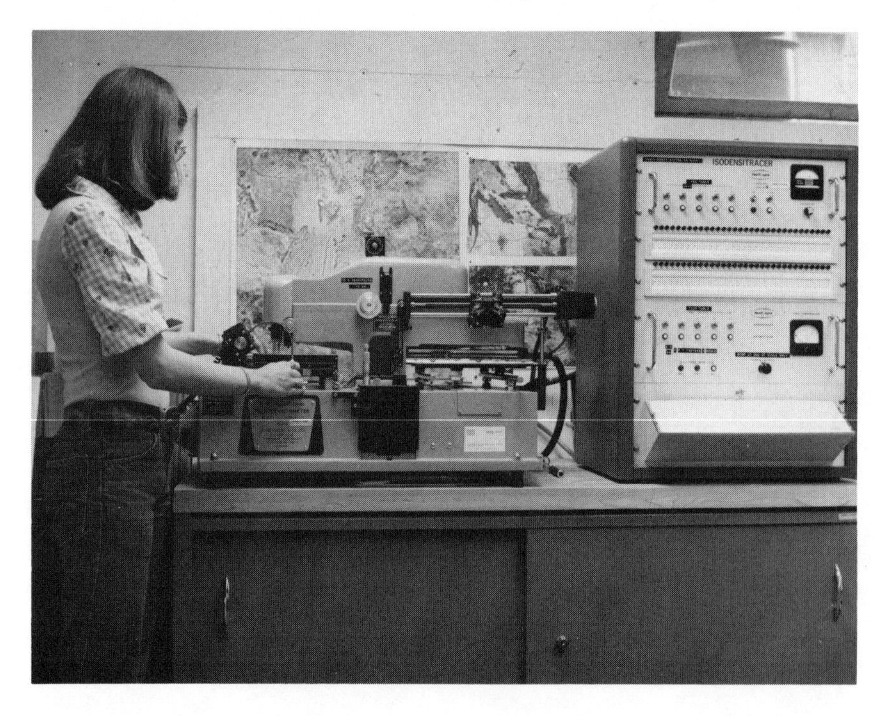

FIG. 11
Joyce Loebl Mark III microdensitometer coupled with a Tech Ops isodensitracer to provide a capability for precise measurement and contouring of film opacity values.

FIG. 12
Video equipment (such as this Spatial Data, Model 401, Edge Enhancer) is available in a variety of systems. These systems can provide a range of image enhancement functions and can be operated in an interactive mode for rapid image analysis.

nique (Wilderman, 1972), a color-additive viewer (Ross, 1973), a multiple-exposure printing system (Kreitzer, 1974), or video equipment.

Stereoscopic Viewing

Most standard photointerpretive techniques are readily applied to either multiband photography or multispectral imagery. One serious drawback to multispectral scanner imagery is that it does not provide the stereo overlap necessary to produce a true stereoscopic model. The scanner does not image an entire scene from a single vantage point as does a camera. Rather, it builds an image, line by line, as the scanner platform progresses along the flight path (Fig. 14). Consequently, the interpreter must depend upon the pseudostereoscopic effect produced by viewing two different bands of a scanner image as a stereo pair, or use side lap of adjacent scanner strips to produce the parallax necessary for stereoscopic viewing (Poulton, 1972). This technique does not form a true stereoscopic model. Instead, it forms a model that appears to sink away from the observer in either direction parallel to the flight path. The model approaches a true stereoscopic model only along the scan line directly beneath the interpreter's apparent vantage point and cannot be used for topographic contouring. It does allow the interpreter to estimate relative relief of adjacent features.

Film transparencies are generally superior to paper prints for purposes of interpretation because they offer higher resolution and present a greater range of brightness values. They also allow greater flexibility in illumination and image combination. Transparency viewing tables are often equipped for variable-intensity illumination so that the interpreter can adjust the illumination for optimum viewing over a broad range of image densities (Fig. 15). These tables can also be equipped for microscopic or stereomicro-

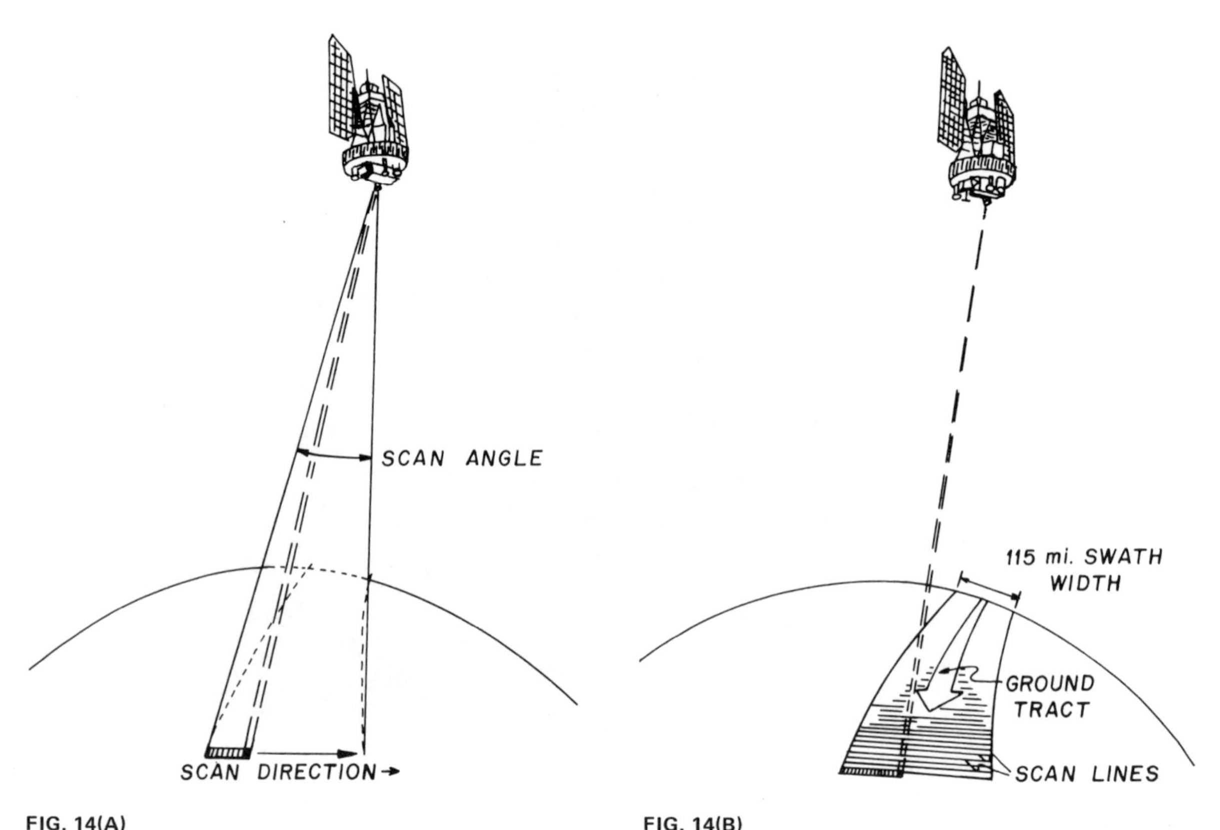

FIG. 14(A)
LANDSAT multispectral scanner system. A scanner "sees" only a very small area at one time but scans the terrain through a moving optical system. Each sweep of the optics records information for a single scan.

FIG. 14(B)
The scanner "builds" the image line by line as the platform moves along the ground track.

scopic image viewing, features that become particularly important when working with small-scale satellite imagery.

The ideal arrangement for the geologist is a device that projects a stereoscopic view of a standard or enhanced image on a base map (Fig. 16). If the equipment will accept image transparencies, it is possible to take advantage of the greater resolution of the transparency. Such instruments may have capacity for image enlargement, and can be equipped to plot information accurately on a topographic or planimetric base map. A number of instruments of this type are available.

GEOLOGIC APPLICATIONS

Geologic Mapping and Map Editing

Photographs and images (especially space photographs or images) have major advantages and disadvantages when used for geologic mapping. The photo geologist is still limited in his ability to identify rock units and map complex detail. Therefore, interpretations must be tied to field studies for lithologic control and verification of results. The most useful photograph or image for geologic interpretation is usually a color or color infrared image of the highest resolu-

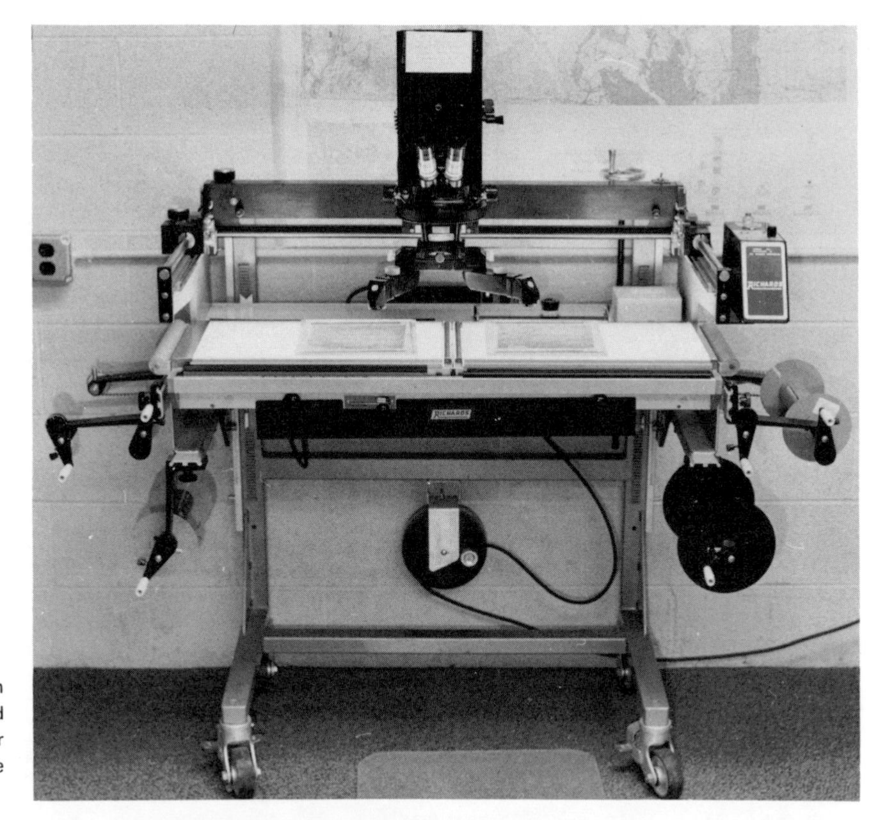

FIG. 15
Richards MIM-3 interpretation table equipped with a Bausch and Lomb zoom stereoscope for stereoscopic viewing of image transparencies.

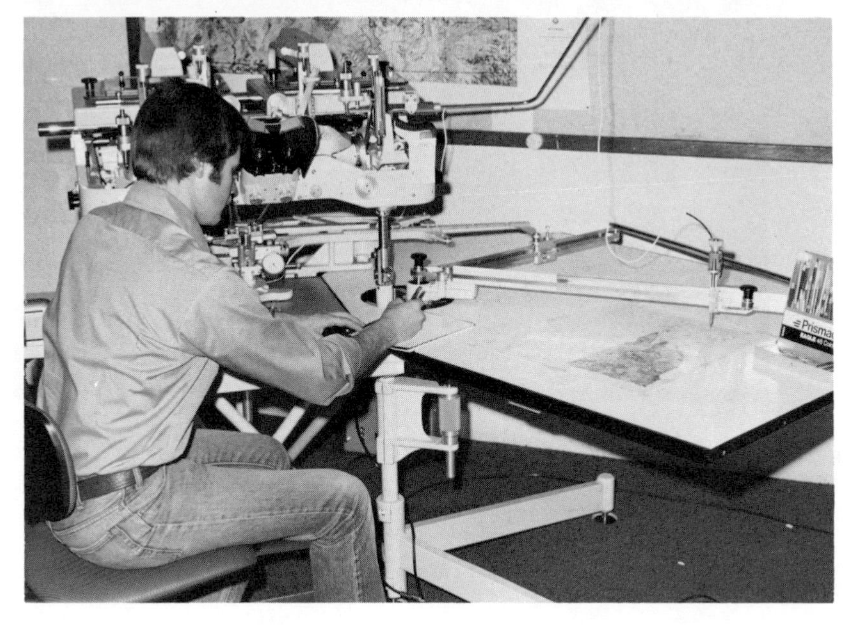

FIG. 16
Kern PG-2 stereo plotter used for direct transfer of information from a projected stereo model to a base map.

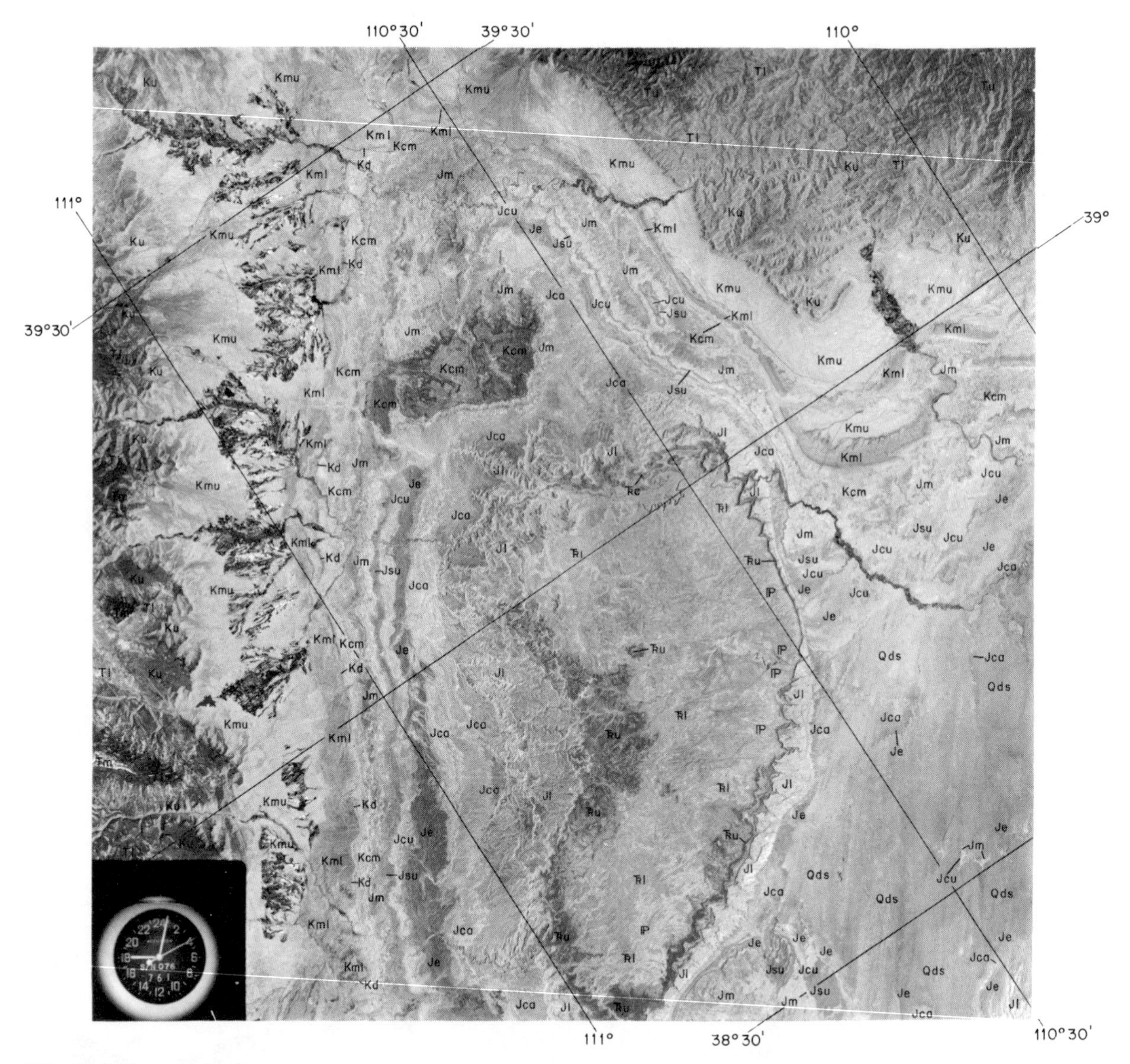

FIG. 18 (*above and right*)
Skylab S-190B image of the San Rafael Swell in Southeast Utah. Contrasting rock units and structural features are so obvious that the annotated image is useful as a geologic map.

SKYLAB-3 S-190B FRAME 81-014
10/29/74 R.W. MARRS

LEGEND

TERTIARY—QUATERNARY
Qds DUNE SAND AND AEOLIAN DEPOSITS
Tu UPPER TERTIARY UNITS
Tm MIDDLE TERTIARY UNITS
Tl LOWER TERTIARY UNITS

CRETACEOUS
Ku UPPER CRETACEOUS UNITS
Kmu UPPER MANCOS FM.
Kml LOWER MANCOS FM.
Kd DAKOTA FORMATION
Kcm CEDAR MOUNTAIN SHALE

JURASSIC
Jm MORRISON FORMATION
Jsu SUMMERVILLE FORMATION
Jcu CURTIS FORMATION
Je ENTRADA SANDSTONE
Jca CARMEL FORMATION
Jl LOWER JURASSIC UNITS

TRIASSIC
ꞦU UPPER TRIASSIC UNITS
Ꞧl LOWER TRIASSIC UNITS

PENNSYLVANIAN
IP PENNSYLVANIAN UNITS

0 10 20 km
0 10 20 miles
N

tion. Fortunately, much of the United States has been covered by high-resolution aerial photography as part of the NASA Earth Resources Aircraft Program and as aircraft support for LANDSAT and Skylab studies. These surveys normally employ metric cameras that obtain high-resolution photographs in both color and color infrared. This coverage is usually obtained with 60-percent overlap to allow stereoscopic viewing.

No geologist will claim that color is diagnostic of rock type; but, in a given region, a particular rock unit often displays a characteristic color over a large outcrop area. Color transparencies and prints are favored by most geologists because they show rocks in a mode to which the geologist is accustomed; but color infrared photography is generally just as useful and is often sharper, because the hazy, blue-band information is filtered out. Rocks that show a strong variation in iron content exhibit a greater tonal contrast in the infrared. Because the color infrared image combines the red (which shows low reflectance for vegetation) with the infrared (which shows high reflectance for vegetation), vegetation is notably enhanced. Thus, rock units that support characteristic vegetation assemblages are readily mapped. Once the geologist is familiar with the color shift (Fig. 17), he can map as well or better with color infrared as with color photography.

Photogeologic mapping is not a substitute for detailed field studies, but in areas of sparse vegetation, photogeologic maps that have been carefully field checked may be as accurate as a field map. The cost of the photogeologic map is usually much less than that of a map obtained by surface mapping. Formations defined by field geologists can sometimes be recognized and mapped by photointerpretation, but the photomap units do not ordinarily correlate exactly with the established stratigraphic units. Where the units do correspond, a rectified aerial photograph may show formations in such detail that a properly annotated photograph may serve as a geologic map (Fig. 18).

Space photography or imagery has the advantage of the synoptic view, but is limited by lower resolution. LANDSAT color infrared composites have the lowest spatial resolution of the products made available through NASA Earth Resources Programs (~ 70 m). Skylab S-190A resolution varies from 30 to 75 m on the different bands; and Skylab S-190B color, color infrared, and high-resolution black and white photography is effective for definition of objects as small as 10 m. This sets obvious limitations on the use of these images for geologic mapping. LANDSAT composites are useful for mapping at a scale of 1:500,000 or 1:250,000 (Figs. 19 and 20), whereas Skylab S-190A may be suitable for mapping at 1:125,000 and Skylab S-190B at 1:125,000 (Fig. 21) or 1:62,500. Most small-scale geologic maps (1:250,000 to 1:500,000) are prepared by reduction and generalization of available large-scale maps supplemented by reconnaissance maps and photogeologic studies. The image data base provided by the satellites yields more than adequate resolution for small-scale maps, and the final product easily meets regional map standards. Because of the extremely broad coverage, compilation of small-scale maps is very time consuming. Typically, state geologic maps take years to assemble and may be seriously out of date at the time of publication.

The space image shows the synoptic view of a given area without the prejudice of prior mapping or information definition. The earth's surface is displayed as it really appears, with none of the basic information misinterpreted or omitted. Consequently, a mosaic of

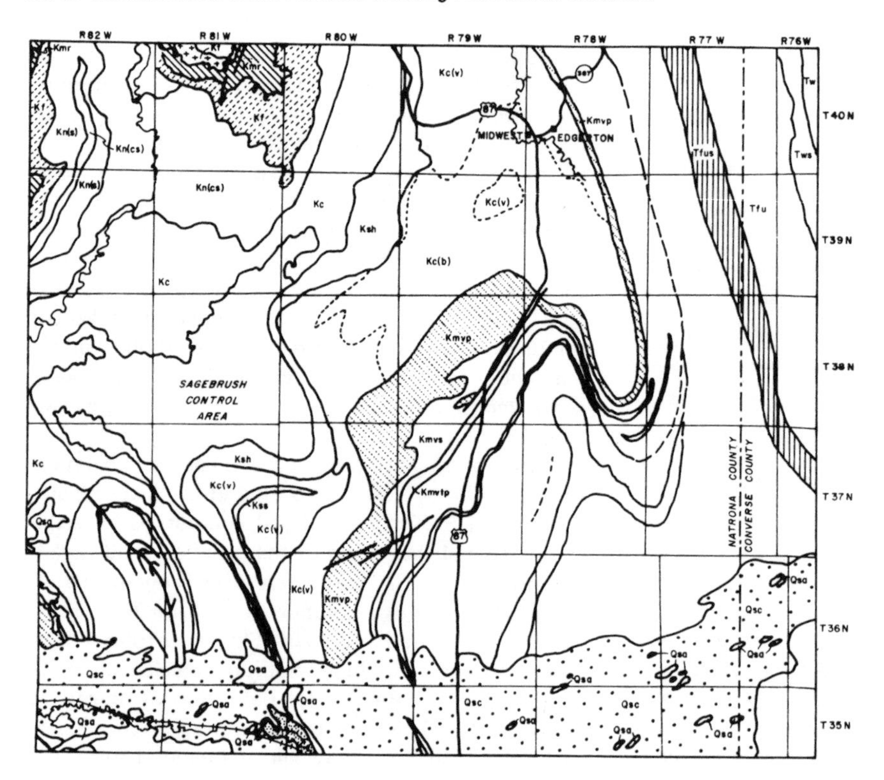

FIG. 19
Photogeologic map prepared from LANDSAT-1 color composite 1030-17235. This area lies north of Casper, Wyo.

satellite images with appropriate annotation can be as useful or more useful than a map and does not become obsolete as quickly as a map.

Because of the limited resolution, mappable units defined on a space image are not always the same as those mapped in the field. For example, two thick shale sequences (defined as separate formations) may be separated by a thin sandstone that is not resolved on the space image. They may, therefore, show as one unit. Likewise, any formation contact that is not marked by a distinct change in physical characteristics of the rock is difficult or impossible to define by photointerpretation. Conversely, intraformational changes in rock type may be selected as a photomap boundary even though they do not represent a formation boundary. Some users of remotely sensed data have referred to this problem and suggested that the subdivisions mappable on the imagery be called "remote-sensing units." Remote-sensing units can be used to make useful reconnaissance geologic maps, but the lack of correspondence to existing maps sometimes makes it difficult to use image interpretations as supplements to existing maps.

Despite its limitations, the space image offers the geologist a product with a number of unique uses. Subtle lithologic variations are often apparent on space images because of reflectance changes related to gradual changes in rock composition. Therefore, it is sometimes possible to use the images in the study of regional facies changes. Regional geologic maps may not show these facies changes because of the manner in which some formations are named or because lithologies may be classified by age rather than lithologic type. Photogeologic interpretation of a LANDSAT image or Skylab photograph gives the geologist a firsthand familiarity with the regional geology of an area that he cannot readily obtain through most other methods of map compilation. In addition, the time required to prepare a reconnaissance map is far less than that required for compilation by other methods. The geologist may also acquire new information through interpretation of the space image or correct errors in earlier compilations. Thus he may use the space imagery as a tool for map editing or to define and map critical geologic and physiographic features.

A common fallacy concerning geologic mapping in the United States is that the country has been adequately mapped (especially at scales of 1:250,000 or larger). Some maintain that there is no longer any domestic need for such mapping and that satellite imagery is of little value to geologists working in the United States. Neither of these attitudes can be supported. In addition to having large unmapped areas (Fig. 22), much of the reconnaissance and large-scale mapping is out of date or inaccurate. If we take the U.S. Geological Survey estimates (Kinney, 1960) that only 25 percent of the country has up-to-date mapping at a scale of 1:62,500, this leaves 75 percent large-scale mapping that can probably be significantly improved by use of space images.

Lineations

Photogeologic lineations are lines chosen from images on the assumption that they originate through geologic processes. If the lineation originates through geologic processes, it is often a fracture (fault or joint) or a linear expression of a fracture system or shear zone. When the geologist can prove (normally by field checks) that these lineations are fractures, the photogeologic lineations may become geologically or economically significant. Faults recognized by this method are of potential economic value because they serve as a prospecting guide for fault-controlled groundwater (Goetz et al., 1973), metallic mineral

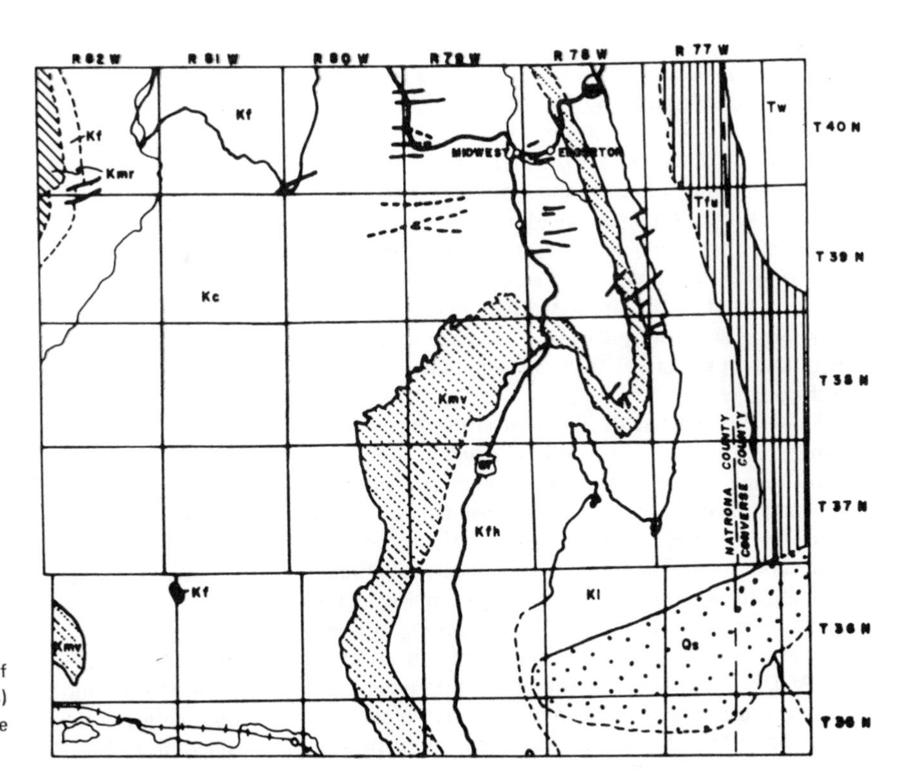

FIG. 20
A portion of the geologic map of Wyoming (Love et al., 1955) showing the known geology in the area of Fig. 19.

deposits (Rowan, 1973), petroleum (Collins et al., 1973), and in engineering environmental problems.

The synoptic view of the space image allows the geologist to see lineations of regional extent that cannot be recognized by any other technique; but the relatively low resolution of the image makes it difficult or impossible for him to distinguish geologic features from those of nongeologic origin. For example, study of lineations seen on LANDSAT images of Wyoming shows that a significant, but variable percentage of the photo lineations are of nongeologic origin. In parts of some mountain uplifts approximately 80 percent of the lineations are fractures or joints (Parker, 1972), but in some basin areas (Blackstone, 1973) virtually no lineations can be proved to be geologic in origin.

The low resolution of LANDSAT may cause the interpreter to erroneously identify other linear features as potential geologic lineations (e.g., graded roads, highways, fence lines, tornado swaths, contrails, railways, fire lines, and such geomorphologic features as linear blowouts and long, transverse stabilized dunes). Most of these features can be recognized as nongeologic simply by use of high-resolution aircraft images or Skylab 190B or 190A photographs.

Short and Lowman (1973, p. 12) draw particular attention to lineation studies done by Isachsen in New York and Weidman in Montana. Their studies illustrated that many linear features are not of geologic origin. For example, Isachsen shows that less than one third of the linear features mapped in New York State are strictly structural in nature, and that many known structural features cannot be identified on the imagery.

These problems with lineation study using space imagery are not new to photogeologists, who have found many photolinear features such as alignments of topographic structure. These problems are more severe for the space image because of the lower resolution and greater number of potentially significant linear features apparent on a single image. Three factors have a serious effect on the mapping of linear features: resolution, sun elevation, and sun azimuth.

LANDSAT is sun synchronous so that images are ordinarily generated in the morning with a sun azimuth that establishes a bias toward northeast-striking

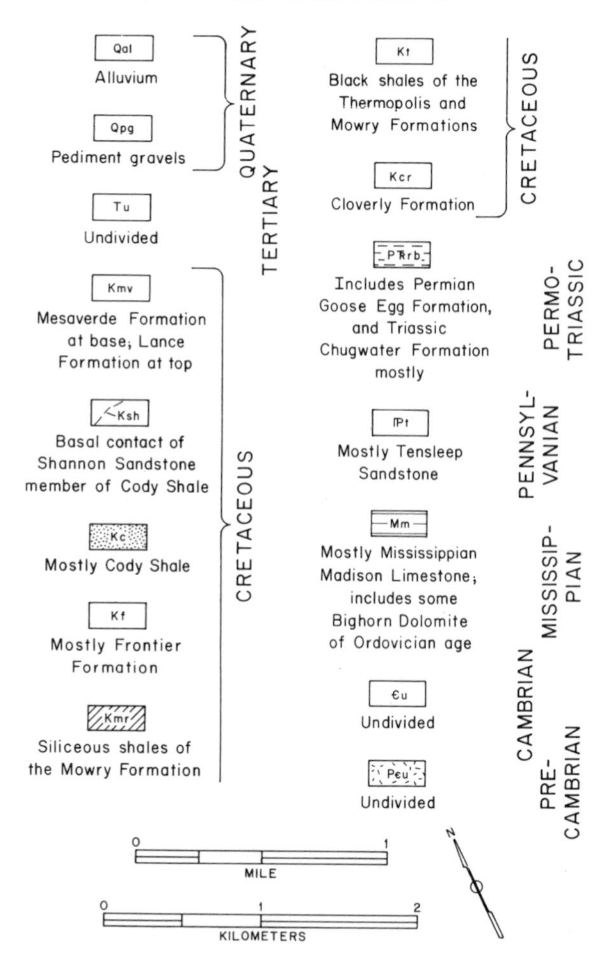

FIG. 21 (*above and right*)
Reconnaissance geologic map of the Horn area from Skylab S-190B photograph and Skylab S-190A color and color-infrared photographs. Area outlined in black was also mapped from aerial photography.

linear features, whereas Skylab passes give different perspectives and may result in the emphasis of a completely different set of lineations (Fig. 23). Rose diagrams in Fig. 23 show that there are not only more

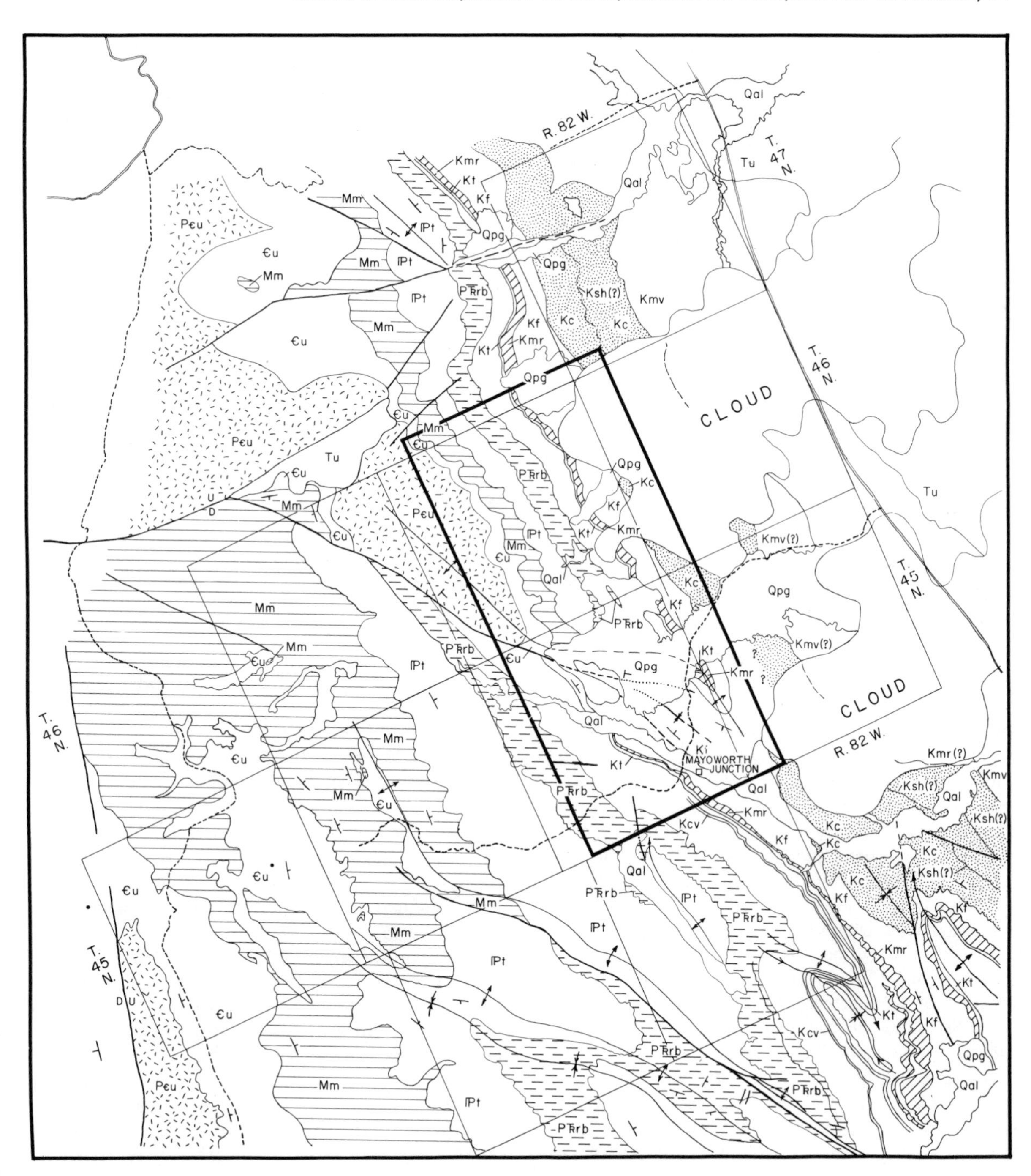
R. 82 W.
T. 47 N.
T. 46 N.
T. 45 N.
CLOUD
CLOUD
R. 82 W.
MAYOWORTH JUNCTION
T. 46 N.
T. 45 N.
Qal
Tu
Qpg
Kmv
Kmr
Kt
Kf
Kc
Ksh(?)
Kmv(?)
Kmr(?)
Kcv
Ki
P₸rb
Pt
Mm
Єu
PЄu

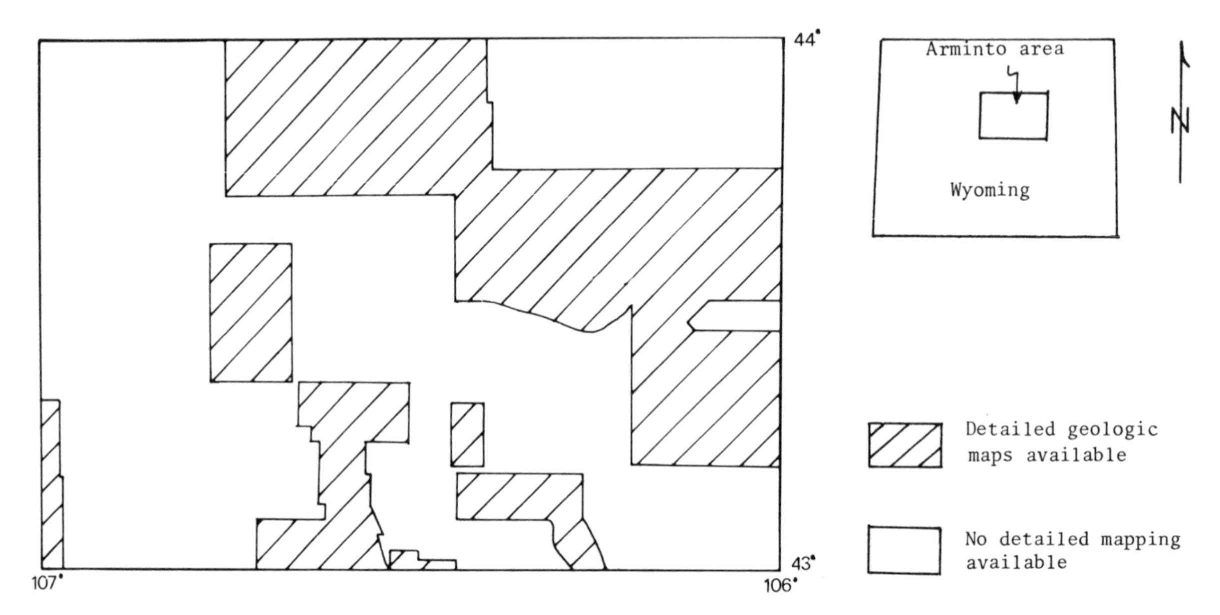

FIG. 22
Index showing areas where detailed geologic maps are available within the Arminto region of central Wyoming.

lineations recognized on the Skylab S-190A photograph, but that the major set strikes north–northwest rather than northeast, as is the case for LANDSAT. This is a sun azimuth effect, because the major linear features are approximately perpendicular to the sun azimuth in both examples. This sun azimuth effect is pronounced because many of the lineations are expressed topographically, and shadowing due to topography is greatest orthogonal to the direction of illumination. Sun elevation will also play a role in lineation mapping, because low-angle illumination causes distinct shadowing where lineations are topographically expressed. In the given example (Fig. 23), a few more lineations might have been mapped at the 49° sun elevation of Skylab than at the 54° sun elevation of LANDSAT, but this cannot be quantified because of the variation in azimuth.

An area in the Wind River Mountains east of Fremont Lake, Wyoming (Fig. 24) was mapped using aircraft images (C-130 infrared photographs). Comparison of Figs. 23 and 24 show clearly the great advantage of high-resolution aerial photographs in lineation mapping. The number of lineations mapped is several times greater than the number mapped from LANDSAT or Skylab. A "woods for the trees" relationship was noted, because some major lineations located on Skylab were overlooked on the aerial photograph until the interpretation was rechecked. The regional lineations were present.

Perhaps the ultimate value of images taken at different altitudes will depend simply on the care geologists take to verify the features. Statements that "northeast-striking linear features are more common than previously supposed" have little meaning without supporting data. Only when the linear features are verified and found to be new faults or extensions of known faults can they be considered both economically and scientifically significant.

Space imagery may also be of value in defining patterns of linear features over large areas and in the recognition of large-scale circular features. Buried salt domes and reefs may be recognized by trends in linear features peripheral to the bodies, and major subsurface discontinuities may be recognized by changes in linear patterns in overlying rocks.

LANDSAT has demonstrated a special facility for use in detection of circular as well as linear features. Most of these are volcanic or intrusive in nature, and many have been recognized for the first time on LANDSAT. One of the first new structures of this type identified from LANDSAT is a 25-mile diameter ring of hills just northwest of Reno, Nevada. The circular feature encroaches on the east front of the Sierra Nevada Mountains. The feature has not been fully explained but may be a ring complex of volcanic origin that has subsequently been disturbed by fault-

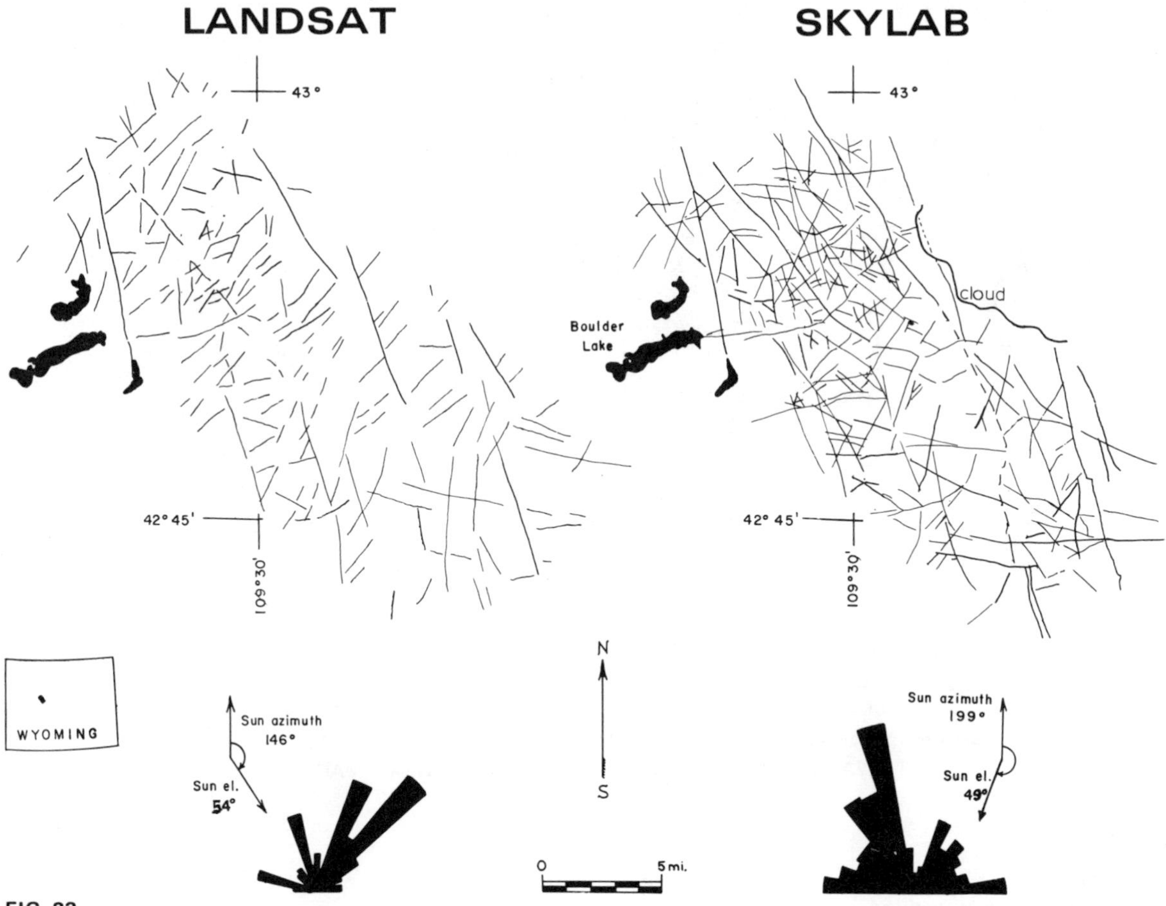

FIG. 23
Comparison of linear features detected in a portion of the Wind River Mountain Range, Wyo., using LANDSAT imagery and Skylab S-190B photography. The rose diagrams show that a bias is established by illumination, causing the interpreter to "see" more lineations that lie approximately perpendicular to the sun azimuth.

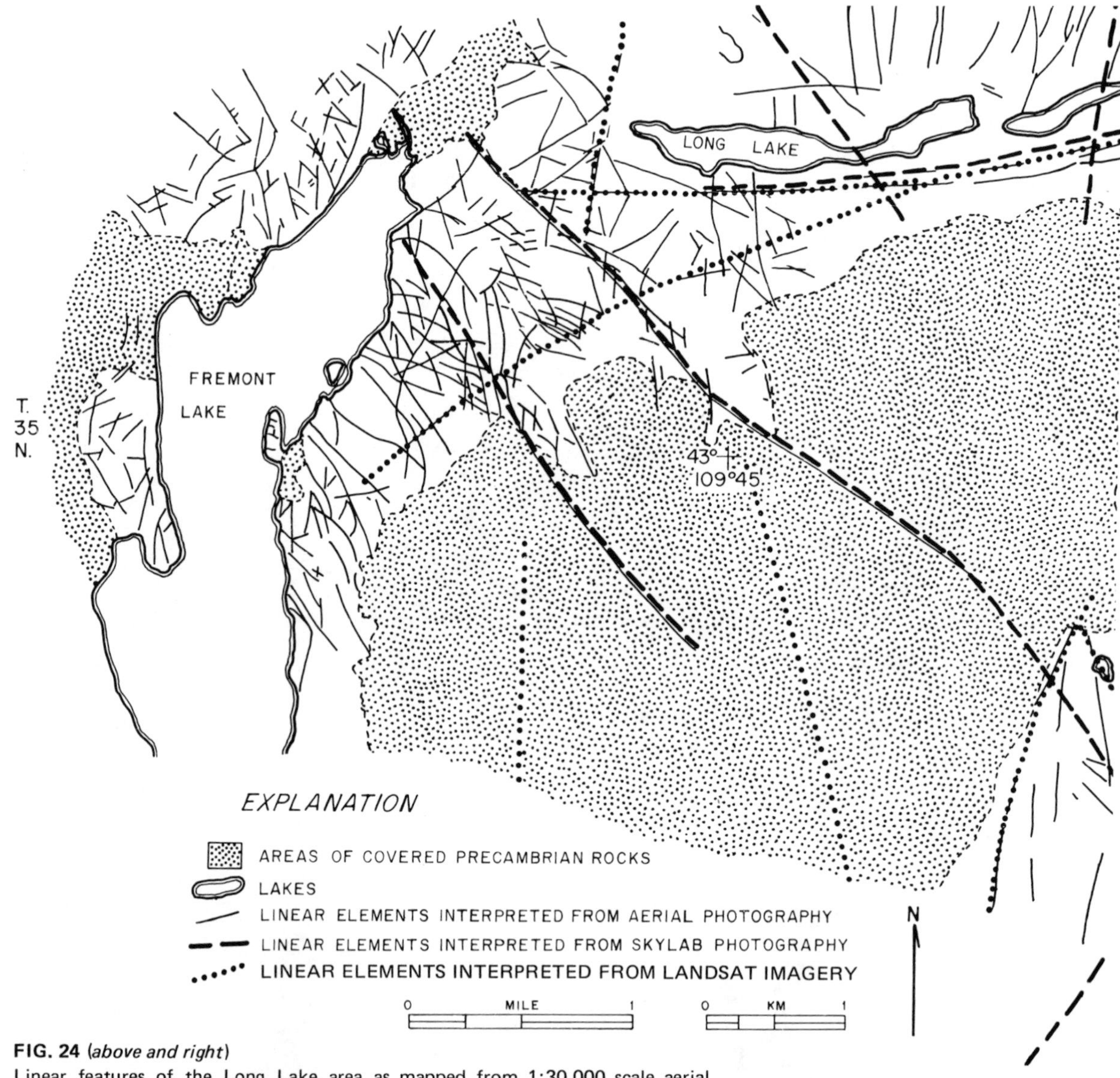

FIG. 24 (*above and right*)
Linear features of the Long Lake area as mapped from 1:30,000 scale aerial photography (NASA Mission 184, roll 11).

ing. Another example (Fig. 25) is an even larger elliptical feature northeast of Crater Lake, Oregon. Preliminary field studies indicate a concentration of small volcanic structures (cones and vents) along the forested rim of the large feature. A ring dike system would explain this alignment.

Often, structures such as these are subtly expressed on existing maps but have failed to be specifically delineated as distinct surface patterns. Consider the portion of the geologic map of Arizona shown in Fig. 26. An elliptical outcrop pattern is evident when pointed out, but generally has not been

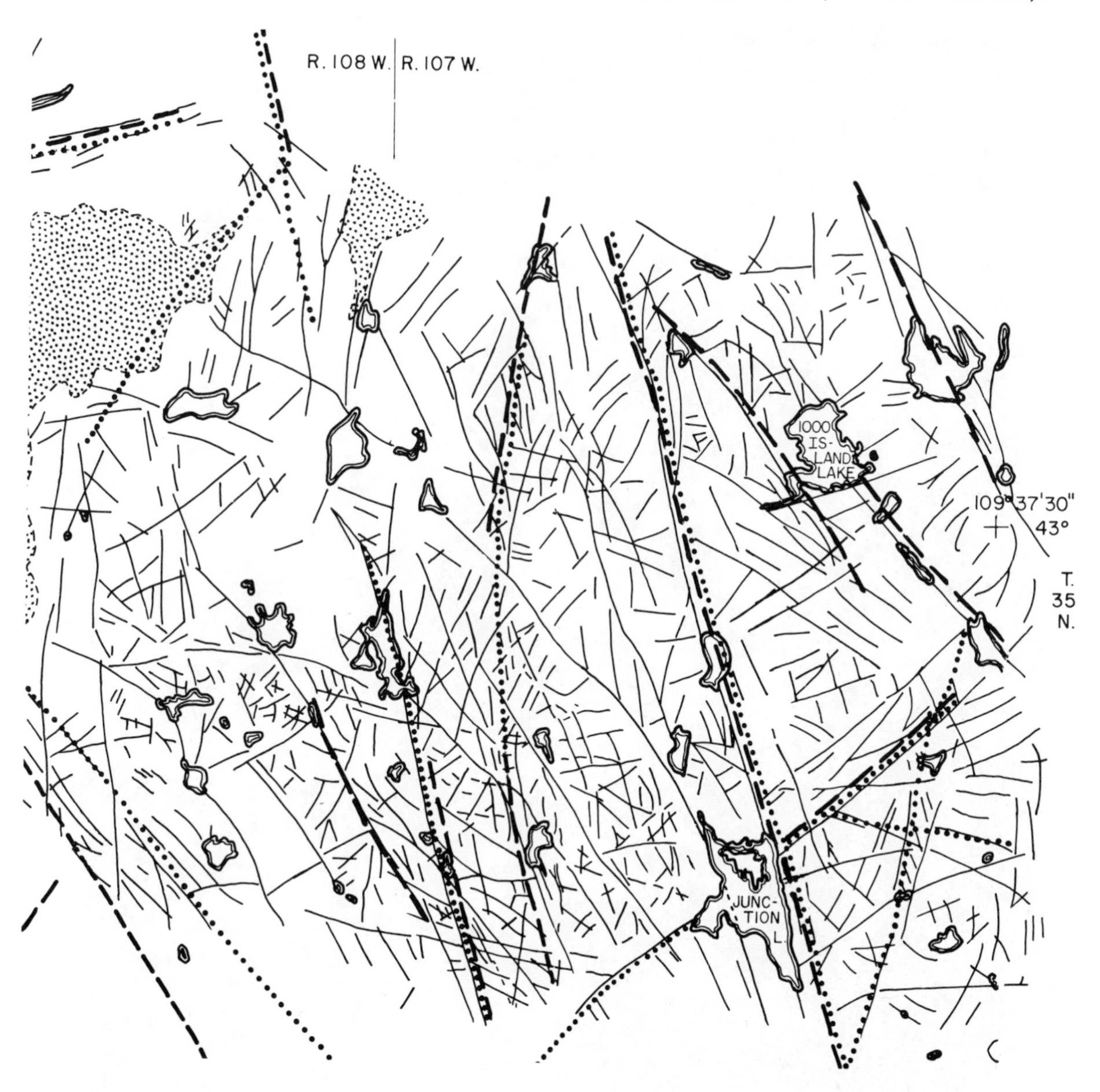

recognized by geologists. The LANDSAT view of this area (Fig. 27) defines a prominent drainage and topographic pattern that aids detection of the corresponding patterns on the map.

A few circular features have more exotic origins. Planetary-oriented geologists have learned to "think impact" whenever a structure with a near-perfect circular outline is seen. On earth, more than 70 structures, ranging from a few hundred feet to over 60 miles in diameter, have now been identified as recent or ancient meteorite craters. Many of these were first spotted on aerial photos.

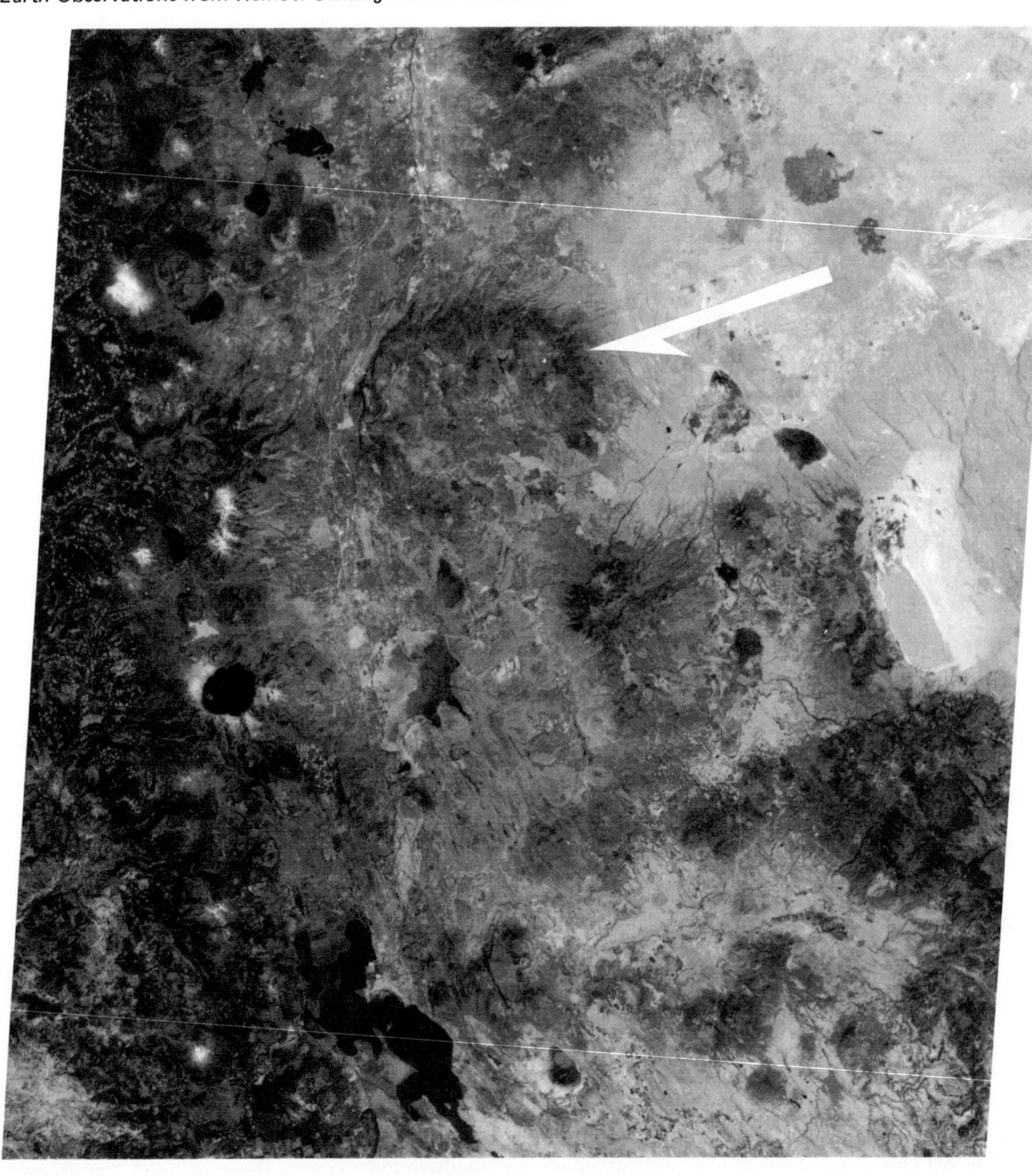

FIG. 25
LANDSAT MSS-5 image of south-central Oregon showing Crater Lake (lower left) and Newberry Volcano (top center). A prominent elliptical structure (arrow) is apparent between Crater Lake and Newberry Volcano.

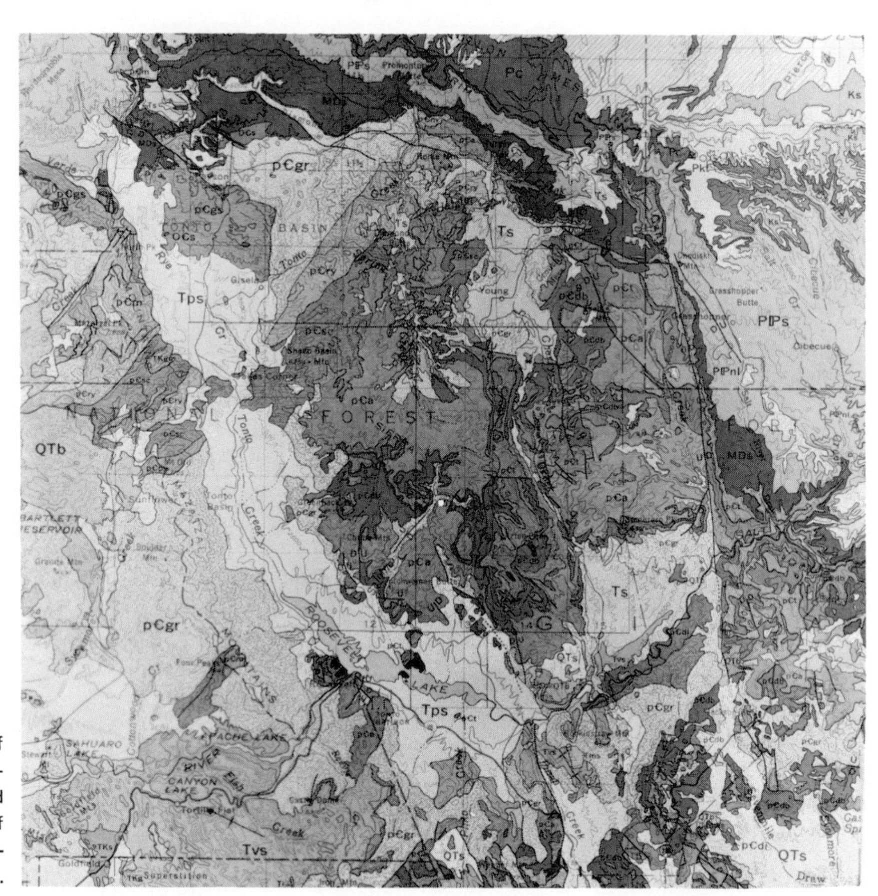

FIG. 26
A section of the geologic map of Arizona showing a region northeast of Phoenix. Tonto Creek and the Mogollon Rim are the chief components of the elliptical pattern apparent in the geologic map.

LANDSAT is expected to be a powerful instrument for detecting new, large, impact craters. Many of the known structures show up clearly on LANDSAT imagery. But, to date, just one new crater has been found with LANDSAT. Also, a structure in Brazil, some 400 miles west–southwest of Brasilia, was recently identified as an old crater because it contains shock-metamorphosed rock materials. This feature, called the Araguainha Dome, had been listed as having a diameter of about 13 miles. As seen from LANDSAT (Fig. 28), this old crater is nearly 25 miles in diameter, with two distinct dark "rings" comprising rock units that support thicker stands of trees in the semijungle.

LANDSAT mosaics are ideal for interpretation of the tectonic framework of large regions of the earth's crust. Lineaments that continue for hundreds of miles can be integrated into a unified network which reflects the influence of fractures in an ancient basement or results from stress systems developed from more recent tectonic movements. In a broad context, diversities of topography and surface geology may be filtered out, and a new synthesis of structural data can be defined. Geologists are even now building new structural models based on LANDSAT image interpretation.

Tectonic Maps

A tectonic map is normally prepared by summarizing the information on large-scale maps to be reduced

FIG. 27
Part of a LANDSAT-1 image covering the same area represented by the map in Fig. 26. The elliptical pattern is strongly expressed in the topography.

to the small-scale map. Consequently, the final product of such an effort has greater reliability. Interpretations of aerial images may be used to fill gaps between available maps, and space images may be used to detect regional structural features that are not apparent in the compilation or to detect features that are so subtle that they were overlooked in field studies.

Where tectonic maps have not been prepared and where mapping is limited, the space image is of even greater value. Structural elements, such as closed anticlines, domes, intrusive bodies, folded mountain belts, fault zones, regional joint patterns, and other fracture systems, may be readily mapped. The interrelationships among structure, topography, vegetation distribution, and solar illumination may enhance the structural elements so that subtle relationships are disclosed. The combination of snow and low-angle illumination are particularly effective for enhancing topographically expressed structures (Fig. 29). In some areas (e.g., Gulf of Oman, the Afar, Afghanistan) space imagery has brought about a fundamental resynthesis of the entire tectonic framework of a region.

In many areas the space image or a mosaic of images may simply be annotated to show the major structural features (Fig. 30). Newly discovered features may be designated as such. These space-image "discoveries" must then be checked through fieldwork or by interpretation of high-resolution images (photography, radar, imagery, etc.) Once proved, the new features may then be added to a final tectonic map. Thus, the space image has two basic applications in tectonics: (1) compilation of new tectonic maps in poorly mapped areas, and (2) updating and improving existing maps through recognition of new features and filling gaps between maps.

GEOMORPHOLOGY, QUATERNARY, AND GLACIAL GEOLOGY

Deposits of Quaternary age and of glacial origin may develop characteristic landforms, and certain landforms may be typically covered with specific types of deposits of Quaternary age. Thus, a landform map, or geomorphic map, may include information regarding deposits of Quaternary age or deposits of glacial origin, and vice versa. Because there are surficial deposits and surface landforms, aircraft and spacecraft images are ideal for mapping them. Furthermore, many landforms and glacial and Quaternary deposits support a characteristic vegetation assemblage that allows the interpreter to use color infrared and color images to great advantage.

Surprisingly, many geomorphologists have not taken full advantage of aircraft and spacecraft images to map and study these features. Perhaps this is partially because recent emphasis has been on more detailed and quantitative methods of study, and geomorphologists may have the opinion that images are

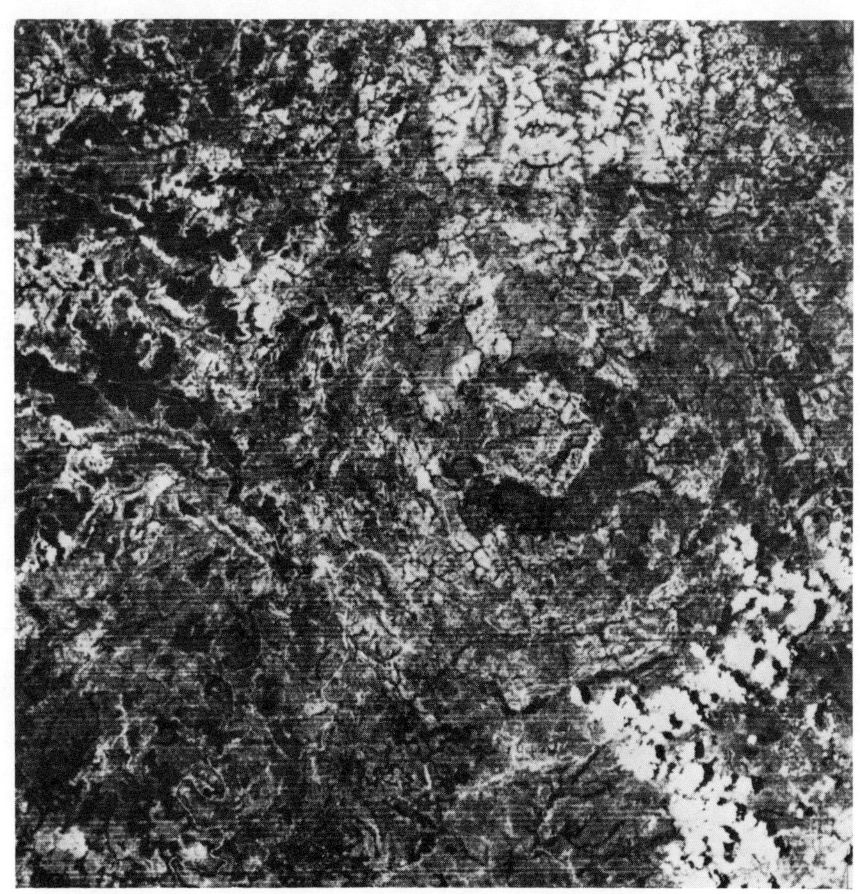

FIG. 28
Enlarged portion of a LANDSAT-1 image showing the double-ring structure of the Araguainha Dome (dark tones represent heavy vegetation). This structure has been identified as an ancient meteorite impact crater.

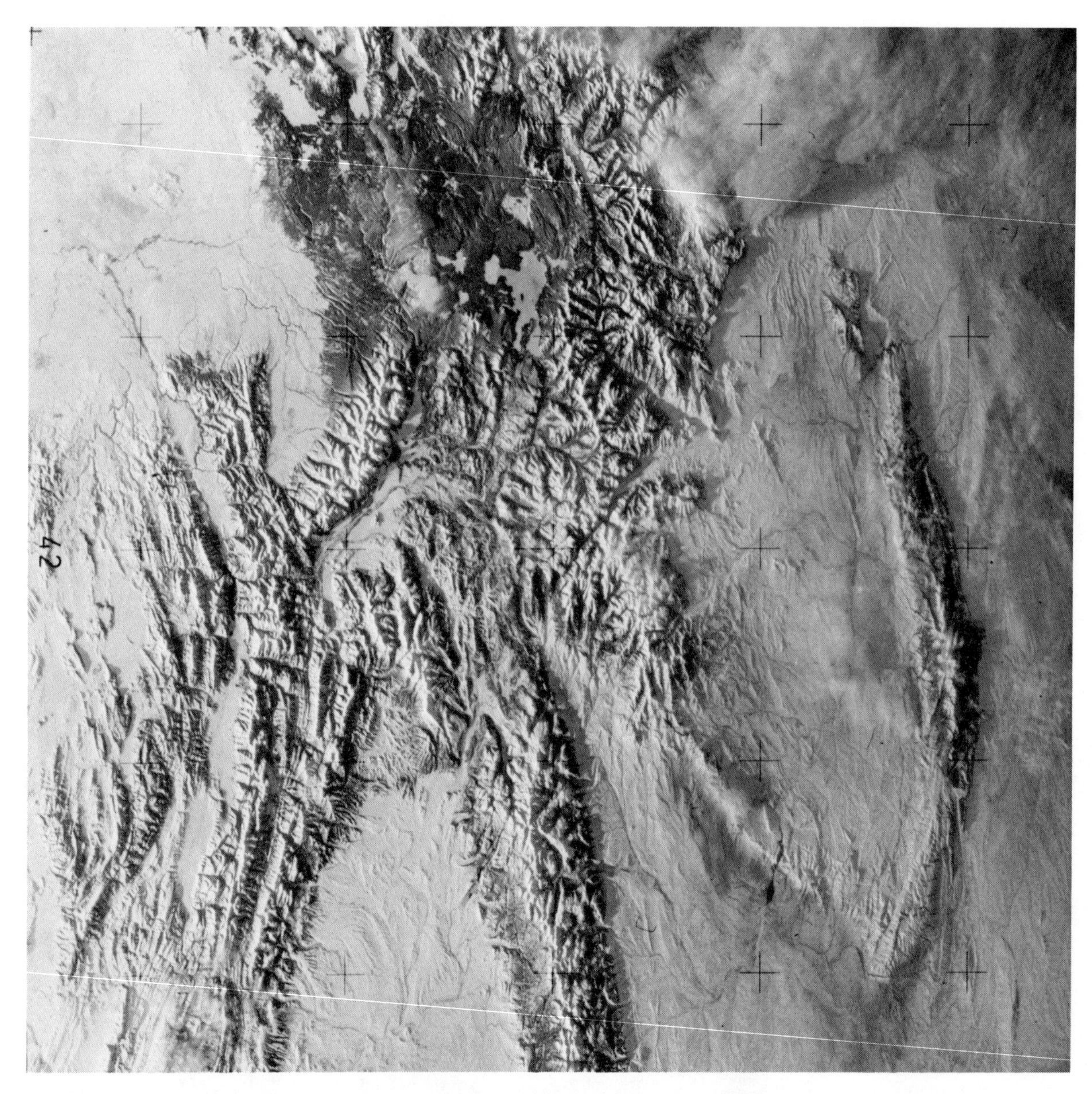

FIG. 29
Near-vertical view of the snow-covered northwest corner of Wyoming as seen from the Skylab space station in earth orbit. A Skylab-4 crewman used a hand-held 70-mm Hasselblad camera to take this picture. The low-angle illumination and oblique aspect help to enhance certain topographic elements. Yellowstone Park and Yellowstone Lake appear in the top center, the Wind River Mountains in the bottom center, and the Wyoming Thrust Belt in the lower-left corner of the photograph.

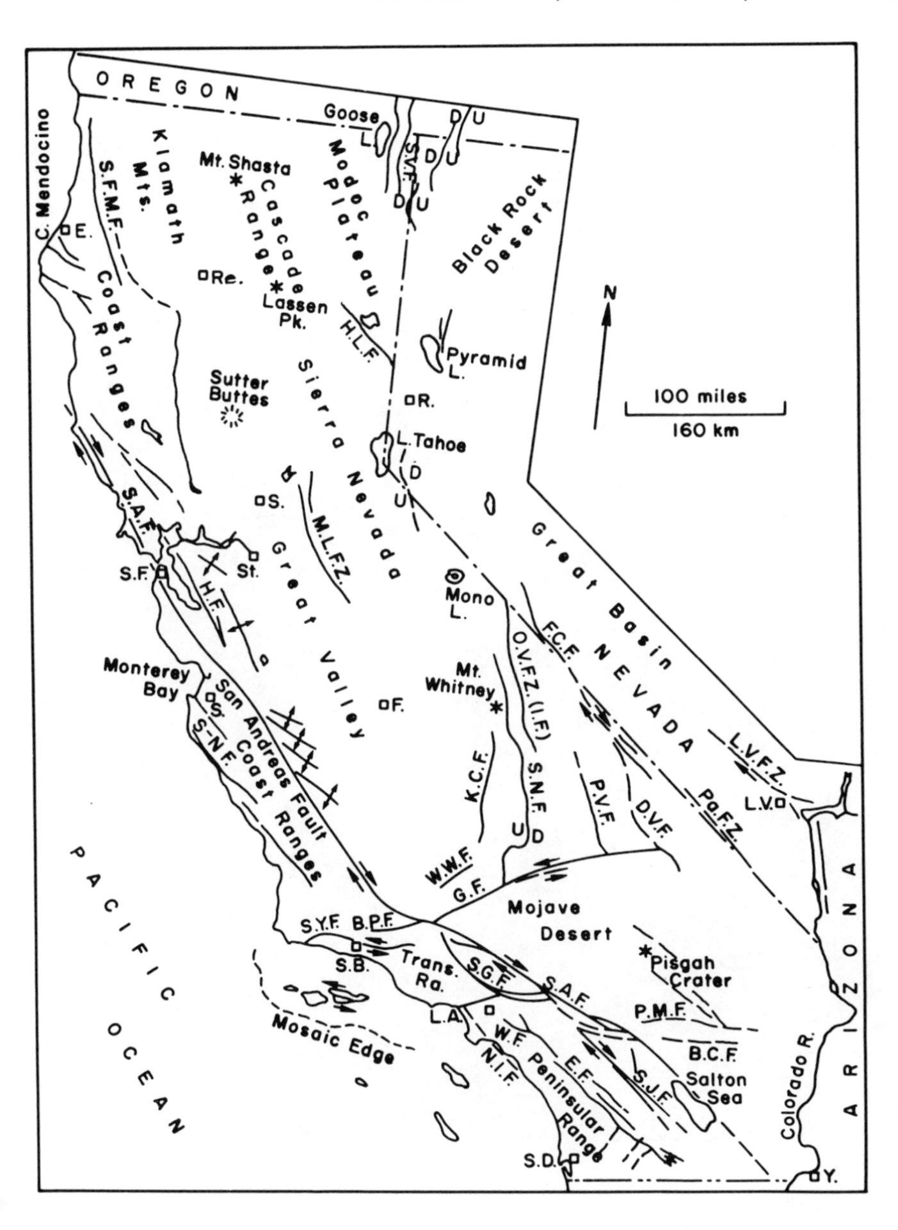

FIG. 30(B)
Geologic structure, California and west Nevada. Based on General Electric Co. LANDSAT-1 mosaic, Paul D. Lowman, Jr., Goddard Space Flight Center, November 1974.

not suited to quantitative study. Maps of landforms, Quaternary deposits, and deposits of glacial origin are relatively uncommon compared to geologic maps and tectonic maps; yet the solution to many of our pressing environmental problems is dependent on the availability of this information. For example, in compiling the new geologic map of Wyoming (J. D. Love, personal communication, 1973), less information was available for the deposits of Quaternary age than for any other unit.

Remote-sensing images can be used in a number of ways to construct useful maps of surficial features and deposits. Apollo and Gemini photographs taken at oblique angles and under different sun-azimuth and sun-angle conditions provide a useful perspective for landform analyses. The same is true for hand-held photographs taken by astronauts on the Skylab missions (Fig. 31). Several adjacent physiographic provinces can be inspected in a single view. The oblique photographs may substitute for or supplement quasi-relief maps.

Depending on the side lap of adjacent image strips, LANDSAT imagery can be viewed pseudostereoscopically. Also, much Skylab photography and aerial photography has the proper overlap for stereoscopic viewing. In the stereoscopic presentation, landforms

FIG. 31
View across southern Wyoming and Colorado taken by a Skylab-2 crewman with a 35-mm hand-held camera. The Colorado Front Range is seen in the center of the picture, the Medicine Bow Mountains and Laramie Basin in the right foreground, and the Denver Basin (Platte River Valley) in the left foreground. The San Luis Valley and San Juan Mountains can be seen in the distance (top center).

are seen in greater detail than in the nonstereoscopic photographs. LANDSAT images have the additional advantage of worldwide seasonal coverage, which allows the geomorphologist to take advantage of vegetative or snow-cover enhancement of landforms as seen at different times of the year (Fig. 32). Skylab photographs also offer this seasonal comparison capability for a few areas. Repetitive aircraft coverage can be flown as required, but the costs can become prohibitive.

It is now plausible to consider classification schemes tailored to regional representation of landforms as evident on small-scale imagery. With selection of suitable symbols, maps showing the distribution of all major and many smaller landform types can be generated for political units (states or countries), entire continents, or the whole world (Short et al., 1972). Such maps will provide new insights into regional geomorphology and into the many processes that shape and alter the land surface.

Aerial photography has been used quite successfully for mapping glaciated terrain, but the use of space imagery for mapping either Quaternary deposits or deposits of glacial origin has been limited to a few areas. One excellent example of the use of LANDSAT imagery in mapping glacial deposits is that of Breckenridge (1973), who made a regional map of glacial deposits of western Wyoming. Using LANDSAT imagery, he was able to map major glacial features, flow directions, and the maximum extent of the Wisconsin stage (Fig. 33). Studies of active glaciers in Alaska and the western United States with LANDSAT-1 imagery yields new information on glacial processes and helps establish relationships between glacial deposition or erosion and the climate or other controlling factors.

McKee's study of sand dunes (McKee et al., 1973) is one of the better examples of mapping Quaternary deposits with space imagery (Fig. 34). McKee has selected a number of test areas around the world in which he finds large areas of aeolian features (sand seas). These features have a characteristic morphology that reflects the interaction of climate with the surface materials. Such studies should eventually provide a basis for better understanding of aeolian deposition and allow for more thorough interpretation of the aeolian formations that occur in the geologic record.

ENVIRONMENTAL AND ENGINEERING GEOLOGY

Both LANDSAT and Skylab have produced excellent images showing sedimentation patterns in the open oceans or restricted embayments where rivers carry their loads into marine waters (Fig. 35). Sedimentologists are afforded an opportunity to observe modern-day offshore transportation and deposition processes at a viewing scale never before available. Seasonal and long-term changes can be monitored using the LANDSAT repetitive system. Development of new sedimentation models will have obvious practical applications to harbor and shoreline maintenance, ocean waste disposal, marine pollution, and fisheries. When detailed information is needed, aircraft data can supplement satellite data in sedimentological studies.

Thermal scanners have been used to monitor and map thermal pollution (Ruggles, 1969) and to locate warm or cold springs discharging into lakes or the sea (Lee, 1969). The Skylab S-192 scanner, which has a thermal infrared band, will undoubtedly prove helpful in regional studies of the thermal regime of water bodies.

Studies of active glaciers monitored from space should give new impetus to developing hypotheses on the growth and recession of ice sheets and on the likelihood of renewed continental glaciation. Surging glaciers have already been observed from LANDSAT (Fig. 36). Vast regions of permafrost can be inspected for evidence of activities detrimental to the environment or special conditions that could influence engineering activities (Fig. 37). Areas influenced by piedmont or mountain glaciers may reveal new relations between the present ice bodies and the outwash and depositional plains beyond (Fig. 38). LANDSAT images of the Northern Hemisphere, where previous advances of continental glaciers have greatly modified the surface terrain, are disclosing new patterns of morainal deposits, buried topography, and drainage shifts. This new knowledge is of great significance to groundwater hydrologists, construction (foundation) engineers, builders seeking gravel, clays, and other

FIG. 32
LANDSAT-1 image (1171-17072) of northwest Nebraska and southwest South Dakota taken in January 1973. The snow cover enhances the geomorphic pattern of drainages and dune fields.

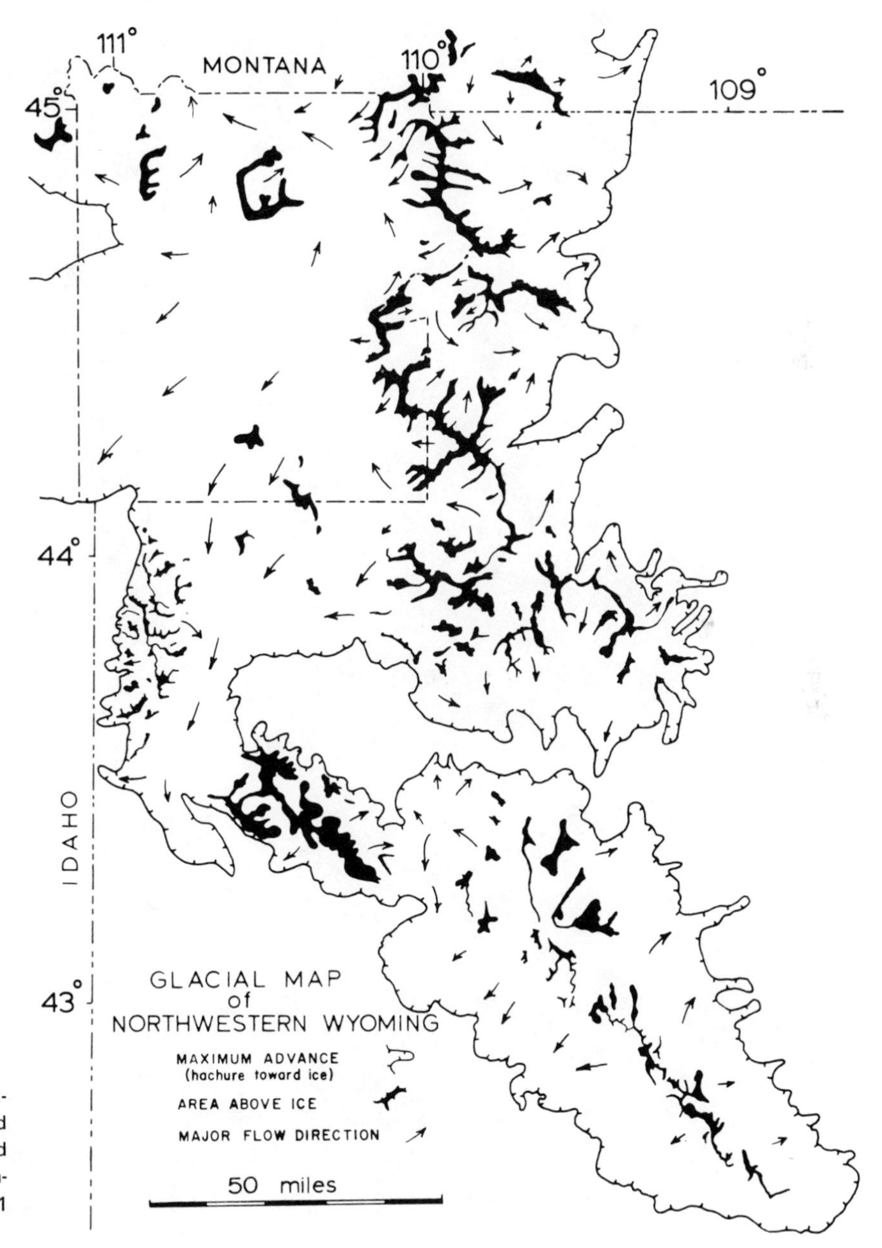

FIG. 33
Map showing the maximum extent of Wisconsin glaciation and ice flow directions, compiled from interpretations of glacial features detected on LANDSAT-1 imagery.

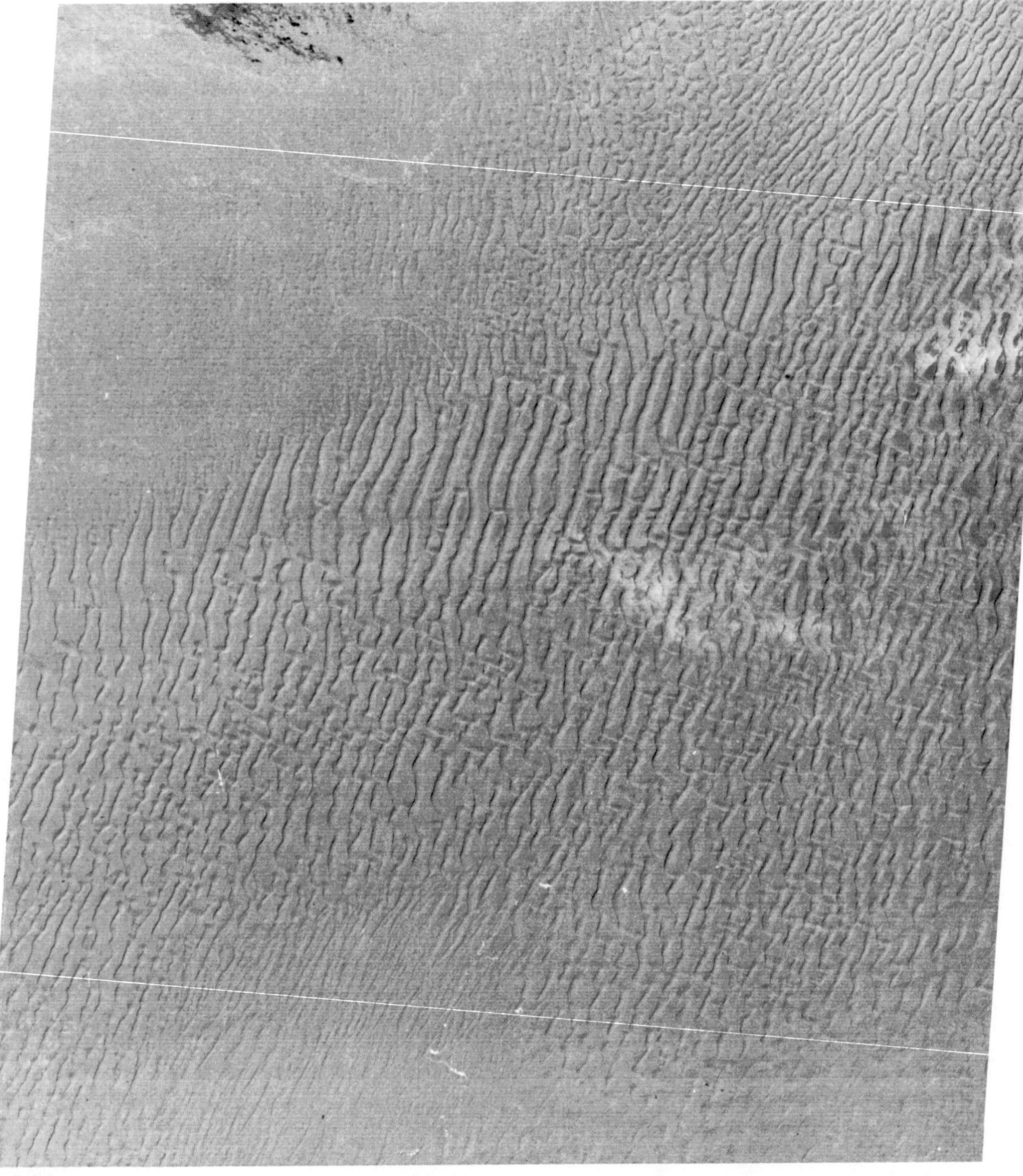

FIG. 34
Distinctive dunes in a "sand sea" in the Takla Makau Desert of western China in an LANDSAT image taken in November 1972.

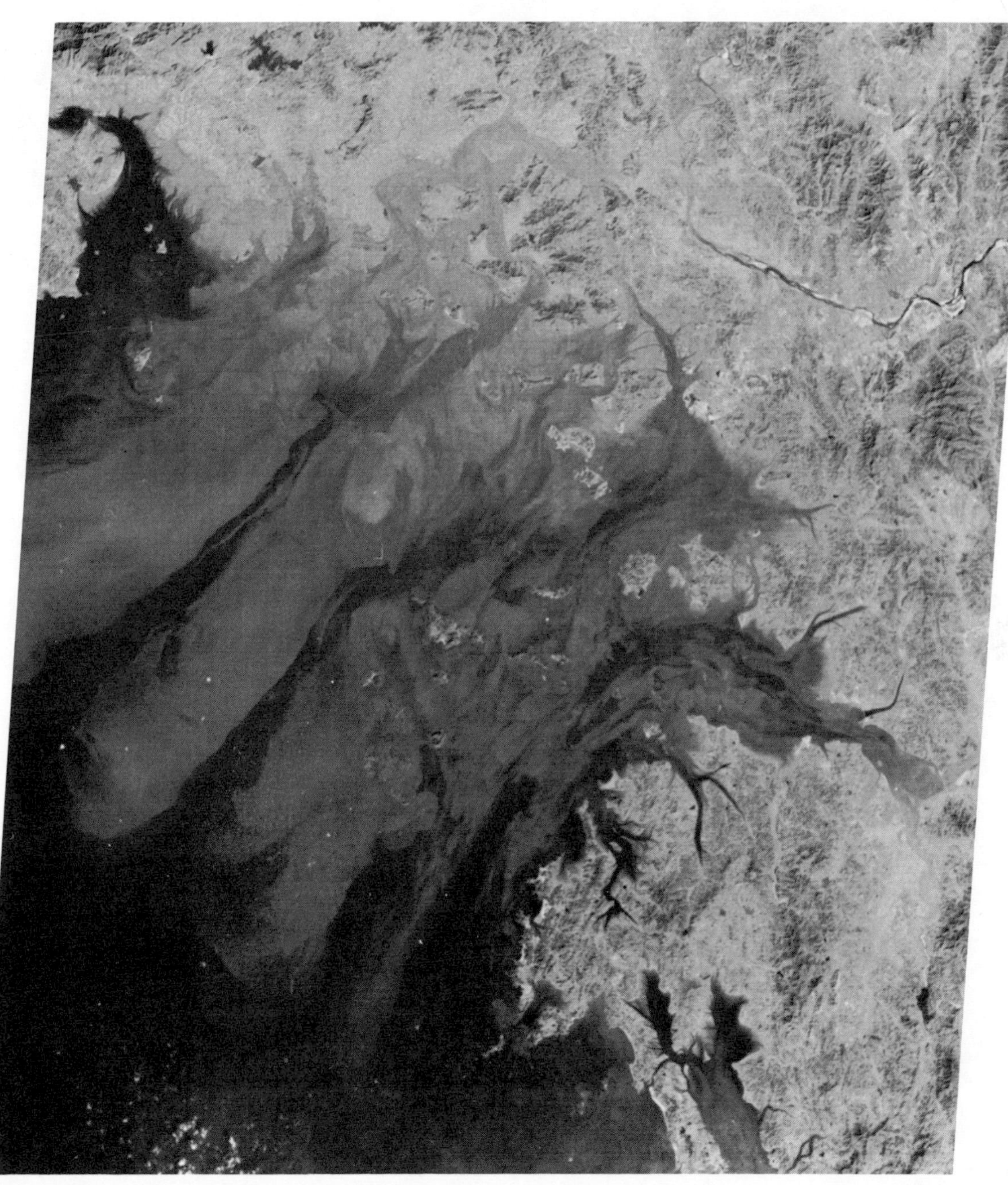

FIG. 35
Sediments from the Han River northwest of Seoul, Korea, being carried into the Yellow Sea, as seen from LANDSAT.

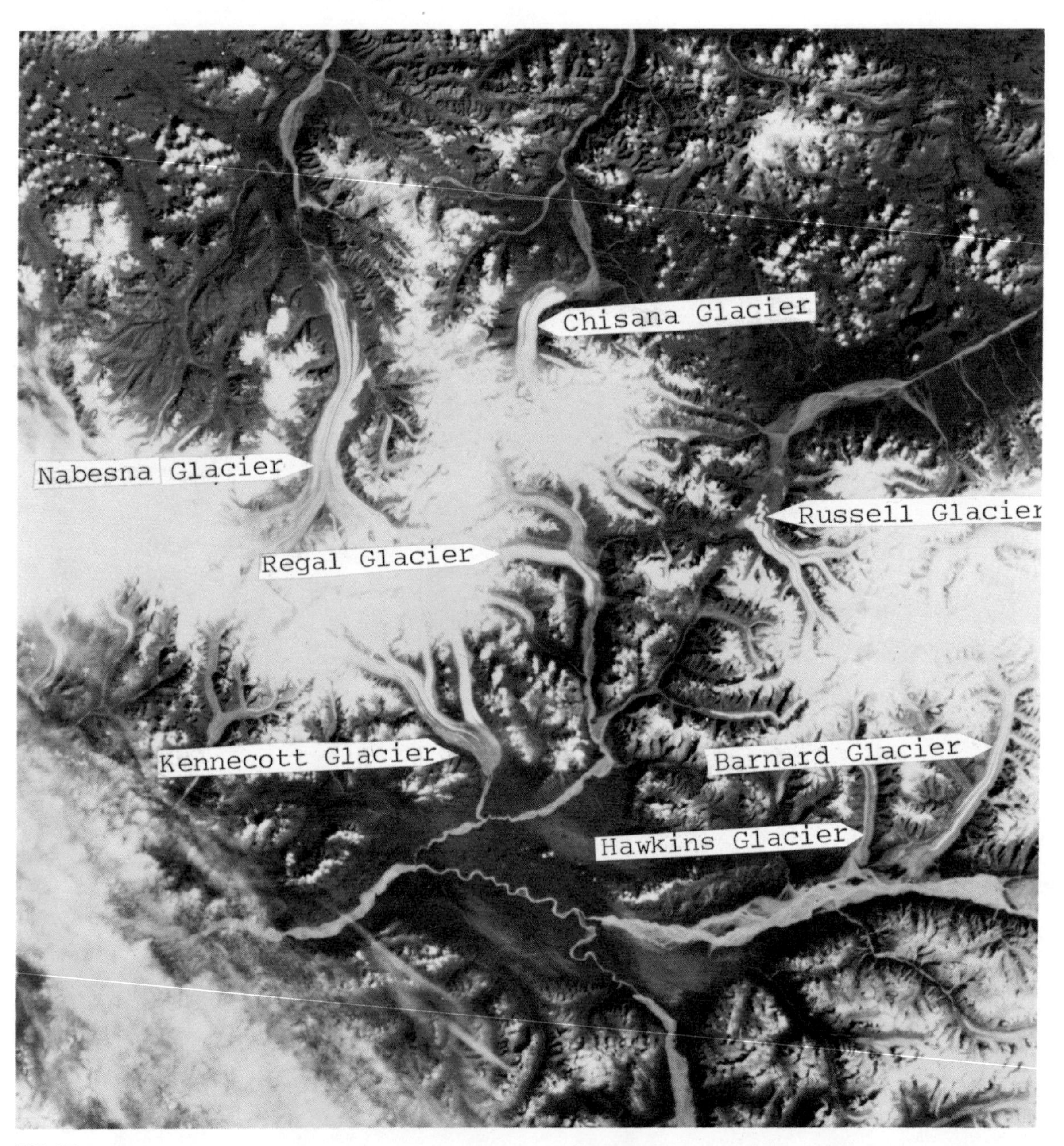

FIG. 36

Image (LANDSAT 1026-20220) showing a portion of eastern Alaska, including the Wrangell Mountains. Of the many glaciers in this area, at least one (Russell Glacier) exhibits the crenulated morainal pattern typical of surging glaciers.

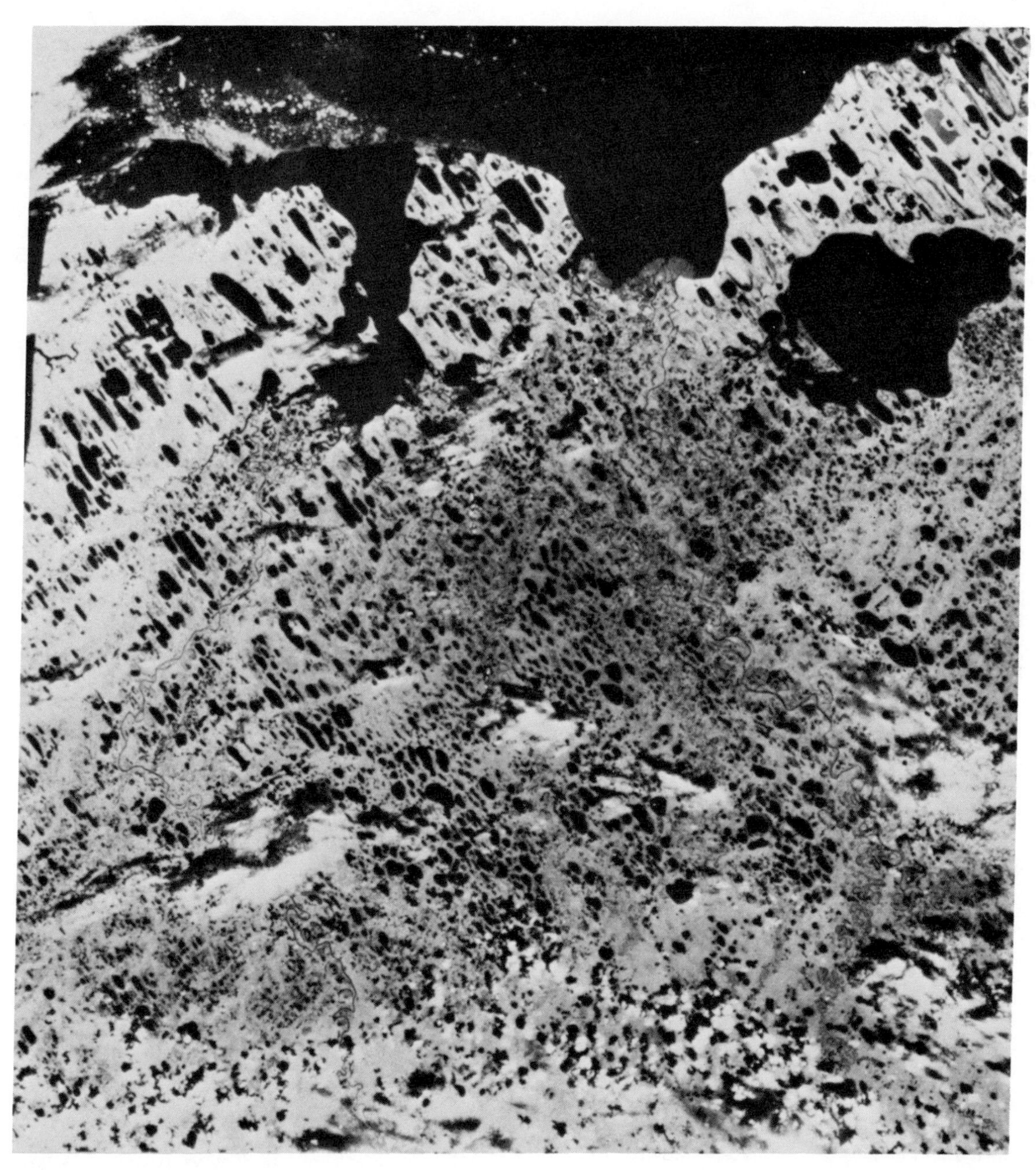

FIG. 37
Aligned glacial lakes south of Point Barrow, Alaska, as imaged by LANDSAT (1006-21510). At least some of these lakes may be thermokarst depressions related to permafrost effects.

raw materials, land-use planners, those concerned with recreation, and farmers. The year-round surveillance of sea ice in the polar regions (Fig. 39) provides a continual record of the status of these frozen wastes to meteorologists concerned with weather-making conditions and shippers seeking new routes through the Arctic.

Geomorphologists use satellite imagery in studying the size, shape, and development of large sand bodies, which may have ultimate value in understanding petroleum reservoir rocks. These same images are useful to environmental geologists or engineering geologists who may simply seek a regional map of sand deposits for land-use purposes or for construction material. The LANDSAT system allows the observer to monitor individual or grouped sand deposits as they move, to learn about the dynamics of aeolian processes, and to predict encroachments on developed land. Regional mapping of sand seas and stabilized dune fields also gives clues to wind condition through recent history and may be useful in siting wind-powered generators.

Information from remote-sensing platforms is particularly useful as a tool of the environmental geologist. This fact is recognized by some schools that combine programs in environmental geology and remote sensing. Modern environmental study programs combine the talents of geologists, hydrologists, soil scientists, ecologists, plant scientists, planners, and others concerned with environmental problems. One example of a major environmental problem in the United States is the impact of the development of the vast coal resources of the western United States. Large areas underlain by coal will be strip mined, and, in addition to environmental problems associated with mining and transportion of coal, the likely development of coal-fired power plants, coal gasification plants, and coal liquefaction plants further compounds the environmental problems. Fortunately, these areas were covered by LANDSAT and Skylab sensors prior to the period of major development so that an environmental base line is established for monitoring and change detection. Preliminary studies suggest that strip-mining activities can be detected easily and monitored by use of LANDSAT or manned spacecraft (Fig. 40). Regional vegetation maps can be prepared from space imagery, information can be obtained on range quality, maps can be made of floodplains that show natural vegetation versus irrigated farmland, land-use maps can be made, and maps of regional slopes can be prepared by interpretation of drainage pattern and texture. The LANDSAT system is ideal for regional studies of this type because many tasks are dependent on seasonal coverage. Skylab provides stereoscopic coverage and additional detail, but does not give repetitive coverage. Enhancement techniques are useful in some environmental studies. Certain tasks can be expedited by computer mapping techniques that improve resolution or make automatic classifications (Sellman, 1973). Land-use mapping of small urban areas is best accomplished with imagery showing the area after a light snow (Fig. 41). This shows the clear advantage of seasonal coverage. Skylab has the advantage of higher resolution in both S-190A and S-190B sensors and stereoscopic coverage (useful in some studies such as regional slope estimation and classification), but coverage is limited. Perhaps the main value of Skylab photographic sensors is to illustrate that the resolution of the S-190B sensor (Fig. 42) is adequate for virtually all regional mapping tasks. Virtually all needs of environmental mapping and monitoring teams could be met with a repetitive spacecraft system of Skylab resolution. Whether such a system would have an overall cost advantage remains to be seen, but there is no doubt that existing space imagery is far less costly to the user than comparable coverage with aircraft.

MINERAL EXPLORATION

Mineral exploration from remote-sensing platforms may be either direct (tailored to the detection of a

FIG. 38 (*right*)
Glacial-scour-produced "Finger Lakes" in the Kilbuck Mountains of southwest Alaska. This LANDSAT MSS band 7 image enhances differences in the outwash plains drained by the Nuyakuk River. The long linear valley west of the lakes is the westernmost extension of an arcuate strike-slip fault that continues across Alaska into Canada.

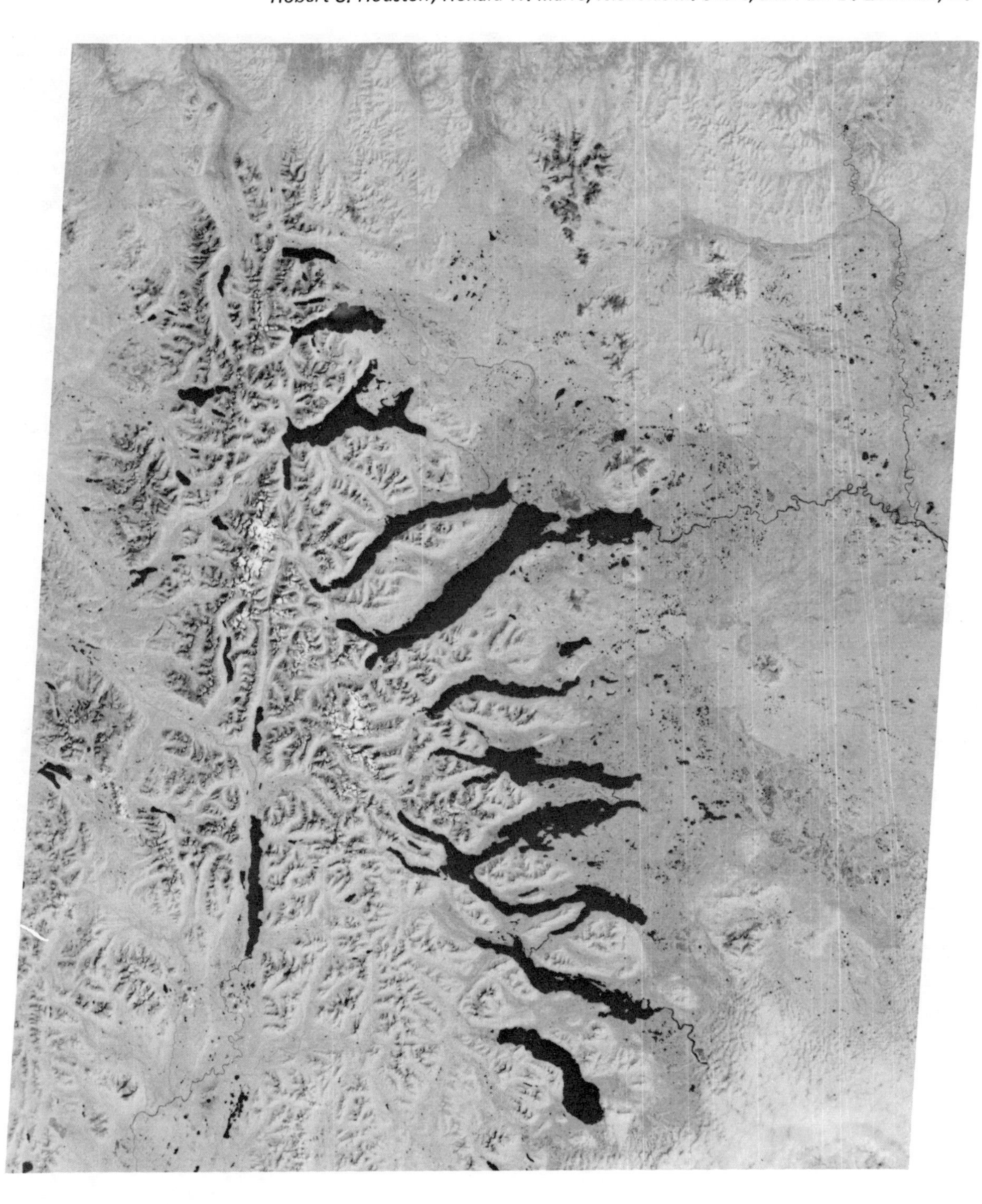

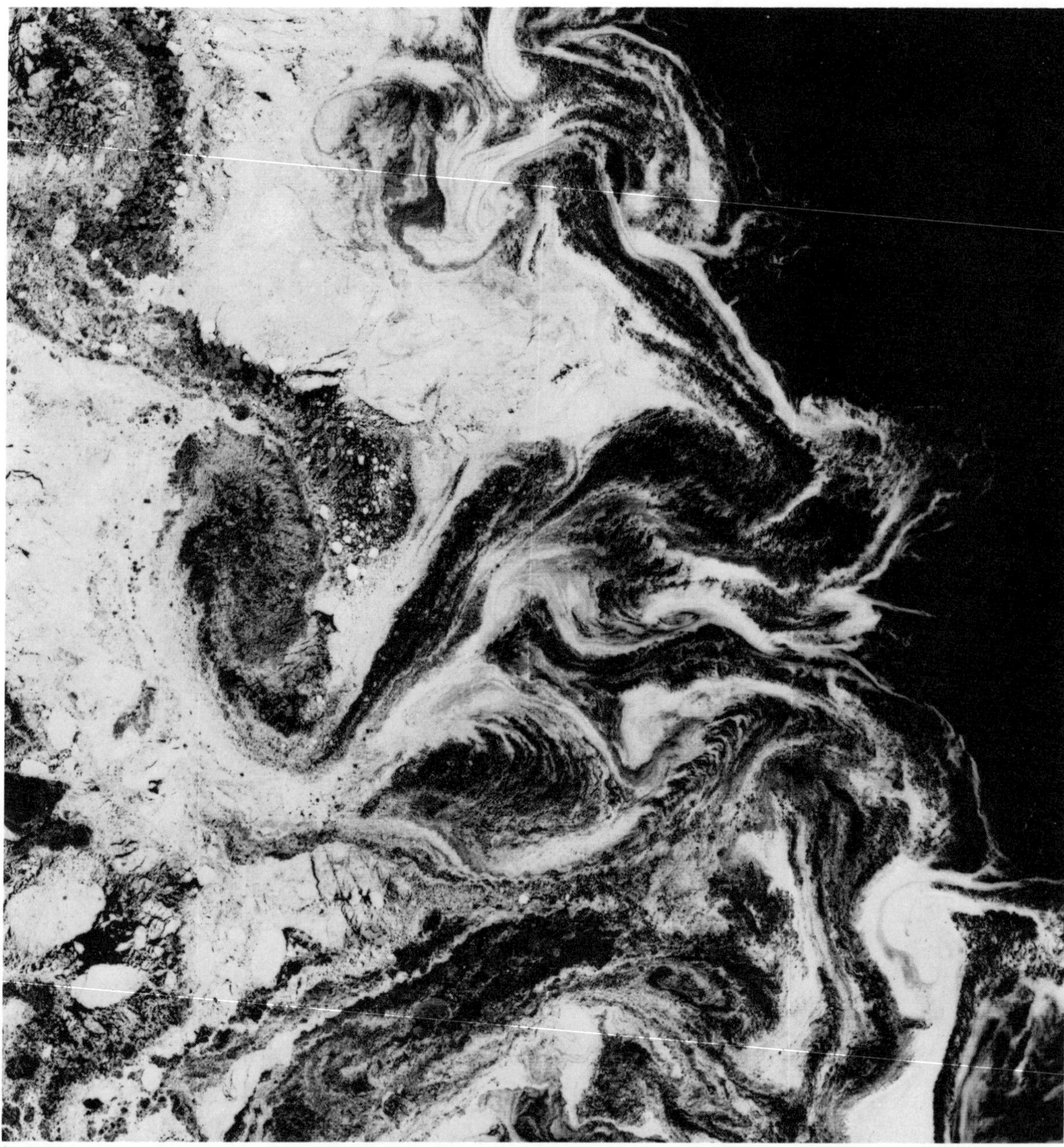

FIG. 39
Sea ice forming in the Greenland sea just off the coast of King Christian Land in eastern Greenland; image taken by LANDSAT in early October 1972.

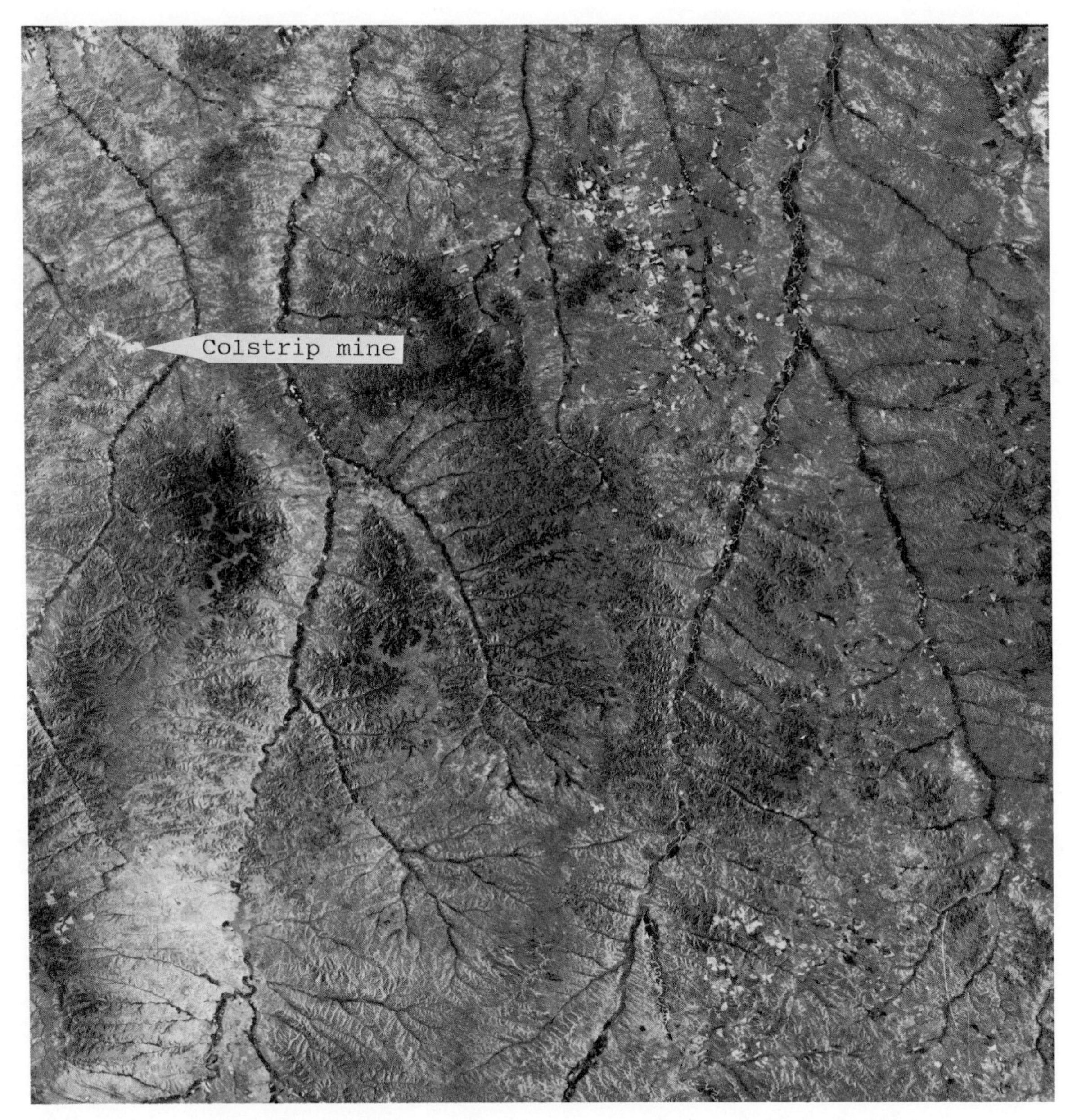

FIG. 40
Skylab photography such as this S-190B coverage of the Powder River Basin, Mont., can be used to map disturbed areas and to monitor mining and reclamation activities.

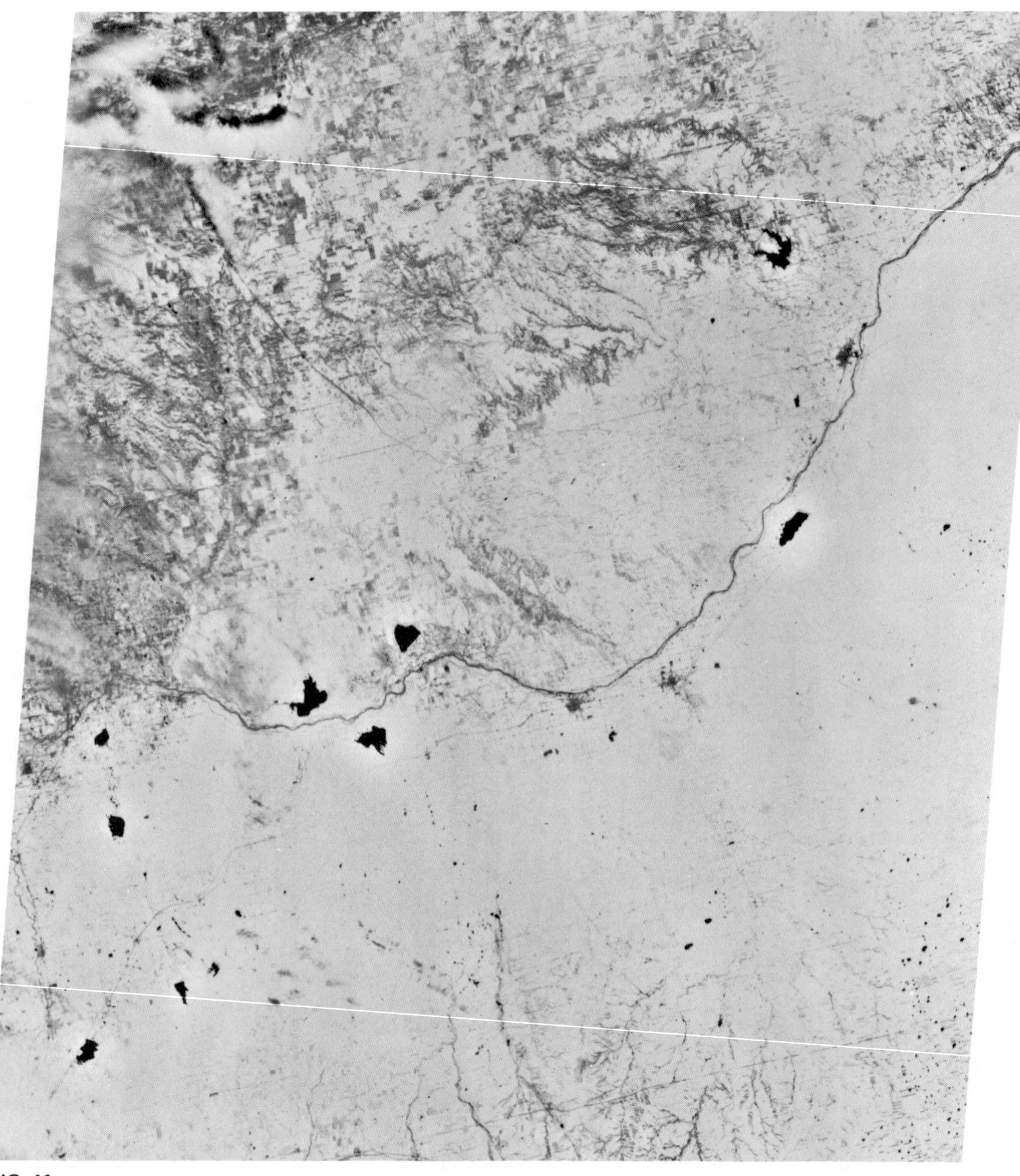

FIG. 41
LANDSAT-1 band 7 image (1261-17085) of the Platte River Valley in northeast Colorado. Small towns and transportation routes contrast sharply with surrounding countryside, which is shrouded in snow.

FIG. 42
Skylab S-190B photograph of western South Dakota showing the obvious potential of space imagery for land-use studies. Fields, forests, and rangeland are easily discerned, and even the street pattern in Rapid City (bottom center) is apparent.

property that is diagnostic of a given mineral resource) or indirect (geologic information is acquired by remote-sensing techniques and used to develop a mineral exploration program). Direct exploration techniques include mapping with an airborne magnetometer to search for magnetic minerals or sulfides associated with magnetic minerals, mapping with an airborne gamma-ray spectrometer to search for radioactive ores, and mapping by use of electrical prospecting techniques, which detect or induce and record eddy currents in conductive ore bodies. These techniques have been used in mineral exploration for several decades and can be considered standard airborne exploration tools. The literature on these methods is extensive (Keller, 1966; Adams, 1970; Nettleton, 1971) and well known to exploration geologists; it will not be reviewed here. Other direct methods include optical remote-sensing systems for use in remote detection of vapor and fluorescence, and techniques to identify rocks by use of the infrared emission spectra. Optical systems developed for detection of air pollutants have been successful in detecting iodine vapor, a halogen that may be emitted from some ore deposits (Barringer, 1969). Some ore minerals, asphaltites, and organic compounds are fluorescent, and optical techniques have been developed by the U.S. Geological Survey (Hemphill, 1968) that may allow detection of these materials from remote-sensing platforms.

A system previously discussed is that developed by Lyon and Patterson (1969) for rock identification, which uses an aircraft-mounted spectrometer and radiometer to record the emission spectra of rocks in the far infrared. This system takes advantage of the shift in the silica absorption band with silica content in the far infrared. Characteristic spectra of reference samples are compared with those of the target or unknown as determined by the spectrometer. Under optimum conditions, rock types and contacts can be mapped along the path of the spectrometer. As suggested earlier, a system of this type may not have general applicability to geologic mapping because of the large number of variables that affect rock spectra, and because rock types and contacts are mapped only along the line path of the spectrometer. The system might be tailored to a specific area where distinctions among exposed rocks are important in exploration. The major question is whether or not such a system has cost benefits over standard photographic techniques. These techniques, unlike the standard geophysical techniques, are not proven exploration methods, but they have potential and their development should be monitored by exploration geologists.

Associated with many ore deposits are alteration haloes that are much larger than the deposit itself. In some cases these haloes show a marked color anomaly or an anomalous concentration of elements that may be detected by remote exploration techniques. Color anomalies may be detected by interpretation of color or color-infrared photography or by use of false-color images produced by band combination of multispectral data. A typical alteration halo may have a large area of reddish discoloration that results from oxidation of iron. Depending on the size of the altered area, it may be detected by use of aerial photography or Skylab color and color-infrared photographs and LANDSAT color-infrared images. The color shift from shades of red in color positives (Fig. 43) to shades of yellow in color-infrared positives (Fig. 44) is unfamiliar to many geologists, but once the observer becomes familiar with the color shift, the haze penetration of color infrared makes these images better for color anomaly detection.

Among the most promising remote-sensing exploration methods are techniques that have been developed to enhance or distinguish these altered zones from other objects on a photograph or image. Where ferric or ferrous iron are abundant in altered zones, it may be possible to take advantage of iron absorption bands in the near infrared to distinguish iron-rich and iron-poor rocks. This can be done by ratioing procedures developed by Rowan (1973), Vincent (1973a), and Goetz and Billingsley (1973). Goetz of the Jet Propulsion Laboratory and Rowan of the U.S. Geological Survey have made reflectance measurements on limonite and have concluded that this mineral has a spectral response different from most other common minerals. Although LANDSAT bands are not ideal for enhancement of this mineral, these investigators have found that the spectral response of limonite is enhanced by composites of specific band ratios (i.e., LANDSAT MSS bands 5 to 4, 4 to 6, 7 to 6). Each ratio represents a new image, which, like individual MSS bands, can be used to construct color composite images. In this manner, Goetz and Rowan produce false-color composites by the combining of computer-generated ratio images (Fig. 45). The method has been tested at Goldfield, Nevada, and the enhanced color anomalies were verified as gossans (Rowan, 1973). The method did not produce a completely unique enhancement because one other nongossan rock unit was also enhanced. The false-color ratio composite represents a marked improvement over the standard infrared color composite; further experimentation may prove this to be a major breakthrough in direct remote sensing of altered areas.

Indirect exploration may involve any use of remote-sensing techniques that aid the explorationist and lead to a mineral discovery. The development of spacecraft-mounted sensors adds a new dimension to exploration geology, just as it does to other geologic studies. The exploration geologist may use available imagery in stages—the multilevel sensing concept. For example, he may recognize major linear features of possible economic interest using LANDSAT, Skylab, Gemini, and/or Apollo photographs and then examine key areas with aircraft images and ground surveys. He may find major color anomalies with standard or enhanced space imagery and verify or disprove their value with aircraft and ground surveys. Currently, the chief value of space imagery for mineral exploration may be its use as a tool for targeting anomalous areas for examination by other methods of exploration.

One example of the multilevel sensing approach in mineral exploration is the use of space imagery to map greenstone belts in Wyoming, the isolation of particularly interesting units from aircraft images of the greenstone belts, and the discovery of iron-rich formations by field check of these units. Another example of this approach might be the recognition of mafic and ultramafic complexes by interpretation of color-infrared satellite images, isolation of specific prospects by use of aircraft-mounted sensors, and ground verification or rejection.

It is reasonable to suppose that new oil prospects may be recognized in unmapped areas through a subtle expression of structural features as seen from the space perspective. These may also be verified by aircraft survey and ground study.

One of the most significant indirect exploration tools for some areas may be radar imagery. The ability to penetrate clouds in areas of almost continuous cloud cover allows delineation of topography and some structure. Such information has obvious utility for the explorationist.

Vegetation stress is readily detected using infrared images because of the reduced reflectance of stressed or diseased vegetation in the near-infrared range. A multilevel approach may also be useful here, since abnormally high concentrations of some metals are known to cause plant stress (Cannon, 1964).

Successful exploration programs are often maintained through careful use of complementary exploration tools. Combinations of several remote-sensing techniques, or the use of all information currently available from different sensing levels, may be the key to the solution of many exploration problems.

HYDROLOGY

Surface-water resources can be studied, mapped, and monitored by a number of remote-sensing techniques, and information on groundwater resources may be derived indirectly. Mapping of impounded surface water is straightforward. Infrared images are particularly well suited for this task because they show water bodies and water-saturated soils contrasting sharply with other materials. This striking contrast is the result of the strong attenuation of near-infrared energy by water. The repetitive coverage provided by LANDSAT is exceptionally well adapted to mapping and monitoring surface-water resources. Although the system has resolution limitations for some studies, users have reported that water bodies as small as 2 acres can be recognized (Work and Gilmer, 1973). Additional detail of drainage patterns may be obtained from Skylab S-190A and S-190B photographs or infrared aerial photography.

Snow-line information aids in estimating snowpack in mountain areas. Repetitive coverage is essential in monitoring the changing snow lines; again, the LANDSAT system has a considerable advantage. Some confusion may occur in mapping snow lines because of similar appearance of snow and cloud cover. This can be avoided by pseudostereoscopic viewing in areas where side lap is adequate. If the snow-line information can be successfully correlated with water content in storage areas, spacecraft may augment or substitute for ground surveys that are used to predict water available for agricultural, municipal, and industrial use.

LANDSAT also provides a tool to monitor floodwaters and related problems (Hallberg et al., 1973). With timely coverage, water damage associated with major storms can be assessed and recovery monitored.

The long-term monitoring of surface-water resources by spacecraft and correlation with climatic factors may ultimately enable scientists to predict the effect of climatic changes on the surface-water regime. For example, storage of water in lakes and reservoirs may be assessed over large areas, and predictions of actual storage capacity and supply can be based on complete, factual information rather than on small samples. Predictions can be refined as monitoring continues through several seasons.

Direct sensing of groundwater resources is generally impractical except for shallow aquifer surveys using induced polarization systems. But where groundwater is fault or fracture controlled (Walker, 1972), the imaging systems can be effective exploration tools. For example, in the Powder River Basin of Wyoming, the major, deep aquifer is the Madison Limestone. The outcrop of this aquifer exhibits a paleokarst topography that can be recognized on space photographs and aerial photographs. The Madison Formation is also cut by major fracture systems that can be recognized on these photographs. Initial studies suggest that the groundwater in the Madison aquifer is fault controlled; thus, space and aircraft photographs may be a major aid in locating the principal productive horizons. Another study of fracture control of aquifers is that of Goetz et al. (1973) who have used LANDSAT imagery to identify and prospect for fault-controlled aquifers in the Grand Canyon area of Arizona.

CONCLUSIONS

Aircraft have been successfully used as sensor platforms for a number of years. Now earth resources satellites provide coverage capabilities far beyond those of aircraft. Development of these new platforms has been paralleled by development of new sensors and new and better interpretive methods. The new sensors can be used in a variety of ways to give the geologist more information or provide him greater flexibility in his analyses.

Synoptic coverage from orbital altitudes is one of the basic advantages of the satellite sensors. It allows the geologist to recognize and interpret interrelationships between regional features. Nonphotographic sensors are particularly helpful in providing information for solution of special problems (e.g., radar imagery for penetration of cloud cover, thermal infrared for temperature anomalies). They allow the geologist to circumvent some limitations of photography while providing new information about the earth's surface.

Much of the power of the new remote-sensing techniques lies in manipulation of the data by special processing or enhancement procedures. Some sensors record data in digital form so that the data are readily processed by computers. Computer handling of the data allows for rapid preprocessing (noise removal, atmospheric correction, etc.) and for application of various combinations or comparison functions. For some applications, the computer can be used to automatically classify surface features and produce maps; but geologic applications have generally proved too complex for automatic classifications. Computer processing methods are viewed as aids to geologic interpretation, which can help the interpreter see subtle variations, make accurate spectral and spatial comparisons, define patterns, or classify certain data sets.

Other remote-sensing instruments record information only in image format, but provide flexibility for analysis by segregating the information into discrete spectral bands (multiband photography, multispectral imagery). The images representing different spectral bands can then be compared or combined through a number of optical techniques to enhance specific contrasts. Other optical techniques improve contrast on specific features (edge enhancement, density analysis, etc.) presented by a single image, and can be used as aids in interpreting standard photography as well as multispectral imagery.

Despite the added capability provided by the new techniques in remote sensing, standard aerial photography remains the most useful single tool for geologic remote-sensing studies. But aerial photography usually contains more information than is realized in the interpretation. The interpretation techniques and enhancement procedures developed for use with the new sensor data can help the geologist glean more information from his aerial photography, his satellite imagery, and from any nonstandard sensor data that he may use in solving special problems.

REFERENCES

Adams, J. A. S., 1970, *Gamma-Ray Spectrometry of Rocks,* Elsevier, New York, 295 p.

Anuta, P. E., S. J. Kristof, D. W. Levandowski, T. L. Phillips, and R. B. MacDonald, 1971, Crop, Soil, and Geological Mapping from Digitized Multispectral Satellite Photography, *7th International Symposium on Remote Sensing of Environment,* v. III, University of Michigan, Ann Arbor, Mich., p. 1985–2016.

Avery, T. E., 1968, *Interpretation of Aerial Photographs,* Burgess, Minneapolis, Minn., 324 p.

Barnes, J. C., and C. J. Bowley, 1973, Use of ERTS Data for Mapping Snow Cover in the Western United States, *Symposium on Significant Results Obtained from ERTS-1,* NASA/Goddard Space Flight Center, New Carrollton, Md., p. 855–862.

Barringer, A. R., 1969, Remote Sensing Techniques for Mineral Discovery, *Proceedings of the 9th Commonwealth Mining and Metallurgical Congress,* Applied Science Publishers, Ltd., Essex, England, p. 649–690.

Benson, C. S., 1973, Snow Cover Surveys in Alaska from ERTS-1 Data, *Symposium on Significant Results Obtained from ERTS-1,* NASA/Goddard Space Flight Center, New Carrollton, Md., p. 1593–1595.

Bjornsen, R. L., 1968, Infrared Mapping of Large Fires, *Proceedings of the 5th Symposium on Remote Sensing of Environment,* Ann Arbor, Mich., p. 459–464.

Blackstone, Donald B., Jr., 1973, Geology of Photo Linear Elements, Great Divide Basin, Wyoming, *NASA Tech. Rept. NASA-CR-133634,* 14 p.

Breckenridge, R. M., 1973, Glaciation of Northwestern Wyoming Interpreted from ERTS-1, *Symposium on Significant Results Obtained from ERTS-1,* NASA/Goddard Space Flight Center, New Carrollton, Md., p. 379–386.

Buckley, B. A., 1971, Computerized Isodensity Mapping, *Photogrammetric Eng.* v. 37, no. 10, p. 1039–1042.

Cannon, H., 1964, Geochemistry of Rocks and Related Soils and Vegetation in the Yellow Cat Area, Grand County, Utah, *U.S. Geol. Surv. Bull. 1176,* 127 p.

Collins, R. J., et al., 1973, An Evaluation of the Suitability of ERTS Data for the Purposes of Petroleum Exploration, *NASA Tech. Rept. NASA-CR-132980;* available from National Technical Information Service, Springfield, Va., Report E73-10646, 19 p.

Dellwig, L. F., 1969, An Evaluation of Multifrequency Radar Imagery of the Pisgah Crater Area, California, *Mod. Geol.* no. 1, p. 65–73.

Goetz, A. F. H., and F. C. Billingsley, 1973, Digital Image Enhancement Techniques Used in Some ERTS Application Problems; 3rd ERTS Symposium, Dec. 10–14, 1973, Washington, D.C., sponsored by NASA/Goddard Space Flight Center, Greenbelt, Md.

_____, F. C. Billingsley, D. P. Elston, I. Lucchitta, and E. M. Shodmaker, 1973, Application of ERTS to Geological Problems on the Colorado Plateau, Arizona. *Proceedings of 3rd ERTS Symposium,* NASA/Goddard Space Flight Center, Greenbelt, Md.

Greene, G. W., R. M. Moxham, and A. H. Harvey, 1969, Aerial Infrared Surveys and Borehole Temperature Measurements of Coal Mine Fires in Pennsylvania, *Proceedings of the 6th International Symposium on Remote Sensing of Environment,* Ann Arbor, Mich., p. 517–525.

Hallberg, G. R., B. E. Hoyer, and Albert Rango, 1973, Application of ERTS-1 Imagery to Flood-Inundation Mapping, *Symposium on Significant Results Obtained from ERTS-1,* NASA/Goddard Space Flight Center, Greenbelt, Md., p. 745–754.

Hemphill, W. R., 1968, Remote Detection of Solar Stimulated Luminescence, paper presented at 19th Congress of International Astronautical Federation, New York, no. AS-156, 7 p.

Houston, R. S., F. W. Zochol, and S. B. Smithson, 1974, ERTS-1 Applied to Geologic Mapping in the Dry Valleys, *Antactic J.,* v. 9, no. 3, p. 68–76.

Keller, G. V., 1966, *Electrical Methods in Geophysical Prospecting,* Pergamon, New York, 519 p.

Kinney, D. M., 1970, Mapping, *in* Earth Science in 1969, *Geotimes,* v. 15, no. 1, p. 12–23.

Kreitzer, M. H., 1974, Direct Additive Printing; *Photogrammetric Eng.,* v. 40, no. 3, p. 281–285.

Lee, Keenan, 1969, Infrared Exploration for Shoreline Springs at Mono Lake, California, Test Site, *Proceedings of the 6th International Symposium on Remote Sensing of Environment,* Ann Arbor, Mich., p. 1075–1100.

Love, D., et al., 1955, Geologic Map of Wyoming, Department of Geology, University of Wyoming.

Lowman, P. D. Jr., 1974, Geologic Structures in California, NASA/Goddard Space Flight Center, Greenbelt, Md.

____, and Tiedemann, H. A., 1971, Terrain Photography from Gemini Spacecraft, *Final Geologic Report, X-644-71-15,* NASA/Goddard Space Flight Center, Greenbelt, Md., 75 p.

Lyon, R. J. P., 1970, The Multiband Approach to Geologic Mapping from Orbiting Satellites: Is It Redundant or Vital? *Remote Sensing Environ.,* v. 1, no. 4, p. 237–244.

____, and J. W. Patterson, 1966, Infrared Spectral Signatures—a Field Geological Tool, *Proceedings of the 4th Symposium on Remote Sensing of Environment,* Ann Arbor, Mich., p. 215–230.

____, and J. W. Patterson, 1969, Airborne Geological Mapping Infrared Emission Spectra, *6th International Symposium on Remote Sensing of Environment,* v. 1, University of Michigan, Ann Arbor, Mich., p. 527–552.

McKee, E. D., C. S. Breed, and L. F. Harris, 1973, A Study of Morphology, Provenancy and Movement of Desert Sand Seas in Africa, Asia, and Australia, *Symposium on Significant Results Obtained from ERTS-1,* NASA/Goddard Space Flight Center, New Carrollton, Md., p. 291–304.

McLerran, J. H., and J. O. Morgan, 1964, Thermal Mapping of Yellowstone National Park, *Proceedings of the 3rd Symposium on Remote Sensing of Environment,* Ann Arbor, Mich., p. 517–530.

Merifield, P. M., 1964, Some Aspects of Hyperaltitude Photography, Report no. 1, contract NAS5-3390, Lockheed-California Company, Burbank, Calif.

Mier, M. F., 1973, Evaluation of ERTS Imagery for Mapping and Detection of Changes of Snowcover on Land and on Glaciers, *Symposium on Significant Results Obtained from ERTS-1,* NASA/Goddard Space Flight Center, Greenbelt, Md., p. 863–875.

Miller, L. D., 1966, Location of Anomalously Hot Earth with Infrared Imagery in Yellowstone National Park, *Proceedings of the 4th Symposium on Remote Sensing of Environment,* Ann Arbor, Mich., p. 751–769.

Moxham, R. M., and A. Alcaraz, 1966, Infrared Surveys of Taal Volcano, Philippines, *Proceedings of the 4th Symposium on Remote Sensing of Environment,* Ann Arbor, Mich., p. 827–843.

National Aeronautics and Space Administration, Principal Investigations Management Office, 1973, *EREP Investigator's Information Book,* NASA/Johnson Space Center, Houston, Tex.

Nettleton, L. L., 1971, *Elementary Gravity and Magnetics for Geologists and Seismologists,* Society of Exploratory Geophysicists, Tulsa, Okla., 121 p.

Parker, R. B., 1972, An Evaluation of ERTS-1 Imagery for Mapping of Major Earth Fractures and Related Features, Report E72-10349; available from National Technical Information Service, Springfield, Va., 7 p.

Pecora, W. T., 1967, Surveying the Earth's Resources from Space, Proceedings 27th Annual Meeting, American Congress on Surveying and Mapping, Washington, D.C., 7 p.

Poulton, C. E., 1972, The Advantages of Side-Lap Stereo Interpretation of ERTS-1 Imagery in Northern Latitudes, *in* W. A. Finch, ed., 1973,

Proceedings of 1st ERTS-1 Symposium, NASA/Goddard Space Flight Center, Greenbelt, Md., p. 157–161.

Raines, G. L., and K. Lee, 1974, Spectral Reflectance Measurements, *Photogrammetric Eng.,* v. 40, no. 5, p. 547–550.

Ross, D. S., 1973, Simple Multispectral Photography and Additive Color Viewing, *Photogrammetric Eng.,* v. 39, no. 6, p. 583–591.

Rowan, L. C., 1972, Near-Infrared Iron Absorption Bands: Application to Geologic Mapping and Mineral Exploration, *4th Annual Earth Resources Program Review,* NASA, Manned Spacecraft Center, Houston, Tex., p. 60-1 to 60-18.

____, 1973, Iron-Absorption Band Analysis for the Discrimination of Iron-Rich Zones, *NASA Technical Report NASA-CR-130788;* available from National Technical Information Service, Springfield, Va., Report E73-10338, 12 p.

____, et al., 1973, Mapping of Hydrothermal Alteration Zones and Regional Rock Types Using Computer Enhanced ERTS MSS Images, Third ERTS Symposium, Dec. 10–14, 1973, Washington, D.C., NASA/Goddard Space Flight Center, Greenbelt, Md.

Ruggles, F. H., 1969, A Thermal Study of the Connecticut River Estuary, *Second Annual Earth Resources Aircraft Program Status Review,* NASA, Manned Spacecraft Center, Houston, Tex., p. 41-1 through 41-12.

Sabatini, R. R., Rabchevsky, G. A., and Sissala, J. E., 1971, *Nimbus Earth Resources Observations,* Technical Report 11, 9690-8, Prepared for NASA/Goddard Space Flight Center, Greenbelt, Md., 256 p.

Schlosser, M. S., 1974, Television Scanning Densitometer, *Photogrammetric Eng.,* v. 40, no. 2, p. 199–202.

Sellman, Buzz, 1973, Land Resources Survey for the State of Michigan, *Symposium on Significant Results Obtained from ERTS-1,* NASA/Goddard Space Flight Center, Greenbelt, Md., p. 1083–1090.

Short, N. M., and P. D. Lowman, 1973, *Earth Observations from Space: Outlook for the Geological Sciences,* NASA/Goddard Space Flight Center, Greenbelt, Md., no. X-650-73-316, 115 p.

____, V. V. Salomanson, and N. H. MacLeod, 1972, Global Mapping of the Earth's Land Surface Using ERTS Data: A Planning Document, NASA/Goddard Space Flight Center, Greenbelt, Md., X-650-72-317.

Smedes, H. W., H. J. Linnerud, L. B. Wodaver, Ming-Yang Su, and R. R. Jauror, 1972, Mapping of Terrain by Computer Clustering Techniques Using Multispectral Scanner Data and Using Color Aerial Film, *4th Annual Earth Resources Program Review,* vol. III, NASA, Manned Spacecraft Center, no. 05937, Houston, Tex., p. 61-1 to 61-30.

Specht, H. R., D. Needler, and N. L. Fritz, 1973, New Color Film for Water Penetration, *Photogrammetric Eng.,* v. 39, no. 4, p. 359–369.

Vincent, R. K., 1973a, Spectral Ratio Imaging Methods for Geological Remote Sensing from Aircraft and Satellites, *in* Abraham Anson, ed., 1973, *Symposium on Management Utilization of Remote Sensing Data,* American Society of Photogrammetry, Sioux Falls, South Dakota, p. 377–397.

____, 1973b, Ratio Maps of Iron Ore Deposits, Atlantic City District, Wyoming (abs.), *Symposium of Significant Results Obtained from ERTS-1,* NASA/Goddard Space Flight Center, Greenbelt, Md., p. 42.

Walker, A. S., 1973, Geologic Evaluation of Remote Sensing Imagery of the Mesabi Range, Minnesota, *Proceedings of the 8th International Symposium on Remote Sensing of Environment,* Ann Arbor, Mich., p. 1137–1146.

Watson, R. D., and L. C. Rowan, 1971, Automated Geologic Mapping Using Rock Reflections, *7th International Symposium on Remote Sensing of Environment,* vol. III, University of Michigan, Ann Arbor, Mich., p. 2043–2053.

Wilderman, W. E., 1972, Preparing Color Enhancements with a Diazo Machine, *in* Colwell et al., 1972, *An Integrated Study of Earch Resources in the State of California Based on ERTS-1 and Supporting Aircraft Data,* National Technical

Information Service, Springfield, Va., Report E73-10027, p. 9.3.

Work, E. A., and D. S. Gilmer, 1973, Preliminary Evaluation of ERTS-1 for Determining Numbers and Distribution of Prairie Ponds and Lakes, *Symposium on Significant Results Obtained from ERTS-1,* NASA/Goddard Space Flight Center, Greenbelt, Md., p. 801–805.

Yost, Edward, and Sandra Wenderoth, 1971, Multispectral Color for Agriculture and Forestry, *Photogrammetric Eng.,* v. 37, no. 6, p. 590–604.

7

EXPLORATION FOR FOSSIL AND NUCLEAR FUELS FROM ORBITAL ALTITUDES

Nicholas M. Short

Nicholas M. Short is with the Earth Resources Branch, NASA, Goddard Space Flight Center, Greenbelt, Maryland.

A pressing need to find new methods and approaches in exploring for oil and gas and other sources of fuels no longer requires demonstration or proof. Any American who experienced interminable waits in long gas lines in the 1973 winter of discontent, or reacts to his ever-rising monthly utility bill, or simply reads the depressing reports of still-impending energy crises in his local newspaper will not require any convincing to become alarmed about the outlook for the future. Many responses to these problems have been advanced—conservation, pressures on our foreign supplies, and accelerated development of solar, aeolian, geothermal, and fusion energy sources lead the list.

An acceleration in exploration is another obvious response. Tried and true techniques will be applied with increased vigor. But innovative and even unconventional techniques must also be devised, tested, and put into operation as soon as they are declared to be practical and productive.

One such novel techniques has moved across the horizon of possibility and looms now as one of the more promising approaches in an expanding search for fuel sources and other raw materials relevant to a rising worldwide demand for more energy. The approach is simply to use orbital space platforms—both automated satellites and manned space vehicles—from which to gain a new perspective of the earth's surface by means of standard and/or specialized remote-sensing techniques. This approach is actually an outgrowth of many years of aerial photography as applied to geologic mapping and mineral exploration. Its important new (although not unique) advantages over the commonly used conventional aerial systems are the following:

1. Greatly expanded synoptic view provided by individual images owing to a ten- to hundredfold increase in operating altitudes coupled with high-resolution imaging sensors.
2. High frequency of (repetitive) coverage resulting from the numerous orbital passes available during spacecraft missions of months' (manned) to years' (satellites) duration.
3. Use of a variety of sensors extending through several regions of the electromagnetic spectrum,

including multispectral imagers, that acquire coherent data simultaneously.

4. Ease with which orbital remote-sensing data can be transmitted or later converted into a digital mode, allowing their further treatment by a wide range of computer-processing techniques involving enhancements, selective information extraction, comparison of repetitive scenes, etc., to be made on a large volume of information.

Practical remote sensing from space platforms began almost with the launch of the first rockets that carried recoverable photographic equipment. Photographic data recording wide areas of the earth's surface continued to be gathered on nearly all manned missions from Mercury and Gemini through Apollo. Weather satellite systems, such as TIROS and Nimbus, returned useful data on the land and sea surfaces in addition to the atmospheric data for which they were designed. Drawing upon years of experience in remote sensing from aircraft, the designers of orbital remote-sensing platforms placed the first "sophisticated" multispectral sensor system on LANDSAT-1, the Earth Resources Technology Satellite launched in mid-1972. Soon thereafter the Skylab manned laboratory began to acquire an even broader range of remotely sensed data, using film cameras, multispectral scanners, an infrared spectrometer, and several radiometers. Future systems now on the "drawing boards" will incorporate this growing wealth of experience with a new generation of varied sensors in space. Thus will emerge ever more versatile research satellites and, in all likelihood, on-line or operational satellites (such as the LANDSAT Follow-on series) and spacelabs (including shuttle-serviced stations).

In keeping with the author's current base of experience, this chapter will be confined almost exclusively to results from the LANDSAT program pertinent to exploration for oil and gas and, very briefly, for uranium deposits. The remainder is divided into three parts: (1) a review of achievements to date in relevant aspects of general geologic studies from LANDSAT, (2) a survey of reported accomplishments oriented specifically toward exploration for energy sources (with emphasis on petroleum), and (3) an evaluation of the prospects and limitations of the space platform approach to fuel exploration.

Before proceeding through these four sections, it is appropriate to summarize as a "preview" the present status of exploration for fossil and nuclear fuels from space. Simply synopsized, it must be stated that, as far as NASA officials can determine, no previously unknown petroleum or uranium deposits have yet been discovered directly (and probably indirectly) from interpretation of images or other forms of remotely sensed data obtained either from satellites or from visual and/or instrument observations made by astronauts. However, "rumors" have reached the author and others of considerable interest in this approach by oil companies and other exploration-minded enterprises. It is in the nature of the petroleum industry to be secretive about successes with new techniques that score in finding oil and gas—witness the delay of several years in making seismic prospecting generally available throughout the industry after a few companies had verified its value. What has been brought to light by LANDSAT in particular that provides invaluable adjunct or supplementary data to the exploration scientists are these accomplishments: (1) updating and refining rock unit boundaries and distributions on small-scale maps, (2) recognition of hitherto unsuspected lineations, especially regional straight to circular or arcuate fracture systems, that could control or influence the location of oil traps or uranium concentrations, (3) better definition of often subtly expressed geomorphic "anomalies" that bear some relation to subsurface structures, and (4) an *apparent* surface manifestation of alterations of rock, soil, or vegetation tied to escaping hydrocarbons or to redistribution of elements associated with shallow uranium deposits. According to the growing convictions of those geologists now working with NASA's Earth Resources programs, it is just a matter of time before an oil field or a nuclear ore deposit is found through the use of significant data obtained from LANDSAT or some other spaceborne sensing system as an integral part of the exploration program.

PROGRAM RESULTS*

General View

The LANDSAT-1 spacecraft was launched on July 23, 1972, from NASA's Western Test Facility in California.† Both the return beam vidicon (RBV) and the multispectral scanner (MSS) began to transmit image data on July 26, 1972. Owing to a switching circuit problem, the RBV has been shut down since early August 1972. As of October 1, 1974, the MSS has imaged more than 100,000 scenes covering greater than 85 percent of the earth's land surface. About 30 percent, on average, of these scenes are largely cloud free and well illuminated. Coverage of all North America has been continuous, but coverage of other continents became severely limited by tape recorder difficulties since March 1973.

The reader is urged to consult the three-volume proceedings of the second (March 5–9, 1973) and third (December 10–14, 1973) *Symposia on Significant Results from the Earth Resources Technology Satellite*‡ for detailed treatment by the investigators, program managers, agency representatives, and invited speakers of the major findings of LANDSAT-1 for the various discipline groups comprising the Earth Resources Program.

Value of Synoptic Coverage

ERTS provides a remarkable sequence of uniformly illuminated, essentially planimetric vertical views of the earth's surface that cover very large areas (~12,500 square miles) in an image obtained from the 570-mile orbital altitude. Like the Gemini and Apollo pictures, these images are invaluable because of their synoptic aspects; that is, a single scene (approximately 115 miles on a side) surveys a wide variety of terrain, geology, and land-use types under near-optimum viewing conditions that emphasize the contextual relations of these surface features. Unlike the Gemini and Apollo pictures, or those from Skylab, the LANDSAT images meet the necessary conditions for being readily joined together in mosaics. This greatly increases the synoptic character of this imagery and extends the assessment of contextual relationships to regional and even subcontinental proportions. The resulting mosaics are surprisingly close to being orthographic, as can be determined by comparing both individual images and composite mosaics to such projections as Albers equal area or Lambert conformal.

LANDSAT-1 has now acquired enough cloud-free imagery to allow assembly of vast areas of earth in mosaics. Some spectacular examples have already been made public. A black and white (red band 5) mosaic of the entire state of Oregon is reproduced here (Fig. 1), and many other states individually as well as the entire continental United States have been similarly mosaicked. Color mosaics are completed for the eastern United States from Maine to Florida, the entire western coastal United States, Florida, Louisiana, Michigan, Wyoming, and Montana (among others), and for several regional sections. Foreign areas available in color now include Italy, parts of west Africa, all of Iran, the Red Sea area, and Yemen.

Beyond the esthetic and technical achievements associated with such mosaics, the scientific merits alone—particularly in geology—are a sufficient justification for their production. Comparison of small-scale geologic or physiographic maps of large areas with LANDSAT mosaics reveals at once the remarkable utility of the latter for presenting regional interrelationships among major structural or land-form units. Such mosaics are especially suited to lineaments analysis where uniform lighting serves to highlight trends of continuous, often deep-seated fracture zones.

*Most of the observations on program results have been extracted from the NASA-Goddard Space Flight Center X-Document X-650-73-316 entitled *Earth Observations from Space: Outlook for the Geologic Sciences,* by N. M. Short and P. D. Lowman, Jr., October 1973.

†A second LANDSAT was successfully launched on January 22, 1975. Both satellites remained operational as of July 1, 1976.

‡Volume I of each symposium proceedings is available from the Government Printing Office; inquiries about availability of Volumes II and III can be made there.

FIG. 1
Uncontrolled photomosaic made from LANDSAT images acquired in 1972 showing the entire state of Oregon and parts of the surrounding states of California, Idaho, and Washington. All images were produced from red band 5. (Courtesy Oregon State University.)

Other Special Advantages of LANDSAT

Geologists experienced in using radar imagery look forward to the day when an active microwave system is flown on a spacecraft. One property of radar, that of cloud penetration, offers a distinct advantage in regions that are habitually overcast. Another important property, at least of some of the microwave bands now used, is the "apparent penetration" of heavy foliage that allows radar to "see" the terrain beneath an extensive tree canopy. It is surprising to note that under some conditions the images made from the LANDSAT infrared (IR) channels (bands 6 and 7) will provide a rendition that has a "quasi-radar" aspect. This is graphically displayed in Fig. 2, where the thick jungle in southern Venezuela appears

FIG. 2(A)

Red band 5 LANDSAT image taken on March 9, 1973, over the Orinoco River basin; the river divides eastern Columbia from southern Venezuela. The Llanos, a grass-covered plains, appears on the west side; a thick jungle-like forest covers the terrain on the east side.

FIG. 2(B)
IR band 7 LANDSAT image of the same scene shown on the opposite page. Some of the details in the Llanos are emphasized by the stronger contrast. The dark jungle cover seen in the band 5 image now acts as though it has been "penetrated," revealing the underlying terrain, here consisting of Precambrian igneous and metamorphic rocks.

almost uniformly black in the green and red bands (4 and 5) but seemingly is "stripped" to bare ground in the IR bands. In reality, the close association of canopy profile with ground topography is being revealed when the foliage is examined in the highly reflective infrared. The crystalline rock terrain can be broadly defined and differentiated in the IR band image.

Some stereo capability exists for LANDSAT images where sufficient side lap is maintained. However, most of each scene cannot be viewed in this way. Using the multispectral scanner data stored on the digital tapes, members of the U.S. Geological Survey remote-sensing facility at Flagstaff, Arizona, have developed a reprocessing procedure that produces a pseudorelief or three-dimensional effect. One phase of that procedure involves a change in the contrast ratios of gray levels in a given band; this selective darkening and lightening acts much like airbrush shading used to give an appearance of relief to a topographic sketch map of mountainous terrain. An example of the final product is shown in Fig. 3.

Specific Geologic Applications

It should be kept in mind that LANDSAT is basically an extension of aerial photography to large-area mapping. Aerial photographs in the past have been used chiefly as aids to the geologist in preparing or revising various kinds of geologic maps. The techniques and approaches of photointerpretation developed prior to availability of multispectral imagery from space remain the major tools by which geologists extract information from LANDSAT data. Because of the decrease in resolution (by factors ranging from 5 to 100), certain types of information found in aerial photos are inherently unrecoverable from LANDSAT, but such deficiencies are offset by the synoptic overviews that provide hitherto unobtainable information. Thus, some tasks will still be done better from aircraft, but others may be done best, and even exclusively, from space platforms.

One measure of the success of LANDSAT in geology is the extent to which new information has been acquired. While it is still premature to assign a dollar value or cost-effectiveness rating to the results reported so far, it is clear that significant benefits are gradually accruing from LANDSAT data to the specific applications outlined in Table 1. In time, accomplishments in each of these fields will be translated into economic payoffs as new mineral deposits and oil prospects are discovered and engineering projects are undertaken because of these scientific and technological advances flowing from LANDSAT and other programs that couple space-acquired data with conventional ground exploration methods. The outlook is especially promising for certain parts of the world where LANDSAT images represent the first detailed surface coverage of regions never before surveyed or mapped beyond a reconnaissance level.

Some of these applications are reviewed in the following sections.

TABLE 1
Applications of LANDSAT to Geology

Map editing
- Boundary and contact location
- Stratigraphic and/or "remote sensing" unit discrimination
- Scale-change corrections
- Computer-processed "materials" units maps

Landforms analysis
- Regional or synoptic classification and mapping
- Thematic geomorphology (e.g., desert, glacial, volcanic terrains)

Structural geology
- Synoptic overviews of tectonic elements
- Appraisal of structural styles
- Lineaments (and "linears") detection and mapping
- Metamorphic and instrusion patterns
- Recognition of circular features

Lithologic identification
- Color-brightness (spectral reflectance) classification
- Ratio techniques
- Photogeologic approach

Mineral exploration
- Reconnaissance geologic mapping
- Lineaments trends (especially intersections)
- Surface coloration (blooms and gossans)
- Band ratio color renditions

Engineering and environmental geology
- Dynamic geologic processes (sedimentation and coastal processes, sea ice, active glaciers, permafrost effects, landslides and mass wasting, shifting sand seas, land erosion)
- Strip mining, surface fractures—mine safety
- Construction materials

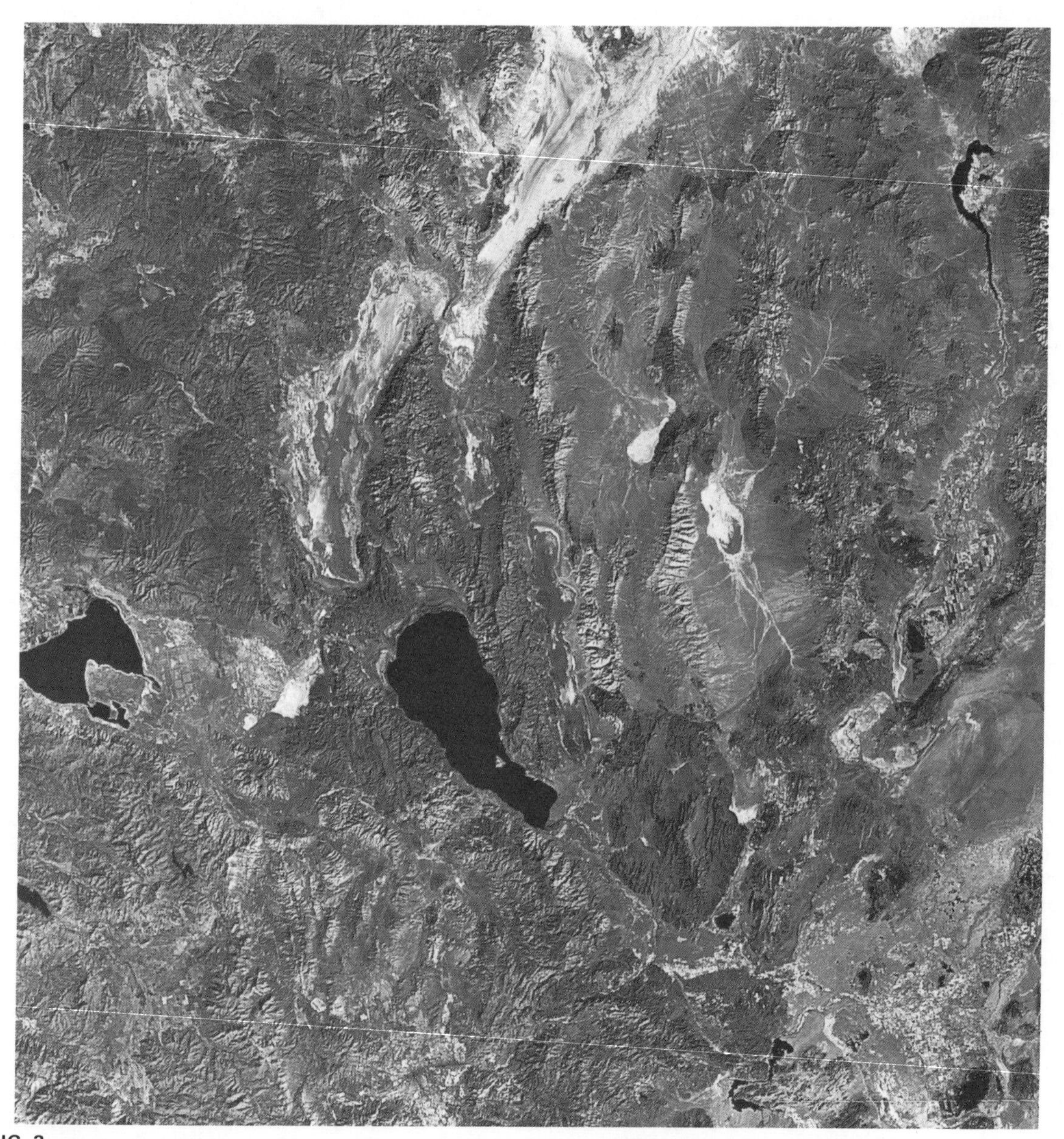

FIG. 3

Computer-reprocessed rendition of the band 7 LANDSAT image taken on November 22, 1973, by the multispectral scanner as the spacecraft passed over western Nevada (Reno in lower left, Pyramid Lake in center, Black Rock desert at top). The pseudorelief effect is the result of contrast stretching. (Courtesy U.S. Geological Survey.)

Map editing. To some extent, LANDSAT data can be used to make new maps, but these would not be equivalent to those produced from aerial photographs. To construct a standard geologic map, it is normally necessary to recognize stratigraphic units and sequences at the *formation* or even *member* level and, to a lesser extent, to discriminate among the major lithologies in the area. For mapping from aerial photos, this requires recognition of unit boundaries and definition of differences among units on the basis of rock color and/or surface weathering effects, topographic and/or geomorphic expression, soil associations, and characteristic vegetative cover, among other criteria. Most units depicted on large-scale maps range in thickness from a few tens to a few hundred feet at most. However, because of resolution limitations, most *stratigraphic* units (defined from ground studies by criteria that usually require close-up or even hand-specimen examination) cannot be recognized and separated in LANDSAT images along the same boundaries selected for mapping purposes. Several ground-distinguishable units with similar reflectance properties (termed remote-sensing units) might group or blend into a single discernible unit in an LANDSAT image that may or may not have a meaningful stratigraphic and/or lithologic significance. When examined in the field, some remote-sensing units actually correspond to single stratigraphic units, but others are comprised of several stratigraphic units having similar reflectance.

Nevertheless, under suitable conditions geologic maps of considerable usefulness have already been produced from LANDSAT imagery. The example from Wyoming shown in Fig. 4, when compared with the published map of the same area, indicates that new units having a field-checked reality were defined from the LANDSAT images even though the contacts among these or previously known units may not be as precisely located as those in the ground-based map version. In some instances, mapping from LANDSAT can be made more accurate by referral to coverage from several seasons, as effectively illustrated in Fig. 5. This makes use of the repetitive (18-day cycle) aspect of LANDSAT coverage from its near-polar orbit.

A more immediate application of LANDSAT images lies in map editing or revising of previous small-scale maps. In regions of the world where rock exposures are sharply defined (mainly in deserts or other areas of low vegetation), the correspondence of LANDSAT viewed surface geology patterns with those in the maps is almost self-evident (Fig. 6A and B). But close comparison of image to map frequently points to serious discrepancies in the map version. Reality resides with the LANDSAT image.

Lithologic identification. Identification of rock types from aerial or space platforms has long been a goal that consistently remains elusive. The high hopes that at least the major rock groups could be recognized with presently used remote-sensor data have met with varied success. Depending on the experience of the interpreter and his awareness of the rock types known to occur in the imaged area, the photogeologist frequently has been able to correctly identify basalts, granitic rocks, some metamorphic types, limestones, shales, and sandstones. This ability will, of course, decrease considerably as resolution becomes too poor to single out individual lithologic units. However, in some geologic terrains, generally homogeneous rock units are exposed over wide enough surfaces to produce distinguishing tones and patterns in LANDSAT imagery, as illustrated in Fig. 7.

It is not likely that remote sensors operating from space will ever achieve a high degree of reliability in rock-type identification. Unlike laboratory methods, such as x-ray diffraction in which unique solutions to component mineral identity result from fundamentally different combinations of atomic structure, there is little that most remote-sensing devices can measure that is exclusive to any given rock type. In the spectral range scanned by the LANDSAT MSS, the only rock properties directly measured are color and brightness; indirectly, derivative properties such as relative erodability (expressed topographically), surface stains, soil associations, structural response, vegetation preferences, etc., are taken into account in making identifications. However, it is not possible to set up a meaningful working classification of rocks based primarily on typical colors and relative brightness (most classifications are built from mineral as-

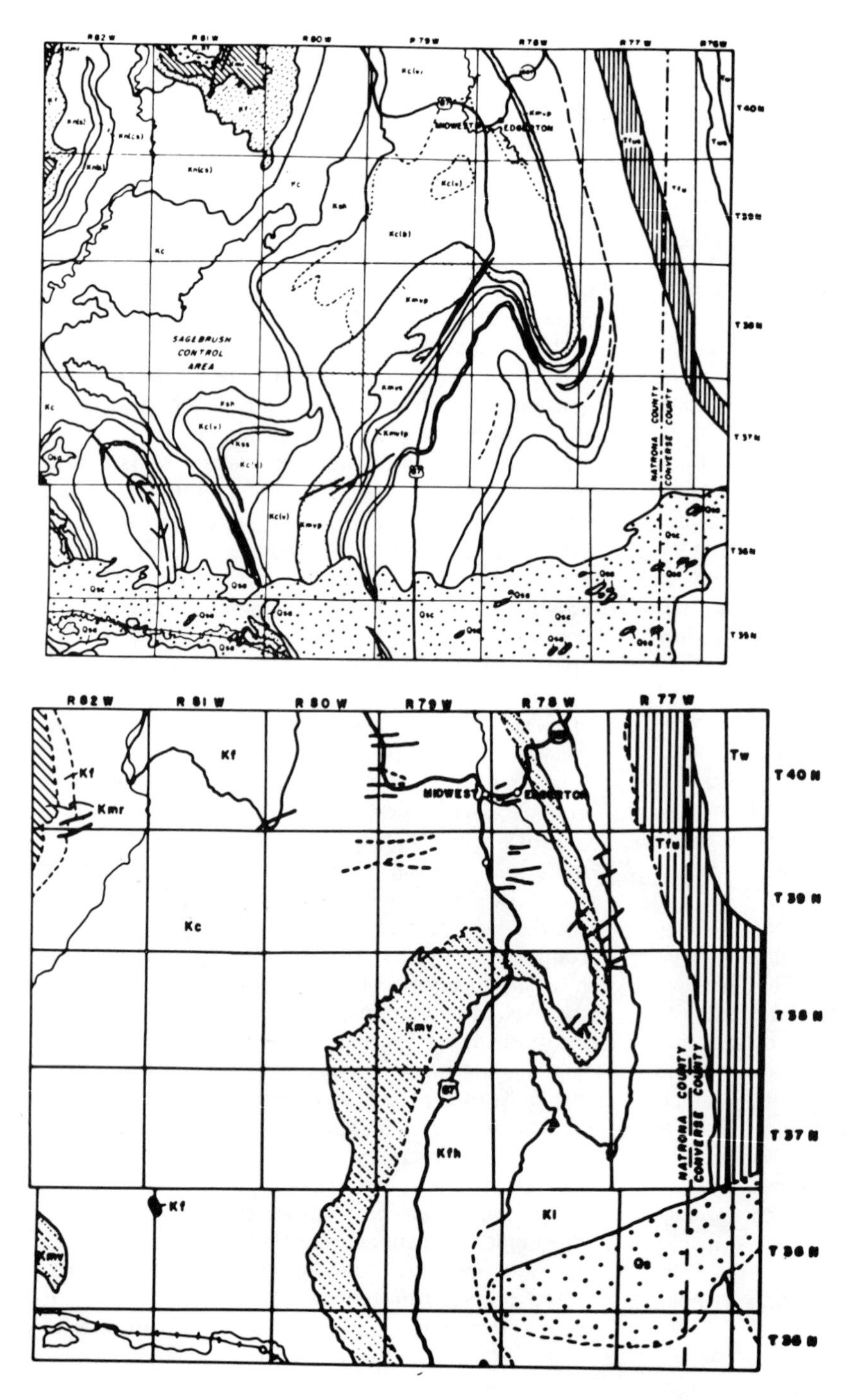

FIG. 4
Details of the southeast quarter of the map of the Arminto area in central Wyoming as prepared from LANDSAT imagery (top) compared with the equivalent area taken from the state geologic map (bottom) published in 1955. (Courtesy R. S. Houston, University of Wyoming.)

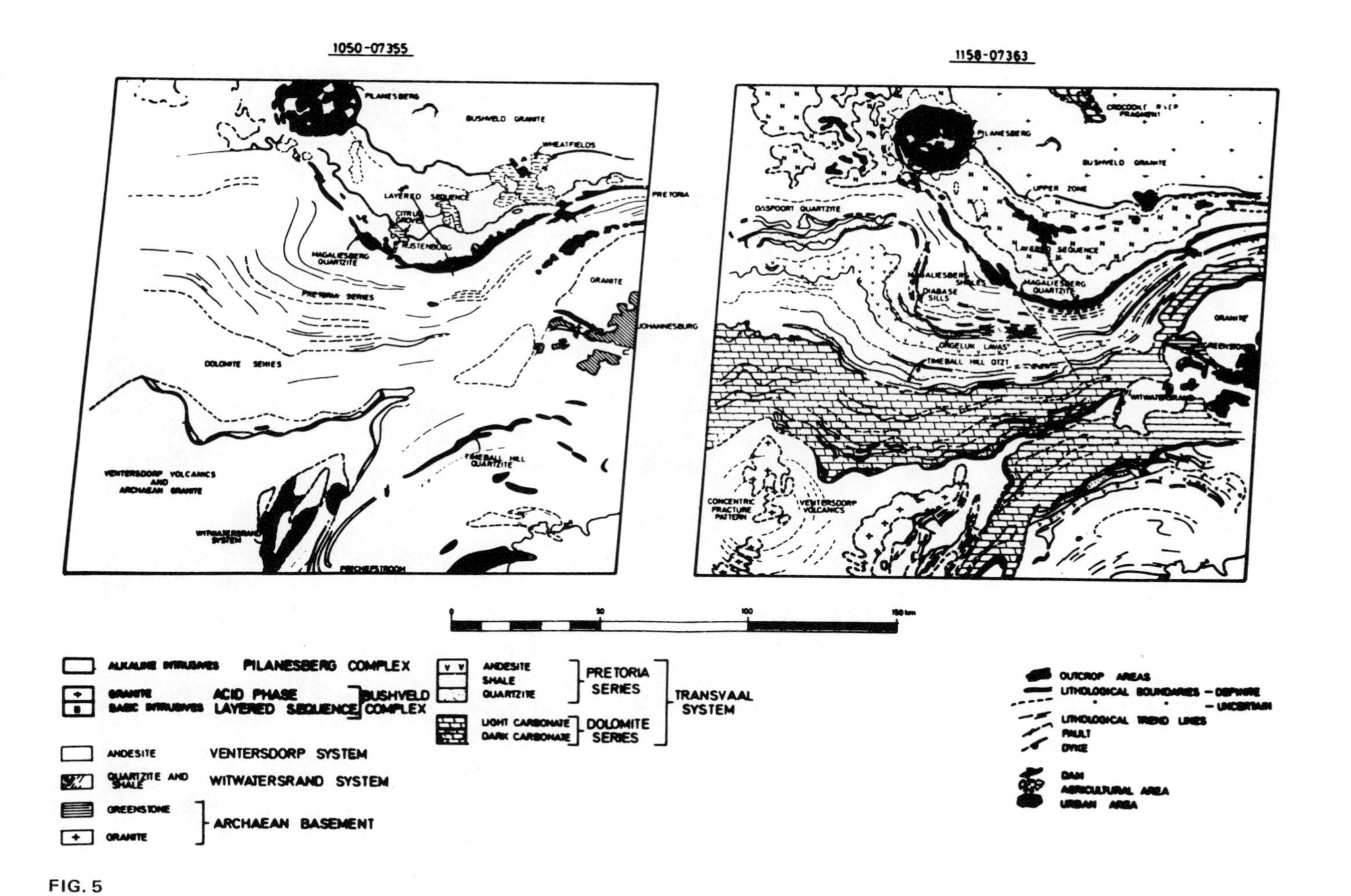

FIG. 5
Sketch maps of geologic interpretations made from LANDSAT images of part of South Africa west of Johannesburg acquired in early September 1972 during the dry season (left) and late December 1972 during the wet season (right).

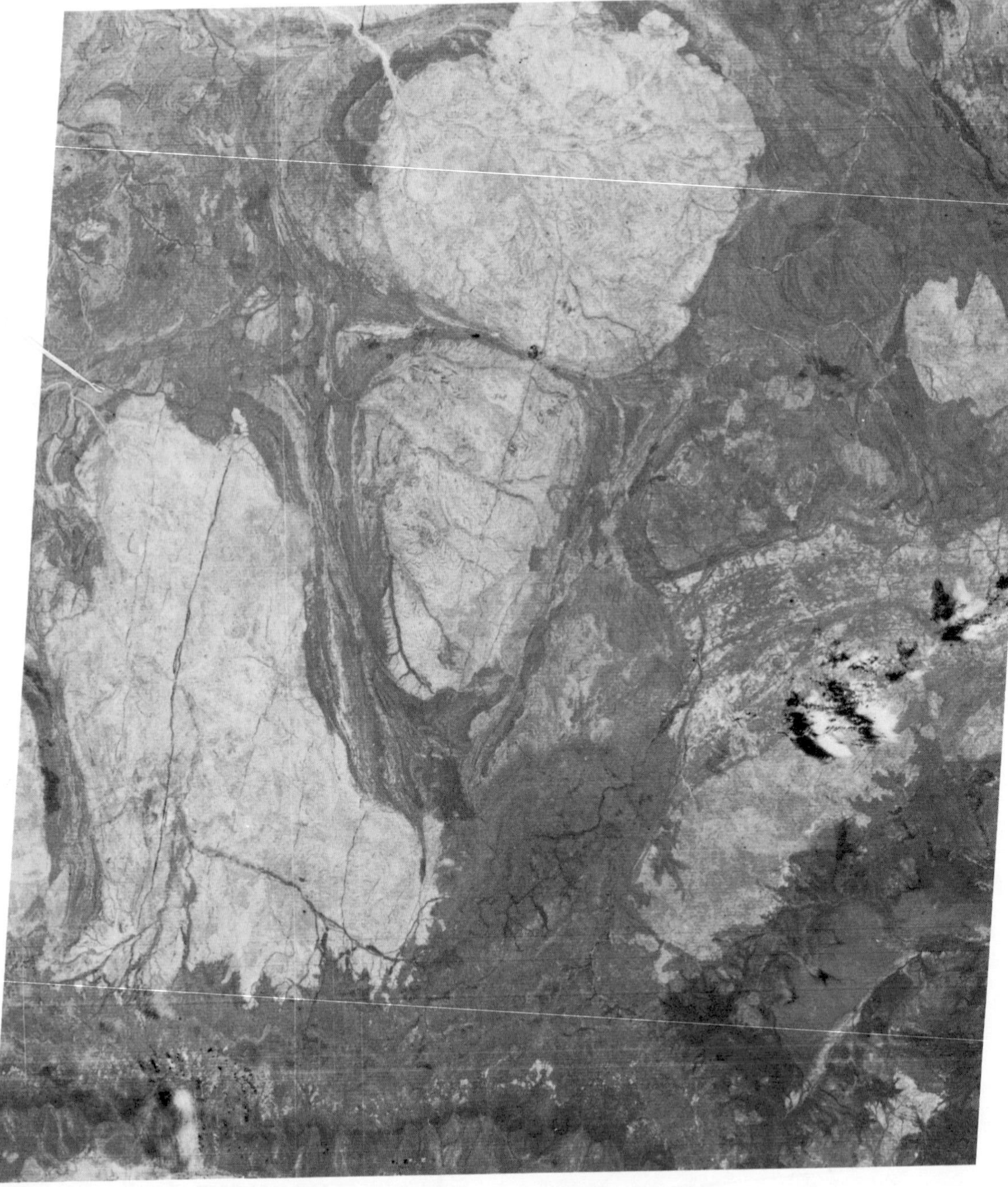

FIG. 6(A)
LANDSAT red band image of a region in western Australia just south of the northwest coast near Port Hedlund showing several large igneous plutons cutting into metamorphic rocks (mantling bands).

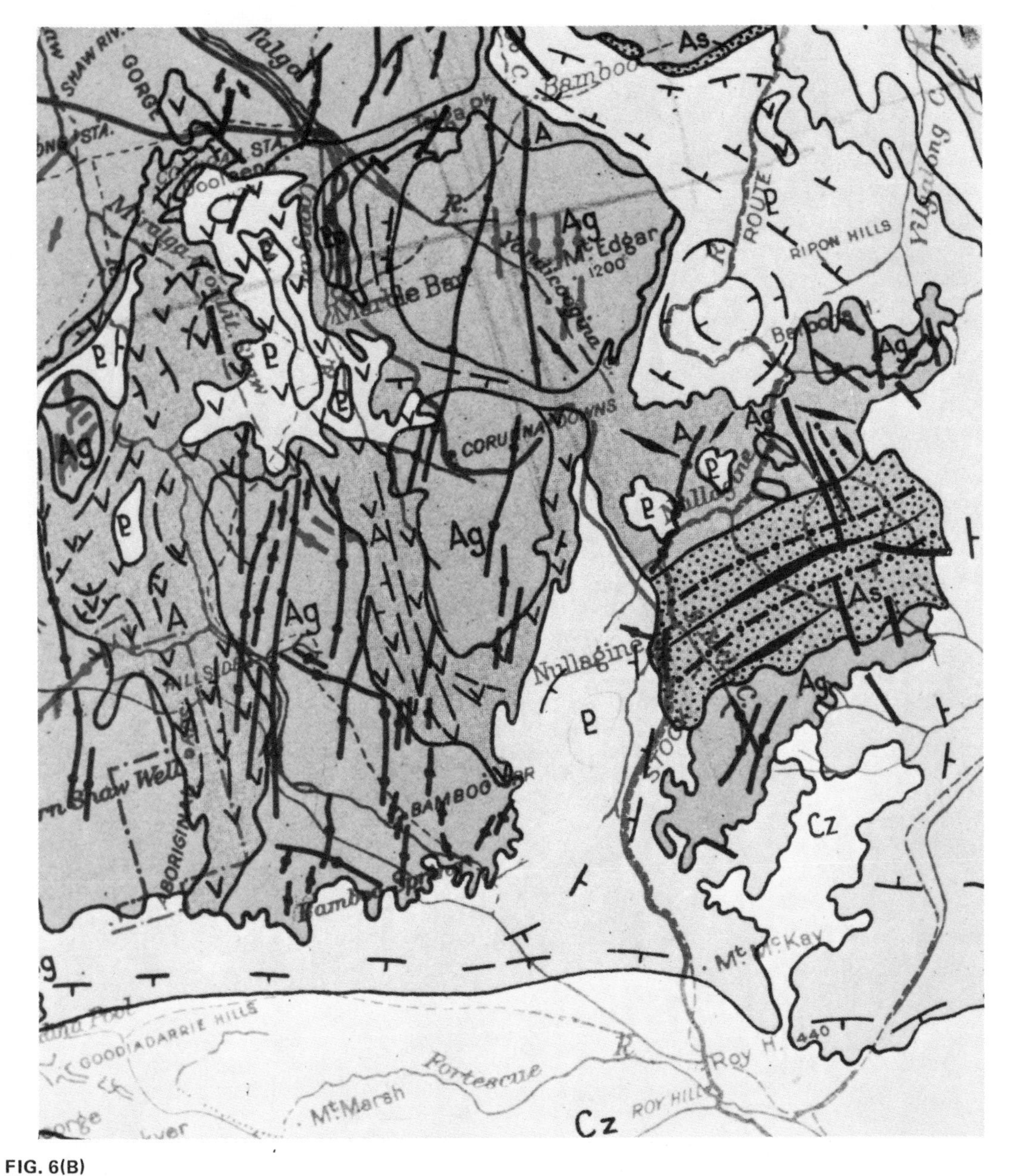

FIG. 6(B)
Part of the tectonic map of Australia that includes a good bit of the area imaged in Fig. 6a. Ag, As, and A refer to Archean granitic and metamorphic rocks; P denotes Proterozoic metasediments; Cz relates to Cenozoic sediments. Close comparison between mapped unit boundaries here and those in the LANDSAT image reveals that improvements in the precise locations of these boundaries can be made from the LANDSAT view.

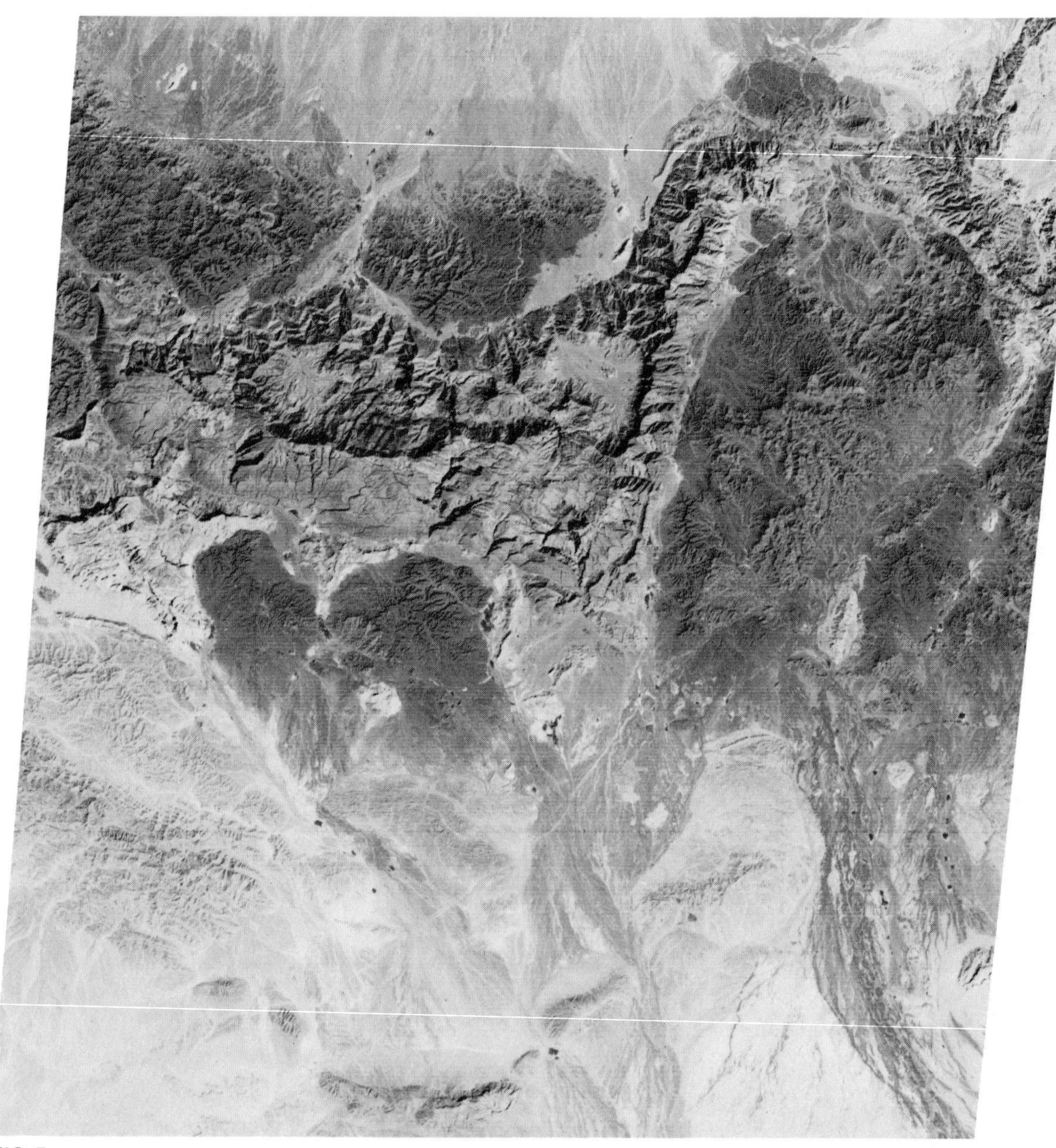

FIG. 7
Oman Mountains in Oman along the Gulf of Oman in the northeast sector of the Arabian Peninsula. The darker areas are mostly ophiolites and serpentines. Several folded belts of Paleozoic and Mesozoic carbonate-shale sequences lie between the basic igneous intrusives; one near top center has been breached to expose Paleozoic dolomites. Tightly folded radiolarites and other sedimentary rocks are evident in center left.

semblages, textural aspects, and field relationships). Thus, granites and schists, sandstones and limestones, shales and slates, and other lithologically or genetically dissimilar rock pairings may have roughly the same colors and brightnesses. Conversely, one given rock type may have many color variants, for example, green, red, buff, gray, and black shales, or white, dark-gray, buff, and red limestones.

Structural geology. Experience with earlier space imagery had disclosed the exceptional value of synoptic imagery for displaying extended structural elements such as closed anticlines, domes, and intrusive bodies, folded mountain belts, fault zones, regional joint patterns, and other fracture systems in their regional context. In arid regions, especially, the surface expression of structurally disturbed parts of the crust was often better revealed in the images than in maps of the same areas. The interplay among underlying structure, topography, vegetational distributions, and solar illumination commonly enhanced the appearance of structural elements, so that subtle relationships not apparent on the maps were made to stand out. New lineaments of considerable magnitude and extent were picked out in the images, because their breadth and continuity were commonly overlooked on the ground where only small, localized effects of a segment exposed discontinuously from one outcrop or topographic expression to the next were insufficient to manifest the "whole from the parts." In some areas of the world (e.g., Gulf of Oman, the Afar, Afghanistan) space imagery has brought about improved understanding or even fundamental resynthesis of the tectonic framework. LANDSAT has broadened these observations to sections of the globe never before imaged in detail from space. Four outstanding LANDSAT views of complexly folded and faulted parts of the crust are documented in Fig. 8A–D.

As expected, the principal output so far from examination of LANDSAT images for new structural information is the recognition of numerous linear features, ranging in length from 1 to 2 miles up to several hundred miles. Almost every LANDSAT image having notable geologic content is marked also by occasional to frequent linears. In the first rush to report significant results, investigators usually equated these linears with structural features such as faults, joints, or inclined strata. Many of these interpretations have stood the test of field checking. But others have been abandoned when individual linears were found to be lighting artifacts, spurious alignments of diverse ground features, or man-made objects; the degree to which a region is vegetated represents another factor that influences the apparent occurrence of linears.

The geologic studies of the state of New York by Y. W. Isachsen provide a case in point. At the March 1973 ERTS Symposium, Isachsen displayed a mosaic of the state that had been analyzed for structural features. His efforts concentrated on the Adirondack and Catskill mountains (Fig. 9), where large crustal fractures were well known and mapped, in part because of their control on regional topography. He has since presented an updated LANDSAT-based map (Fig. 10) of linears of all kinds observed in the eastern half of New York. His first appraisal of these linears had indicated that (1) many of those already recorded from geologic studies were recognizable, (2) other known ones which should be visible did not show up, and (3) still others prominent in LANDSAT images were completely new features not recorded on any maps. However, after careful field checking and examination of maps and photos, he concluded that less than one third of the new linears are likely to be strictly structural in nature. Many undiscerned faults and lineaments fail to be expressed in LANDSAT images because of unfavorable illuminations and/or degradation effects in the third- and fourth-generation images used.

Another study reported by N. H. Fisher and his colleagues in the Australian LANDSAT investigations is particularly instructive. Several test areas that have been thoroughly studied and mapped over the years on the ground and from aerial photographs were chosen for comparison with the information content extractable from LANDSAT. In the case of linears, it was found that only 30 percent and 10 percent of the previously known faults and major fractures were recognized in the 1:1,000,000 and 1:250,000 scale LANDSAT images, respectively, covering the same area. Furthermore, many of the LANDSAT-identified linears were new and did not coincide with the field-

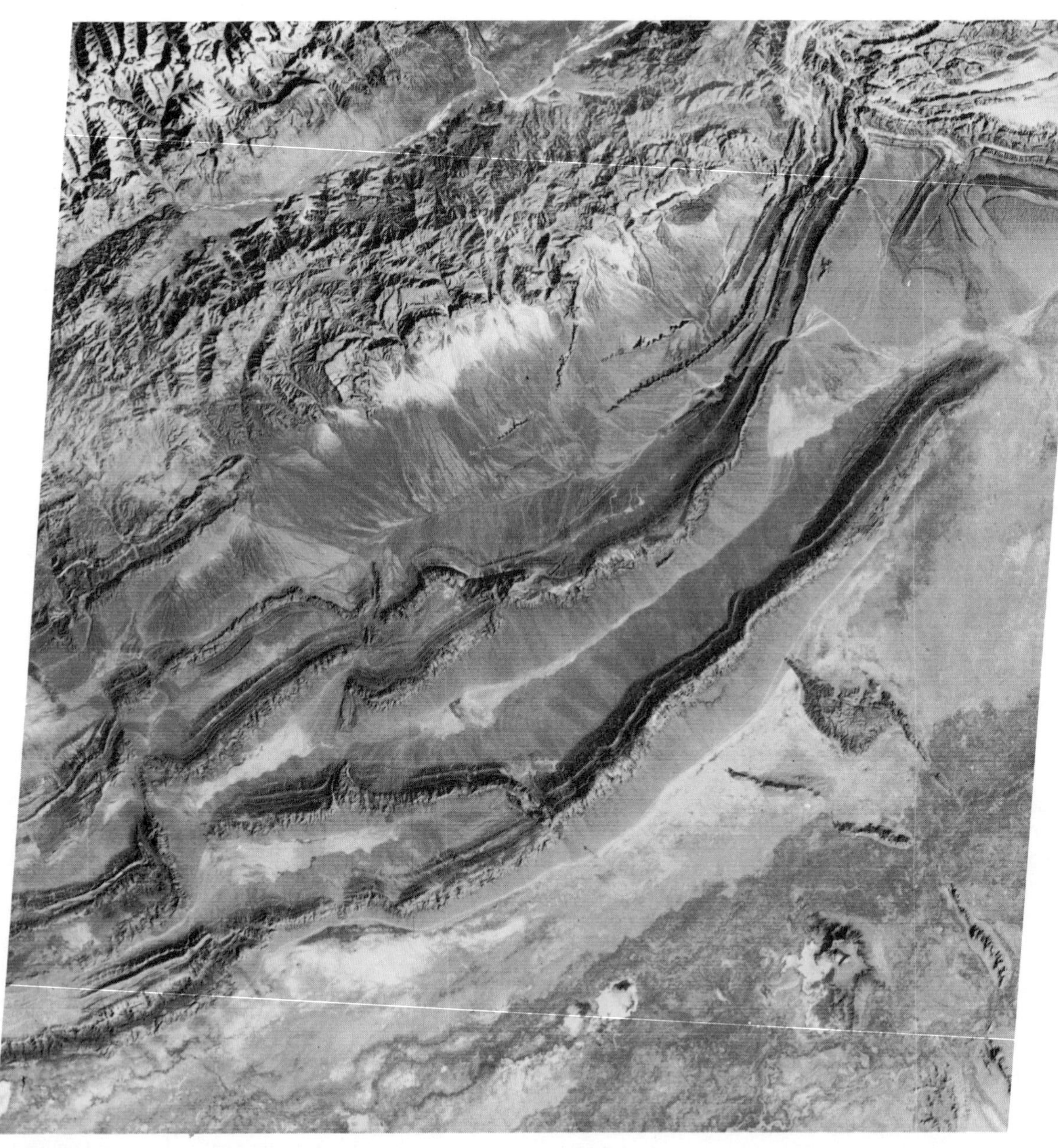

FIG. 8(A)
Broad folds, offset by faults, of the K'op'ing Shan, a series of ranges set against the edge of the Takla Makan desert of western Sinkiang Province in westernmost China. The higher mountains to the north are part of the Tien Shan, which passes along the border with the Kirgiz Republic in the USSR.

FIG. 8(B)
Part of the folded Appalachian Mountains, the Blue Ridge, the Piedmont, and the dissected Appalachian Plateau in western Virginia (Roanoke appears in the right center) and West Virginia.

FIG. 8(C)
The Kuruk Tagh mountain range, a complex of Precambrian igneous and metamorphic rocks and infolded Ordovician rocks in the Sinkiang Province of China. The great east–west wrench or tear fault can be traced for more than 300 miles. Lake Baghrash lies to the north.

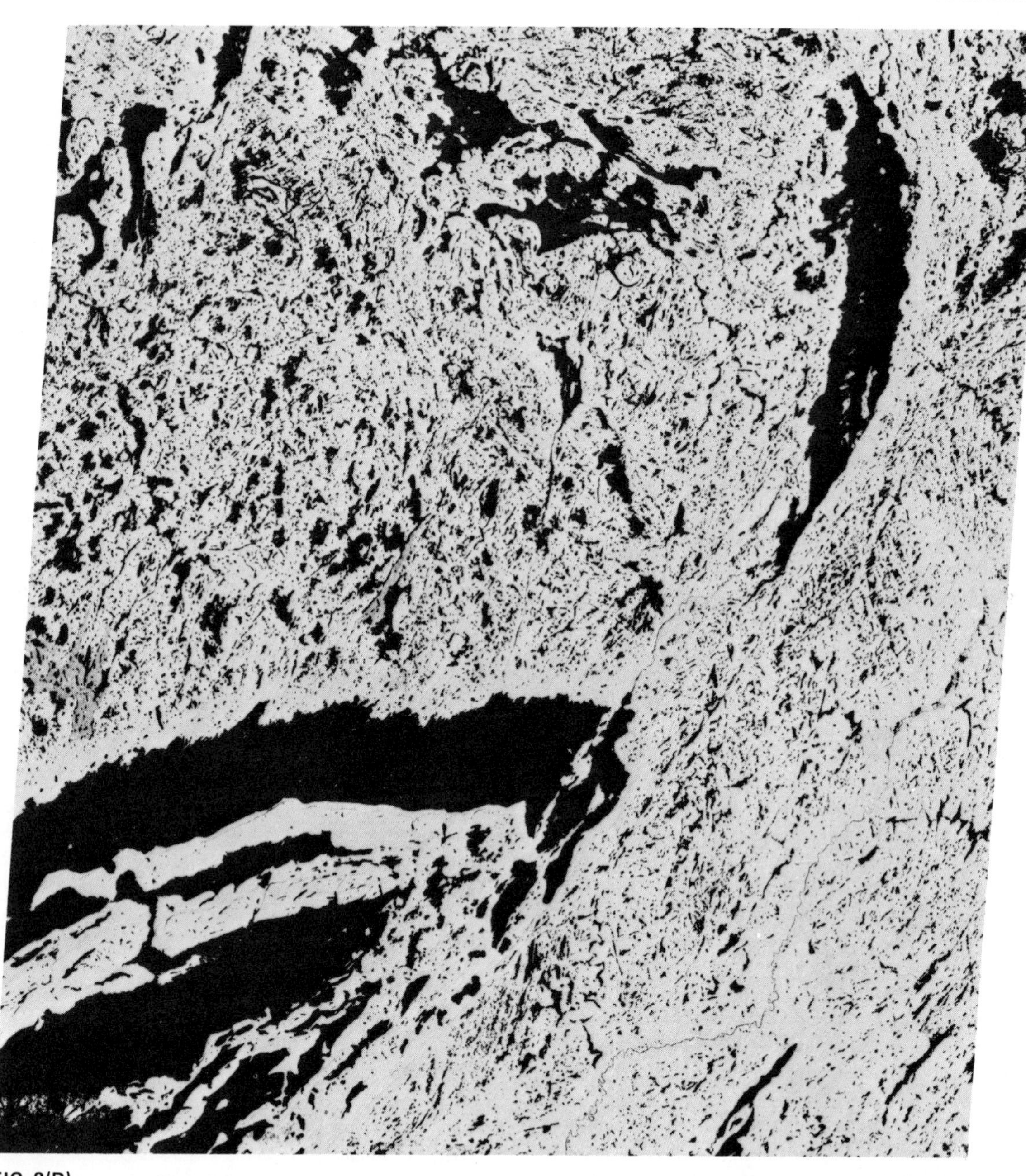

FIG. 8(D)
Band 7 (IR) LANDSAT image of the eastern end of the Great Slave Lake in Canada's Northwest Territories. Glacially scoured fractures in the Precambrian crystalline rocks and glacial lakes in moraines are filled with water in this scene; the water appears very dark in this infrared image and thus tends to delineate these structural and geomorphic features.

mapped lineaments. Also, radar-detected lineament patterns generally were not compatible with the LANDSAT linear sets. This discrepancy between LANDSAT and ground truth, while it seemingly casts doubt on the validity and reliability of the space imagery as a discriminator of structural features, tends to be reduced when it is realized that the mapped lineaments (1) were often defined by criteria discernible only on the ground (e.g., fault gouge, slickensides), (2) included many short lineaments below the maximum length (1 to 2 miles) detectable from LANDSAT imagery, (3) often consisted of close-spaced sets counted as single linears in LANDSAT scenes, and (4) did not suffer from the bias of one-time-a-day coverage at mid-morning. A related Australian study disclosed an additional effect of operator bias: the same interpreter picked out (or missed) different linears when examining the same image at intervals several months apart, and several interpreters tended to produce different, subjective linears maps from the same image.

The Wind River Mountains of Wyoming provide a dramatic indication of the rapidity with which a mapping effort can be accomplished using LANDSAT imagery. R. Parker of the University of Wyoming has been mapping in the high country of this range for 5 summers, a task carried out on pack mule and "shanksmare" in the grand tradition. His labors led to the map produced in Fig. 11, left. After receipt of LANDSAT imagery covering this range, he completed the map shown in Fig. 11, right, in just 3 hours. Although this map should be rated as "preliminary" because most of the lineaments have not been verified, some confidence in the correctness of identification is afforded by field checks at several localities where evidence of fracturing was then obtained. Still, a note of caution has been added to this work following receipt of some Skylab imagery of the same area. A linears map made from the Skylab scene is compared with the LANDSAT version in Fig. 12. The difference in numbers of linears detected results from the higher resolution of the Skylab metric camera. The differences in orientation of prevailing linears is almost certainly due to the times of day when the data were acquired; the morning sun direction from LANDSAT favors enhancement of northeast sets, whereas the afternoon sun-illuminated images from Skylab emphasize north to northwest sets.

LANDSAT has shown a special, almost unique facility for calling attention to circular as well as linear features. Most of these are volcanic or intrusive in nature and many are newly recognized. The circular or arcuate markings traced in Fig. 13 are thought to be fractures in the country rocks overlying a series of intrusives in central Colorado. Similar curved linear features, consisting of segments that commonly encompass a full 360°, are in places associated with porphyry copper deposits in many parts of the world. Again, these may have developed over stocks and diapirs, but many are believed to be fractures within the initial subsurface or "roots" zone of volcanoes long since eroded below their cones or craters.

LANDSAT mosaics are ideal for getting a perspective on the tectonic framework of large regions of the crust (Fig. 14). Thus, through-going lineaments that continue for hundreds of miles can be integrated into a unified network that reflects the influence of fractures in an ancient basement or results from stress systems developed from more recent plate tectonic movements. Seen in a broad context, where diversities of topography and surface geology tend to be filtered out, a new synthesis of structural data can emerge. Various investigators are even now building revised structural models from LANDSAT for their regions of interest. Some first results of one such effort applied to all the conterminous United States are depicted in Fig. 15, although no follow-up verification of the existence of heretofore unrecognized lineaments has been carried out as yet.

Mineral exploration. Work to date has pinpointed two potentially useful ways in which LANDSAT data could help to locate conditions favorable to the concentration of metals and other mineral materials. First, the recognition of new crustal fractures and, especially, intersections in lineaments systems improves the probability of finding ore *if* one believes in

FIG. 9 *(right)*
LANDSAT mosaic showing the entire Adirondack Mountains of eastern New York. The St. Lawrence River appears at the upper left, Lake Champlain at the upper right, and the Mohawk River near the bottom. (Courtesy Y. W. Isachsen.)

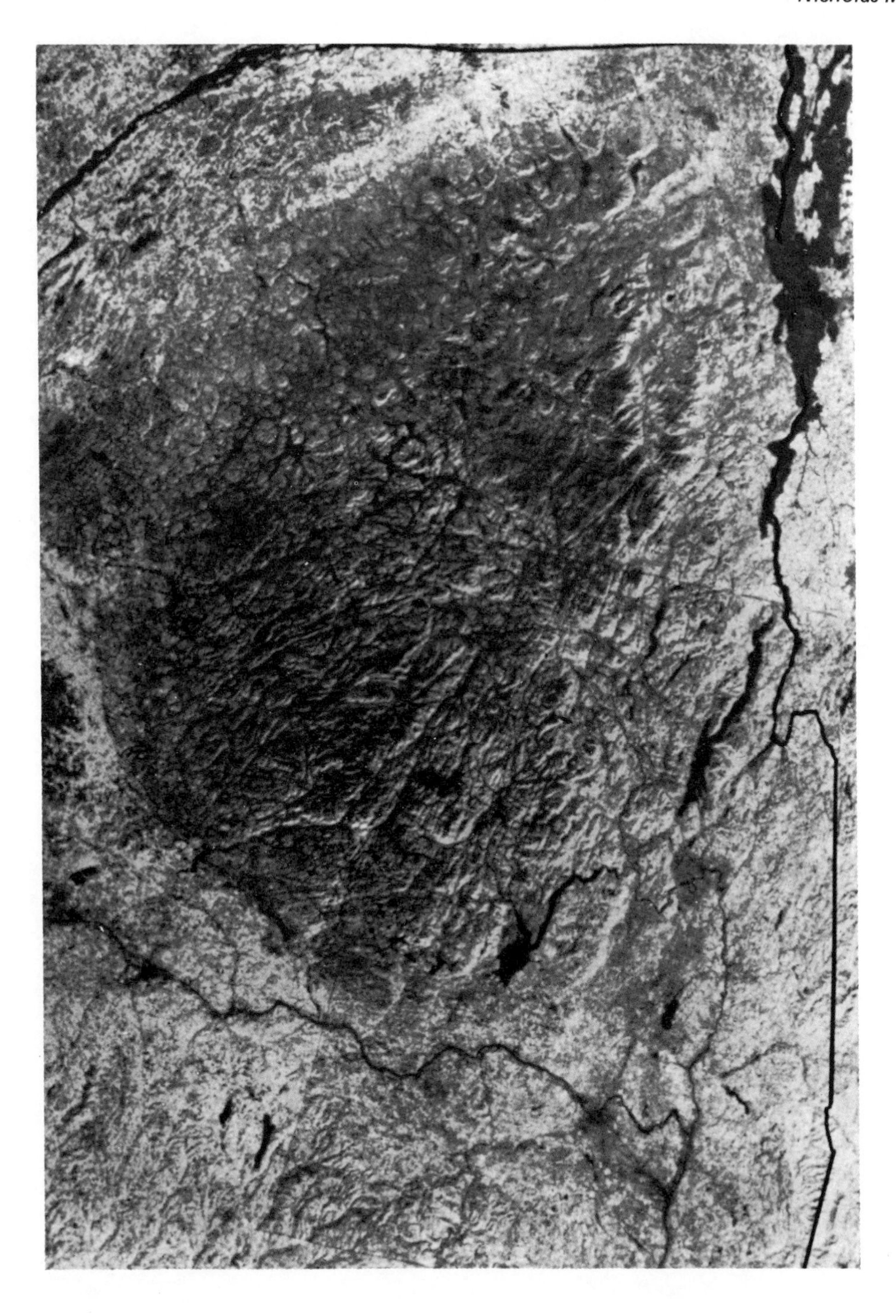

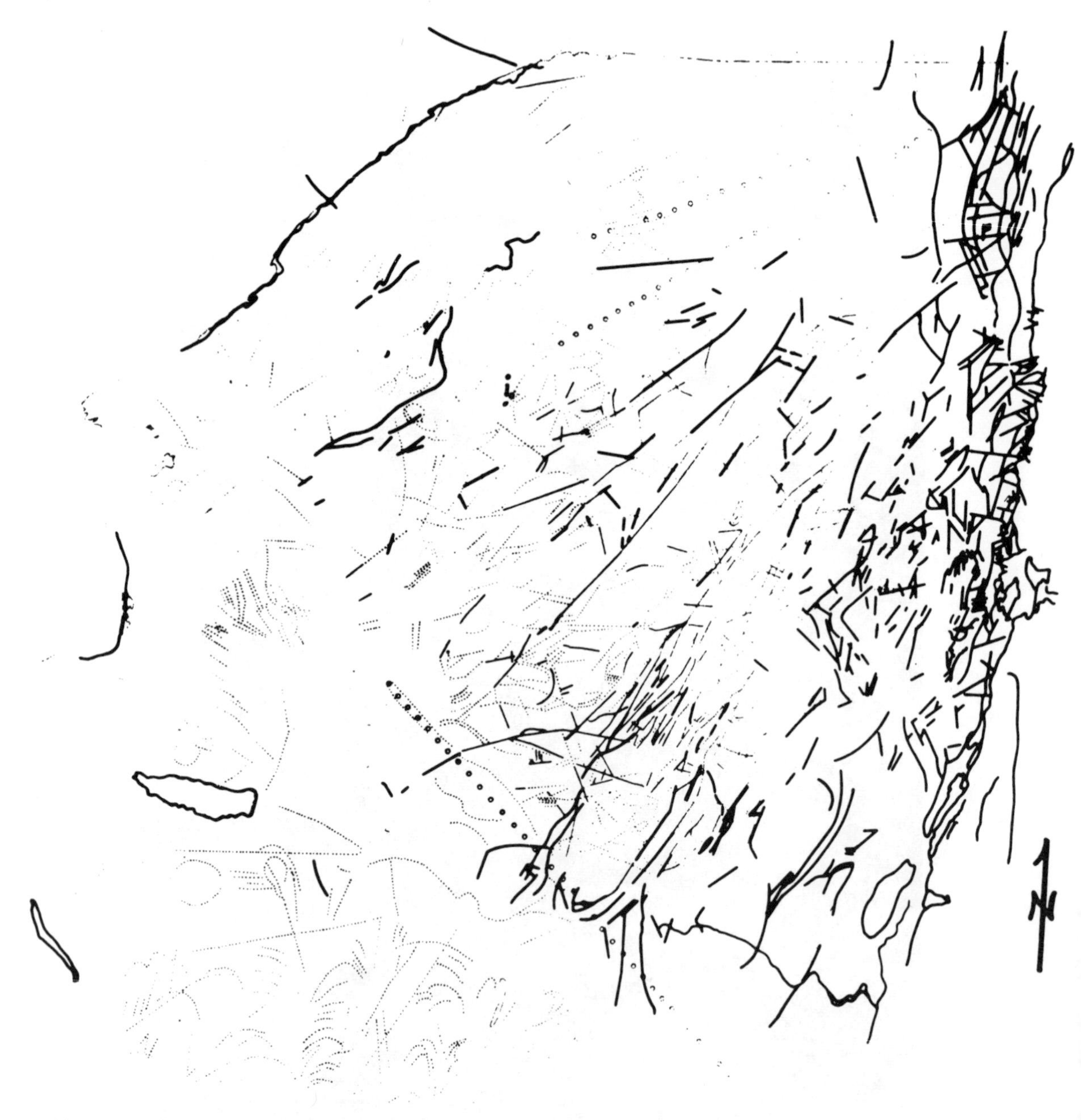

FIG. 10
Sketch map of previously mapped lineaments (solid lines) and new "linears" (dotted lines), not all of which have proved to be rock fractures, observed in LANDSAT images of eastern New York. (Courtesy Y. W. Isachsen.)

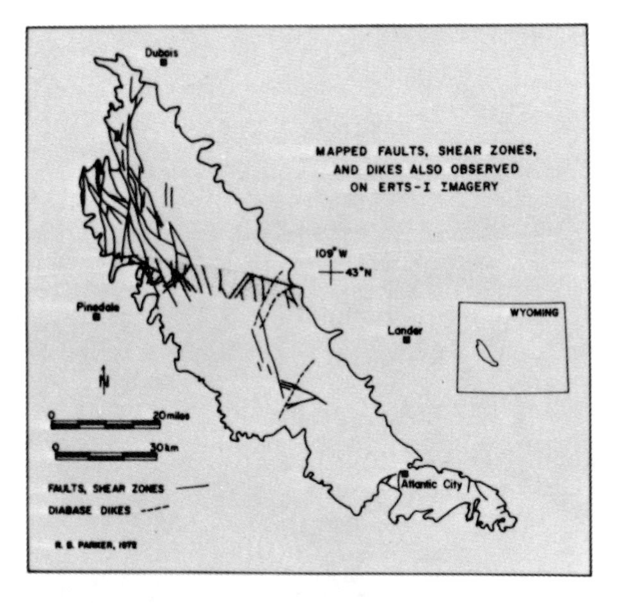

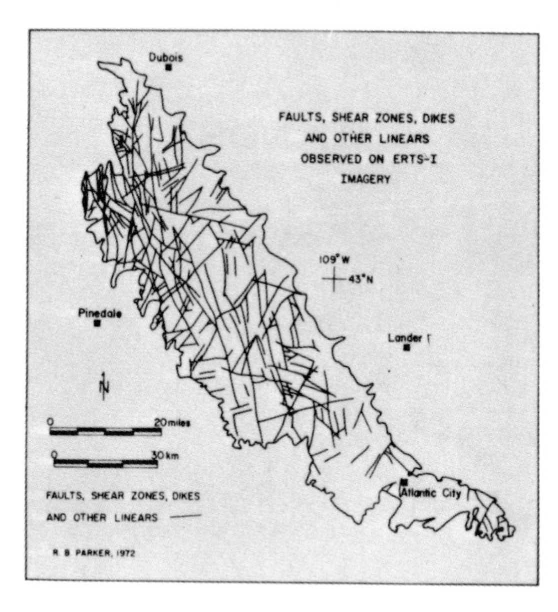

FIG. 11
Interpretation of major fracture systems in the Wind River Mountains of western Wyoming made by R. B. Parker of the University of Wyoming from field studies prior to 1972 (left) and then updated by analysis of a single LANDSAT image (right).

the commonly held view that such fractures control localization of mineralizing solutions. Each new fracture or intersection provides new targets for exploration. Point intersections, particularly, represent a significant narrowing in on promising zones of concentration so that exploration of vast areas can be greatly compressed. An example of this approach has already been presented in Fig. 12.

Second, many shallow mineral deposits give rise to distinctive surface stains (gossans and blooms) caused by alteration or secondary enrichment. If broad enough, some of these stains should be detectable as color-brightness anomalies—subject to the caveats raised in the section on lithologic identification. A simple test of this capability would be to look at LANDSAT imagery for any evident visual (tonal) differences around known mineral deposits that single them out from their surroundings. Caution must be maintained in examining active mining areas to avoid confusion between surface conditions at man-made workings (excavations, mine dumps, dried-up lakes, etc.) and natural stains present before exploitation.

A. F. H. Goetz (Jet Propulsion Laboratory) and L. C. Rowan (U.S. Geological Survey) have developed a computer-based method for enhancing LANDSAT imagery to bring out the effects of surface alteration; usually subtle accumulations of hydrated iron oxide (iron rust or gossan) associated with sulfide deposits are emphasized. From field and laboratory measurements with a reflectance spectrometer, these investigators have confirmed that limonite has a spectral response quite unlike that of most other common minerals. This response can be made more sensitive to small differences when several ratios of different LANDSAT MSS band pairs are calculated. Each resulting ratio represents a variable signal, which, like individual MSS band analog signals, can be used to construct photo images. The individual ratio images are passed through an optical processor using color filters to produce a color composite. Filter combina-

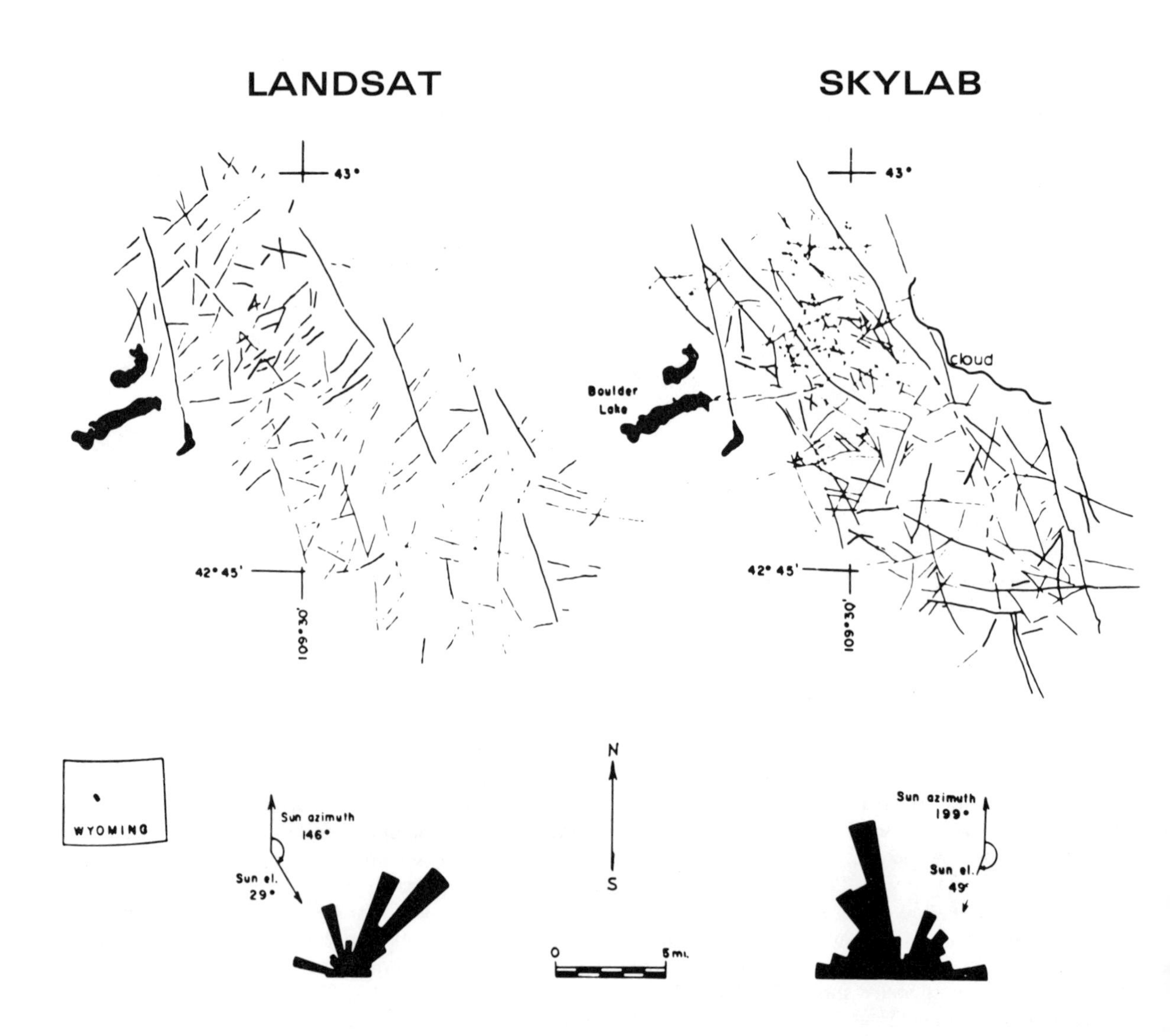

FIG. 12
Comparison of linear features mapped using LANDSAT images and Skylab S-190B photographs of a part of the Wind River Mountains. The rose diagrams indicate the sun-azimuth bias introduced by acquiring the images in the morning and the photographs in the afternoon. (Courtesy R. S. Houston, University of Wyoming.)

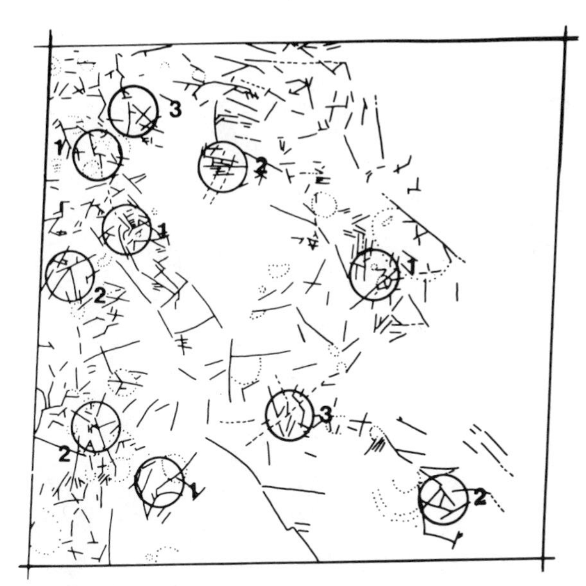

FIG. 13
Tracing of straight and curved or arcuate linear features recognized in LANDSAT imagery on a winter LANDSAT scene over the central mineralized district of Colorado. The area is south of Denver; Colorado Springs appears in the right center. Higher frequencies (or densities) of linears have been circumscribed. (From S. Nicolais, 1973.)

tions have been found that cause the tones or gray levels representing limonite in three different ratio images to appear a yellowish-brown, much like that of iron stain. Different rock types, vegetation, etc., also take on distinctive hues; in particular, clay alteration products also indicative of certain ore deposits can be made to take on a characteristic set of colors. Goetz and Rowan have tried out their method on a LANDSAT scene of central Nevada that includes the Goldfield mining district (gold and silver accompanied by iron sulfide). Prominent yellow-brown color patterns are observed around Goldfield (principally in a ring or aureole that roughly outlines that alteration zone surrounding the underlying intrusive) and other areas where surface iron stains were known before. The color composite has now been field checked from the air and ground during which many of these color anomalies were verified. Insofar as gossan often indicates mineralization (including uranium ores), this enhancement technique, if it bears up under further testing, may well prove to be a major breakthrough in mineral prospecting.

SPECIFIC RESULTS IN FUEL EXPLORATION

Both direct and indirect information bearing on the search for gas, oil, and uranium* can be gleaned from LANDSAT data. Some types of direct information are obvious to anyone familiar with the use of aerial photographs in geologic exploration. Most have been alluded to previously, along with certain reservations as to their accuracy and utility. Among those guides to the presence of underground fuel sources recognizable at the surface are (1) lithologic units (reservoir rocks or host beds) whose subsurface extensions elsewhere can be inferred, (2) folded structures that persist at depth, (3) relatively short lineations and fracture or joint sets that may localize mineralization or trap petroleum below, and (4) generally longer regional linear systems that afford clues to basement trends responsible for controlling structural and/or stratigraphic traps. Previous illustrations provide examples of the ability of LANDSAT to detect and define such features. Another relevant example appears in Fig. 16, which shows how readily LANDSAT can pick out some of the classic salt domes in the Gulf Coast from which petroleum products have been recovered.

Still another example is that of the actual presence of oil at the surface. This is a rare condition on the land (tar sands and pits being the exception), but natural seeps on the seafloor give rise to surface slicks analogous to the better publicized oil spills. Geologists at the Conoco Oil Company Research Laboratory in Ponca City, Oklahoma, working with the NASA aircraft facility at Johnson Spacecraft Center in Houston, have demonstrated the detectability of marine natural seeps as these are manifested on the ocean surface. Images obtained through narrow-band blue filters on photographic cameras during aircraft flights

*Coal and oil shale have been excluded from this review because their abundance in the United States makes exploration unnecessary at the time.

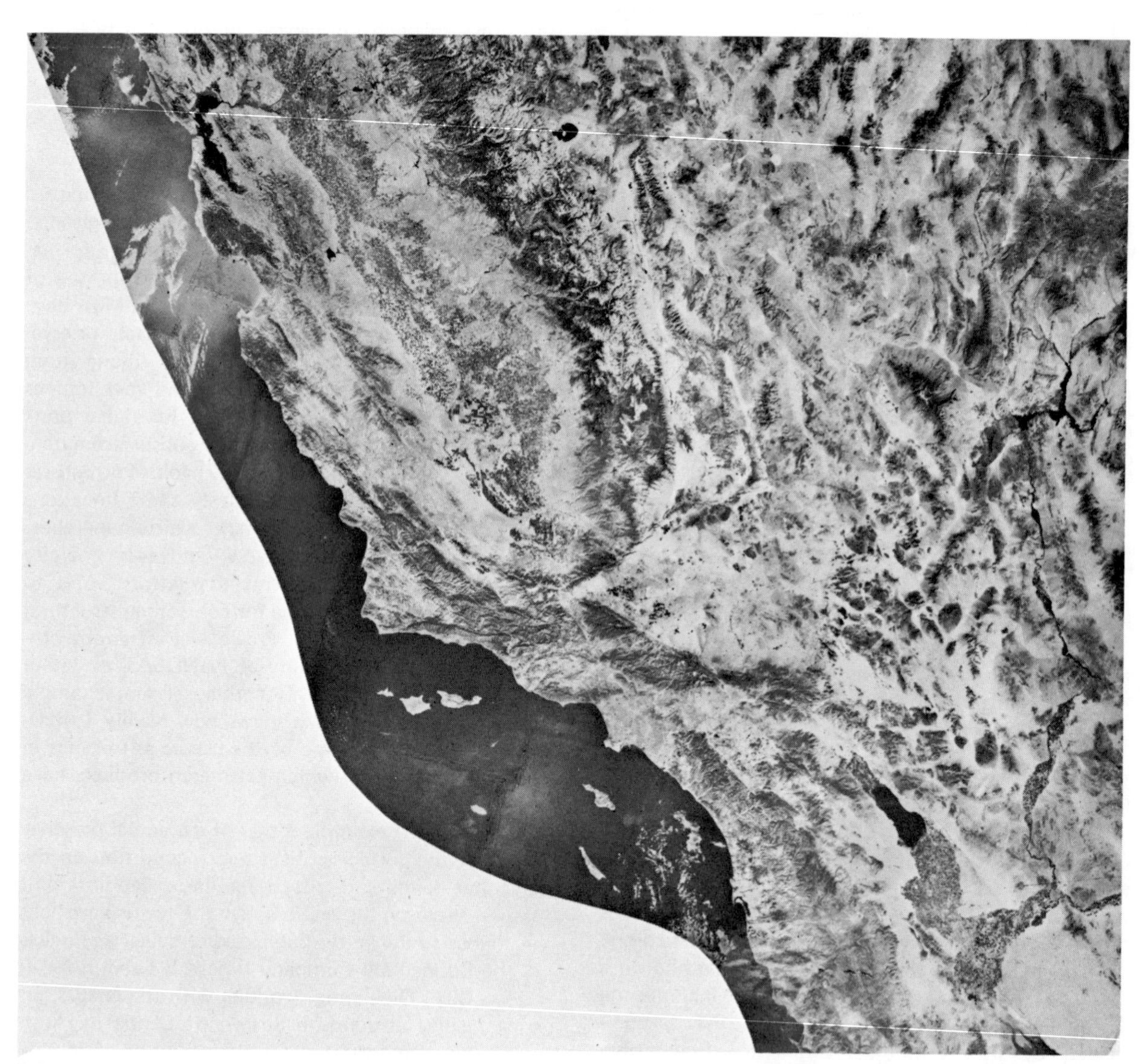

FIG. 14

LANDSAT band 5 mosaic of much of the southwestern United States, including central and southern California, southern Nevada, and small parts of Arizona; northern Mexico around the mouth of the Colorado River is also shown. This is part of the mosaic of the United States prepared by the Soil Conservation Service.

FIG. 15
Sketch map outlining the United States on which W. D. Carter of the United States Geological Survey, Reston, Va., has plotted the major or regional straight and curved linear features that he could recognize in a preliminary examination of the LANDSAT mosaic of the entire United States.

over known seeps in the Gulf of Mexico readily display the outlines of the surface collection of oil (Fig. 17), but film exposed to longer wavelengths fails to define the same spots.

As examples of diverse indirect information types, one can cite (1) possible effects of petroleum on vegetation—a contemporaneous phenomenon, and (2) recognition of modern-day sediment distribution patterns (Fig. 18) in coastal, estuarine, or lacustrine waters, from which come new insight into the accumulation of source beds or reservoir units in depositional basins of the geologic past.

Of the 72 LANDSAT-1 NASA-approved investigations in geology, perhaps more than half provided useful information of immediate interest to a petroleum geologist looking for principles or demonstrations of the exploration capabilities of space imagery. Some of the same information is applicable to the geologist searching for uranium. A few of the investigations dealt specifically with suggested applications to fuels exploration as a secondary concern or byproduct of their main objectives. Only one investigation was completely dedicated to the use of LANDSAT data for petroleum exploration; no investigation

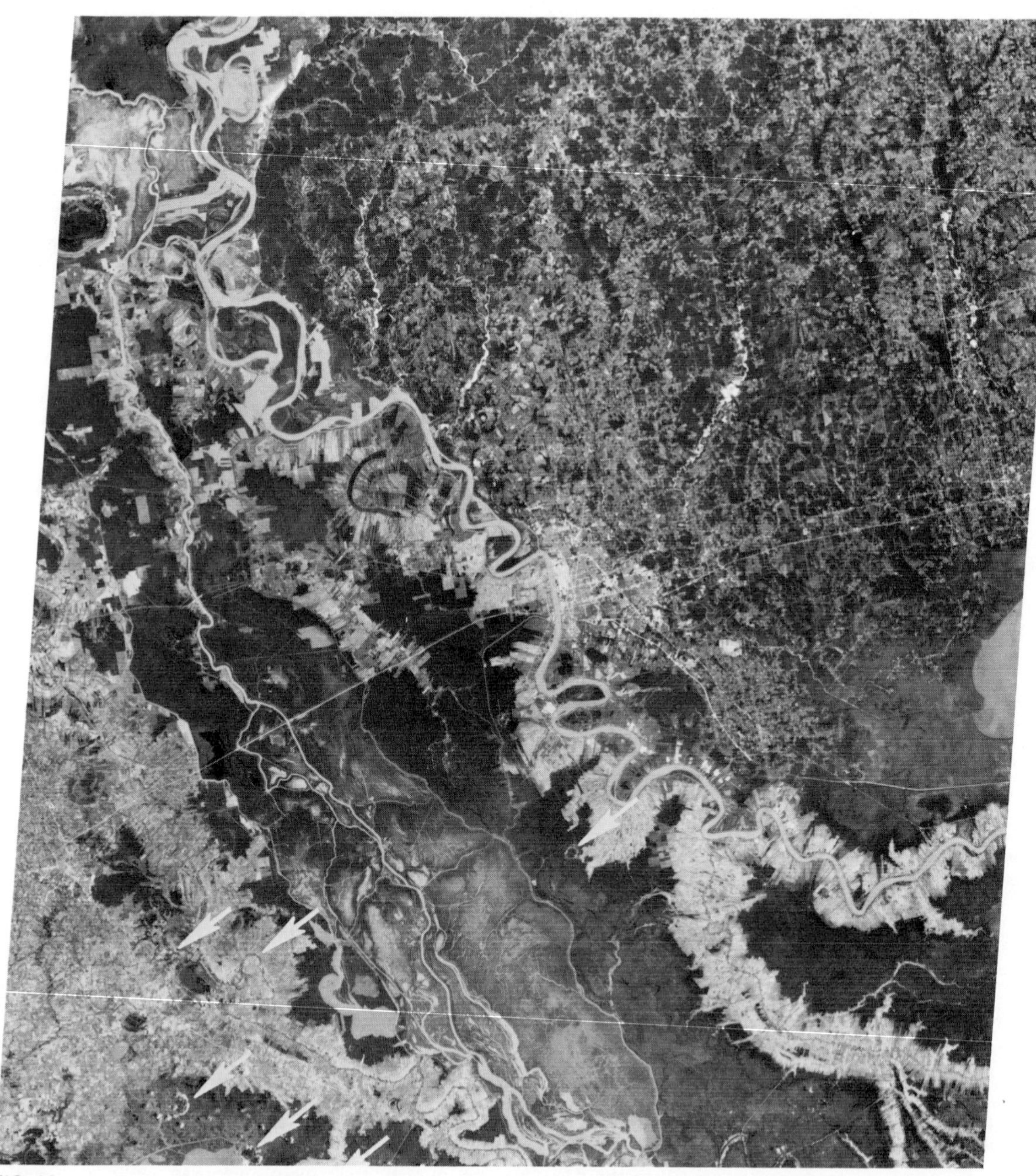

FIG. 16

LANDSAT image of southern Louisiana (just west of New Orleans) and southwest Mississippi (top) through which flows the Mississippi River past Baton Rouge and the Atchafalaya River to the west. Numerous salt domes occur in this region; several, easily identified in this band 5 image, are indicated by arrows.

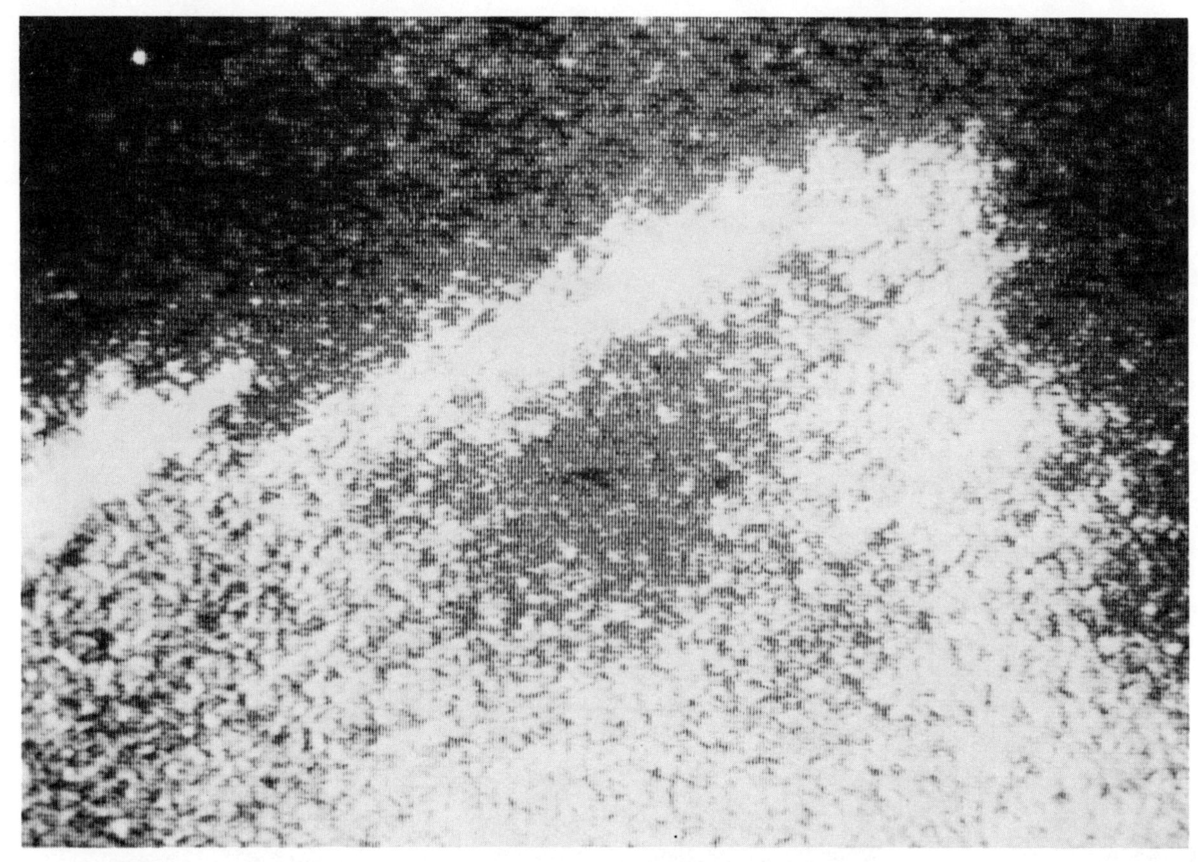

FIG. 17
Black-and-white version of a color-enhanced photograph of the surface "slick" formed from a natural oil seep in the Gulf of Mexico. The photo was made during a NASA aircraft overflight over known seeps to test methods for detection of oil on the sea surface by remote sensing. As shown here, best detection was achieved with a No. 35 filter (near UV-blue) over the KA62 camera loaded with panchromatic film. (Courtesy J. A. Eyer, Continental Oil Company.)

considered the question of prospecting for uranium from space.

The sole LANDSAT investigation in petroleum exploration was conducted by R. J. Collins, Jr., president of the Eason Oil Company of Oklahoma City, and his colleagues. Under his direction, the field and interpretive work was carried out by F. P. McCown, L. P. Stonis, and G. J. Petzel of the company and J. R. Everett of the Earth Satellite Corporation, Washington, D.C.

Their philosophy underlying the objectives of the study followed a sound and proper approach: concentrate on a major oil basin already in production as though it were an unknown or virgin territory earmarked for the first phases of exploration. This outlook tends to reduce the bias of familiarity, but still retains the eventual opportunity to compare their findings with established ground truth, that is, the location of productive fields and the data on the geologic factors responsible for the petroleum accumulations.

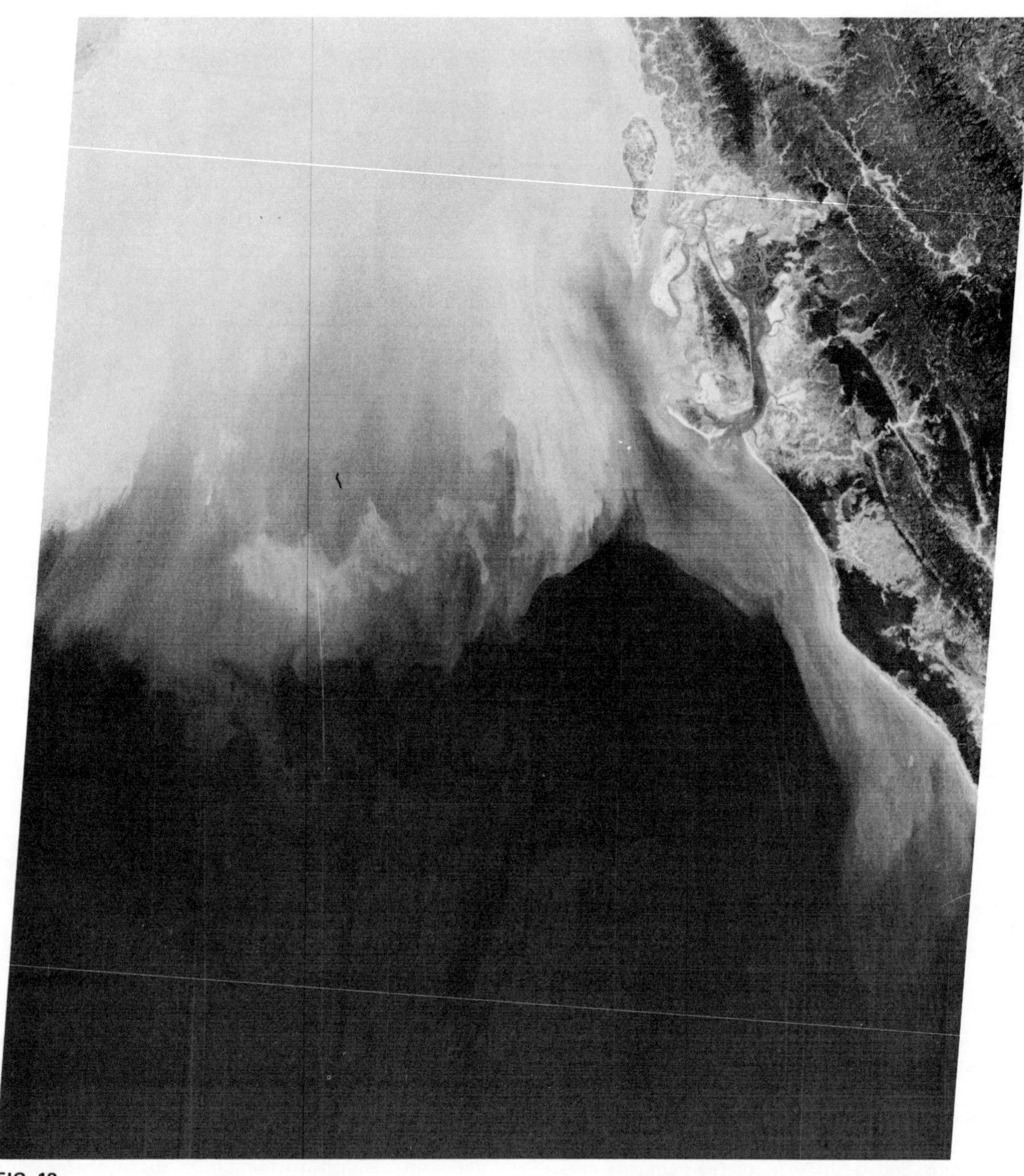

FIG. 18
Sedimentation pattern of discharge zone of the Ganges River into the Bay of Bengal south of Bangladesh, LANDSAT band 5.

The choice of test area was dictated in part by company interests and proximity, but was also based on a set of conditions that offered an optimum appraisal of the potentialities of exploration from space. Thus, the Anadarko Basin of southwest Oklahoma and the Texas Panhandle was selected for detailed analysis. This is one of the older oil-producing regions in the United States, having been developed in the 1920s. More than 25 major producing fields occur in the structural traps within the basin and many more are found in stratigraphic traps.

The basin consists of a west–northwest trending subsidence trough that was filled with more than 45,000 ft of epicontinental sedimentary rocks from the early Cambrian through the Pennsylvanian. Maximum sedimentation took place in the Pennsylvanian when about 11,000 ft of clastic rocks were deposited. Subsequent to this the basin underwent strong deformation, more or less contemporaneous with that in the Ouachita Basin farther east, leading to a steepening of the southern basin flank accompanied by extensive basement faulting that carried crystalline rocks upward at the Wichita Mountains. Following rapid erosion, about 3000 ft of Permian red beds, saline deposits, and carbonates was laid unconformably over the now asymmetric basin—these units are only slightly deformed. Thin deposits of Tertiary sediments cover the western part of the basin.

The modern-day surface, therefore, consists of flat-lying Permian and younger sedimentary rocks that show little or no direct evidence of the subsurface conditions around the structural traps (except for several fields along the south limb). Most traps lie below 2000 ft, but many occur at depths of 8000 to 12,000 ft, and, more recently, production has reached 17,000 ft and deeper. The present surface is used both for farming and ranching. Scattered woods mixed with shortgrass are interspersed with farmlands on the eastern side. This gradually gives way westward to country used primarily for grazing in open, rolling long-grass prairie and sagebrush. The test area, then, is characterized as low to high plains on which several different vegetation covers have developed on soils derived from near-surface rocks that have no involvement with the more deeply buried and "masked" producing zones. This type of petroleum environment provides an exacting test of the capability of LANDSAT to "sense" clues to hidden petroleum reserves.

The Eason Oil investigators have compiled one of the most comprehensive and analytical reports in the geology phase of the LANDSAT program. They have thoroughly documented the variety of techniques used in their study. They have demonstrated the degree of reliability to which small-scale mapping can be done from space imagery and have shown the value of multiseasonal coverage in separating and identifying surface units. They have also evaluated the presumptive cost effectiveness of the LANDSAT approach to petroleum exploration—major savings in the initial (or reconnaissance) stages of exploration are indicated and further savings are suggested for later stages, for example, a potential reduction in the number of seismic lines that may otherwise have been planned, which could then be eliminated over areas rated as unfavorable for accumulation. However, in the remainder of this survey of their results, we shall confine our attention to the two most promising observations that could lead directly to the discovery of petroleum.

First, as did most other investigators the Eason Oil group gathered a great deal of new information on surface linears. This is almost self-evident in comparing Figs. 19A and B. The increase in known linears as defined in LANDSAT images is impressive, so much so that it raises a question of doubt in any petroleum geologist (by nature a skeptic made so through the experience of many dry holes) as to their validity. The immediate reaction is to wonder why only a few of the linears have been found in the field or through subsurface drilling. One obvious answer is that they have, but the policy of proprietary use practiced by the oil companies has prevented the linears from being recorded on published maps. However, the investigators discount this as a prime explanation. Another answer resides in the definition of linears; many may be nongeological and only a fraction might be fracture zones, faults, or surface expressions of basement lineations. The investigators again counter this by claiming that almost everywhere they field-checked roadcuts along the linears path they found some verification of a recorded linear, even though that same linear was usually "invisible" in aerial photos or by visual inspection from an aircraft. The

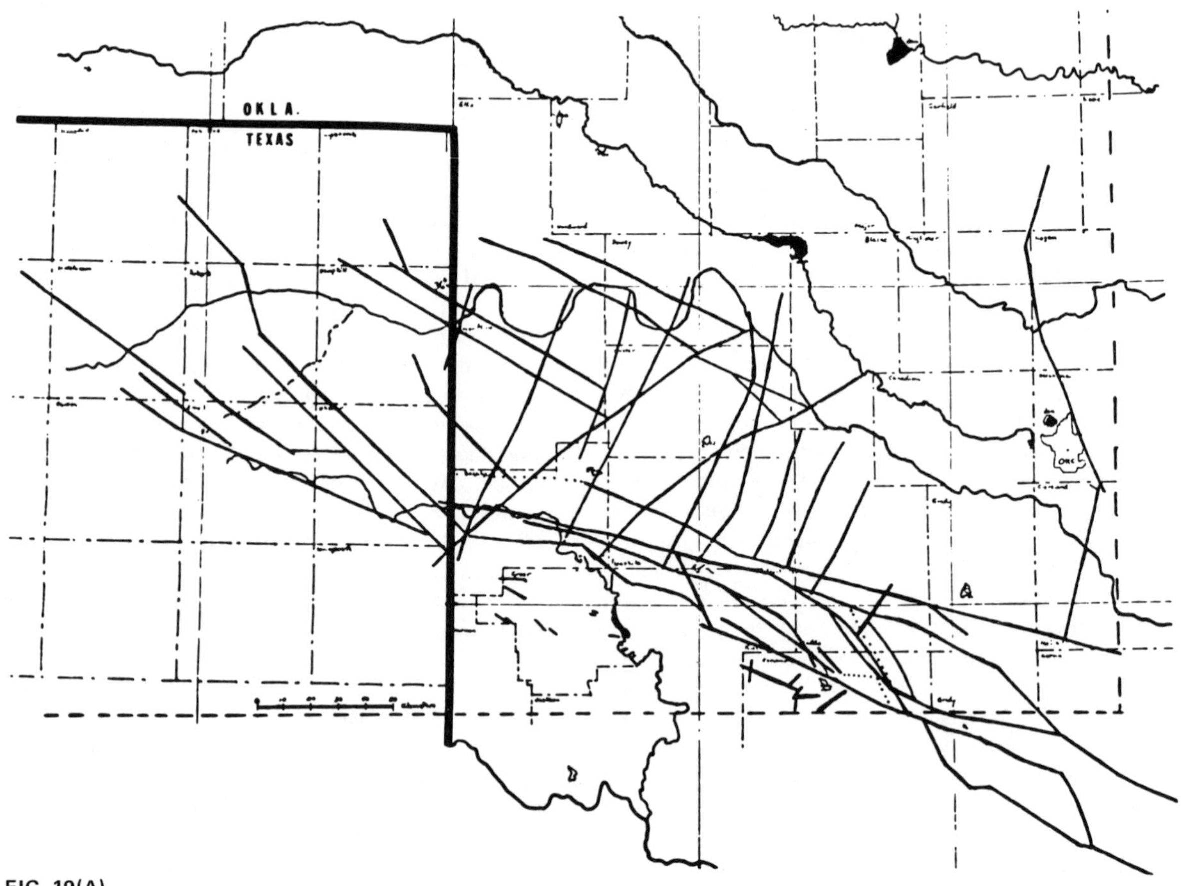

FIG. 19(A)

Tracing on a geographic base map of the major known and hypothesized faults and linears expressed at the surface over the Anadarko Basin. (From Eason Oil Company final LANDSAT report.)

question remains open but, suffice to say, the likelihood that many of the linears actually exist as structural features seems reasonably high in the Anadarko Basin. The importance of this in oil finding is that (1) more oil traps can now be looked for, (2) the linears themselves—if they are fracture zones—can be loci of increased porosity, especially at intersections, and (3) the patterns revealed by the linears should better define regional stress fields and trends of structural adjustment.

The second important result concerns several types of surface anomalies defined by the investigators. A few individual examples can be seen in the LANDSAT image reproduced in Fig. 20, although most are revealed only by inspection of transparencies on a light table. Most of these anomaly types have been categorized as *closed anomalies,* in that they can be circumscribed by a boundary line. One type clearly is due to drainage patterns, some of which represent adjustments to subsurface structures, such as an anticline over which younger strata have unevenly settled. Another type is topographic, caused in some instances by differential erosion. Still another comes under the heading of geomorphic and is based on

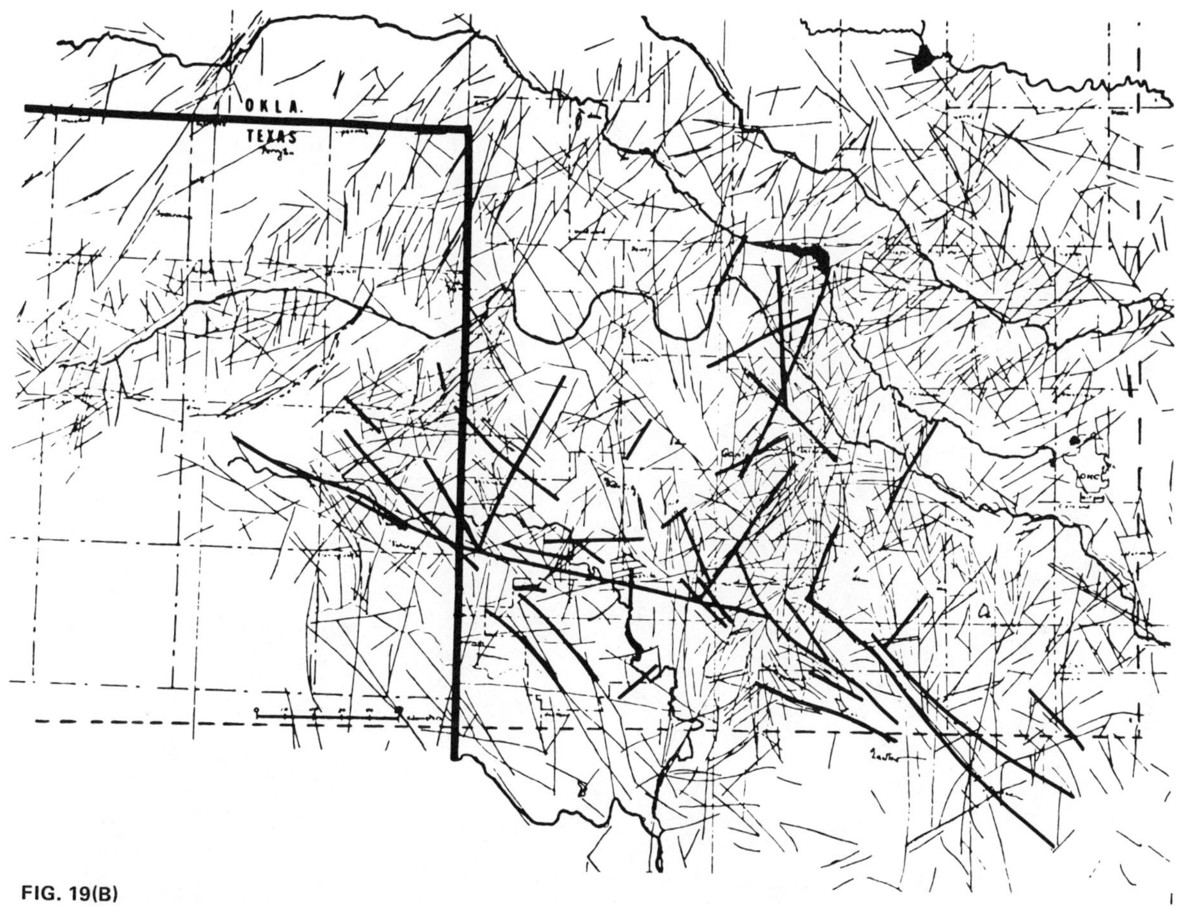

FIG. 19(B)

Linears recognized in LANDSAT imagery covering the Anadarko Basin and surrounding areas in Oklahoma and Texas. Heavy lines emphasize ERTS linears coincident with or extending from linear features plotted in Fig. 19A.

subjective interpretations by the investigators. Less definitive, but nevertheless recognizable in the imagery, are the two classes termed tonal and textural anomalies. Tonal anomalies are expressed as unaccounted-for differences in gray level, whereas textural anomalies appear as streaked, mottied, or rough patterns in the imagery. These may have a variety of causes, from cultural activities, or as peculiarities in soil, rock, or vegetation cover, through geological effects of uncertain nature, to unusual reflectances of unexplained origin. Figure 21 presents a map of the large or more conspicuous closed anomalies over the Anadarko Basin. Many of these, however, show no positive correlation with, or superposition on, underlying oil and gas fields.

Two classes of anomalies yield a high correlation with the location of producing areas. One is the geomorphic–topographic type. The other is the so-called *hazy* anomalies (see Fig. 20), described by the investigators as resembling a "blurring" or "smudging" of the photo image. A map of their distribution in the basin appears in Fig. 22. The extent to which

FIG. 20
LANDSAT band 5 image covering part of western Oklahoma and some of the Texas Panhandle. The meandering Canadian River and the North Canadian River are the principal drainage features. "Hazy" anomalies are particularly evident in the bend of the Canadian River near the center of the image.

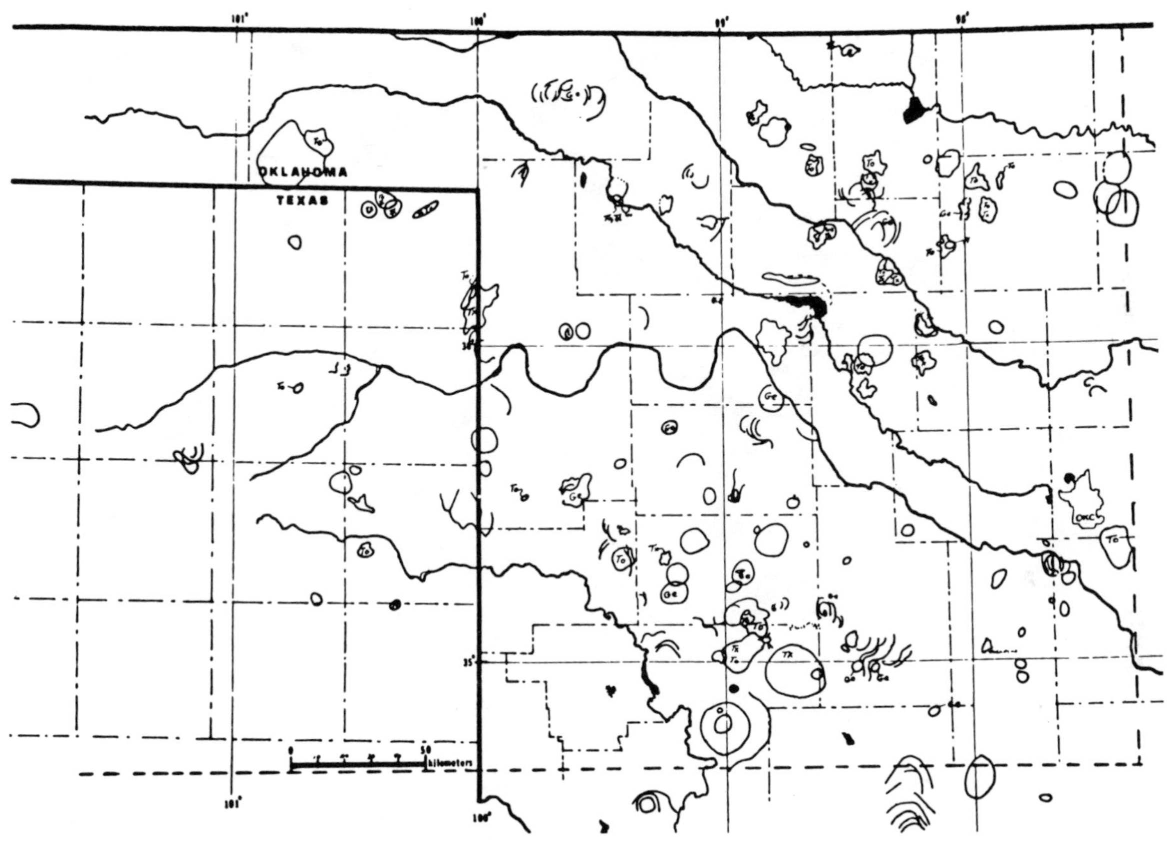

FIG. 21
Sketch map showing the traced outlines of many of the closed anomalies recognized in the Anadarko Basin during the Eason Oil Company LANDSAT investigation. To refers to tonal anomalies, Tx to textural anomalies, and Ge to geomorphic anomalies. (From Eason Oil Company final LANDSAT report.)

both classes correlate with known fields is outlined in Table 2 along with data on the other anomaly types. Some of the hazy anomalies occur over fields producing from structural traps as much as 8000 to 14,000 ft down; anomalies over stratigraphic traps are rare. The correlation is so strong that one must accept some kind of relationship between a surface effect and the presence of petroleum beneath, even though the nature of that relationship is not yet settled and is still open to suspicion.

The investigators note that the hazy anomalies are best seen in dry-weather fall images and are also well displayed in wet spring imagery. They are best developed in grazing lands on the west but are less easily found in farmlands. Almost invariably, when visited on the ground, they coincide with local areas of sandy soils that, in places, even have formed into dune-like deposits. They also tend to associate with the Ogallala formation or with Plio-Pleistocene terrace deposits.

Already these hazy features have engendered considerable debate as to their nature, cause, and value as

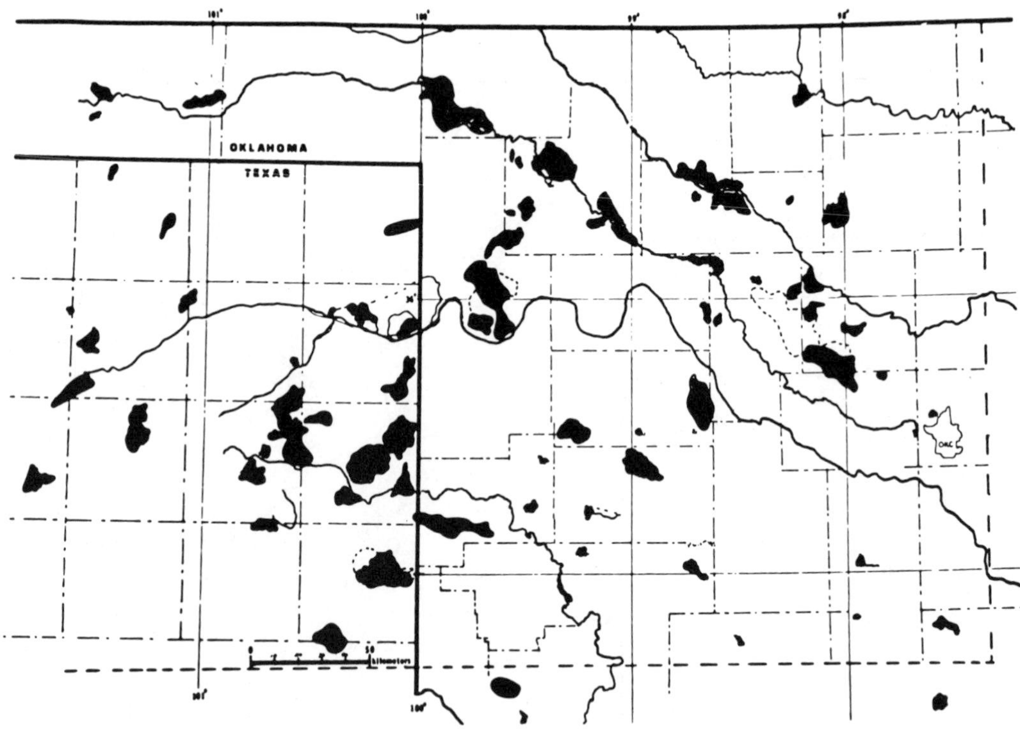

FIG. 22
Sketch map of hazy anomalies in the Anadarko Basin study area. (From Eason Oil Company final LANDSAT report.)

TABLE 2
Closed Anomalies

Survey 1 (Fall)	*Geomorphic, Tonal, and "Hazy" Anomalies*	
76	Total anomalies	
59	Producing fields	
11	Nonproductive structures	
6	No coincidence	
33 of 37	Geomorphic	Anomalies coincide with field or structure
33 of 35	"Hazy"	
0 of 4	Tonal	
Survey 2 (Fall and Spring)	*"Hazy" Anomalies*	
57	Total anomalies	
42	Producing fields	
6	Nonproductive structures	
9	No coincidence	

a guide to petroleum. Two diverse views will be discussed here; the opinion of the present author is intertwined with those of other supporters or critics.

One piece of evidence is given as a computer processed enlargement (Fig. 23) of the LANDSAT view of the most prominent hazy anomaly shown in Fig. 20, that in the bend of the Canadian River near Webb, Oklahoma. This rendition is remarkably similar in details to the view presented in an RB-57 high-altitude aerial photo. Both show extensive drilling and pumping sites along a network of roads. Ground inspection indicates the presence of stabilized sand dunes and very sandy soils. A hypothesis to explain this, and other hazy anomalies elsewhere, simply holds them to be consequence of man's activities in extracting the oil, such as damage to the terrain by countless vehicles and bulldozers plus possible effects

FIG. 23
Black-and-white version of color image photographed from the TV monitor screen of the Image 100 computer-interactive processing system at the General Electric Space Sciences Laboratory in Beltsville, Md. This scene shows a bend of the Canadian River in western Oklahoma enlarged to a scale of approximately 1:30,000 from the original scale of the same LANDSAT image shown in Fig. 20; however, this rendition is made entirely from the LANDSAT computer-compatible tape of the April 6, 1973, pass over Oklahoma.

from escaping hydrocarbons, which may have occurred prior to current land use and restoration regulations. However, the explanation does not adequately account for the extensive sand deposits unless it is assumed that so much vegetation has been stripped off that the area(s) converted to a sand waste during the dust bowl days (or at a later time).

A competing explanation is more exciting and desirable to those seeking to show the utility of orbital exploration. This postulates that hydrocarbons have leaked or migrated up and out from the trap and upon reaching the surficial layers introduce chemical and/or botanical changes. This idea is old; the notion of telltale alterations by escaping gases or fluids has long been advocated despite meager proof. One variant holds to geochemical changes in bedrock or soil resulting in bleaching or staining to produce color anomalies. Another envisions the escaping substances as reacting with the country rock to induce distinctive metasomatic products. A third considers the hydrocarbons as capable of moderating surface vegetation, either by killing off plant life and therefore accelerating erosion, or by "fertilizing" the soil and thereby stabilizing growth. This last view might seem to apply to the sand-rich hazy anomalies, if toxic hydrocarbons have damaged the grasses and sage, allowing an increase in loss of the fine fractions from the soils, or if beneficial hydrocarbons have fostered thicker vegetation that causes the sand to pile up.

So far, no decisive data have been gathered to explain the specific nature of these anomalies. However, T. J. Donovan of the U.S. Geological Survey reports in the March 1974 issue of the *American Association of Petroleum Geology Bulletin* (v. 58, pp. 429–446) the results of some highly relevant analyses at the Cement field near the axis of the Anadarko Basin. This field occurs within a long doubly plunging anticline that, unlike most in this basin, has prominent surface expression. Exposed bedrock consists of Permian red sandstones and gypsum beds. In the field in many places, the sandstone has been bleached to a yellowish-brown to white color (Fig. 24). This is ascribed to the reducing action of hydrocarbons carried in expelled reservoir pore waters. The gypsum beds are converted to low-porosity carbonate rocks. Sandstones are also recemented by migrating carbonate solutions. Both the metasomatized gypsum and these sandstones tend to form more resistant rocks that stand above the terrain as butte-like outliers.

Such effects have been observed before but their relationship to hydrocarbons has never been firmly established. However, Donovan has determined the carbon and oxygen isotope compositions (Fig. 25A and B) of the Permian sandstone and has been able to explain the extremely anomalous isotopic compositions—a deficiency of ^{13}C and exceptionally high ^{18}O—by a plausible geochemical model dependent on the reducing action of CH_4 and other hydrocarbons. He has also found anomalies at six other localities overlying producing fields.

Strangely, the Cement field itself is detected only with difficulty in the LANDSAT imagery. This is unexpected in view of the prominent color and rock-type changes noted at the surface. Special enhancement techniques (e.g., those developed by Goetz and Rowan) may be needed to discriminate these surface alterations in some kinds of terrain or surface materials. However, Donovan's results look highly promising as a general guide to petroleum accumulation if it is ultimately proved that (1) reservoirs leak, (2) the escaping products reach the surface, (3) these products interact with surface materials to bring about discernible changes, and (4) the changes are detectable from space.

In summary, I shall state briefly that some of the observations made at the Anadarko Basin, together with the results of Goetz and Rowan, have a direct bearing on prospecting for certain kinds of ore deposits. Like most mineral deposits, uranium ores tend to be localized by fractures; hence any new information on linears will be helpful in exploration. One type of uranium deposit (the sedimentary roll deposits of Wyoming and Colorado are excellent examples) in most instances is accompanied by secondary hydrated iron oxides released from the original source rocks or the eventual host rocks in the migrating solutions that concentrate the ore. This gossan is commonly located at the surface over the roll front. Present exploration methods include a search for iron staining of soils or

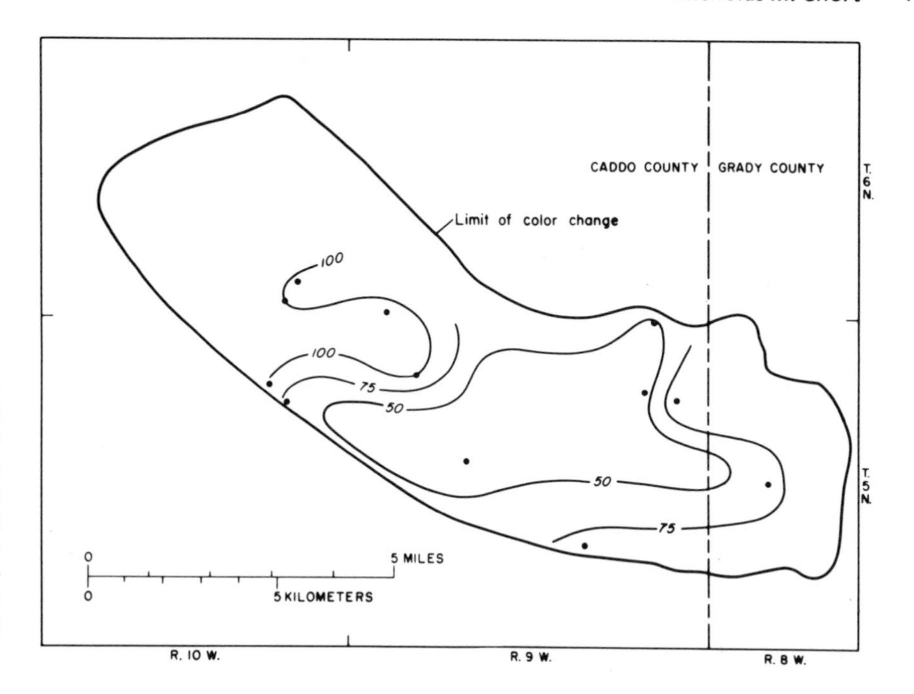

FIG. 24
Contoured values of total iron content (in parts per million; each number should be multiplied by 100) within the "bleached" areas of the Rush Springs sandstone (Permian) exposed at the surface over the Cement field, Anadarko Basin, Oklahoma. Solid dots represent sample locations. (Courtesy T. J. Donovan, U.S. Geological Survey.)

surficial bedrock, and in some instances the limonite alteration products are widespread enough to be detectable from aircraft (and by presumption from space). The band ratio method that worked in the Goldfield, Nevada, district has now been tried on several areas of known uranium roll deposits, producing definite indications of iron oxides associated with the rolls.

POTENTIAL FOR PETROLEUM EXPLORATION FROM SPACE

Too little work has been done to date to allow a definitive prediction about the outlook or prospects for success of petroleum prospecting from satellites or other orbiting platforms. However, enough has been learned already from the LANDSAT investigations in geology, and particularly those carried out by Eason Oil Company and by Goetz and Rowan, to permit a summary to be made of the positive factors that offer real promise for the approach. These are the following:

1. Improved perspective of exposed surface structural expressions.
2. Subtle indications of subsurface structures through drainage control (circular or offset stream patterns) and other geomorphic anomalies.
3. Direct indications of linears or circular features that can be related to local or regional fractures and lineations, which also influence the localization of petroleum.
4. Recognition of distinctive tonal or hazy anomalies, especially through enhancement techniques, that are eventually proved to be caused by or related to natural petroleum occurrences.
5. Association of geobotanical anomalies with the

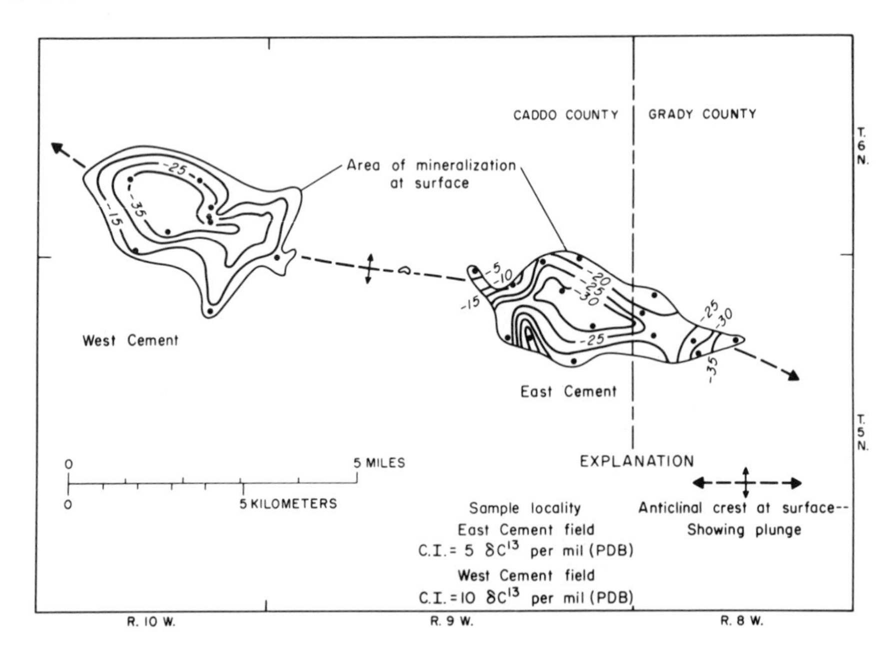

FIG. 25(A)
Variations in carbon-isotope composition within the area of strong carbonate mineralization of the Rush Springs sandstone at the Cement field. (Courtesy T. J. Donovan, U.S. Geological Survey.)

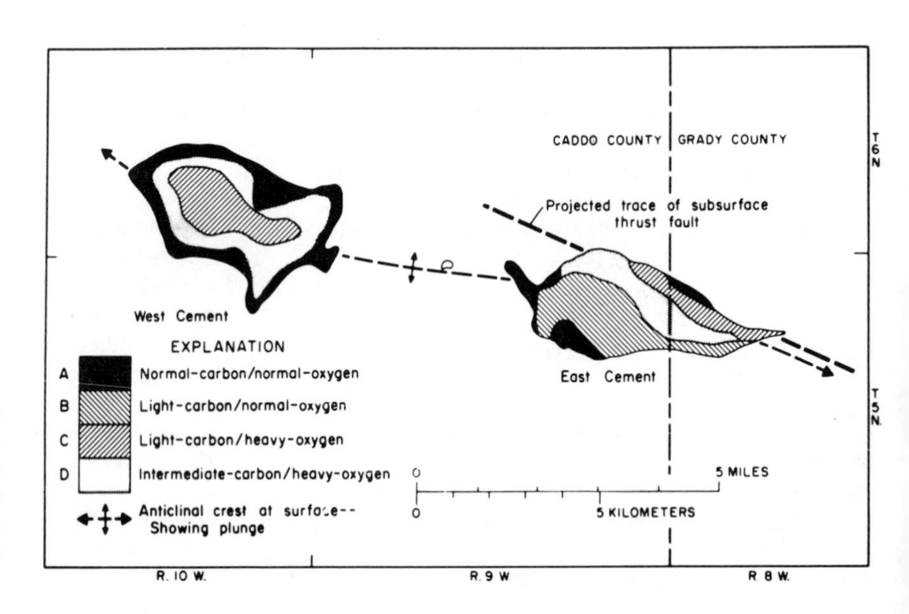

FIG. 25(B)
Distribution of different types of carbonate cement in the Rush Springs sandstone based on the relative proportions of different carbon and oxygen isotopes. (Courtesy T. J. Donovan, U.S. Geological Survey.)

subsurface presence of petroleum; this has not yet been established.

Enough experience is at hand to suggest a suitable working procedure for the analysis of LANDSAT imagery in search of guides or clues to petroleum. The procedure should begin with a standard photogeologic interpretation of LANDSAT transparencies and paper prints leading to maps of surface materials units, linears, and various types of closed anomalies. This should be accompanied by preparation of appropriate color composites. Special processing should follow; this would include edge enhancements (by optical, electronic, or computer techniques) to bring out more information on linears and band ratioing (with color output) and other computer-based data-reformating methods.

At this point it is wise to reiterate emphatically that remote sensing from aircraft and/or spacecraft is not a "magic black box" method that will supercede the many tried and true field and instrument methods used over the years in petroleum exploration. Remote sensing is still just another "tool" in the workbox of skilled, intuitive exploration geologists. It will be put to its best use when it is considered as another data source to be integrated with field mapping, subsurface data, and methods that can detect geophysical, geochemical, and geobotanical anomalies.

A word of caution is in order due to the limitations already set forth by doubting critics (which include from time to time even some of us who are actively trying to apply remote sensing to petroleum exploration). Among these are the following:

1. Surface indications of oil and gas are relatively uncommon and most have probably already been detected in the more accessible parts of the world.
2. Today most new oil discoveries are being made primarily through use of geophysical methods and by drilling, inasmuch as the most decisive data now needed relate to deeply buried subsurface conditions that generally have poor surface expression.
3. The precise role of linears in localizing oil has not been fully established, and depends on time of formation, depth to producing zone(s), cap-rock characteristics, etc.
4. Geological complexities (unconformities, glacial cover, etc.) unrelated to petroleum accumulation frequently mask oil and gas traps.
5. The hypothesis that hydrocarbons can escape to the surface and cause recognizable alteration effects is unproved; the rate of escape may be less than the rate of surface erosion or other removal factors.
6. It is difficult to assess man's role or that of vegetation in producing (or obscuring) tonal and hazy anomalies.

SUMMARY

The results of the Eason Oil study have been sufficiently encouraging to prompt NASA to continue and expand its efforts to evaluate the feasibility of exploring for petroleum (and, to a limited extent, uranium) from LANDSAT and Skylab data already on hand and from satellites or manned missions planned for the future. The philosophy behind this decision is as follows:

1. As part of the mandate in its Earth Resources program, NASA has an obligation to demonstrate the value of remote sensing from aircraft and space platforms and to make the resulting information public.
2. At this time, at least, NASA must do this "proofing," as the oil companies are likely to be reticent in revealing their own conclusions.
3. For a start, it is important that the Anadarko results previously reported by thoroughly checked and understood.
4. Positive, or at the minimum encouraging, results should be obtained from no less than five other petroleum provinces or basins, preferably from a variety of surface terrains, before it is reasonable to argue for general acceptance of orbital remote

sensing as a valuable adjunct to petroleum exploration.

Evaluation of investigation results from LANDSAT-1 leads to the conclusion that this satellite can do many of the same prime tasks in geological applications done in the past four decades with aerial photography. A broad challenge has been thrust upon geologists to more fully exploit and extend these accomplishments. The ever-rising demands for more energy sources and related raw materials places petroleum exploration near the top of the list of priority applications of space-acquired data that must still be demonstrated and implemented.

8

THE ROLE OF REMOTE SENSING FOR ENERGY DEVELOPMENT

John E. Johnston
Frank J. Janza

John E. Johnston is Deputy Chief for Coal Resources with the Office of Energy, U.S. Geological Survey, Reston, Virginia. Frank J. Janza is a Professor of Electrical Engineering, teaching remote sensing at California State University, Sacramento, California.

PART A
REMOTE SENSORS, THEIR FUNCTIONS, AND METHODS OF DATA ENHANCEMENT

In this decade remote sensing (RS) has captured the imagination and attention of scientists, engineers, and many other people in government and industry. The term "remote sensing" is generally restricted to the detection and analysis of radiant energy (emitted or reflected) in some part of the electromagnetic (EM) spectrum. Sensing of phenomena, such as gravity, magnetism, and force fields, are here excluded, as is the detecting of odors and chemicals. An abundance of literature describes various types of sensors operating in the EM spectrum, and methods of recording, measuring their response, and displaying selected data. Instruments and techniques have been designed, built, and described that increase the accuracy and reduce the time involved for data reduction, analysis, information extraction, and detection of change. Sensors and ancillary equipment are used in the laboratory and field or from aircraft and spacecraft. In the development of energy, laboratory use of remote sensing and associated techniques should not be ignored.

It is not the purpose here to describe in depth each component and all the various RS systems and their "trade-off" values in terms of effectiveness or efficiency. A summary of sensors and ancillary equipment is included as background necessary to evaluate "data use technique" applications to energy development. The prime purpose is to describe problems of obtaining energy from various undeveloped natural sources and to point up remote sensing and relate technology that could play a role in energy development. The feasibility of several sensing systems and techniques has been experimentally demonstrated. However, not until problems of energy development that are seemingly unrelated become clearly focused will RS requirements be defined for the development of energy; this subject will be discussed in Part B.

Background on Remote Sensing

Remote sensing is the technology of studying objects from a distance by measuring and recording energy from one or more segments of the EM spectrum. The applications of RS techniques specifically for energy development have been underway for less than a decade.

The early history of the RS field is covered in Chapters 1 and 2 of the *Manual of Remote Sensing* (1975); subsequent chapters provide background, theory, and specifics of remote sensing, sensors, and systems. Only a summary of the elements useful to energy development is presented here.

A RS system may include one or more sensors for collecting spectral data, a data-storage device, such as film or magnetic tape, and devices for measuring, analyzing, formulating, and displaying the spectral data. Remote-sensing systems, components, and peripheral special-function devices may be classified on many bases.

A few introductory questions should be asked before presenting pertinent details on remote sensing. What makes remote sensors useful tools? This question leads to a more definitive set of questions:

1. What interaction mechanisms of phenomena transform or modulate the EM carrier so that they convey information about the surface or subject?
2. What sensors are available; what are their capabilities and limitations?
3. What regions of the total EM spectrum are utilized?
4. What surfaces or subjects are sensed?
5. What disciplines are covered?

PRINCIPLES OF REMOTE SENSORS AND NATURE OF DATA COLLECTED

Interactions of Electromagnetic Radiation Utilized in Remote Sensing

The numerous kinds of remote sensors that are currently employed and which are herein discussed in part utilize electromagnetic radiation (EMR) collected from the land, water, and atmosphere over the EM spectrum from microwaves to x-rays. Electromagnetic radiation is propagated from an initial source and transmitted to a remote sensor. During the transmission, the EMR undergoes absorption, reradiation, reflection, scattering, polarization, and spectrum redistribution.

The changes in the EMR from source to sensor can be subtle or pronounced, depending upon the interactions that take place with the media or materials involved. These changes provide data for the analyst or photointerpreter about the properties of these media on materials. They become the heart of the RS analysis, since they supply vital information that can be identified with the temperature, moisture, texture, and electrical properties of whatever has been surveyed. Cause, and, particularly, effect are primary concerns, for herein lies the foundation for interpretation of data collected by the remote sensors.

The remote sensor is a transfer device that changes the collected EMR at the input aperture to a usable form at its output. This output is generally processed electronically or chemically to provide data on the subject as analog or digital records on magnetic tape, imagery (photographs, scan-line prints), or visual displays (TV monitors or oscilloscopes).

The parts of the EM spectrum available for remote sensing are given in Fig. 1. Remote sensors have been developed to operate in either narrow or broad spectral bands throughout this vast spectrum. This chapter, however, is necessarily limited in its scope to techniques and equipment found in the ultraviolet (UV), visible, infrared (IR), and microwave spectrums.

EMR sources. We are not primarily concerned with the generation of EMR in this discussion, but rather with the manner in which it is utilized in remote sensing. It is important, however, to identify whether the EMR generation is passive or active, since the interaction phenomena can vary considerably. The sun, for example, could be stated as our greatest "passive" source, radiating a very intense and broad spectrum of EMR. In addition, the moon, the earth, and its atmosphere are passive broad spectrum sources of considerably less intensity.

The "active" sensors are largely man-made and can be thermal or radio EMR generators. Lasers and radars are examples of such generators that are now actively utilized as remote sensors. Active EMR, from which extremely high power densities can be generated over very narrow spectral bands, is contrasted with that obtainable from passive systems, which generate power densities that are considerably lower and have much broader spectral bands. Active sources are usually monochromatic (single-frequency) generators from which direct control can be maintained over the intensity. The intensity from these generators can be orders of magnitude greater than the material background radiation.

EMR coefficients, ratios, parameters. Much research and experimentation in remote sensing are conducted to determine the changes or interactions that are meaningful in the application of remote sensing to energy development, geology, agriculture, meteorology, oceanography, and other disciplines. These changes in the EMR are sometimes called "signatures," which correlate the sensor outputs with such surface variations as temperature or moisture change.

The sun, as a source of EMR, provides an entree to the subject of interactions, since it is common knowledge that it irradiates the atmosphere and the earth's surface, where both, in turn, absorb or redirect the EMR. The redirected radiation is mildly or appreciably affected by the electrical and physical properties of media, and knowledge of these effects is vital to the identification of signatures.

The EMR spectrum of the sun at the outer limits of the earth's atmosphere is considerably different than that detected at sea level. Solar EMR is subject to gross and selective absorption and scattering by the atmosphere. The EMR incident at the earth's surface is largely absorbed by the ground, thus raising its temperature. The longer wavelength radiation propagates below the surface to be absorbed or reflected back by the interfaces between differing media; however, much of the incident radiation is directly reflected from the ground surface. These reflections can be intense or diffuse, or both, depending on the wavelength of the EMR. When the EMR is broken up into undefined groups of rays or wave fronts, it is said to be scattered.

Remote sensors are made to respond to EMR instantaneously or to varying degrees, such as with time-averaging techniques. Short-time variations in the incident or reflected radiation can be detected and measured. It thus becomes apparent that the usefulness of remote sensing and its further development hinge on the understanding of instantaneous and long-time EMR interactions due to reflection, scattering, and absorption. The latter terms are stated as simple ratios or coefficients, but more often as involved discrete or statistical mathematical expressions, which are functions of frequency, angle of incidence, polarization, temperature, etc. Special phenomena, such as diffraction and refraction, are also included.

Specific examples of EMR interactions. The important goal then with remote sensors is to correlate their output with the subject being surveyed. To achieve this goal requires accurate knowledge of the effects on the collected EMR caused by the subject and of the anomalies introduced by the sensor itself. In the analysis of these effects, which may be in the form of simulated models, first-order approximations are established for the coefficients of reflectance, emittance, absorption, and transmittance. Such approximations, though valuable, leave much to be desired for a complete picture of the signatures.

The well-known equation for the Fresnel voltage reflection coefficients uses the ideal assumption of a perfectly flat interface of infinite extent between two homogeneous and isotropic media. Such conditions in nature are rare indeed. Consequently, to show how these basic coefficients need to be modified to fit real conditions encountered in the application of remote sensing, a more direct and useful approach needs to be taken by the researcher. The analysis of a number of actual experiments using remote sensors is essential.

By studying the theory and techniques of selected examples in the visible, infrared, microwave, and x-ray regions, the variations in these expressions are revealed and are correlatable with observed or measured values.

Each researcher generally selects an appropriate model to show the differences and refinements necessary in the common definitions of reflection, scattering, and absorption for application of remote sensing to specific parts of the EM spectrum.

Modeling. The tools of modern science—the computer, modeling, statistical analysis, information theory, etc.—have been relied upon heavily in building up libraries and sources of data for RS applications. Of these, modeling and the computer are playing important roles in the science of remote sensing. Interpretations of RS observations can be conducted at two levels of sophistication:

1. Interpretation of image display or recorded analogs.
2. Digital computer analysis based on "automated" processing techniques.

In both cases the observations are compared to preconceived standards: in the former case the previous experience of the interpreter plays the major role in his recognition criteria; in the latter case, statistical standards are commonly based either on training sets carefully selected by the investigator, or on the presence of distinct populations that can be studied by cluster analysis. But the relationship between RS data and variables of interest for resource study is greatly complicated by both the number and the magnitude of the factors that affect the recorded signal. These factors encumber the application of empirical techniques based on either manual or machine processing.

Mathematical models, derived from fundamental physical laws, provide powerful techniques to accomplish the following:

1. Examine correlations between variables.
2. Test assumptions about suspected major factors.
3. Discover and analyze unsuspected factors.

An effective approach in the development of these models is to start with observations of simple areas, analyze the results in terms of simple models, and iteratively increase the complexity of the models and sites in a systematic manner.

Source parameters. The parameters relating to the ground are, in part, both structural and textural. Information about the gross and fine details of the ground surface is important in the analysis of RS data. Even more important is information about the electrical parameters of the surface, how they vary with location and time, and their relationship with ground or soil properties. *The key parameters common to the visible and infrared regions are reflectance, p, absorptance, a, emittance, e, and transmittance, t.* These parameters are functions of the wavelength, angle of incidence, and polarization, and are related to units of power.

Power and voltage parameters. The term reflectivity, or voltage reflectivity, *R,* is used in the common terminology of longer-wavelength radar. The relationship between the reflectivity and the reflectance, *p,* is mathematically expressed as follows: $|R^2| = p$. Reflectivity is related to the dielectric properties of the substance, its conductivity.

Atmospheric parameters. The atmosphere is also described by the parameters identified above, except that textural properties are no longer appropriate; rather, properties such as stratification, layering, and turbulence are considered.

Platform and sensor parameters. Another set of parameters associated with the platform and remote sensor, and important in modeling and in the consequent interpretation of results, is geometric; these parameters establish the spatial relations of the remote sensor and the ground or region being surveyed. Generally, spherical coordinates are used to show sun, satellite, or aircraft position and the direction of the EMR.

Remote Sensors

Since remote sensors are the heart of the data gathering or measuring system, some of their pertinent characteristics are first reviewed. Numerous remote sensors exist, but the ones listed are those most widely utilized:

1. Photography (including cameras, lenses, filters, and films)
2. Television
3. Infrared scanners
4. Infrared radiometers
5. Infrared imagers
6. Spectrometers
7. Multispectral scanners
8. Radar
9. Microwave radiometers

As noted before, sensors are called active if their energy source is an integral part of the system (e.g., a side-looking aircraft radar, SLAR). Passive systems rely mostly on the reflection or reemission of solar energy from the surface of the object sensed.

Material for the following description of these sensors and systems has been summarized from the literature, particularly from the *Manual of Remote Sensing* (1975) and MacDowall (1972).

Photography. Photography is the most widely used and most highly developed method of remote sensing of earth resources. Extremely high resolution (spatial and spectral) images can be collected, and with well-trained interpreters abundant data and amounts of information are available from photography. Photographic film is one of the most compact information storage methods presently available; as such, the camera can collect huge quantities of data easily and concisely.

Photography has constraints not inherent in other sensors. One constraint deals with quantitative analysis. With film, obtaining quantitative information on radiation levels is rather difficult since the film density is a critical function of a large number of factors including film type, exposure parameters, and the method of developing and processing. Even without the ability to obtain quantitative data, a tremendous amount of useful information is still available from photographs. In view of the many advantages of photography, a considerable effort is justified in retrieving the photography taken by unmanned spacecraft. Presently, it is necessary to resort to electrooptic sensors to receive imagery taken by such spacecraft. A compromise in spatial resolution is therefore necessary.

Three basic types of cameras, frame, slit, and panorama, have been developed. The first of these takes a single frame of the field of view with an open–close type of shutter. The frames are generally overlapped, about 60 percent, to provide a stereoscopic effect. By flying over a ground area with a number of sweeps of the aircraft, the total area can be scanned. This method is very good for mapping and producing aerial mosaics; however, if camera platform velocities are high, image motion compensation (IMC) is required; otherwise, the scene is smeared.

The slit camera focuses a strip of the area of interest on a narrow slit behind which the film is placed. As the aircraft is flown the film is advanced at a rate proportional to the velocity–altitude ratio of the aircraft. This system requires no IMC; however, the film rate must be very accurately controlled to obtain good imagery.

The third camera type is the panoramic. With this type of camera, the field of view of the lens is swept transversely to the flight path of the aircraft; coverages of 180° are common. The image is then recorded on a film that is swept synchronously with the lens sweep. By sweeping the camera so that the coverage is contiguous, very wide swath strip maps can be built up. Large amounts of information can be collected rapidly with these cameras. However, they are not suitable for mapping, mosaics, or other purposes requiring structures shown normal to the surface (orthonormal).

Photographic cameras operate in the visible and near-infrared (IR) portions of the EM spectrum. At these wavelengths, EMR has no cloud penetrating capabilities; as a result, photography can be used only on cloudless days.

Photographs can be taken on a number of different films including black-and-white (B&W), black-and-white with IR sensitivity, color (in which radiation is recorded in three spectral bands, generally red, green, and blue), and false color (where infrared, as well as the visible bands, are assigned different colors). Regular IR color film has the infrared band in red or shades of red depending upon the reflectances in the viewed scene. Color photography generally adds information to the imagery obtained, which makes qualitative interpretation somewhat easier.

The utilization of IR radiation in false color gives two advantages over photographs taken in the visible band. The first is the ability to penetrate haze, which can and frequently does attenuate shorter-wavelength radiation, and the second is its sensitivity to reflectance changes of vegetation. Since chlorophyll reflects strongly in the IR region, living vegetation shows up strongly in false-color photography.

The resolution of a camera is limited by the size of the lens and its aberrations. The ultimate angular resolution is given by $\theta = \lambda/D$, where D is the diameter of the lens in wavelengths, λ is the wavelength, and θ is the angle of acceptance in radians. The angle

θ also determines the ability of the lens to separate two point sources or objects, and determines the spatial resolution. Achievable performance is limited by the elimination of aberrations over the entire wavelength range of interest.

Another consideration in the resolution of a photographic system is the film resolution. The film grain size must be chosen so that its spatial resolution is higher than the lens spatial resolution; otherwise, a degradation in the overall resolving power will result. Generally, fine-grained films can be selected with sufficiently high resolving power so that lens resolution is the controlling factor.

Although photography is an extremely useful and widely employed method of sensing earth resources, its usefulness as a space sensor is limited by the difficulty of either digitizing the film-recorded data or returning the processed hard copy to earth. Development of an unclassified system for returning photographic film from unmanned spacecraft would be most valuable.

Television systems. One of the most commonly used instruments for space imagery and observation is the television (TV) system, which is popular because of small size and weight, lack of moving parts, and the ease with which images can be transmitted to earth. The TV system also has very good low-light-level capability and resolution.

The television camera forms frame images of the field of view by sweeping an electron beam in a horizontal-line (raster type) pattern across a photosensitive layer onto which the image has been focused. The scanned image is thus converted to analog form as a varying voltage with respect to time. The analog voltage scan lines are telemetered to a ground station, where a reverse process produces a television picture of the subject. The image quality, spatial and spectral, depends on the type of tube used in the system. Three types of tubes are common in spacework: image dissectors, vidicons, and return beam vidicons. Other imaging tubes, such as image orthicons and plumbicons, are available but not generally used for spacecraft. The three main types of tubes differ significantly from each other; a brief description of each type follows.

The image dissector (ID) is a camera tube with a photoemissive layer for its photocathode imaging surface. The optical image focused on this layer causes an emission of electrons that is directly proportional to the intensity of the incoming light. This pattern of electrons is then accelerated down the tube and focused at a plane in which the aperture to an electron-multiplier section is placed. The electron image is then swept over this aperture in a raster pattern by means of deflecting coils, and a highly amplified current output at the photomultiplier results, which is proportional to the light intensity striking the photocathode in the image or resolution element observed. The size of a resolution element, or the smallest area resolved, is determined by the size of the aperture and can be as small as 0.0013 cm in diameter. The electron-multiplier output is then used to modulate the beam of a cathode-ray tube (CRT), which is swept across the screen sychronously with the sweep of the image, thus building up a corresponding image in relation to the light intensity from the area under observation.

The vidicon imaging tube operates in a somewhat different manner. A photoemissive layer is not present but is replaced by a photoconductive layer onto which the image of interest is focused. Unlike the photoemissive layer in the ID tube, the photoconductive layer allows storage of the image for relatively long periods of time (0.1 s or more), thus allowing shuttered operation of the instrument. The vidicon operates by sweeping the beam from an electron gun in a raster pattern over the photoconductive surface.

The photoconductor is a material whose conductance varies with the intensity of light incident upon it. When the electron beam is swept across the surface, current flows through the layer to a signal electrode, which is between the photoconductor and the face of the tube. The amount of current flowing in the beam is then proportional to the intensity of light incident upon the photoconductor. This varying current flow is used as the signal to intensity-modulate the monitor CRT in much the same way as in the operation of the image dissector.

The remaining image tube applicable to remote sensing is the return-beam vidicon (RBV), which operates in somewhat the same manner as the vidicon. However, instead of obtaining a signal from the current conducted through the photoconductive layer, the electron beam is reflected from the photoconducter (the amount of reflection is proportional to the charge on the layer) and returned to the aperture of an electron multiplier, where low-noise preamplification is obtained. The signal output from the multiplier stage is then used to modulate the CRT. This type of tube generally has a higher ratio (signal to noise) and better resolution than the vidicon-type tube.

The output signals of the tubes can be used to intensity-modulate a CRT (as was described for the ID tube) and produce an image in real time of the object observed. Most of the TV systems are intended for space use; therefore, the signals from the tubes are digitized and telemetered to earth along with synchronization information. These data are then used to produce imagery or are stored on magnetic tape for later processing.

The wavelengths at which these tubes operate are generally within the visible spectrum. Exact spectral bands are determined by the type of photocathode used. There are a number of major classes of photocathodes depending upon their spectral response, and any one can be used as the photosensitive layer in the image tube. Graphs of the spectral sensitivities of the various photocathodes are readily available from companies that manufacture imaging tubes. Since the systems operate in the visible spectrum, they depend upon scattered light for their illumination and are, therefore, generally limited to daytime use, although some low-light systems capable of night use have been developed. Also, since cloud penetration is not possible at visible wavelengths, the systems can only be used in clear weather. Thus, the use of this equipment is severely limited when compared with thermal IR systems or microwave systems.

The resolution of a TV instrument is limited by two parameters, the actual resolution capabilities of the imaging tube in lines per millimeter and the focal length of the optics. The resolution capability of tubes is determined by measuring the modulation transfer function (MTF) (the amplitude modulation of an input signal as a function of the frequency of the signal), or by using simple bar charts. Resolution of imagery tubes may vary widely from 10 to 100 lines per mm. If the focal length of the optics is sufficiently large, the resolutions may be made very small. The angular resolution is given by $R = 1/R_T f$, where R_T is the resolution of the tube in lines per millimeter and f is the focal length. In practice, the focal length is limited by physical considerations; among these are the size of the optics, which is generally limited in spacecraft use, and the field of view desired. Adequate earth coverage from a satellite becomes very difficult if the field of view is made too small.

It is interesting to note that multispectral TV systems described have relatively low spectral resolutions and ranges from 10 nm with the Wisp ocean sensor to 100 mm for the Earth Resources Technology Satellite (LANDSAT) multispectral system. These are very coarse spectral resolutions when compared with that possible with the spectrometer devices being developed, and are of the same order of magnitude as obtained in multispectral scanners.

Television has proved highly valuable, especially in weather satellites, for producing real- and near-real-time imaging of the earth's surface. These instruments have proved to be very small in both size and weight and rugged enough for reliable space use. The advantage that they possess of having no moving parts also increases their reliability and usefulness in remote sensing from satellites. At the same time, it must be remembered that TV systems can be used only in earth-pointing and stabilized satellites. Malfunctions of the moving parts in the attitude and stabilization systems will degrade the performance of a TV system.

Infrared Systems. *IR scanners:* Infrared scanners are a group of instruments that produce strip maps of emissions from the earth from the IR portion of the EM spectrum. The IR region lies between the red end of the visible and the millimeter wave region, from approximately 0.75 to 1000 μm. Radiation in this area can be divided into two parts, scattered radiation from the sun and atmosphere, and self-emitted radiation from the earth or object. Most scattered radiation is in the region from 0.75 to 4 μm; where the radiation from the sun is stronger than the self-emissions. In the region beyond 4 μm, the emitted radiation is stronger for bodies at typical earth temperatures. Thus, if a system observes IR radiation during the day in the short-wavelength region, it is principally measuring scattered radiation from the sun; if it operates in the long-wavelength region (longer than 4 μm), it measures emitted radiation. At night, of course, the radiation observed in any region is emitted rather than scattered.

The amount of radiation collected by a sensor from an object is a function of its temperature, its emissivity, and the angle at which it is observed. Any body at a temperature above 0° emits a wide spectrum of EMR, where the radiant power and peak wavelength are functions of temperature. For bodies at normal earth temperature, about 300°K, the wavelength peak lies at approximately 10 μm. The total radiated energy increases as the body temperature is raised. Thus, an object at a particular temperature will emit a given total amount of radiation, and a slightly cooler object will emit a smaller amount of radiation. Objects can thus be distinguished one from another by differences in radiated energy. These parameters are functions of their absolute temperatures and emissivities and can be used, for example, to distinguish between healthy and diseased vegetation, ice and water, and types of vegetation, and to delineate thermal currents in water.

This identification is aided or complicated by two factors. The first is the emissivity of the object, that is, its ability to radiate energy. In the IR region, emissivities can range from as low as 0.7 to as high as 0.98, where a perfect emitter (ideal blackbody) has a emissivity of unity. Since the ability of objects to radiate is not uniform, it is possible for two objects at the same temperature to appear to be at different temperatures, and vice versa. The radiation can be expressed as a radiometric temperature, $T_B = eT$, where e is the emissivity ($0 \leqslant e \leqslant 1$) and T the real or absolute ground temperature. The remote sensor functions on the radiometric temperature, which, in turn, is related directly to radiated EM power. Another factor is the large change in object temperature that can take place from day to night. Objects with a small heat capacity will change temperature rapidly, whereas those with large heat capacities (such as water) will change very slowly; thus, it is normal for a body to appear cool relative to its surroundings during the day and hot during the night.

Infrared radiation has a relatively short wavelength, somewhat less than the diameter of cloud drops, and therefore cannot penetrate cloud cover. It can, however, penetrate haze somewhat better than visible light owing to its longer wavelength. Infrared sensors therefore have an advantage since they can operate at night, can record emitted radiation, and do not require reflected daylight. Because atmospheric components, particularly water vapor and carbon dioxide, have very strong absorption bands in the IR, there are only a limited number of windows in the IR spectrum—the spectral passbands—in which this radiation can be observed. This absorption curve is given in Fig. 1, where it can be seen that eight major regions of the IR spectrum are present where the atmosphere is transparent; the 3- to 5-μm and 8- to 15-μm bands are the spectral regions most commonly used for IR measurements.

The infrared scanner mode of operation is quite simple; a small field of view (resolution element), typically between 1 and 5 milliradians (mrad), is swept across a path transverse to the aircraft flight path by means of a rotating 45° mirror. The radiation from each ground resolution element is detected and used to intensity-modulate a glow tube or CRT, which is swept synchronously with the mirror sweep. The output is then correspondingly focused onto a film and recorded as a scan line. As the aircraft flies forward, a series of lines are built up on the film, and a strip map is produced of the terrain beneath the flight path. Most scanners have facility to also record the data on magnetic tape, which can make data handling and processing easier. Infrared systems must have a detector capable of measuring small changes in temperature in order to obtain good temperature resolution. This can only be done if the detector collects sufficient energy from a resolution element to produce a signal larger than noise signal. The noise equivalent temperature is defined as the change in object temperature required to produce a signal-to-noise ratio of 1 and is the measurement of a system's temperature resolution detecting capability.

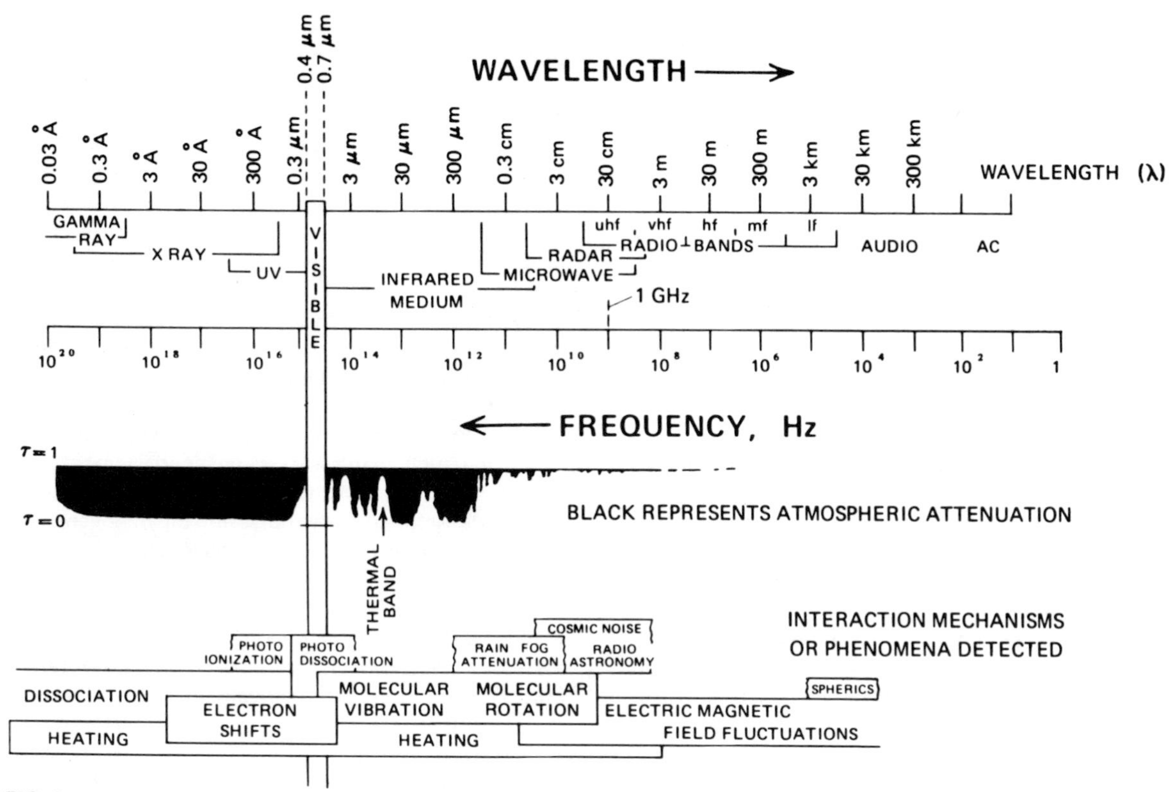

FIG. 1
Electromagnetic spectrum.

The spatial resolution, which is the other parameter of importance in such scanners, is determined by the ratio of d/f, where d is the size of the detector and f is the focal length of the optics. To obtain very fine resolution, a large f or a small d is required. However, the size of the detector that can be produced with existing technology is of the order of 0.1 mm; hence, the resolution in such systems is limited by the detector size. Furthermore, if the focal length is made very large, the field of view is very small and, hence, the amount of radiation collected is decreased, as is correspondingly the sensitivity. As a result, the temperature sensitivity also decreases. Designs of IR systems are restricted by this trade-off between temperature sensitivity and spatial resolution. Typical spatial resolutions lie between 1 and 5 mrad, and noise equivalent temperatures between 0.1 and 0.5°K.

All IR scanners are equipped with calibration sources in order that absolute measurements of temperatures can be made. These sources vary with the system, from that of a blackbody at a constant temperature to looking at skylights, but the basic principle is the same, that is, measuring an object radiating a known amount of energy and using this to establish the radiation levels emitted from the subject.

Various means are used in IR instruments to correct for aircraft motion, because variations from a straight and level flight path result in image distortions. Some of the equipment corrects for these errors before recording the signal (such as the H.R.B. Singer instruments) and others correct after the signal has been recorded on magnetic tape (such as the Daedalus scanner). Another distortion that comes into IR scanner imaging is an apparent roll-off of the terrain at the edge of the strip map. This is caused because the scanning is performed by a rotating mirror, which moves as a linear function of angle; therefore, at long distances, a larger section of terrain is swept per unit of time than immediately below the aircraft. This condition results in distorted imagery, which looks as if it were part of a cylinder. This distortion can be corrected after recording on magnetic tape or by scanning the mirror in a nonlinear sweep. The latter method is used by Texas Instrument's equipment.

Of the 18 infrared scanners studied, the types covered represent most of the unclassified scanners available.

The large number of available IR scanners indicates the usefulness of this region of the spectrum for earth resources work. A great deal of effort has gone into this field, and IR imagery has proved extremely useful for a large number of earth-observation tasks.

Infrared radiometers: In this chapter, infrared radiometers are considered to be nonscanning instruments that measure the radiant flux of IR radiation from an object or area. By comparing the energy thus collected with the energy from a blackbody source at a known temperature, the absolute temperature of the body can be obtained quite accurately. These instruments are, therefore, very useful for measuring temperatures of extended objects.

During the day, radiation with a wavelength of less than 3 μm is mostly scattered sunlight. Radiometers generally work at wavelengths longer than this, typically in the 8- to 14-μm atmospheric window (sometimes referred to as the thermal window). In the latter wavelength range, only emitted radiation is measured, and therefore the object temperature can be determined with a radiometer; however, radiometers can also operate in the short-wavelength optical region to measure scattered radiation.

As with IR scanners, the results of measurements are aided as well as complicated by the emissivity of the object being observed and by atmospheric absorption. If the emissivity of the object or terrain being observed is known, corrections in the data can be made; however, the errors introduced by the atmosphere are somewhat more difficult to correct. To make corrections for the atmospheric effects of water vapor and carbon dioxide, the atmospheric temperature and humidity profiles must be known. It is impossible to make these corrections to temperature from routine atmospheric radio soundings. It is thus possible to remotely measure the temperature at surfaces of known emissivity to an accuracy of about 1°K.

Because this is an IR system, it has day and night capabilities; however, it cannot penetrate dense cloud cover and is therefore limited in its use to cloud-free weather.

Infrared radiometers have been used widely for measuring such features as atmospheric temperature profiles and surface water temperatures. Because of their wide fields of view, however, they are generally limited to measuring fairly large scale features.

Infrared radiometers have a number of advantages for remote sensing uses since they are small, lightweight, and relatively simple instruments, and results are easily recorded and can be readily interpreted. Furthermore, they are relatively inexpensive instruments compared with other remote-sensing hardware.

Infrared imagers: As the name suggests, these instruments form a frame-type image of the IR radiation emitted or scattered from the area under observation. They differ from IR scanners in that they scan in two directions (i.e., line and field scan), thus forming an image without the requirement of platform motion; scanners, on the other hand, sweep only in one direction and require platform motion to produce the strip-map imagery. Imagers operate by sweeping a small-resolution element (determined by the detector area and the optical focal length) across the area to be covered in a raster-type pattern by means of two orthogonal mirrors. For example, in the case of the AGA System 680 there are two rotating prisms; in the case of the Dynarad 201, these are oscillating mirrors. The signal recorded at the detector for each resolution element is then used to intensity-modulate the output beam of a CRT, and the CRT beam is swept synchronously with the sweep of the resolution element in much the same way as the television system operates. The incremental temperature range to be covered can be selected manually; the entire temperature range is picked and the increment can be 1° or higher. Any temperature outside this range is recorded as either solid black or white. On the CRT screen an image of the intensity of IR radiation in the frame is then formed. This image can be either photographed, observed directly, or, in the case of the Dynarad 201, stored digitally on magnetic tape for later processing.

Imagers, IR scanners, and radiometers have the same problems of atmospheric absorption, object emissivities, and thermal and spatial resolutions, because the wavelength of light detected by all these systems is the same. In general, temperature sensitivities (noise equivalent change in temperature) and spatial resolutions are comparable to those of the scanners. Both the Dynarad 201 and Barnes T101 possess the capability of an angular resolution of 1.7 mrad, with a temperature sensitivity of about 0.1°K. Dynarad also specifies a much smaller resolution of 0.12 mrad; however, the temperature sensitivity for this resolution is not known.

These instruments are generally intended for qualitative rather than quantitative imagery; consequently, calibration sources are not included in the equipment. Absolute temperatures thus cannot be obtained as accurately as with scanners or radiometers. Thermal contouring can also easily be done with these systems by using the isotherm mode. With this method of operation, the desired temperature and a small band (as low as 0.5°K) around it are selected, and any object within this temperature range is recorded as a saturated white. This allows easy identification of isothermal areas. Most of the systems will also give color images; however, this requires time integration of approximately 10 s per frame in most cases and is, therefore, too slow for aircraft use or spacecraft platforms.

Infrared imagers are ideal for real-time qualitative detection of thermal anomalies. A considerable amount of work has gone into the development of these sophisticated instruments. Although they are not designed specifically for aircraft, they can be adapted without much difficulty.

Spectrometers. Every material emits a unique pattern or spectrum, because, under given conditions, the amount of radiation emitted from an object is recorded as a function of wavelength. The reflection spectrum is also a function of the illumination. Then, if the spectrum of an object can be observed and measured, it can be compared with spectra of known materials and identification can be made. This is the basic principle used in the application of spectroscopy to remote sensing. It should be noted that the technique is founded on the science of reflection spectroscopy, which is still in the development stage. In the case of remote sensing from aircraft or satellites, the problem is complicated by the effects of the atmosphere through which radiation must pass to get to the system. The atmosphere produces backscattered radiation, which raises the noise levels significantly as well as absorbing radiation at certain wavelengths, thereby directly affecting the spectrum being observed.

Spectrometers, the instruments that observe such spectra, come in many different types, ranging from prism and grating systems, in which the light is dispersed by an element and the resulting spread-out spectrum is detected on a photographic film or a moving detector, to interferometers using either interference filters or Michelson-type systems. The output of the interferometer types is more complex to interpret, since it records at the output the interference pattern of all wavelengths of the radiation at once as a function of path-length difference. A spectrum is obtained by taking a transformation (Fourier) of the output by the use of an optical processor. Various means of coding input and output can be used to make these instruments more effective in both dispersion and light efficiency.

Two different properties are of concern in spectrometers; these are resolving power, that is, the ability to separate one wavelength of light from another, and its ability to collect and use light. The two properties are interdependent, and in general, as the resolving power is increased, the efficiency of the instrument diminishes. Of the four types of spectrometers considered here, the prism is most limited in its resolving power. Grating spectrographs can be built to have large resolving powers, and the resolving power of interferometric types is limited only by the size of the equipment.

There are two ways in which the efficiency of the instrument can be improved. The first way is by multiplexing, that is, observing the light in more than one wavelength at the same time. This is automatically done with the Michelson Interferometer since it measures the intensity of all wavelengths. In the other, spectrometer methods can be used to produce multiplex spectra. The gain in signal is proportional to $\sqrt{n}$, where n is the number of spectral elements sampled at a given time. This is the method being used by the York University Fabry-Perot spectrometer to produce a gain in signal strength. This method has the disadvantage that the output is not in a spectral form but must be analyzed to produce the desired display.

The advantage to be gained is the "throughput," that is, the total amount of light that the instrument can collect. The throughput is proportional to $A\Omega T$, where A is the area of the collecting aperture, Ω is the solid angle viewed by the spectrometer, and T is the transmission of the system. For a given area, Ω is a constant; however, it can vary widely from one device to another. For a given resolution, the throughputs of prism gratings and interferometers are in the ratio 1:10:1000.

As a result, it can be seen that, for low-light-level work or short exposure times, an interference spectrometer with multiplexing and high-order dispersion is desirable.

Multispectral scanners. The multispectral scanner is a highly developed, complex instrument designed to measure the radiant flux from a large number of spectral bands scattered from the surface of the earth. Light radiation from the visible spectrum from each small resolution element observed by the optics is dispersed and recorded in each spectral passband. In other remote-sensing devices in which the radiation is measured in a single passband, objects are recognized on the basis of shape, texture, and their environmental relationships to other objects. The multispectral scanner represents the first combination of the power of spectral and spatial recognition, with the result that much more data are available for identification procedures. Sophisticated statistical interpretation techniques are now being applied to such data. Multispectral imagery hopefully will make possible detailed and high-speed identification of materials and objects.

In general, multispectral scanners can be divided into two sections: a scanning section, which includes the collecting optics, and a spectrograph section, which includes the detectors. The first section is very similar in nature and operation to the IR scanners, and the second section behaves as a spectrometer. The scanning section consists of a rotating 45° mirror and optics that focus the radiation from the resolution element onto a field stop. The field stop serves as the entrance slit to the spectrometer section. The light is then collimated, passed through the dispersive element, and focused on an array of detectors (generally linearly analyzed), where each detector collects a separate band of the light spectrum. The amplitude variations of the wavelength of light collected by each detector are then stored on separate tracks on magnetic tape. As the scanner sweeps, the area under observation changes, and hence multispectral information on the terrain beneath the aircraft is obtained. From this imagery a large number of spectral bands are produced.

Because the dispersion is performed after the light passes through the field stop, each detector sees light from the same resolution element. As a result, the registration difficulties inherent in TV multispectral cameras do not exist for these scanners. A spectrum of each ground resolution element is then obtained. This spectral signature can be processed and used to analyze the imagery for various different types of material. Multispectral scanners generally cover a wide spectral range, typically from 0.4 to 2 μm; however, the spectral resolution is not high, usually between 10 and 50 nm. As a result, detailed spectral analyses, such as are available from spectrometers, are not possible. Spectral resolutions in these scanners is approximately the same as those of TV multispectral systems.

Spatial resolution is comparable to that of IR scanners and, as with the IR system, it is limited by detector size and focal length. Resolutions of about 3 mrad are typical for multispectral scanners; however, the Nimbus medium-resolution IR scanner (MRIR) has a much lower angular resolution of 50 mrad, and the ERTS MSS has a very much higher resolution of 0.07 mrad.

The instruments are generally calibrated using a number of different calibration sources. These include blackbodies, skylight, integrating spheres, and other sources. These calibration methods allow accurate measurements of absolute temperature and radiance of the object under observation. As is the problem with all remote sensors, the atmosphere can produce serious effects on the data collected. Radiation levels are affected by both black-scattered light from the atmosphere and by absorption of radiation by atmospheric components. Correction for these effects is a very complicated problem, and such correction cannot yet be done entirely satisfactorily.

Direct imagery is not generally made from the data collected by these instruments. The amount of information is so large that interpretation is extremely difficult; as a result, data are generally digitized and statistically analyzed by a computer.

Radar. Radar is an active system that emits EMR in the radio-frequency range from 0.3 to 40 gigahertz (GHz). This radiation, either pulsed or continuous, illuminates the target, and the scattered or reflected radiation is received and displayed. By measuring both the transit time of the radiation and the amplitude, phase, and polarization of the back-scattered radiation, information can be obtained about the range and physical characteristics of the targets. If then the direction in which the antenna is aimed is accurately known, a map of these scattered characteristics can be developed. Since the system operates at radio frequencies, the map obtained contains new information not available from the IR systems. Radar is capable of penetrating snow and, if the wavelength is long enough, ice and soil to a limited depth. Furthermore, the penetration and scattering from these objects depends on the moisture content. These systems are also capable of penetrating clouds and rain; they therefore have all-weather, day and night capabilities.

One main disadvantage of radar is its relatively poor spatial resolution, which is the result of the long wavelength at which it operates. Furthermore, the systems require large antennas, complex electronics, and a great deal of power to operate. Since the angular resolution is proportional to λ/D, where λ is the wavelength and D is the diameter of the antenna, better resolution can be obtained by employing either a larger antenna or shorter wavelength. With the use of short wavelengths, the system is limited by its lack of ability to adequately penetrate clouds and rain at very short wavelengths (Fig. 3). Therefore, to increase spatial resolution, larger antennas are generally used.

Side-looking airborne radar (SLAR) has a very large antenna and is the type of system generally used to obtain fine-angular resolution. By mounting the antenna so that it is stationary and looking to the side of the aircraft in a direction perpendicular to its line of flight, a very long antenna (about 7 m) can be constructed. The size of the antenna is limited by the difficulty of stabilizing a very large structure in flight.

A very narrow beam of radiation is sent out from the antenna in the direction of flight or along the track. Since range resolution is a matter of timing the return signal, very high resolution can be obtained perpendicular to the line of flight of the aircraft by the use of a narrow pulse. The azimuthal (along-track) resolution is determined by the beam width of the antenna. In this way, by using long radar antennas and short wavelengths, antenna beam widths of the order of 2 mrad can be obtained. The resolution transverse to the plane, however, is partly a function of the range at which the signals are detected. The resolution is consequently relatively poor at long ranges.

The image is formed by focusing the light from the trace of a CRT on a strip film. The CRT electron beam is intensity-modulated by the amplitude of the radar echo, and the motion of the beam is determined by the range of the target. Thus, as the beam sweeps across the CRT, a strip of light is made corresponding to the radar intensity of various points on the ground as the range increases from the aircraft. The film is then moved along at a rate proportional to the velocity/height (v/h) of the aircraft and a strip map of the terrain to the side of the aircraft is formed.

Two methods are used to improve the resolution of radar systems. The first, called chirping, is used to improve range resolution; the second, the technique known as synthetic-aperture radar, is used to improve along-track resolution.

Since range resolution is obtained by timing the flight of radar pulses, the resolution is ultimately limited by how short a pulse can be produced at the antenna. To overcome this limitation, the chirp technique is used. The transmitted pulse is swept over as large a frequency range as is practical. This information can be used to produce increased range resolution. If the received pulses are then passed through a suitable filter that allows all frequencies transmitted in the pulse to come out at the same time, a very short processed pulse length can be produced in comparison to the transmitted pulse, thus increasing the range resolution.

To increase the along-track resolution, a longer antenna is needed. This is produced artificially by observing the reflection from all targets in the broad beam of a shorter antenna at a number of different points along the aircraft flight path for as long as the target is in the antenna beam.

The radiation received from each target is then recorded in both phase and amplitude. The result is later manipulated by an optical processor, which transforms the phase (or Doppler) histories to produce the final imagery. Because the signals received from the target are recorded over a long distance, the synthetic-aperture radar acts as a very long antenna, thus greatly increasing the resolution. The resolution of the synthetic-aperture radar is essentially independent of range, because at long range the target is in the antenna beam longer; hence returns coming from it are recorded over a longer distance. The theoretical resolution limit of a synthetic system is equal to half the length of the antenna. A typical antenna of this type ranges from 1 to 10 ft; however, the processing of the resulting phase histories can improve the resolution obtained. Synthetic-aperture systems (SAR), however, require sophisticated methods of maintaining the antenna in a straight line during overflights. This signal-phase detection technique coupled with the chirped method of measuring range produces images of very high spatial resolution.

Thirteen of the listed radars have been used, or are intended for use, in earth resource applications. Because of their all-weather, day and night capabilities, they are especially valuable for surveillance that requires coverage at all times. Furthermore, they offer the opportunity of observing scattering from terrain in the microwave band that promises to unveil information not readily available from shorter-wavelength radiation.

Microwave radiometer. A microwave radiometer collects and measure the flux of microwave radiation incident upon its antenna. The amount of radiation obtained is limited by the system bandwidth and the size of the area at which the antenna looks, or the beam width. The noise-like radiation collected appears as an average voltage level at the filtered output of the detector, and is generally read out for recording purposes as an equivalent radiometric temperature, that is, the temperature of a blackbody source that would produce the same amount of noise in the bandwidth of the system. The equivalent temperature is determined by comparing the level at the antenna with that of a reference resistance held at a constant temperature. Comparison is carried out by rapidly switching from one source to the other and measuring signal differences; thus the change in antenna equivalent temperature is recorded.

The equivalent temperature recorded is not equal to the temperature of the area observed. Since it is a passive system, a large number of factors contribute to the radiation levels, thus complicating interpretation. Some of these factors are (1) actual temperature of the body, (2) emissivity of the body, (3) sky reflections, (4) polarization of the radiation, (5) angle of observation of the body, and (6) noise from the radiometer itself.

The amount of radiation emitted by an object is a function of both its temperature and emissivity. For an ideal blackbody emitter, which has an emissivity of unity, the spectral radiance (L) in the microwave region is given by $L = 2kT/\lambda^2$, where k is Boltzmann's constant, T the temperature, and λ the wavelength. Thus, a relatively small linear temperature dependence exists compared with the fourth power of temperature dependence found in IR regions. In the microwave region, however, materials have large variations in emissivity, that is, the ability to emit radiation. These emissivities may vary from 0.41 for liquid water to almost 1.0 for ice; thus, for a water surface with ice floating in it the water appears very cold and the ice very warm. Other materials have different emissivities generally between these two figures.

If the surface of an object is very smooth, it can also reflect radiation from the sky and from space. Thus, smooth water will reflect the sky (or cloud) temperature. The reflected energy of course depends on the angle of reflection, on the reflectivity of the surface, and on the polarization of the radiation. Thus, a smooth water surface will reflect a very low sky temperature. However, if it is choppy, the reflectivity and emissivity are changed, and the apparent temperature can increase or decrease considerably. Thus, interpretation requires a comprehensive knowledge of instrumental, environmental, and propagational factors.

Radiation is also dependent upon the angle of incidence at which the object is observed. It is generally found that, as the angle increases from the vertical or normal, a drop in apparent temperature is apparently due principally to the reflective quality of the emitter. However, in some cases a rise in apparent temperature may be found as the angle is increased, for example, over rough seas. Another effect is the polarization of the radiation detected. Significant differences exist in emission when observed in vertical and horizontal polarization.

Other remaining factors in microwave radiometry that need consideration, as in all airborne or satellite sensors, are atmospheric absorption, reemission, and scattering. For wavelengths longer than 3 cm, the air is quite transparent; however, for shorter wavelengths significant absorption by water vapor and oxygen components occurs. Such gases also reemit radiation at similar frequencies, thus making interpretation even more difficult. Reemitted radiation can, however, be used to advantage if the observer wishes to analyze the components of the atmosphere.

As with radar, the spatial resolution of the microwave radiometer depends on the wavelength of the radiation and the size of the antenna. Thus, for fine spatial resolution, a large antenna and a high frequency are desirable. However, atmospheric absorption becomes excessive if the frequency is too high. The antenna also must be sufficiently small to be carried on an aircraft. Also, the radiation bandwidth must be large for good temperature resolution when signal integration time is limited.

The temperature sensitivity of microwave radiometers can be increased merely by integrating the signal over longer periods of time. This is a great advantage for ground-based systems; however, the dwell time cannot be very long for airborne or satellite systems. As a result, integration times in aircraft radiometers are generally kept below 1 s. Nevertheless, very good temperature sensitivity can be obtained with these systems, and the only problem is one of interpretation of results.

Although the passive system has many difficulties in interpretation that can be eliminated in active systems, passive systems have an advantage due to their size and power requirements, since they do not require transmitting equipment. Passive systems also produce a good deal of information not available to active systems, since they measure self-emissions from the targets rather than scattered radiation.

The field of passive microwave radiometry has been extensively developed only recently. Its equipment is less sophisticated than found in other more highly advanced RS systems; however, it has the potential to develop much information concerning soil, moisture, ice, water interfaces, vegetation, and other parameters relevant to earth resource measurements.

Sources of Information

As shown by the previous sections, a large number of remote sensors are available throughout the visible, IR, and radar region for energy development applications. For the user of such sensors, the difficult question is which, either singular or plural, of the numerous types should be applied to a specific problem. Because of long usage, the first tendency is perhaps to apply photographic remote sensors. The main thrusts in current remote-sensing applications consequently have been made using photography and its close counterpart, the imaging sensors, such as the vidicons or, more completely, the television systems.

It is imperative that designers analyze and compare

the capabilities of the remote sensors in the three major regions of the EM spectrum—visible, IR, and radar—to prevent biasing the future development of remote sensors with photographic techniques in order to fill certain voids in our present capabilities. An analysis of essential information from the typical remote-sensing survey is essential to form a basis for such design considerations. Such an analysis requires that priorities be assigned to the most useful or important information yielded by the sensor. The priorities need to be based upon general survey requirements if any comparisons are to be made at all. Certainly, there are many exceptions where special experiments are designed for obtaining specific information. With these limitations, the question is what is the most desired feature of an image taken in the visible, IR, or radar regions? For viewing purposes the image has been processed into either a black-and-white or color photograph.

Spatial resolution. Invariably the user will ask for greater spatial resolution or the ability to separate objects or geometric terrain features. A tremendous amount of information is contained in such spatial detail either directly, by correlation, or by inference. However, spatial resolution is locked up in the angular resolution of the remote sensor, previously expressed as $\theta = \lambda/D$. (Every circular aperture or lens has a cone of acceptance that collects the EM radiation—the larger the lens, the narrower the cone. By passing a plane through the axis of the cone, the apex angle is θ. It is a planar angle as compared with the solid cone angle.) For example, in the visible band (0.4 to 0.7 μm), with the angular resolution at a wavelength of 0.5 μm and a lens about the size of the human eye (0.5 cm = 10,000 wavelengths), the angular resolution is 0.1 mrad. Thus, this lens at 100 m can distinguish point sources only 1 cm apart. Such resolution is phenomenal. In comparison, at a radar wavelength of 10 cm and for the same angular resolution, an antenna 1000 m in diameter would be required, equivalent to the length of about 10 football fields. The conclusion is that the shorter the wavelength, the greater the spatial resolution possible for a given size of lens, aperture, or antenna.

Material identification. In the analysis of an image, the identification of the observed material probably has equal or higher priority than spatial resolution, and can best be established by separately considering each application. Present remote sensors are only partially capable of identifying various earth materials and then only to a limited degree. Vegetation, land, ice and snow, and water as general classes can be identified. As a special example, further progress is required before the moisture content of soils can be determined remotely. Considering another case, present sensors do not directly and positively provide discrete data on whether the material sensed, for example, is granite or clay.

Correlation techniques are being presently tested on spectral signatures of minerals and vegetation with the aid of computers. Known spectral signatures of a substance taken with a spectrometer are correlated with spectra from a number of unknowns. Correlation coefficients obtained by a computer of 60 percent or over are usually acceptable as an identification of a substance.

EM penetration. The pentrability of EM radiation into a material media is significant, since further information can be made available from subsurface reflections and interactions. For example, the potential of determining soil moisture on an area-extensive basis depends heavily on the penetrability of longer microwaves. The penetrability of short wavelengths in the visible and IR regions is superficial, and the reflected or diffusely scattered EMR comes from the surface. Penetrations greater than 100 μm are not available for subsurface material identifications.

For the much longer wavelengths, 1 mm to 10 m, the penetrability of EMR is vastly improved. Thus, the promise of realizing ground surface and subsurface information from longer-wavelength remote sensors has accelerated new investigations.

The absorption of EM radiation by the various constituent gases and water vapor of the atmosphere is both undesirable and desirable in remote sensing, depending on whether information is collected about the earth or its atmosphere.

Various gas molecules are set into resonance at specific frequencies in the EM spectrum and thereby absorb energy. Sophisticated measurement and analysis techniques have been developed that provide air pressure and temperature profiles of the atmosphere using either a down-looking (from satellites) or up-looking (ground-based equipment) remote sensor.

Such pressure and temperature profiles are valuable to the meteorologist and the RS data user; consequently, special remote sensors have been designed in the visible and IR regions to obtain them.

Absorption. Whereas absorption spectra are of value for collecting meteorological and weather information, they do degrade to different degrees the images from photographic and IR sensors. The transmission windows available throughout the EM spectrum are presented in Fig. 1. In some regions, the transmissivity is almost unity, or all the radiation passes through the atmosphere without attenuation. The IR region contains a number of high transmissivity windows. It is informative to contrast the transmission characteristics in the microwave region with those in the IR region. For wavelengths longer than 3 cm, the atmosphere is nonabsorptive or transparent. Some absorption and scattering are still prevalent for heavy rains.

Table 1 is a summary of the pertinent features of both passive and active sensors for the parts of the EM spectrum discussed and clearly shows the tradeoffs between spatial resolution, atmospherics, penetration, and material identifications.

CONVERTING SENSOR DATA TO USEFUL INFORMATION

Remote-sensor systems provide data in either real time or in some stored form such as film or magnetic tape. We are primarily concerned with the stored forms of data and the techniques available to improve the extraction of information from them.

The film exposed by a camera is developed and processed into B&W or color photographs for viewing. Generally, the film negative or positive is retained after development and used directly in viewing monitors, where certain areas can be enlarged, color enhanced, or filtered to aid in the photointerpretation.

Photographic film is an excellent way of storing information, and IR and radar RS systems nearly all eventually have their output data put into such a form. Imaging systems utilizing a vidicon or a scan-

TABLE 1
Remote-Sensor Capability Comparison

Spatial Resolution (mrad)	*Visible 0.001–0.1*	*Near IR (photographic) 0.01–0.1*	*Thermal IR 1.0*	*Radar (active) 10*	*Microwave Radiometer (passive) 10–20*
Material identification	Limited capability through indirect spatial and color correlations	Improved capability through multispectral correlations and reflectance changes	Requires multiple sensors and correlations using thermal information	Limited capability through indirect textural and reflection measurements	Improved capability through the use of thermal, reflectance, and polarization information
Depth of earth or subsurface EM penetration	0.01 μm	0.1 μm	10 μm	10 cm	1000 cm
Atmospheric constraints	Requires clear weather or light hazy atmosphere	Requires clear weather or hazy atmosphere; a number of transmission windows available with limited transmissivity	Requires clear weather, hazy or smoky atmosphere; surface thermal radiation found in 8–14 μm region	Operates in clear, foggy, and cloudy weather and in moderately heavy rain	

ning-type sensor transfer a viewed scene to photographic film for storage and further analysis. Imaging systems are not unique to the visible and IR regions, and are also found at the longer wavelengths. The side-looking radar (SLAR) as well as the synthetic-aperture radar (SAR) both resort to the indirect transfer of the collected EMR by their receivers to provide a corresponding image on photographic film. Even where each scan line of the viewed scene is recorded in analog form on magnetic tape, the final processed product can eventually be transferred to film for analysis. Therefore, the abundance of film data processors and analyzers found on the market should come as no surprise to the prospective user. This equipment is designed to permit the operator to view any area or picture element (pixel) of the film, or a hard copy, to enable the analyst working in a particular discipline to obtain the maximum amount of information. Even the simple operation of enlarging the image to where the spatial resolution of the scene is apparent is very valuable in data analysis.

Techniques

Before reviewing some of the more common hardware available to the experimenter, manager, or scientist, a few of the popular techniques for viewing, modifying, or enhancing an image or film are given, with a brief explanation of what is done to improve the analysis capabilities of the observed scene.

Magnification. The film is placed in a viewer where optics provide one or two possible magnifications, or are continuously variable. The primary function of magnification is to provide ease of viewing the image. The spatial resolution is not improved since it is controlled by the lens and the film grain size, and not the magnification.

Densitometry. Density profiles are derived optically or electronically. By moving a small focused spot in a straight line across a film and collecting the transmitted light through the film with a photodiode, an analog signal showing the density variations in the slice is obtained. The narrower the spot, the finer the density detail. The name "microdensitometer" is given to small-spot equipments. Similar results are obtainable utilizing a scanning electron beam. Density profiles are recorded, which can be analyzed visually or through image-processing methods. By magnifying the image and scanning with a fine spot, microscopic density detail is obtainable.

Restoration. The image, either recorded on film or tape, can be improved by removing the degradation imposed by the sensor, receiver, or the atmosphere. The improvement is generally accomplished by removing the noise background, where noise may be defined as any EMR that degrades the desired image. The sensor lens, the aircraft or satellite collection platform, and the viewer all introduce geometrical distortion to some degree. An inverse mathematical or mechanical correction process provides a means of restoring the scene to that of the original.

Registration. This term applies primarily to the precision with which a multispectral viewer can overlay three or four spectral images. Geometric distortions of sensor lens and platform are key factors in providing good image registration. Registration is important for mapping and obtaining image spatial information. For mapping, registration is vital to photograph normal to the earth and is appropriately called an orthonormal image.

Image transfer. Filmed scenes can be optically corrected for geometric fidelity and projected in registration onto a topographic map. Thus, a true scaled photographic image can be transferred to update a map or plot rapidly and accurately. Such equipment is commercially called a Zoom-Transfer scope. Precision transfer techniques are valuable for conducting surveys and continuing inventories.

Image processing. Various electronic techniques are amenable to imagery that has been scanned by a point-source electron beam so as to provide a voltage time signal. The image is scanned similarly to a typical TV vidicon camera except, the analog signals now can be electronically processed to provide the following: (1) constant density or amplitude slices, (2) contour of maximum density gradients through the differentiation of the signal, and (3) the inversion or negative of the image. By adding a negative and positive image, a differential image is obtained that highlights contours, lineaments, and boundaries of rapid change.

Image enhancement. Image processing provides enhancement through density slicing, differentiation, and inversion. Enhancement is also achieved by quantizing the levels of gray or amplitude of the image, and provides for each quantized level separate level control whereby image gradients can be more precisely analyzed. To aid in the analysis, the levels are enhanced by adding different colors of light. Thus, the fine-amplitude changes in a level are more readily seen. Various colors can be dialed into the quantized levels in available equipments.

Image coding. Images can be separated statistically into areas of commonality through the use of computers and high-powered mathematical algorithms, which involve clustering and other pattern-recognition techniques. With such techniques, inventories can be made of earth resources, such as agricultural lands and forests, through the use of computers. Partial success has resulted in the identification, separation, and inventory of crops—wheat, corn, soybeans, etc.

Hardware

Two general groups of hardware are used for the analysis of RS film data: (1) static types that utilize film or prints directly, and (2) image data processors. In the latter group, the film is scanned electrooptically, and the resulting scanned data are digitized with an A/D converter and routed to a computer, which analyzes the data. Large amounts of data thus can be processed and analyzed.

A large number of regular and special types of equipment are available for RS data analysis. These can be categorized as input and output devices, systems, computers, and software, and are presented in part.

Electrooptical input devices. These input devices consist of a TV scanner, either a vidicon or solid-state semiconductor photosensor array, and a digitizer to provide conversion of images to digital codes suitable for direct entry to a computer or communications system. The film is placed on a light table below the TV scanner for uniform illumination. Scanning heads are available for visible and thermal spectra, and, with the addition of fluoroscopic adaptors, extend the operating range to ultraviolet and x-ray spectra. The scanned image is processed and displayed on a 0.46- to 0.64-m CRT monitor.

These devices are usually real-time scanning analyzers that rapidly read, review, measure, correlate, and reformat image data to produce calibrated displays of selected areas. Multispectral and multilevel distributions and complex gray-scale variations that are visually difficult to detect or separate can be enhanced, discretely displayed, and referenced to the original imagery.

The equipments are usually versatile and directly scan convert, redisplay, and analyze output data from sonars, radars, radiometers, IR thermal imagers, and spectral scanners.

Most analyzers are versatile in the processing phase and provide interactive graphic-gradient displays, color-encoded isodensitometer contours, digital readouts for density quantizations and planimetry, selective-area population or events counting, directing charting, and measurement of motion for loop-film inputs.

The researcher who is new to the use of electrooptic analyzers will find them simple and direct-acting, with self-evident controls, instantly responsive displays, nonambiguous and accurate digital readout capabilities, and requiring no special training or skills.

A unique feature of some analyzers allows computer input of data from standard TV vidicon scan head or magnetic tape recorder to be displayed on a TV screen with a small white movable dot superimposed. Digital *xy* coordinates are generated by moving the spot to the desired location and activating a readout button. Self-calibrating features and a highly accurate TV grating pattern allow assessment of the geometric quality of the video input signal. Automatic tracking is also available.

Another important feature of the electrooptical analyzer is the increase in the number of horizontal TV scan lines allowing near-photographic spatial resolution. Whereas the ordinary TV receiver displays 512 scan lines per frame, the analyzer vidicons have up to 4000 lines per frame.

Electromechanical devices. Photographic plate or sheet film, 0.23 $\times$ 0.73 m, 70 mm or 35 mm, can be scrutinized with a microdensitometer to positioning of ±1 μm (covers a 1.17- × 0.25-m format). Precision microdensitometers usually have the following capabilities:

Positioning:
: motorized *xy* positioning and readout, and 360° rotary staging.

Scan:
: 0 to 4D density range, 25 to 400X magnification, eight main and preslit selectable apertures, eight neutral-density and three color selectable filters, and analog and digital outputs.

Write:
: photographic plate—sheet film, 64 gray levels, white light source, and programmed scans.

The laser is also applicable to the field of microdensitometers. A recently developed laser image processing scanner provides for scanning, digitization processing, and recording of high-resolution imagery, and has an optical table similar to the unit described above.

Image systems. A large number of different image systems are available for mensuration, manipulation, and image classification. A class of film viewers is now available to view the film directly and project the image onto a viewing monitor. Zoom-lens features are available for studying fine spatial features.

Multispectral views are probably the most useful of the image systems. They are used when generally four spectral bands of the same scene, including one in the near IR, are combined into one image. Each spectral band is projected separately and in near-perfect registration onto a common viewing screen. Colored light obtained through red, green, and blue filters can be used for the separate light sources to provide a means of reconstituting the color of the original scene. The amount of color in each spectral band is controlled by a density filter. False-color combinations are readily available to the operator. The very important color-IR images can be projected by adding red light of proper intensity to the IR spectral band, green to the red band, and blue to the green band.

A list of manufacturers producing RS equipments for data processing and analysis is given in the Appendix.

APPENDIX

The following is a list of manufacturers dealing in image processing: hardware, software, and/or services.

Alden Electronic, Westboro, Mass.
Antech, Inc., Waltham, Mass.
Bausch and Lomb, Rochester, N. Y.
Bendix Corp., Ann Arbor, Mich.
Constantine Engineering Laboratories, Upland, Calif.
Comtal Corp., Arcadia, Calif.
Carl Zeiss, New York, N.Y., and Oberkochen, Germany
Carson Astronomical Instruments, Inc., Valencia, Calif.
Cohu, Inc., San Diego, Calif.
Colorado Video, Inc., Boulder, Calif.
Conrae Corp., Covina, Calif.
CBS Laboratories, Stamford, Conn.
Dest Data Corp., Sunnyvale, Calif.
David W. Mann Co., Burlington, Mass.
Dicomed Corp., Minneapolis, Minn.
Digicom, Chelmsford, Mass.
EMR Photoelectric, Princeton, N.J.
E. Leitz, Wetzler, Germany and Rockleigh, N.J.
Earth Satellite Corp., Washington, D.C.
Edo Western Corp., Salt Lake City, Utah
ESL, Sunnyvale, Calif.
General Electric Co., Daytona Beach, Fla.
Gould, Inc., Newton, Mass.
Honeywell Test Instruments, Denver, Colo.
Hughes Aircraft Co., Oceanside, Calif.
Information International, Los Angeles, Calif.
Image Analysing Computers, Monsey, N.Y.
Interpretation Systems, Inc., Lawrence, Kans.
International Imaging Systems, Mountain View, Calif.
Kantronics Inc., Lawrence, Kans.
Mead Technology Laboratories, Dayton, Ohio
NAC Inc., Japan
Optromics International, Chelmsford, Mass.
Perkin-Elmer Corp., Wilton, Conn.
Photo Digitizing Systems, Burbank, Calif.

PhotoMetrics, Inc., Lexington, Mass.
Philco-Ford, Houston, Tex.
Ramtech Corp., Calif.
Recognition Systems Inc., Van Nuys, Calif.
Spatial Data Systems, Goletz, Calif.
Sierra Scientific Corp., Mountain View, Calif.
SEP, France
Technical Operations Inc., Burlington, Mass.
TRW Systems Group, Redondo Beach, Calif.
Zia Associates, Inc., Boulder, Colo.

PART B
FOSSIL ENERGY, AND THE UTILIZATION AND UNDERSTANDING OF REMOTE-SENSING DATA

The energy crisis of 1973–1974 left a residue of vexing problems. The energy crisis has many complex and interdependent facets which are difficult to pose in a format that is fully intelligible to a broad spectrum of people. Ample energy sources could be employed to supply the world's growing needs for the foreseeable future, but what are the problems of developing these sources?

In 1949, Sumner T. Pike, then employed by the Atomic Energy Commission, in an unpublished address to members of the American Association of Petroleum Geologists, pointed out the potential of atomic energy to fill the gap between energy supply and demand as fossil-fuel reserves dwindle toward the close of this century. He stated that the biggest problem and constraint to the development of atomic energy would be the management of the garbage from spent fuel cells. Nearing the end of this century, we find that his statement is not only true for atomic energy waste but also for fossil-fuel and other industrial wastes. We have the added dilemma that no universal economic or social forum exists that can adequately cope with these problems, although in theory the technology has been developed.

Let us examine briefly some of the energy sources that are or could be exploited and indicate the future potentials of several of the major sources before we delve into the basic tasks of developing our fossil energy sources and the roles remote sensing might take.

SOURCES OF ENERGY

Energy sources can be classified or grouped by different criteria. Those extracted from the earth include the fossil fuels, those hydrocarbons that have been buried and altered in the earth's crusts in antiquity, such as, coal, oil shale, sands bearing heavy oil or tar, crude oil and natural gas (methane) liquids, and natural gas. Other energy sources from the earth's crusts are fuel for nuclear fission (^{235}U, ^{238}U), various minerals for synthetic fuels, and geothermal sources (hot plutonic rocks). Another group of energy sources, called *renewable* energy sources, include fusion nuclear fuel, solar energy, wind power, tidal current flows, hydrogen gas combustion, converted vegetative material (peat), wood, manure, garbage, etc.

Background and Priorities

A proposed 5-year federal budget (starting in FY 1975) contains $10 billion allocated to help ease the nation's energy crisis. Of this sum, nearly $2.2 billion was proposed for the substitution or conversion of coal and its by-products for energy roles that oil and gas now fill; $1.84 billion was allocated to exploit renewable energy resources; $1.44 billion was earmarked for developing the conservation of energy and its use; and only $460 million (less than 5 percent) was allocated to increase domestic oil and gas production. Although large, these budget figures are not significant. However, they do represent a proportion of effort that indicates national thinking.

Although the prime concern of this chapter is the development of energy from fossil fuels, it must be recognized that, once a fossil fuel can be economically extracted (mined or otherwise produced for market), it also becomes a raw material source for various chemical end use products. Coal is the source of a large range of chemical products, but at present coal chemical demands are relatively small. However, as more coal is converted to feed stock or gasified, and sources of oil and gas dwindle, the demand could grow at a very rapid rate.

Oil and natural gas, raw material for the petrochemical industry, now provide a significant percentage of the total production. The petrochemical industry started in the 1950s and grew rapidly owing to abundant and relatively cheap oil and gas.

Tar sands and oil shales are more difficult (than gas, oil, or even coal) to extract and convert to usable feed stocks for energy sources, but as these hydrocarbons become marketable, they will meet a dual market for energy and raw material for chemicals.

Coal. Coal was a primary source of energy through the first decades of this century until oil and gas were discovered in progressively larger deposits and their relative low costs of extraction made coal noncompetitive with oil and gas for most purposes. The calculated recoverable coal reserves in this country alone have been projected as a 200-year supply even with doubling and redoubling the demand. At the present, coal appears to be our most dependable near- and intermediate-term energy source. However, developing coal as a major source of energy presents many unique problems, the solution of part of which remote sensing and associated technology may assist.

Oil and gas. Crude oil and natural gas have commanded a prime position in the hierarchy of modern usage of energy resources for the last several decades. They can be produced with the least environmental impact, the "recoverable reserves" have been large, and they are relatively clean burning. Consequently, oil and gas reserves are being depleted at a greater rate than new reserves are being found. Several views have been formulated on the status of the world's oil and gas reserves. The most pessimistic position is that our total oil and gas resources will be gone within a very few decades. Although resource estimates vary by a large percentage, most experts agree that most of the world's large reserves have already been or can be anticipated to be discovered in the near future, and that we consequently cannot depend on adequate supplies in the near future even though world production may temporarily increase over the next 20 years. Regardless of estimates, every effort must be made to discover new oil and gas sources.

The extraction of oil and gas products from oil shales and tar sands is undoubtedly a potential for extending the life expectancy of oil and gas reserves, particularly as technology for extraction is improved. Oil-shale resources are large in this country and tar sands are extensive in Canada. However, these too are expendable energy resources and cannot be depended upon unendingly.

Nuclear energy. Nuclear energy is not considered fossil energy although its development from mined uranium is subparallel to fossil-energy sources; it is discussed only briefly here because it is considered a potential for a vast energy source.

Energy from nuclear fission is thought by many physicists, engineers, and other scientists to be our best long-term energy source. This may be the case, but many compelling arguments have been made refuting atomic energy as a panacea. The billions of dollars spent in the last quarter-century to develop atomic energy has done little to fill the gap between our energy supply and demand. The high number of atomic reactor and refinery accidents suggests that we are risking our very survival for a preciously small percentage of our energy needs, which could be gotten elsewhere relatively more safely and probably keep more people gainfully employed. If breeder reactors are not brought on line soon, our supply of nonenriched fissionable uranium will not be sufficient to sustain the growing requirement. If breeder reactors are brought on line, it is thought that accident risks would be unacceptably high. Another argument is that the energy and mineral resources required to build relatively safe breeder reactors for fissionable atomic energy, and maybe for fusion reactors, would be so great that our net gain in total energy-producing capacity would be nil, or possibly at a cost deficit rather than a cost benefit.

Fusion nuclear energy would appear to be the universal major energy source for the earth; however, realization of this generating capacity is at best several decades in the future and very possibly more than a century away. Although heat is the only foreseeable pollution associated with a fusion system, high thermal anomalies could play havoc with earth environments.

Other sources. Other sources of energy that are being reevaluated are hydropower, geothermal energy, solar energy, and the harnessing of wind and tidal forces. These too have their limitations and impacts.

The ultimate hydroelectric power that can be produced is a function of the number and size of dam sites; thus, maximum development will not add appreciably to our total energy bank.

Geothermal energy must be converted to electricity near well sites located over deep, very hot plutonic rocks. The electricity must be directed to a nearby use terminal (e.g., used for aluminum reduction and processing) or fed into long-distance transmission lines, the latter of which are impractical for island geothermal sources.

The development of solar energy is dependent on low-cost mass production of arrays of solar cell panels. Large arrays require nearly cloud free sites such as the arid Southwest. Glaser (1968) and others suggest that a large solar cell array be placed in synchronous space orbit, where the energy would be converted to microwaves and broadcast to earth. On earth, it would be converted to electric energy for transmission to final users. This concept has evoked considerable interest in some quarters, but safeguarding such an installation in time of international conflict negates the advisability of pursuing development of a large space system at this time.

The development of energy by harnessing wind power through modern windmills may sound a little preposterous, but according to an article by Wade (1974), NASA predicts that windmills "could supply between 5 to 10 percent of the country's total electric power needs by the year 2000." Wade also states that it has been predicted that 2000 windmills could pump 1.5 trillion kWh of electricity into the national grids annually. That figure is almost equivalent to our present total annual production.

Harnessing tidal flows of water for generation of electric energy could add a small percent to our total energy-generating capacity. Tidal power sites will undoubtedly be developed in areas that have a narrow-mouthed bay with high daily tides, and where there is a local shortage in the energy supply. The Netherlands has developed this source of energy to a fine degree.

Present and Projected Energy Status

Fuel production and consumption for 1972 are shown in Fig. 2. Projected energy demands and projected potential energy sources are shown in Table 2.

ENERGY-PRODUCING SYSTEMS

Background

To discuss major energy systems that are based on a particular energy source for human use, the term

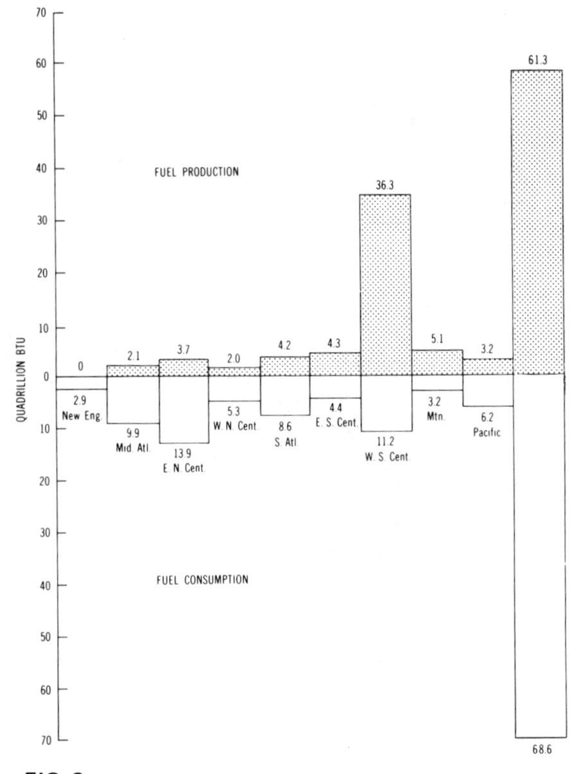

FIG. 2
U.S. fuel production and consumption—1972 (U.S. Bureau of Mines).

TABLE 2
United States Energy Demand and How to Supply It in 1985 [in quadrillions (10^{15}) of Btu]

	Demand			*Independent Supply*
	1971	*1973*	*1985*	*1985*
Petroleum	30.5	–	50	30.1 Btu source (14.4 MM bbl/d)
Natural gas	22.7	–	27	20.0 Btu source (20.0 MMMM ft^3/yr)
Hydro- and geothermal	2.8	–	4	4.0 Btu source (80,000 MW)
Nuclear	0.4	–	12	12.0 Btu source (240,000 MW)
Coal	12.6	–	23	43.0 Btu source (2,275 MM ton/yr)
Shale oil		–		6.0 Btu source (3.0 MM bbl/d)
Total	69.0	76	116	115.11

Source: Modified from G. D. Grayer, A Call for Action, U.S. Energy Independence by 1985, Feb. 1974.

"fate of energy" might be used. Fate of energy in any energy system is the production and expending of energy as work (driving power), heat, and light, or a combination of the three. A list of the most common end uses of energy includes electric drive devices, metals and mineral reduction and refining, mass transit systems, space and other cooling and heating systems, space illuminating, industrial processing, and power for automobiles, airplanes, buses, trucks, ships, and other mobile equipment.

Most energy forms, such as electricity, coal, oil, coal and oil products, and gas, natural and manufactured, require special types of utilizing mechanisms or devices. Some of the most common devices include electric motors and oscillators, electric filament lights, electrolytic cells, electric resistors and/or furnaces, oil, gas, and coke blast furnaces, vapor compressors, gas and oil vapor lights, gasoline engines, and gas turbines.

Each energy system has a unique chain of events

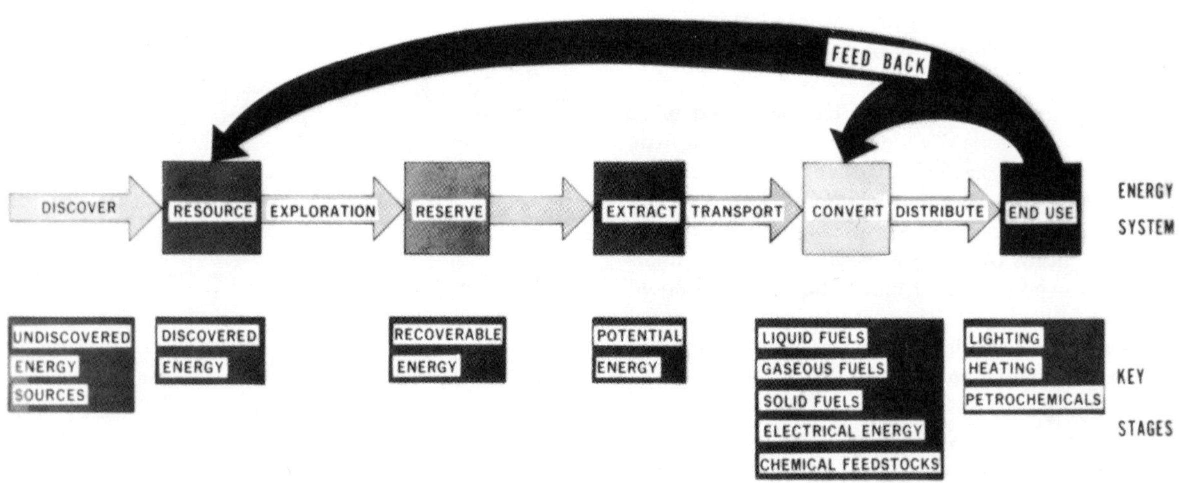

FIG. 3
Parameters in energy systems.

dependent on the basic energy source and the end uses intended for the particular types of energy produced.

The major tasks or key parameters of activities associated with one or more energy systems are diagrammed in Fig. 3. The diagram is over simplified in that few linear routes from a single energy source to its "end use" exist. Each parameter of activities may contain several tasks, and any number of the processing tasks may be required for a particular energy "end use" requirement.

Coal Energy System

Coal does not need to be discovered, per se. Large coal resources are known at least by general locality for many parts of the world, as are the coal beds' positions in the geologic rock columns of that part of the world (Fig. 4).

In utilizing coal there is really a "defining" problem. In the United States, for example, it is estimated that there are nearly 3.5 trillion tons of coal, including anthracite, bituminous, and subbituminous coal and lignite, in beds more than 14 in. thick and ranging in depth from the surface to 6000 ft. Technically, the reserve base of anthracite and bituminous coal beds 28 in. or more thick or subbituminous coal beds 5 ft or more thick, not over 1000 ft in depth, and lignite not more than 120 ft in depth are calculated to be about 434 billion tons. Table 3 shows the general distribution of the reserve base of coal. It is estimated that about half the coal reserve base will be lost in mining; thus, recoverable coal or the actual coal reserve amounts to about 217.0 billion tons.

Prior to the long-term economic extraction of coal, the reserves of a bed being mined must be ascertained, and the characteristics of each particular block of measured reserves should be known in considerable detail.

Coal reserves are classified into several categories such as rank, thickness, and depth of overburden; other important classification categories include sulfur content, organic and inorganic; ash content; coking qualities; trace elements and mineral contents; and detailed petrography. Furthermore, each coal bed should be classified as to precise geographical location, the type of overlying and underlying rocks, the geomorphology of the land surface, the extent and thickness of the coal bed in relation to geological and hydrological settings, the qualities and quantities of surface and ground waters, and the historic and potential land uses in and adjacent to a particular coal. New land-use and mining laws lend special importance to the last category, and better data-collecting and surveying systems are required.

After a coal bed is classified into categories, decisions about methods of coal extraction and plans for "end use" can be made, such as surface mining (stripped), underground mining, or gasification and/or liquefaction in the subsurface or at a surface plant.

Generally, after mining, coal is cleaned and sized in a preparation plant. If the coal is not to be liquefied, gasified, or burned locally for steam generation, it must be transported by rail, barge, truck, or slurry pipeline. Transportation may be to a point of final usage or to a central collection point, where the coal is cleaned and sized and then shipped to a point of final consumption.

Before exploration starts, the collection of data and remote-sensing surveys by high-altitude aircraft and spacecraft can provide enhanceable raw data on many of the larger geologic, geomorphic, and hydrologic features important to coal field classification. These data and surveys can also be used to determine areas of potential in situ usage of coal and to inventory the progress of reclamation and successful revegetation. Remote sensing will be an aid in the planning of transport systems, the selecting of coal conversion sites, and determining the best modes of distributing coal feed stocks and derived energy to the consumer.

Example of Remote-Sensing Application

Background. Examples of remote sensing application to all key parameters (tasks) of coal-energy development have not been documented at this time, and it is probable that remote sensor data and automatic data processing and enhancing (ADPE) will never play a large role in "discovering tasks" or pro-

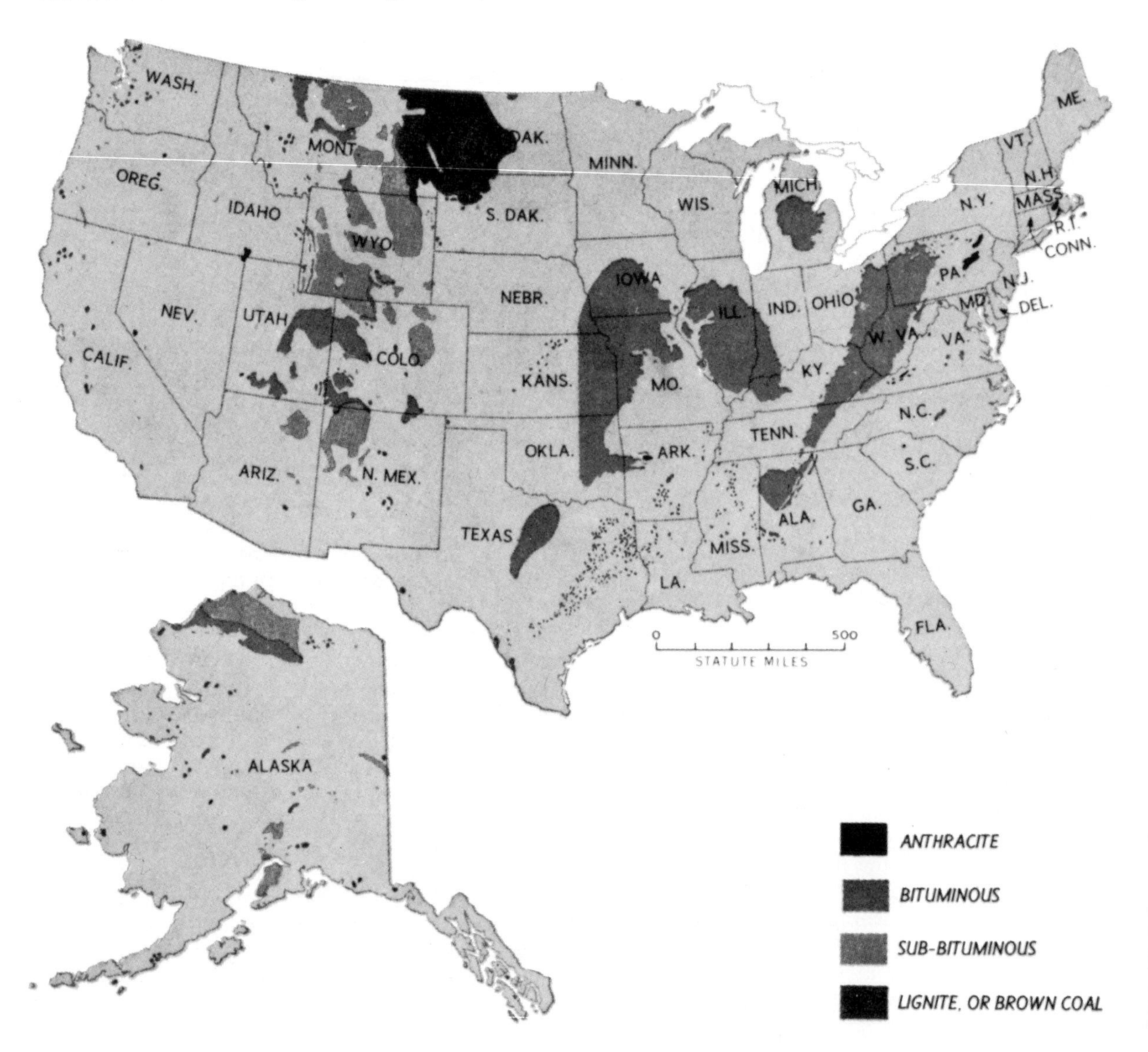

FIG. 4
Distribution of coal fields of the United States by rank.

viding data and information for the tasks of changing coal resources to reserve base coal. These tasks require surface mapping and sample collecting of the coal bed and its associated rocks for analysis of their thickness, physical characteristics, and chemical qualities.

Remote-sensing data have a greater aid potential for extraction, transportation, conversion, and distribution tasks than for proving up reserves.

Strip-mining monitoring. Recently, Rogers et al. (1974), in reporting results from their NASA contract for investigations to determine the utility of LAND-

TABLE 3
Demonstrated Coal Reserve Base[a] of the United States on January 1, 1974, by Area, Rank, and Potential Method of Mining (billions of short tons)

	Anthracite	*Bituminous*	*Subbituminous*	*Lignite*	*Total*[b]
Underground					
East of the Mississippi River	7	162	0	0	169
West of the Mississippi River	c	31	98	0	129
Total	7	192	98	0	297
Surface					
East of the Mississippi River	c	33	0	1	34
West of the Mississippi River	0	8	67	27	103
Total	c	41	67	28	137
Total[b]	7	233	165	28	434

[a]Includes measured and indicated categories as defined by the U.S. Bureau of Mines and U.S. Geological Survey and represents 100% of the coal in place.
[b]Totals may not add due to rounding.
[c]Less than ½ billion tons.

SAT-1 data to detect and monitor areas of strip mining and reclamation, demonstrated the feasibility of using LANDSAT-1 comparative computer tapes (CCT's) as a basis for mapping and monitoring strip mining and reclamation. The methods described are rapid and accurate, and are relatively inexpensive when contrasted to the standard techniques using only aerial photographs and ground teams. It has been stated that coal-stripping and reclamation maps at scales of 1:24,000 to 1:250,000 can be produced from LANDSAT CCT's at one tenth the cost of conventional mapping techniques, but others contest this statement, except possibly for very large or remote areas. Since these maps can be produced quickly and economically, it is now feasible to monitor changes in stripping and reclamation activity at least on an annual basis.

Several areas of investigation by Rogers in Ohio are shown in Fig. 5. Some have been disrupted by coal mining since the early 1800s. Strip mining has been practiced in Muskingum County and in four of the counties to the north and east since the 1910s. The total area of stripping operations in each county was quite large during the period from 1914 to 1947, but is insignificant when compared with the area stripped from 1948 to the present time (1974).

From the earliest days of mining until 1948, little concern was given to the effects of coal mining on the environment. However, reclamation techniques required by 1948 legislation resulted in some grading and planting of trees and forage on spoil banks, although, in some areas, the soil was reported to be too toxic for replanting. Reclamation is proceeding more rapidly and more effectively since stricter laws were passed by the state legislature in 1973.

Strip mining has caused ecologic effects to the extent that large areas have been disrupted and are no longer productive. Such effects include the erosion of bare or sparsely vegetated spoil banks and the discharge of highly mineralized water. Sediment eroded from mined areas tends to silt up streams and reservoirs. Erosion, in turn, leads to flooding, decreased storage area, and the choking off of vegetation. Water that discharges from spoil banks and underground mines is in many places strongly to moderately acidic.

Government agencies commonly do not have current maps or data showing areas that have been disrupted by coal mining operations. Reports available

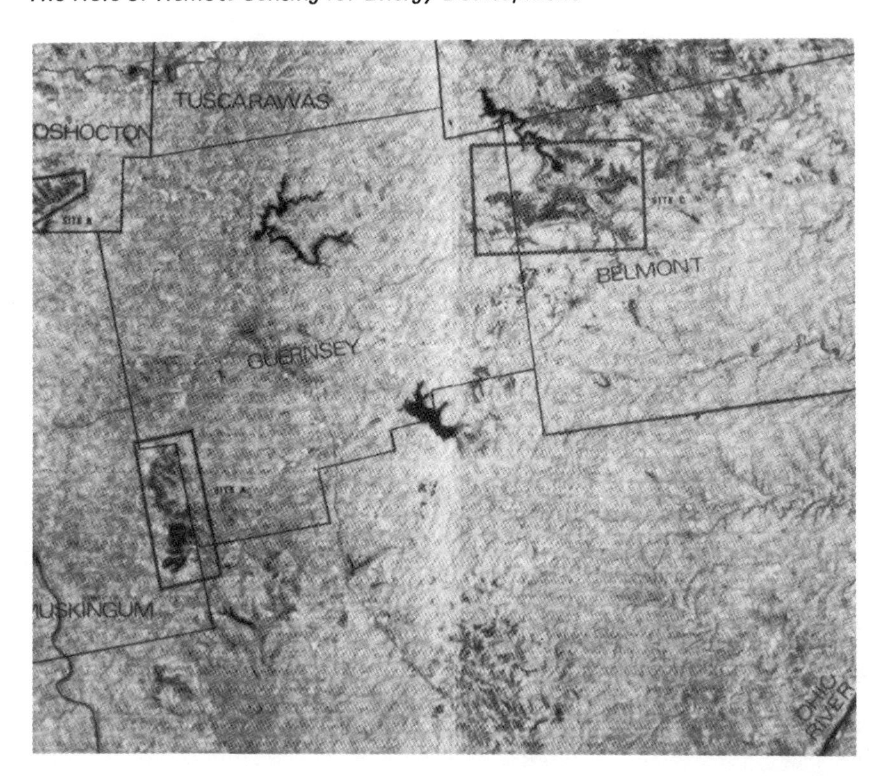

FIG. 5
LANDSAT-1 MSS-7 (IR band): image of Ohio strip mine areas, including Muskingum County site.

to the public may be grossly outdated, inaccurate, and difficult to obtain. This is understandable since on-site examination of the individual mines, and particularly older mines, is impeded by (1) lack of adequate mine maps, (2) nonexistent or blocked access roads for inspection, (3) lack of reliable records, (4) planting along roads that obscures adjacent barren land, and (5) lack of dated aerial photographic coverage. However, it is often possible to recover some of this. Historical data from archived black and white anchromatic aerial photography, taken for crop surveys and/or topographic mapping at an earlier date, are often available.

Utilizing computer techniques and Bendix-assembled equipment, data were enhanced and extracted from LANDSAT tapes of multispectral scanned (MSS) imagery over coal mining operations in southeastern Muskingum County (Fig. 6) on August 21, 1972, and September 3, 1973. Information produced included geometrically correct map overlays showing four earth surface categories of stripped earth, partially reclaimed earth, water, and vegetation (Fig. 6).

The area in Fig. 6 includes a single large strip mine, owned and operated by the Ohio Power Company, in the moderately rolling terrain of southeastern Muskingum County. The mine is nearly 14 km (9 miles) long, and as much as 8 km (5 miles) wide. Aerial photographs indicate that no stripping had occurred in that particular area before 1970. By 1965, however, about 1.5×10^7 m^2 (4,000 acres) had been disrupted; by 1971, strip mining had encompassed almost 4.5×10^7 m^2 (11,000 acres). Reclamation, brought about by stringent legislation enacted in 1973, is proceeding at a rapid rate. Aerial photographs of the northern part of this mine were taken in May 1972. The area was also examined in the course of fieldwork in June 1973. In several parts of the mine, no comparison existed between the landscapes that appeared on the 1972 photograph and the condition that existed only 13 months later. Many of the strip-mine lakes had been filled, much of the area was graded, and various grasses had been planted as part of the reclamation program.

Results of tests of the computer-technique ability to classify each of the four categories are shown on Table 4. Vegetation classification was the most accurate, with a 92 percent correct determination.

Once the total area is classified, the percentage of the area in each category can be determined and then

TABLE 4
Categorization Accuracy Table (units percent)

Category	*1*	*2*	*3*	*4*
Stripped earth	96	4	0	0
Partially reclaimed earth	0	98	2	0
Vegetation	0	8	92	0
Water	0	0	0	100

converted into units of square measurements, as shown in Table 5. The same procedures repeated after a time lapse and compared provide a measure of change in each category, as is shown in Table 6.

Aerial Photo of September 1973
with barren earth
categorized from ERTS Data
of August 1972

Aerial Photo of September 1973
with barren earth
categorized from ERTS Data
of September 1973

FIG. 6
Stripped earth category mapped from 1972 and 1973 LANDSAT data overlaying aerial photograph taken in Sept. 1973. Approximate scale 1:40,000. Same site as Fig. 5.

TABLE 5
Area Printout Table Test Area A, ERTS, Sept. 3, 1973

Category	*Percent of Total*	*Square Kilometers*	*Acres*
Stripped earth	15.54	15.44	3,814
Partially reclaimed	11.86	11.79	2,913
Vegetation	72.08	71.6	17,692
Water	0.53	0.52	129

TABLE 6
Site A Area and Area Changes

Category	*Aug. 21, 1972 Acres*	*Sept. 3, 1973 Acres*	*Difference (1973–1972) Acres*
Stripped earth	2,948	3,814	+868
Partially reclaimed	2,512	2,913	+401
Vegetation	18,657	17,692	−965
Water	433	129	−304

Similar techniques are being employed for a wide variety of other surveys and purposes that require an efficient means of mapping land conditions and temporal changes of surface conditions. With proper point-source data from the mine, more timely methods may emerge for tallying mined and measured remaining mineable coal, or possibly a method for monitoring underground mining by repetitive mapping with enhanced RS data of mines working, but such inductive methods must be tried and evaluated.

Other sources of energy must pass through one or more of the parameters in Fig. 3 as they are developed and move to the energy "end use." For example, solar energy would pass through few system parameters; the resources and reserves do not require discovery or exploration, only collection, extraction, conversion, and distribution.

The other fossil fuels sources of energy have system parameters similar to coal, except that discovery is a large parameter factor for oil and gas. Discovery is also a large factor for nuclear fuel and geothermal sources of energy.

Oil and Gas Energy Systems

The pursuit of new sources of oil and gas is relentless. Nations and corporations are exploring every known sedimentary basin on the globe, on- and offshore. Surface and subsurface surveys using established and experimental geologic, geophysical, and geochemical techniques are being conducted worldwide. Remote-sensing techniques are providing useful information, particularly for increasing knowledge of the geology and structure of sedimentary rocks that may contain petroleum. Johnston (1972), Johnston and Janza (1973), and Withington (1973) demonstrated uses of space imagery for mapping structural trends (lineaments) along and adjacent to the Atlantic coastal plain and continental shelf. The projection of landward structural trends into the continental shelves originated with Hobbs (1904) and was used by Pepper (1954, unpublished report). At the 5th World Petroleum Congress, Johnston et al. (1960) described petroleum potential for Atlantic shelf oil and gas accumulations and estimates of resources that are still acceptable. Brown et al. (1972) demonstrated the presence of sub-bases along the Atlantic shelf.

Saunders (1973) analyzed LANDSAT-1 imagery mosaics over five large areas of North America and delineated known and speculative structural features, such as curvilinears and linears. He determined after plotting these features with the known geology, structural features, and oil and gas occurrences that new areas of potential occurrences should become apparent to a geologist in proportion to his command of the surface and subsurface regional geology. Furthermore, interesting structural areas discerned on mosaics can be compared with available geophysical, geochemical, and subsurface information to further evaluate petroleum potentials, and perhaps select a site for exploratory drilling (wildcatting).

Further steps in exploration surveys, such as mapping, leasing, and drilling, become a set of decision priorities for the exploration group based on economic, land and mineral-right ownership, and available leases.

Remote sensing for energy development subtasks. After oil and/or gas is discovered in a new

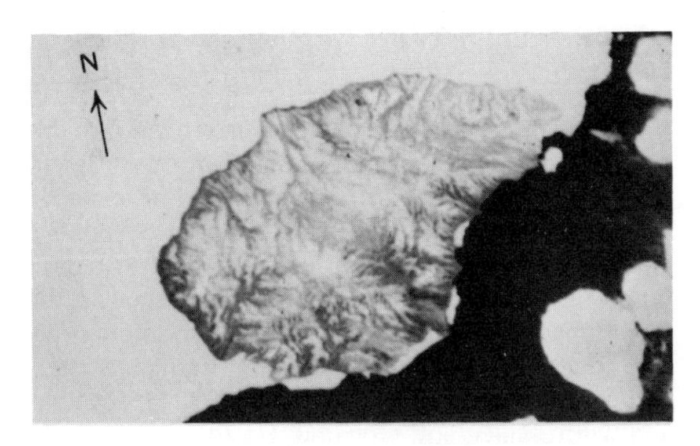

FIG. 7
LANDSAT-1 Image of King Christian Island, Canadian Arctic.

field or province, such as the north slope of the Brookes Range in Alaska or gas in the Arctic (Fig. 7), remote sensing from spacecraft and high-altitude aircraft can provide useful development logistic information, as follows:

1. As a guide to extending exploration for ascertaining potential reserves.
2. For pipeline-route alternatives evaluation based on ecological impact.
3. To plan the most efficient service sites and access routes to these sites and to well sites.
4. To find and survey areas of potential hazard, such as landslides, spring avalanches, floodings, bogs, or permafrost zones, and river mouth silting.

Many of these uses can be accomplished with automatic enhancement and extraction of information from raw remote-sensing data. However, interpretation and field checks are recommended.

Use of Remote Sensing Data for Other Energy Source Developments

As previously stated, numerous tasks are associated with the development of energy sources. The job of extracting fossil fuels and uranium from the earth or using it in place, as with coal gasification, requires several steps of planning, such as, maintaining controls of extraction operations and maintaining an environmentally acceptable rate of reclamation at extraction sites. Timely surveys of mining (extraction) operations are required, and these surveys need to be documented in map form at different scales of detail for mining company tasks, as well as for required functions of local, state, and federal agencies.

AIRCRAFT VERSUS SPACECRAFT IMAGERY

Remote sensing, especially from high-altitude aircraft and spacecraft, provides relatively low cost data surveys for mapping and inventory. Surveying surface features, repeatedly, from both high-altitude aircraft and spacecraft has advantages and constraints; however, the two means should not be compared on a "one or the other basis."

High-altitude imaging can provide the following:

1. Local (intermediate) synoptic view.
2. Choice of time and area of overflight (minus cloud cover).
3. Choice of optimum sensor.
4. Option of stereo photos.
5. Higher ground spatial resolution.
6. Paying organizations may write their own specifications.

Constraints include the following:

1. Need for image mosaicing to view a large area (not synoptic).
2. Cost, amount of data handled, time, and inconveniences are relatively higher than space systems.
3. Imagers require more rectification to match existing map projections.
4. Enhancement of data not fully automatic.

Space imagery can provide the following:

1. Synoptic coverages of more than 10,000 square miles.
2. Savings in time, money, and the amount of data handling.
3. Images in digital form for automatic image processing (AIP).
4. Relatively cheap repetitive overflights.

5. Prediction of time of overflights (minus cloud cover).
6. Near constant sun angle.
7. Nearly matching repetitive image frames.
8. Separate black and white spectrally separated images for color coding and enhancement or optical enlargement.

Constraints include the following:

1. Lack of control over time and location of data collected.
2. Present repetitive systems are government-operated and experimental (LANDSAT-1, NOOA-2).
3. Types of sensors aboard satellites are presently restricted.
4. Ground and spatial resolution that is relatively low.
5. Maximum enhancement by (AIP) methods that are relatively sophisticated and expensive for small areas.

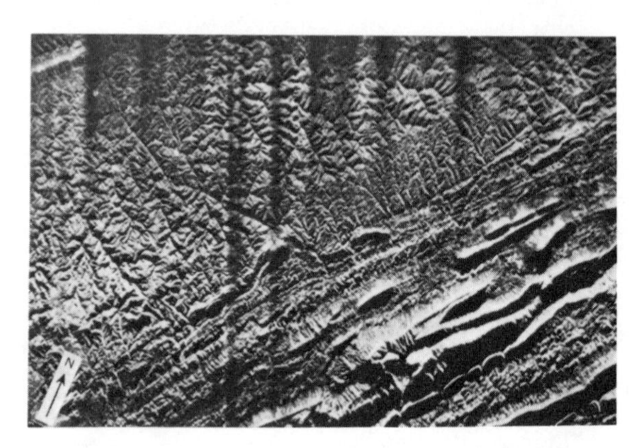

FIG. 8
Radar (SLAR) imagery of part of the Appalachian Plateau and the Valley and Ridge Provinces to the northwest and southeast, respectively, of southeast Kentucky, southwest Virginia and western Virginia. The Russell Fork fault trends southeastward from the northwest, where a small segment of Pine Mountain can be seen; the Canebrake fault trends northwest–southeast in the upper-right corner of the image. The aircraft flew along the northern boundary; note the northwest reflection of radar energy and the deep shadowing on the southeast side of the topography.

OTHER APPLICATIONS OF REMOTE-SENSING DATA

Various remote sensors and their vehicles provide different but complementary data useful to different investigators for different applications. For example, Fig. 8 is a radar (SLAR) image released by the Department of Defense about 1964; because of radar's unique shadowing of the topography normal to the direction of flight, subtle features of the terrain are accentuated in contrast to medium or high-altitude vertical air photographs. Johnston (1972) recognized and with Miller and Englund (1974) subsequently field checked previously unrecognized faults in the southern Appalachians. Similar rock features may prove significant in the discovery of oil, gas, and uranium; in the development of gas and coal produc-

FIG. 9
Black-and-white copy of part of the CIR aerial photograph showing part of the Canebrake fault. The aircraft was flying at approximately 12,000 m over terrain with a mean elevation of 450 m. The camera had a 155-mm focal length lens. The original photographic scale is approximately 1:78,000; conventional photography commonly ranges in scale from 1:20,000 to 1:40,000. (CIR photograph provided by the Rome Air Force Development Group, Rome Air Force Base, Nov. 3, 1971.)

FIG. 10
Canebrake fault in railroad cut 2.4 km northwest of Bishop, Va. Man stands on talus-covered gouge zone with his hand on fault plane.

tion; and in site selection of coal-conversion plants and gasification facilities or atomic energy utility plants. Ryan (1974) and Overbey et al. (1974) analyzed and utilized SLAR for hydrocarbon production siting and satellite imagery for site selection and planning coal gasification projects.

The value of both a synoptic terrain view and surface investigation is illustrated in Figs. 9 and 10. The anomaly of the linear trace of the Canebrake fault cutting across the fiber of a dendritic drainage pattern can be recognized easily by a geologist if the scale of the imagery is optimum, so as to encompass the structure as in Fig. 8. Field studies of the area generally can provide evidence of a geologic fault, as in Fig. 9, where two rock masses have moved relative to each other. A lineament extending several miles to the southeast across the Valley and Ridge Province (shown in Fig. 8) proved not to be a fault when studied in the field by R. L. Miller (Johston et al., 1974). Other lineaments in that image are the results of rock flexures and may not be faults, having appreciable movements. The segment of a LANDSAT image over the same approximate area made January 12, 1973 (Fig. 11) does not show the lineament field checked by Miller. The LANDSAT image also covers an area to the north and west of the SLAR image (Fig. 8). The long prominent scar on the west-central part of Fig. 11 is Pine Mountain, a thrust-faulted block with relief of approximately 2000 ft. The trace of Pine Mountain fault is along the north side of the block trending northeast and southwest. On the south side of the Pine Mountain, about midway, a lineament tends south and then slightly southeast. This lineament was studied from an aerial photograph (Fig. 12) and on the ground; it was also determined to be a fault, which subsequently was named the Coeburn after a small town at its southern end.

Imagery from the NOOA-2 satellite can be obtained in both the visual and a thermal infrared band. Figure 13 is an example of the latter obtained on a relatively cloud free night pass. This synoptic view from beyond the Great Lakes to the northwest and beyond Chesapeake Bay to the southeast provides a

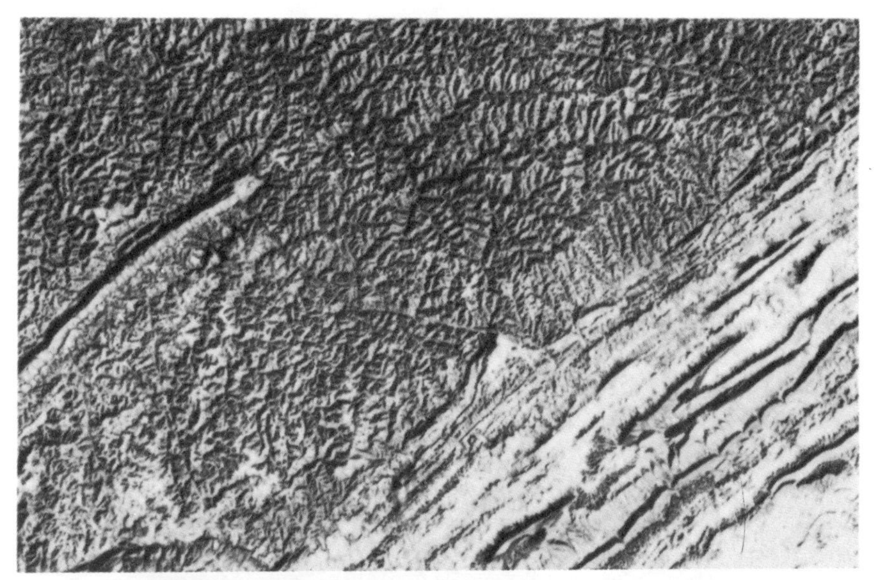

FIG. 11
Segment of LANDSAT near-infrared image showing part of the area in Fig. 8 about 9:30 A.M., Jan. 12, 1973. Note shallow northwest shadowing. Image shows Pine Mountain left center, with the Pine Mountain fault on its northwest flank, and the Russell Fork fault trending southeastward from the end of Pine Mountain. The lineament of the Canebrake fault is in the upper right and that of the north trending Coeburn fault at the lower left. (LANDSAT-1 MSS image 1173-15371, band 7; NASA Goddard Space Flight Center, through the EROS Data Center.)

FIG. 12
Black-and-white copy of a CIR aerial photograph showing a segment of the Coeburn fault. This synoptic view of the region first allowed the authors to identify the lineament. (CIR photograph provided by the Rome Air Force Development Group, Rome Air Force Base, Nov. 3, 1971.)

FIG. 13
Part of a NOAA-2 very high resolution thermal infrared image of part of the United States and Canada, Dec. 3, 1973. (Print 71-18-19-20.)

regional overview of the relationship of several geomorphic provinces. The light tones are warmer.

Geologic analysis of remotely sensed imagery often provides new insights into regional structure. A lineament extending southwest from Long Island marking the fall line between the coastal plain and the Piedmont shows some evidence for a left lateral fault, as shown by apparent displacement of the Delaware and Susquehanna rivers.

Methods of interpreting geology and geography from remote-sensing imagery are both an art and a science. Each individual geologist or organization, knowing its objectives and finished products, must determine how remote-sensing data can be coupled with other information to derive the required maps, reports, or actions (Anderson, 1976).

The basic background for using remote sensing for applications to energy development tasks is a working knowledge of photogeology, a thorough knowledge of the principles of interpretation of topographic and geologic maps, as presented by Dake and Brown, 1925, and an appreciation of the types of information that can be automatically extracted or enhanced from various types of remote sensors.

REFERENCES

Anderson, J. R., et al., 1976, A Land-Use and Land-cover Scheme for Use with Remote Sensor Data, *Photogrammetric Eng.*, p. 28, III. 4.

Averitt, Paul, 1975, Coal Resources of the United States, January 1, 1974, *U.S. Geol. Surv. Bull. 1412,* 131 p.

Brown, P. M., J. A. Miller, and F. M. Swain, 1972, Structural and Stratigraphic Framework, and Spatial Distribution of Permeability of the Atlantic Coastal Plain, North Carolina to New York, *U.S. Geol. Surv. Prof. Paper 796,* 79 p.

Dake, C. L., and J. S. Brown, 1925, *Interpretation of Topographic and Geologic Maps,* McGraw-Hill, New York, 355 p.

Glaser, P. E., 1968, Power from the Sun: Its Future, *Science,* v. 162, p. 857–861.

Hobbs, W. H., 1904, Lineaments of the Atlantic Border Region, *Geol. Soc. Am. Bull.,* v. 15, p. 483–506.

Johnson, J. E., 1972, Remote Sensing as a Tool for Interpreting Structural Geology in the Appalachians (abs.), Am. Assoc. Petroleum Geologists, Eastern Section 1st Annual Meeting, Columbus, Ohio.

____, and F. J. Janza, 1973, ERTS and Remote Sensing's Niche in Geoscience: (abs.), Am. Assoc. Petroleum Geologists 18th Annual Meeting, Anaheim, Calif.

____, R. L. Miller, and K. J. Englund, 1974, Application of Remote Sensing to Structural Interpretations in the Southern Appalachians, *Geol. Surv. J. Res..,* v. 3, no. 3, p. 285–293.

____, James Trumbull, and G. P. Eaton, 1960, The Petroleum Potential of the Emerged and Submerged Atlantic Coastal Plain of the United States, 5th World Petroleum Congress, New York, 1959, *Proceedings,* v. VI, p. 435–445.

MacDowall, J., 1972, A Review of Satellite and Aircraft Remote Sensing Instrumentation, *Proceedings of 1st Canadian Symposium on Remote Sensing,* Ottawa, 2 v., 764 p.

Manual of Remote Sensing, American Society of Photogrammetry, v. I and II, 1975, ASP, Falls Church, Va.

Overbey, W. K., Jr., et al., 1974, Geologic Investigations for Siting and Planning. An Underground Coal Gasification Project (abs.), Am. Assoc. Petroleum Geologists, Eastern Section Meeting, Third Annual Meeting, Pittsburgh.

Richards, G. O., 1975, *Preliminary Guide to Image Processing Equipment and Services,* Electronic Industries Assoc., AIPR 4, 69 p.

Rogers, R. H., L. E. Reed, and W. A. Pettyjohn, 1974, Automatic Mapping of Strip Mine Operations from Space Data, American Congress on Surveying and Mapping/American Society of Photogrammetry Fall Convention, Washington, D.C., 16 p.

Ryan, W. M., 1974, Structure and Hydrocarbon Production Associated with the Pine Mountain Thrust System in Western Virginia (abs.), Am. Assoc. Petroleum Geologists, Eastern Section Meeting, Third Meeting, Pittsburgh.

Saunders, D. F., et al., 1973, ERTS-1 Imagery Use in Reconnaissance Prospecting: Evaluation of Commercial Utility of ERTS-1 Imagery and Structural Reconnaissance for Minerals and Petroleum, Texas Instruments, Inc., for NASA, Goddard Space Flight Center, Greenbelt, Md. 162 p.

U.S. Department of the Interior, 1974, *The EROS Data Center,* Government Printing Office, Washington, D.C., 28 p.

Wade, N., 1974, Windmills: The Resurrection of an Ancient Energy Technology, *Science,* v. 148, p. 1055–1058.

Withington, C. F., 1974, Regional Structures of the Middle Atlantic Coastal Plain as Seen from ERTS-1 Imagery (abs.), Geol. Soc. Am., Northeastern Section, Ninth Ann. Meeting, Baltimore, Md.

9

DIGITAL ENHANCEMENT OF LANDSAT MSS DATA FOR MINERAL EXPLORATION

R. Michael Hord

R. Michael Hord, former Director of the Computer Applications Division of the Earth Satellite Corporation, is currently with the Institute for Advanced Computation, Falls Church, Virginia.

With the advent of LANDSAT (formerly ERTS), the U.S. government has provided the geology community with a source of image data of unparalleled utility. This imagery is now being used by over 300 geological organizations. It is significant, however, that the acceptance accorded LANDSAT data has been largely without the benefit of optimal presentation and exploitation techniques.

THE DATA

Many users are familiar with the LANDSAT film products available from User Services, EROS Data Center (EDC), Sioux Falls, South Dakota. In particular, MSS (multispectral scanner) imagery is available in single-spectral-band black and white or multiband color composites in a variety of scales and in either transparency or print form. Each frame covers a nominal 100 by 100 nautical miles on the ground. Until the launch of LANDSAT-2, each point on the earth was observable every 18 days, but now every 9 days with the two satellites.

Less known but of primary concern here is that these data are also available in digital format on computer compatible tapes (CCTs). This medium, properly processed, is greatly to be preferred in many geological projects over film products for a variety of reasons.

The major advantages of the CCTs are the quality of the data and the accessibility of that data to computer manipulation. The quality advantage is really an aggregate of four advantages:

1. Dynamic range: up to 128 gray levels can be present on CCTs, whereas only 15 to 30 are discernible on film products (particularly important for soils).
2. Precision: the quantitative nature of the medium permits a user to determine a gray-level value exactly.
3. Repeatability: a copy of a copy of a copy of a CCT contains exactly the same data as the original, whereas a fourth-generation photo product is substantially degraded with respect to the original.
4. Resolution: many investigators report that smaller ground details are observable from CCT data than from film products.

These are strong advantages, but the compelling advantages for the geology community derive from the accessability of the data to computer manipulation. Generating a *general-purpose* film product is essentially a compromise among competing requirements from agronomists, hydrologists, foresters, geologists, urban planners, and other user groups and is not optimized for any one. The CCTs lend themselves to custom tailoring by the user for his application in a way film products cannot.

Before discussing this custom-tailoring process, some factual exposition is in order. The LANDSAT vehicles each carry two sensors, the RBV (return beam vidicon) and MSS (multispectral scanner). The RBV acquires data in three spectral channels or bands. The four MSS bands (termed 4, 5, 6, and 7) acquire imagery in the green (band 4, 0.5 to 0.6 μm), red (band 5, 0.6 to 0.7 μm), and infrared (band 6, 0.7 to 0.8 μm, and band 7, 0.8 to 1.1 μm) regions of the spectrum. Each band's image consists of an array of numbers corresponding to the reflectance in that band of a raster of spots on the ground. The array contains 2340 lines, with each line divided into 3240 elements. Each element is termed a pixel, and each is assigned a number from 0 to 127 (from 0 to 63 in band 7).

For $200, EDC will send anyone a frame of this data (the choice of frame is specified by an ID number from a catalog) on four tapes. All four bands of data are supplied, but currently they are not organized with one band per tape. Rather, all four bands are present on each tape for a quarter of the coverage area on the ground. An area measuring 25 nautical miles across track and 100 nautical miles along track is recorded on each tape and termed a panel. Four panels constitute the frame.

PROCESSING

In the years since the launch of the first Earth Resources Technology Satellite, a considerable variety of computer-processing algorithms has evolved. The routine procedure is to use a computer program to read the digits from the tapes, manipulate those digits in accordance with instructions contained in the computer program, and then display the manipulated output digits. The methods of display will be considered later. Often the output will not be displayed immediately but rather stored on another tape, perhaps for another processing step to be performed later. We shall consider here some of the more common processing algorithms, particularly those found to be useful for mineral exploration.

Many algorithms may be applied without intervening human judgment; others depend on the scene content, the specific application, the preferences of the photointerpreter, and the characteristics of the display device.

1. *Reformat:* many facilities prefer to work not with the original CCT format but with a format standard specific to that facility. One common facility format separates the spectral bands onto separate tapes.
2. *Squaring* geometrically corrects the data to equalize the scale factor in the along-track direction with the scale factor in the across-track direction.
3. *Deskew* geometrically corrects the data to compensate for the rotation of the earth during acquisition.
4. *Destripe* compensates for the mismatched sensitivities of the sensor elements in the MSS. Without this compensation the images have the appearance of horizontal "stripes" every six lines throughout the frame.
5. *Contrast enhancement:* a variety of gray-scale adjustment functions are in common usage. Generally, the raw data values are not well suited for display, so each is systematically changed to take full advantage of the capabilities of the display device and the photointerpreter. The simplest of these is the linear gray-scale adjustment, but more complicated functions have been developed and are generally more useful. One example is histogram equalization. This is an adaptive adjustment rule. The computer first ascertains the relative occurrence frequency of the gray levels in a given image. It then computes a nonlinear gray-scale adjustment function, which will pro-

duce an output image characterized by a level histogram.

6. *Edge enhancement:* a family of edge-enhancement operations based on high-pass spatial frequency filters or image derivatives has been shown to improve image interpretability. The resulting picture is most crisp and frequently expedites interpretation.
7. *Enlargement:* digital enlargement by interpolation or another algorithm is often preferred over enlargement by photographic processes.
8. *Density slice:* a special case of gray-scale adjustment, a process that converts all of the pixels with gray level in a specified range to black, while all other pixels are made white.
9. *False color* assigns each gray level to an arbitrary color in order to emphasize gray-level differences.
10. *Ratio:* a multiband process in which the gray level of each output pixel is computed as the ratio of the gray levels of that pixel in two MSS bands.
11. *Eigenvector transform* combines several input bands according to the principal components of the land-cover categories for maximum visual discrimination.
12. *Cloud removal:* using two MSS frames obtained at different times of the same terrain, although each may be highly cloud covered, a single output image without cloud cover can be synthesized.

These then are a dozen of the more common processing options. Each produces a single output image. The images, of course, may be combined into color composites or input to multispectral classification programs, overlaid with interpretation maps and latitude–longitude grids, and, in general, used as ingredients in a thorough exploration effort together with data from other sources.

These represent only a small fraction of the enhancement algorithms that have been developed by teams of geologists, computer scientists, and remote-sensing specialists for application to mineral exploration.

DISPLAYS

After manipulating the data to enhance features of interest in exploration, the analyst faces a display problem before interpretation can begin. Thankfully, image display devices are fairly well developed. A good range of devices is available at widely varying prices.

The first one used by each facility that begins work in this area is the line printer associated with all general-purpose computers. Subimages are displayed as shade prints; each pixel is represented by an alphabetic character chosen to be dark (e.g., M, H) or light (e.g., ., -, =) according to the pixel's gray-level value.

Soon, however, this is recognized as inadequate, and an improved display system is selected. At this point a choice must be made between electronic displays (TV screen) and photographic displays. Both vary regarding color or black and white, and regarding format size, speed, and image quality.

Among photographic displays one may choose among electron beam recorders, laser beam recorders, or more conventional technologies. These have the advantage of producing hard copy that can conveniently be kept for long periods. Electronic displays tend to be faster but unstorable. It should be noted, in this enhancement context, that any compromise on image quality will eventually be regretted.

CONCLUSION

The central message is that a developed technology exists to improve the utility of image data. When decisions involving large amounts of resources are to be made on the basis of information extracted from images, as in the case of mineral exploration, exploiting this technology to maximum advantage is recommended.

GEOPIC EXAMPLES*

The figures described illustrate a few of the more common digital image processing operations.

*Picture credits: R. M. Hord, R. L. Shotwell, R. W. Anderson, and D. Milgram.

A

Example 1: Contrast Enhancement

Figures A and B demonstrate the improvement in overall image quality obtainable through digital linear contrast enhancement compared with the correspond-

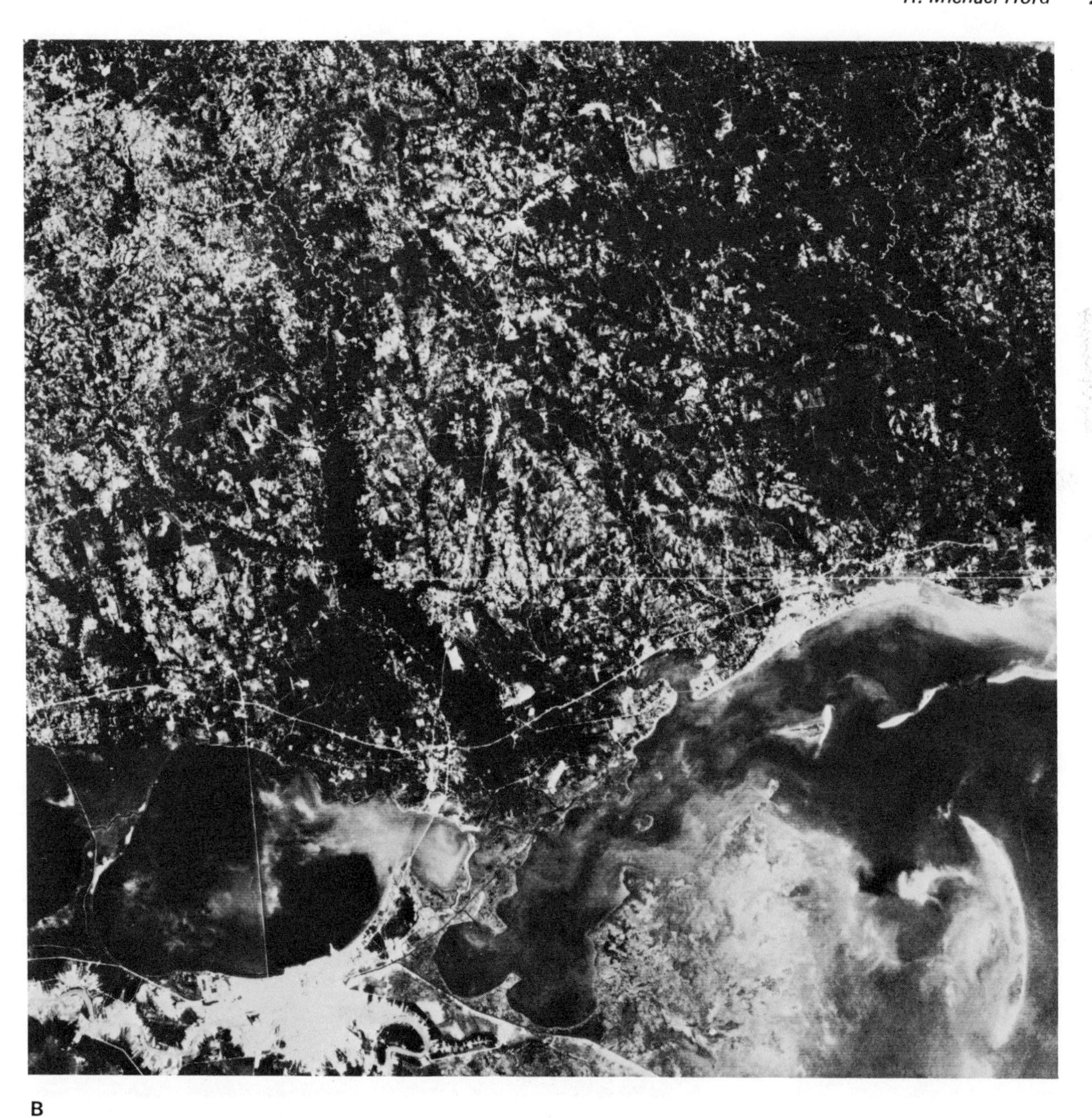

B

ing EDC LANDSAT film product. The image, showing an area around New Orleans, Louisiana, was contrast enhanced to display the water turbidity along the shore. The scale is 1:1,000,000.

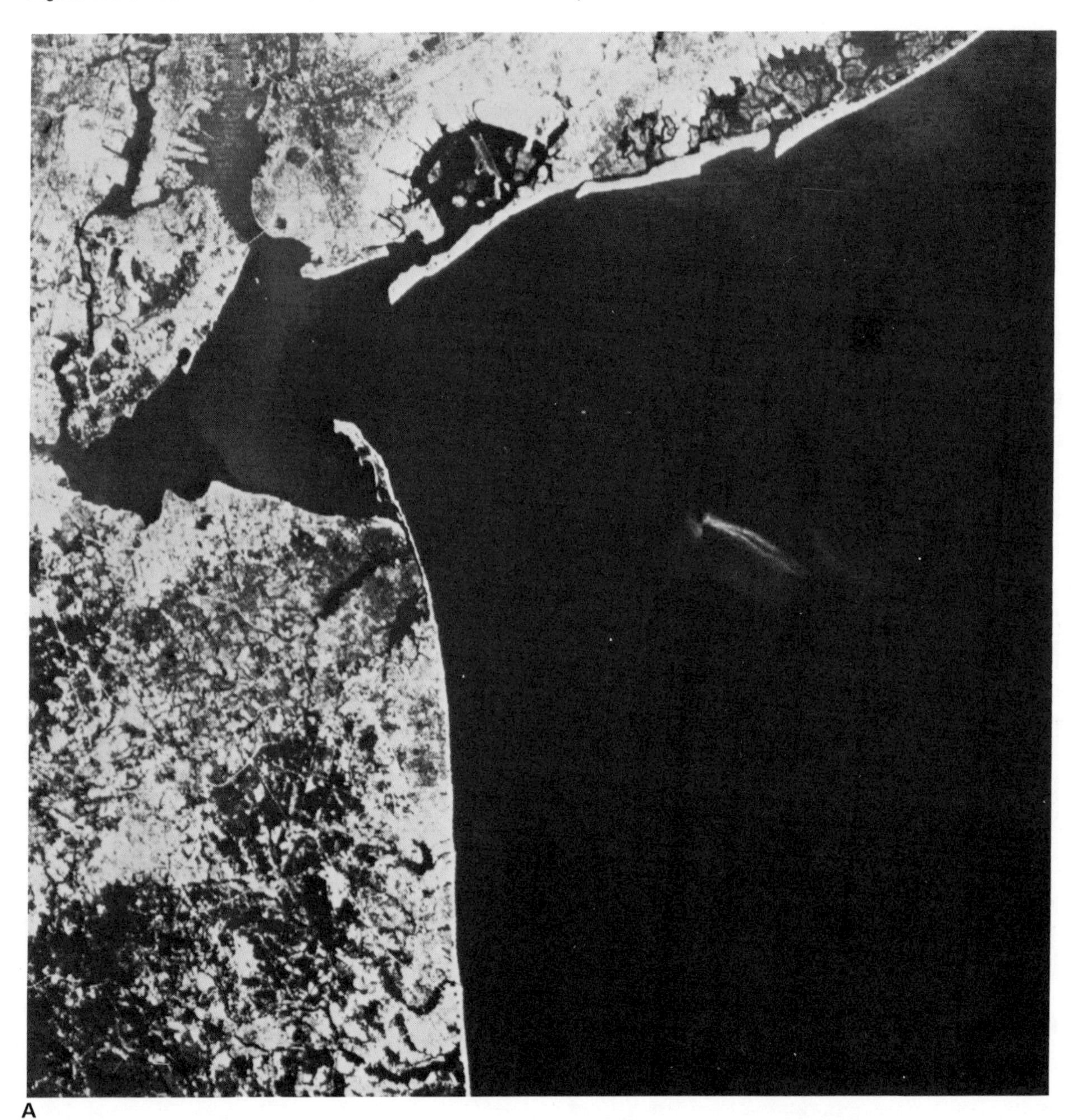

A

Example 2: Density Slicing

Frequently a photo interpreter will be interested in a very specific feature in an image. In this instance it is useful to density slice that image so that only a very narrow range of gray values is displayed. Figures

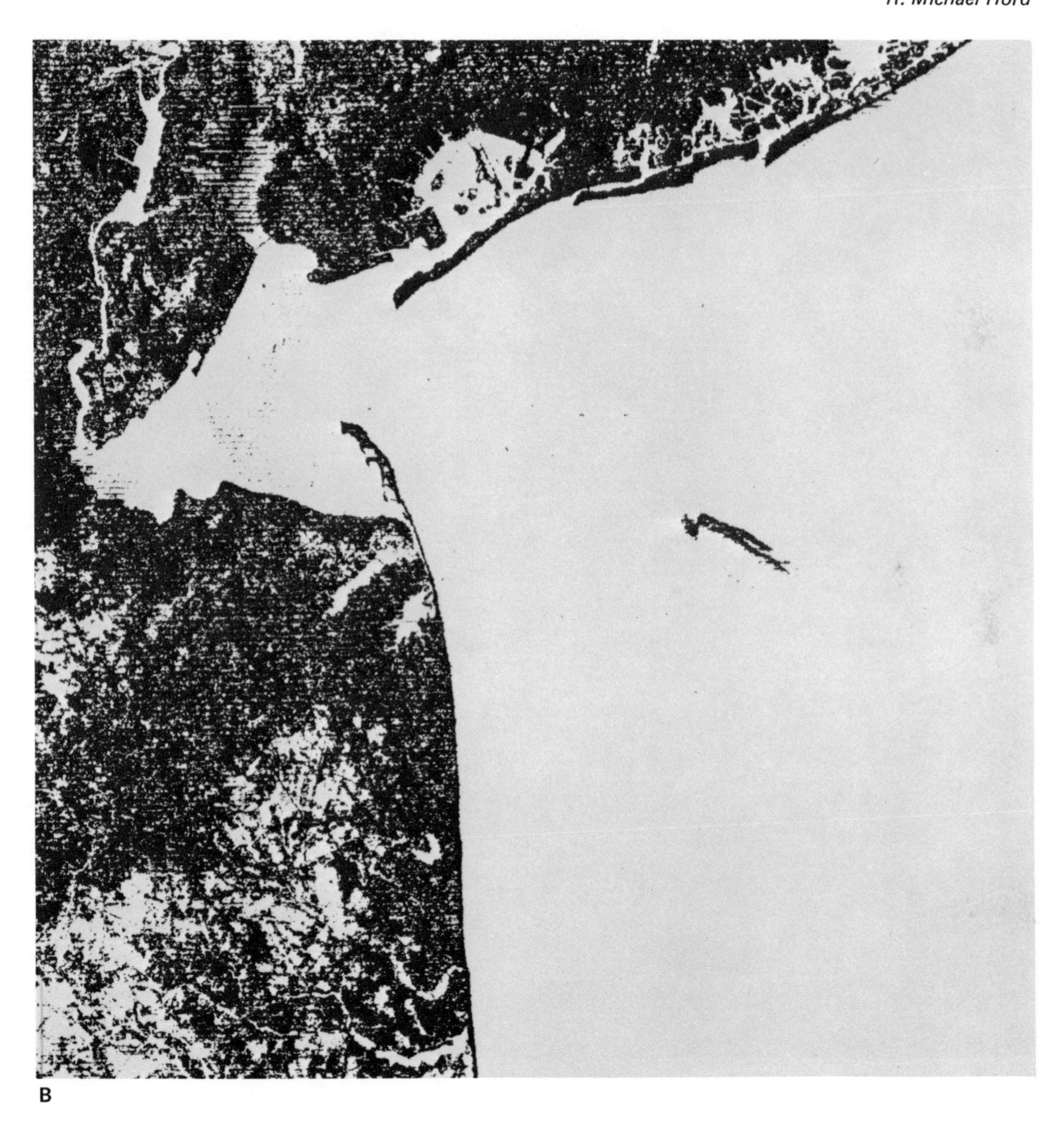

B

A and B demonstrate this technique. The image displays the coastal area around New York harbor. The fresh acid dump seen near the harbor was of particular interest in this example. The scale is 1:400,000.

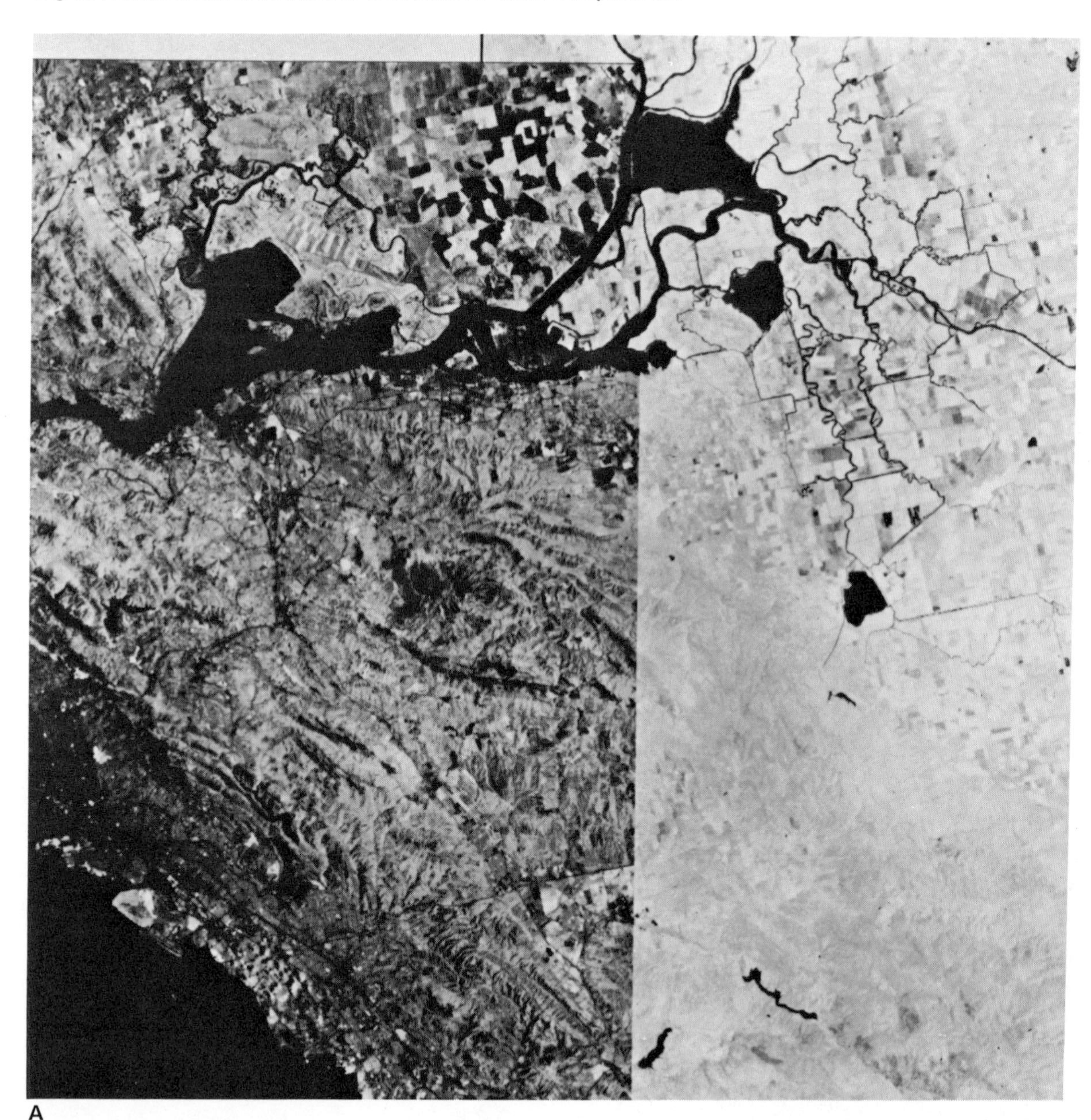

A

Example 3: Digital Photomosaic

Complete area coverage often requires analysis of more than one LANDSAT image. The process of digital photomosaic generation is demonstrated in Figs. A and B in which scenes of the Sacramento, Calif. area obtained on different dates are joined. Figure A is an intermediate product. Figure B illus-

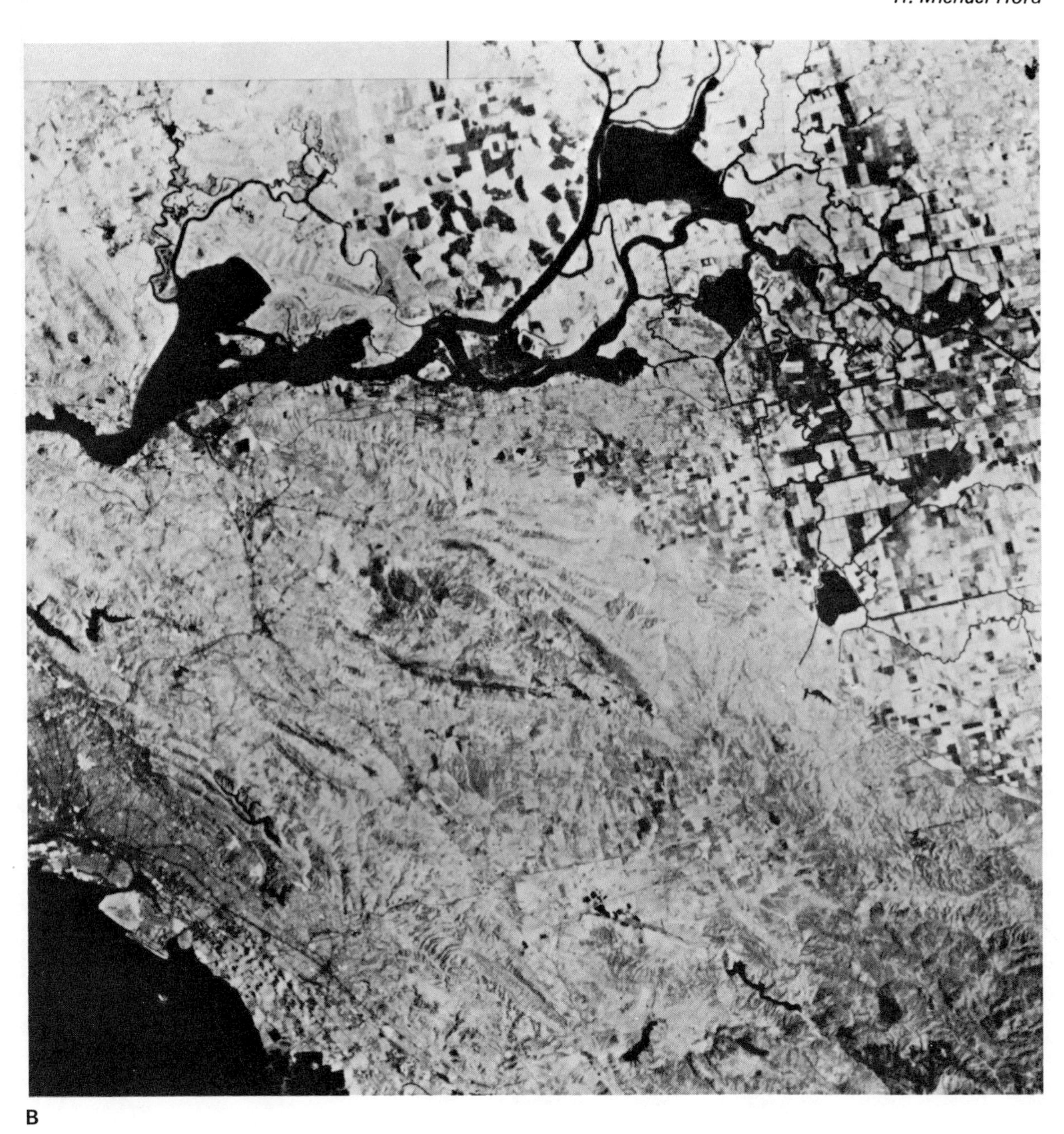

B

trates that the process automatically corrects for radiometric differences between the scenes while maintaining maximal visual contrast. The scale is 1:500,000.

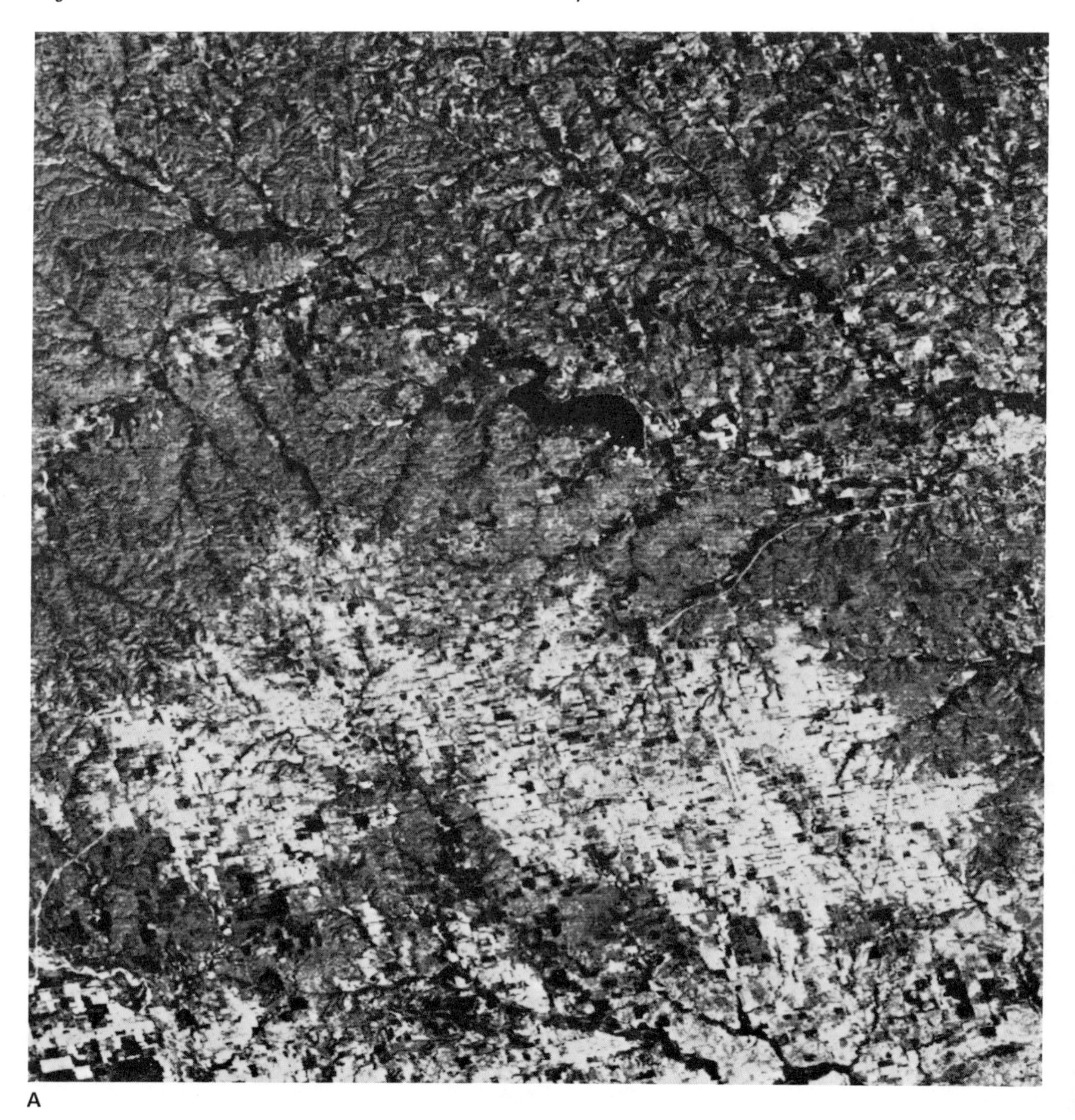

A

Example 4: Logarithmic Ratioing

Selective combination of LANDSAT digital data is performed to foster the analysis of specific features in the image. Figure C demonstrates the effect of loga-

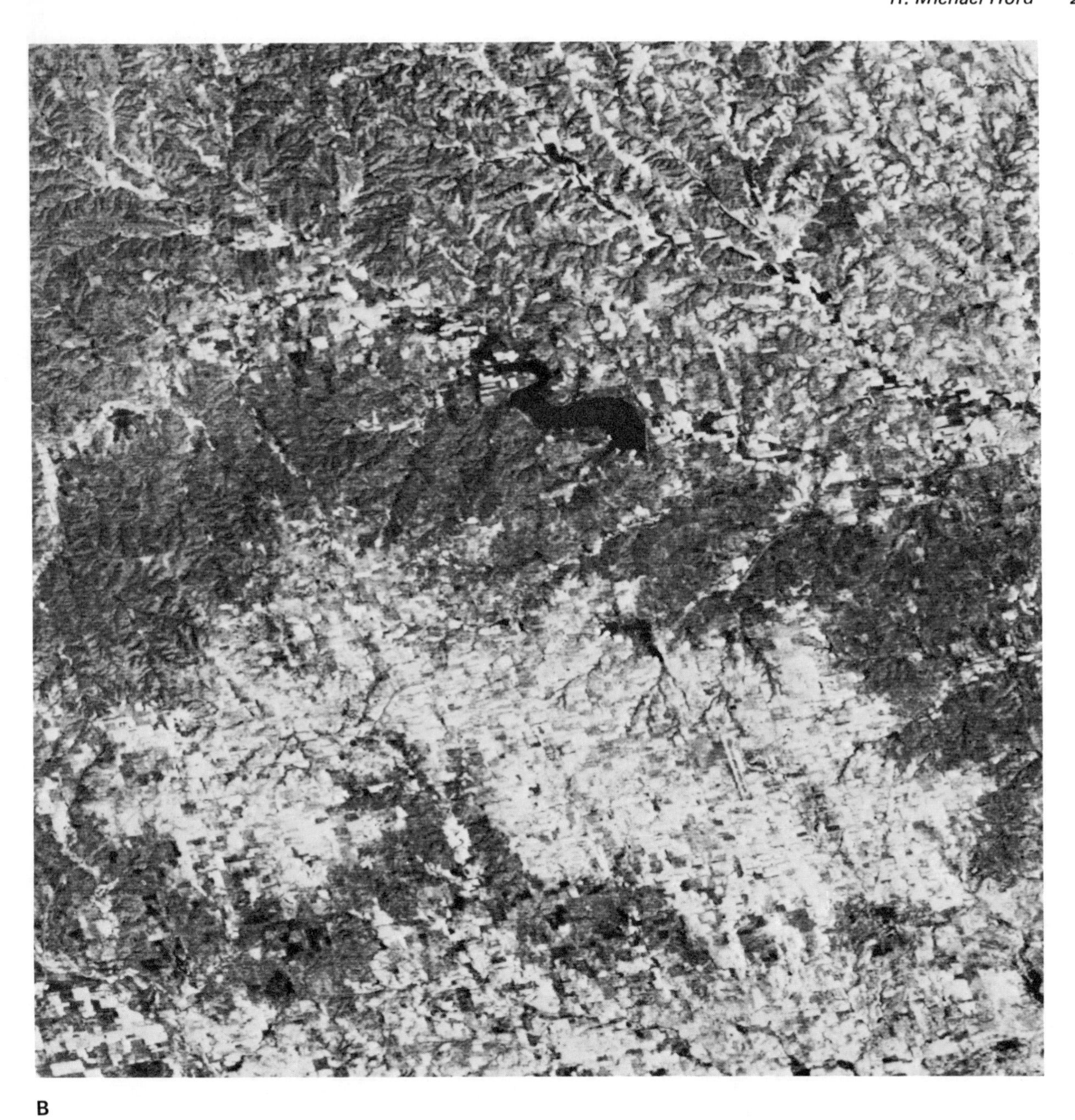

B

rithmically ratioing LANDSAT-1 bands 5 and 7, shown in Figs. A and B, respectively, with the aim of improving the geologic interpretability of a 1973 Oklahoma scene. The scale is 1:400,000.

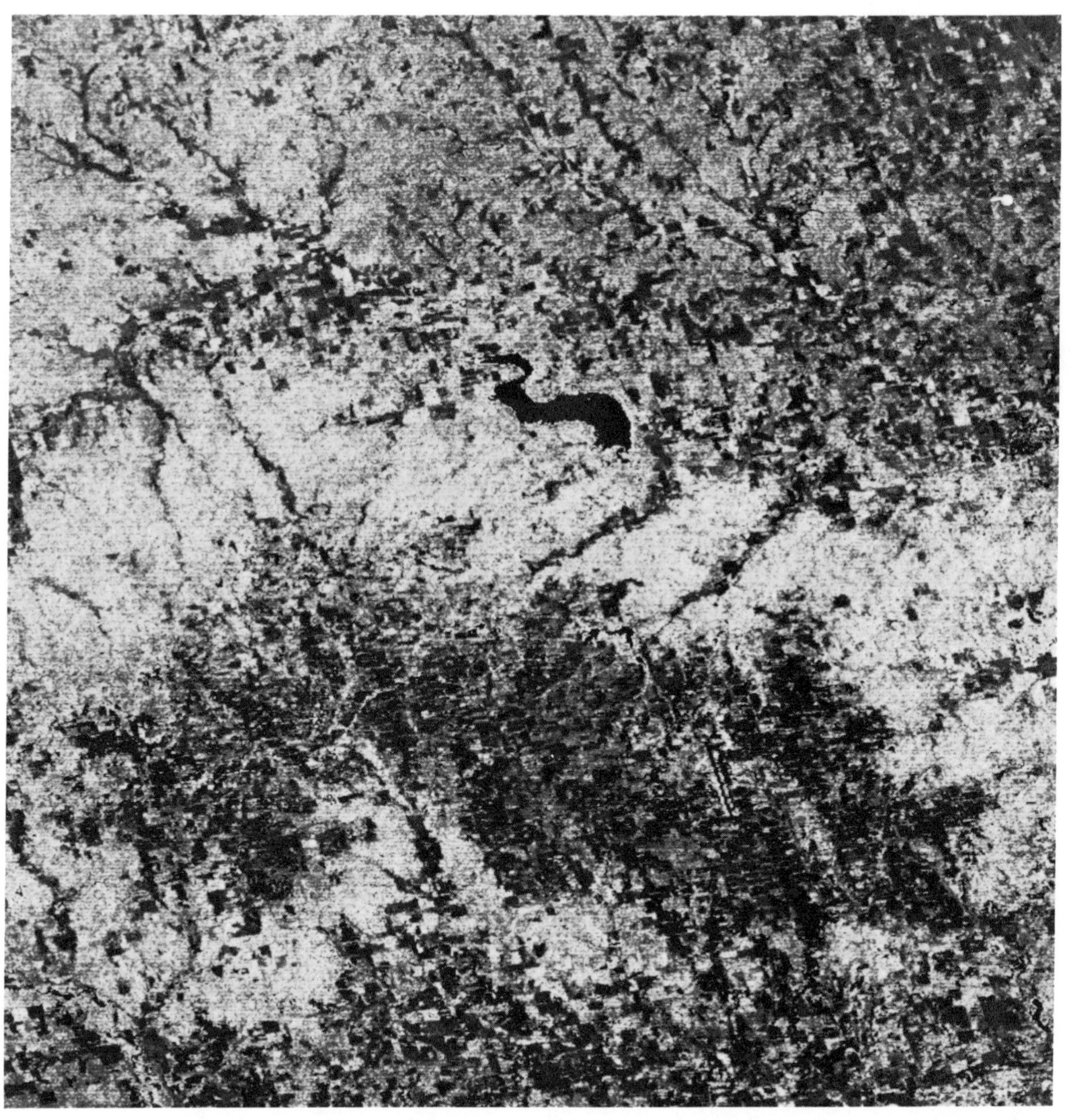

c

A

Example 5: Selected Examples

This example was prepared to demonstrate several processing techniques applied to the same image. Figure A shows an area of Iran at a scale of 1:500,000. Figure B shows a part of the same area

B

that has been digitally enlarged to a scale of 1:250,000. Figure C shows the result of geometrically rectifying the original LANDSAT-1 image. Figure D was processed to digitally enhance edges in this

c

image. Any combination of the above processing techniques may be applied to an image.

D

10

GEOCHEMICAL MAPPING BY SPECTRAL RATIOING METHODS

Robert K. Vincent

Robert K. Vincent is President of the Geospectra Corporation, Ann Arbor, Michigan.

Aerial photography has long been a useful tool for geologists. Camera and film are relatively inexpensive, spatial resolution is excellent, and the services of trained photogrammetrists are available for interpretation of the aerial photos. In contrast, multispectral scanners (MSS) have a short history. Scanner imagery has become widely regarded as a useful tool for geological remote sensing only since the advent of LANDSAT-1 in 1972, when the cost of data collection was substantially reduced by satellites. Accordingly, relatively few trained scientists and technicians have been able to make full use of scanner data. It is quite natural that structural studies (e.g., lineation mapping) have been of great interest early in the history of satellite remote sensing, because they required skills that existed in the geological community prior to the satellite era. However, such studies could probably have been equally well accomplished with space photography instead of MSS images. Although the scanner reduces the need for image mosaicking, compared to photographs, structural studies do not utilize two of the most important advantages of a scanner over a camera: greater spectral range (photographic films are limited to wavelengths shorter than about 1.0 μm, whereas scanners can operate out to 14 μm and longer) and ease of quantitative processing (radiances detected in each single scanner channel can be treated separately for every spatial resolution element in the scene).

Geochemical mapping, where "geochemical" is used in the broadest sense of the word to include gross chemical differences among common rocks and minerals as well as trace-element differences, is a more difficult remote-sensing problem, which does utilize the advantages of a multispectral scanner. For one thing, useful geochemical information exists outside the familiar wavelength regions available to the human eye and the camera. For another, geochemical mapping requires accurate radiometric data (the radiance detected in each scanner channel), which necessitates corrections for spurious environmental effects, such as spatial and temporal changes in solar illumination and atmospheric parameters. In short, there is too much spectral information in the single-channel images of multispectral scanner data for a photointerpreter to be able to extract all of it without the assistance of a computer. Concerning this point, the history of seismic data provides an analogy for geochemical mapping. Photointerpretation of single-channel images can be linked to early seismic techniques, which involved the visual interpretation of seismographs from one or two seismic stations. Although these simpler methods were useful, much more information can be gained from seismic data now that computer data-processing techniques and large arrays of detectors are available.

SPECTRAL EVIDENCE OF MINERALOGICAL COMPOSITION

For optical remote-sensing data to be useful for geochemical mapping, reflectance or emittance features of geologic targets that are diagnostic of chemical composition must exist. In the 0.4 to 2.5-μm wavelength region, spectral reflectance (ρ_λ) is important, because on most of the earth's surface reflected solar irradiance accounts for more energy exiting from the target than naturally emitted radiation. In the thermal infrared atmospheric "window" region from 8 to 14 μm, spectral emittance (ϵ_λ) is more important, because most of the energy coming from a target on the ground is emitted by the target. The wavelength region between 2.5 and 8 μm is mostly absorbed by the atmosphere, except for the 3- to 5-μm region, where the reflected and emittance radiation are more nearly equal. This removes some of the chemically diagnostic information in the 3- to 5-μm region, since $\epsilon_\lambda = 1 - \rho_\lambda$ under equilibrium conditions. The 0.4- to 2.5-μm region is different from the 8- to 14-μm region in another aspect, also. Spectral reflectance features in the former are caused primarily by electronic transitions in transition-metal ions, whereas the spectral emittance features in the thermal infrared region are caused principally by interatomic motions in silicate, carbonate, sulfate, and phosphate minerals.

Figure 1 shows the spectral reflectance features of various minerals *(1–4)* in the 0.4- to 2.5-μm wavelength region. All the particle sizes are $\geqslant$74 μm for each of the crushed samples from which these spectra were measured. Also shown are the spectral bands covered by LANDSAT and Skylab multispectral scanners. For reference, the human eye is limited to approximately the 0.40- to 0.65-μm region, and photographic film is limited to wavelengths shorter than 0.95 μm. Notice that iron oxides (except for magnetite) exhibit a reflectance minimum in the 0.9- to 1.0-μm region, copper sulfides (chalcocite and chalcopyrite) in the vicinity of 0.8 μm, aluminum oxide hydrates (gibbsite) in the regions near 1.5 and 2.4 μm, and carbonate (malachite) in the vicinities of 2.25 and 2.4 μm. The first three types of reflectance minima are caused by electronic transitions in Fe^{3+}, Cu^{2+}, and OH^{1-}, respectively; the carbonate minima near 2.25 and 2.4 μm are caused by overtone and combination tones of the CO_3^{2-} radical and lattice vibrations. Reflectance minima of mafic silicate minerals in the 1.0- to 1.1-μm region are generally caused by the Fe^{2+} iron ion.

Although Fig. 1 shows some significant differences in reflectance among these minerals, the thermal infrared wavelength region perhaps offers more compositional information than the 0.4- to 2.5-μm region, especially for the case of silicate minerals and rocks. To demonstrate this point, Fig. 2 shows a plot of the percentage of SiO_2 versus average emittance in the 6- to 12-μm region and average reflectance in the 0.3- to 2.5-μm region from data by Logan et al. *(5)* for many silicate rocks. Notice that the 6- to 12-μm average emittance correlates much better with the percentage of SiO_2 than the 0.4- to 2.5-μm average reflectance. The thermal wavelength region can have an even greater advantage over the reflective region when both are broken up into narrow spectral bands.

Therefore, important compositional information exists in spectral reflectance and emittance data. How this information can be extracted from aircraft and satellite multispectral data in a reliable manner has been the principal subject of geochemical remote-sensing research to date.

THE CASE FOR SPECTRAL RATIOS

Consider the problem of developing a geochemical mapper for aircraft or satellite use. Desirable characteristics of such a mapper would be that it incorporates the following:

1. Takes advantage of practically all wavelength regions of important geochemical information.
2. Is relatively insensitive to atmospheric and solar

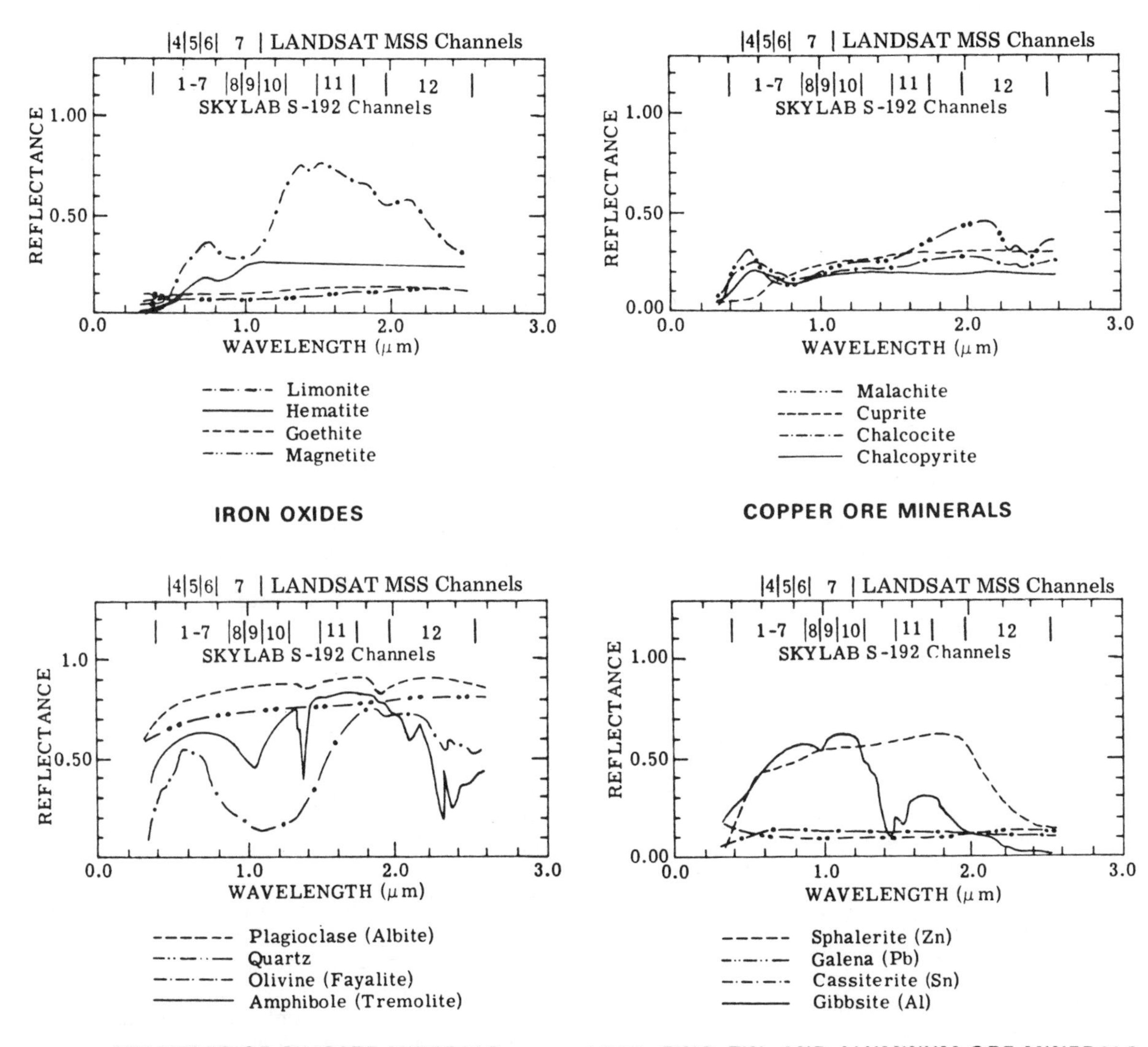

FIG. 1
Spectral reflectance of iron oxides, ore minerals, and silicate minerals. LANDSAT and Skylab multispectral scanner channels are indicated on each graph.

illumination variations, such that data collected at different times and places can be directly compared.

3. Has well-defined limits of accuracy concerning which background targets might be confused with specific geochemical targets of interest.
4. Requires little or no ground truth from the region over which data are collected.
5. Is capable of producing both automatic recognition maps and continuous-tone images for "scannergrammetry" (defined here as photogrammetric techniques applied to scanner data).

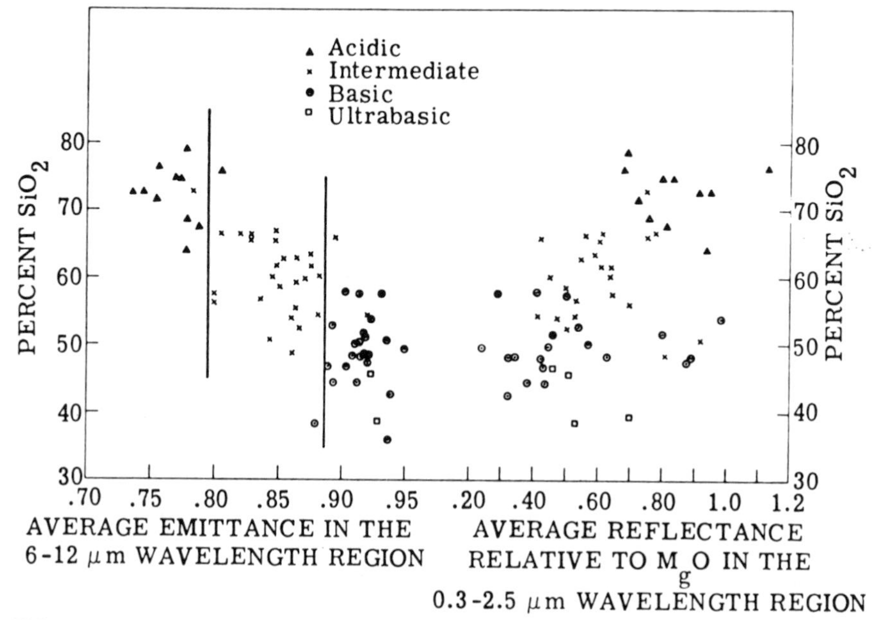

FIG. 2

Comparison of compositional information content of thermal infrared emittance and visible reflective infrared reflectance for silicate rock spectra of Logan et al. *(5)*.

6. Is simple and inexpensive enough to be considered practical for operational use.

Photographs cannot satisfy the first five characteristics and single-channel images, as well as automatic decision theory based on single-channel radiances, do not satisfy the latter three.

One of the best candidates for a geochemical mapper is one based on spectral ratios. A more detailed description of the following procedures can be found elsewhere *(6, 7)*. The calculation of spectral ratios in the 0.4- to 2.5-μm region from satellite and aircraft data begins with the approximate removal of additive atmospheric terms, commonly called path radiance. For low-altitude ($\leqslant$1.5 km) aircraft data, the additive term is usually ignored. Multispectral scanner data are collected on magnetic tape such that the radiance detected in each spectral channel of the scanner can be retrieved for each point (instantaneous field of view) in the scanned scene. The radiance of the darkest object in the scene is subtracted from the radiance of all other points in the scene for each spectral channel. The remaining radiance in each channel is then approximately proportional to the target reflectance. However, multiplicative terms, including atmospheric transmittance, solar irradiance, electronic gain factors, and what can be called a shadow factor (taken to be any spectrally independent suppressor of direct solar illumination within the instantaneous field of view, such as shadowing or sloping away from the sun) are still present. The remaining radiance (after dark object subtraction) in one spectral channel is divided by the remaining radiance in another spectral channel to produce a spectral ratio for each point in the scene. The ratio signal is, therefore, an electronic voltage that is approximately directly proportional to the ratio of reflectance of the target in these two spectral channels. However, the shadow factor has been approximately removed by the ratio, since it is spectrally independent.

If there is one area in the scene for which the spectral reflectance is known, it is possible to determine the proportionality constant between the ratio voltage for a given pair of spectral channels and the actual ratio of spectral reflectances of samples of the target as measured in the laboratory. When this ratio normalization is done, the corrected ratio voltage becomes the actual ratio of reflectances for the target, within the limits of experimental accuracy, which are discussed later. The resulting corrected ratio voltages can then be used as the basis for automatic recognition or for the production of ratio images. To produce a ratio image, the corrected ratio voltage is input to a light source, which is scanned in front of a moving film strip. A high voltage, corresponding to a large ratio, produces a bright spot on the film strip, and a low voltage produces a dark spot. When the ratio image is completed, a gray scale is produced that approximately calibrates the relationship between ratio voltage and film density. In other words, a ratio image is produced in a manner similar to a single-channel scanner image, except that a voltage proportional to a ratio of target reflectance in two spectral regions results.

A geochemical mapper based on these corrected ratio voltages has all the desirable characteristics listed previously. The first characteristic is met by virtue of the fact that a multispectral scanner is the sensor employed. The second desirable characteristic can be provided by spectral ratios, according to recent studies *(8)*. When the magnitudes of one spectral ratio (channel 7 divided by channel 5) of LANDSAT MSS data for a clear August 1972 day and a partly cloudy October 1972 day were compared for a test site near Atlantic City, Wyoming, the dark object subtraction and ratio normalization procedures were found to correct the $R_{7,5}$ ratio voltage within a standard error of 10 percent. The standard errors produced from atmospheric and solar illumination variations for LANDSAT channel 5 (0.6 to 0.7 μm) and channel 7 (0.8 to 1.1 μm) for the same two days were on the order of 50 percent and larger. For both of these estimates, points in the scene estimated to contain $\leqslant$50 percent vegetative cover were used as references, so as to minimize ground changes. Hence, the corrected spectral ratio is at worst more invariant to sky and sun effects than single-channel radiances, and at best can be useful for target discrimination and identification on an absolute basis for some geochemical targets.

The third and fourth desirable characteristics would be fulfilled if the spectral ratios from MSS data could be directly related to reflectance ratios calculated from laboratory data. For LANDSAT, reflectance ratios for the six nonreciprocal spectral ratios possible from MSS data ($R_{5,4}$, $R_{6,4}$, $R_{6,5}$, $R_{7,4}$, $R_{7,5}$, and $R_{7,6}$, where R_{ij} means the ith channel divided by the jth channel) have been calculated for all the rock, mineral, and soil spectra that span the 0.5- to 1.1-μm wavelength region in the NASA Earth Resources Spectral Information System (ERSIS) laboratory data collection *(9–11)*, which is installed at the Johnson Space Center and the Environmental Research Institute of Michigan (ERIM), Ann Arbor, Michigan. Each R_{ij} ratio for LANDSAT was divided into 10 range intervals, each interval containing about 10 percent of the spectral curves in the data collection, and each interval was assigned a digit ranging from 0 to 9. A six-digit LANDSAT MSS ratio code was then used to describe the approximate values of the six spectral ratios possible from LANDSAT data, which permits rapid inspection of the ratio values expected for different rocks and minerals represented in the data bank. These ratio codes have been published *(7)* for 59 soil types (133 laboratory spectra), 89 minerals (197 spectra), 25 rock types (31 spectra), and 18 types of vegetation (18 spectra). Even though the ERSIS collection of data is limited in number and type of rock, soil, and mineral spectra, it has been useful for demonstrating the concept of how laboratory data can be used to make spectral ratios satisfy the third and fourth desirable characteristics. The utilization of laboratory data as training sets has been examined with LANDSAT data and actually executed with aircraft scanner data. A search of the 353 laboratory spectra for "look-alikes" of several iron oxides demonstrates the third desirable characteristic. (*Note:* no spectra of crushed samples with grain sizes exclusively less than 74 μm were included in the 353 curves.) The results are shown in Table 1. It is significant that there are relatively few laboratory spectra with ratio codes similar to the ratio codes of iron oxides. Although it is not realistic to have only one or two samples of each material, this is the best that can

TABLE 1
Results of Searches of Laboratory Spectra for Materials That Are Inseparable from Several Iron Oxides by LANDSAT-MSS Ratio Methods
(Left-to-right order of code is $R_{5,4}$, $R_{6,4}$, $R_{6,5}$, $R_{7,4}$, $R_{7,5}$, $R_{7,6}$)

Subject of Search			*Materials with Which Subject Can Be Confused*	
Name	*Ratio Code*	*Radio Code Range*	*Name*	*Ratio Code*
Goethite (74–250 μm, Minn.)	642311	6, 4, 0–2, 3, 1, 1–2	None	
(250–1200 μm, Minn.)	640312			
Hematite (74–250 μm, Minn.)	888887	8, 8, 8, 8, 8, 7	Loam, Blakely clay type, dry	888887
Limonite (250–1200 μm, Ala.)	889650	7–8, 6–8, 6–9, 5–6, 2–5, 0–2	Loam, Aiken clay, wet	877651
			Loam, Colts Neck type, wet	877630
(74–250 μm, (Ala.)	776520		Loam, Greenville sandy, dry	876631
(250–1200 μm, Ala.)	765542		Loam, Santa Barbara type, wet	876520
			Loam, Tifton sandy type, wet	876641
			Loam, Hamakua heavy type, wet	777651
			Loam, Ookola clay, wet	776642
			Sand, Colts Neck loamy, dry	876641
			Sand, Colts Neck loamy, wet	887640
			Yellow sand, wet	765542
			Loam, Aiken clay, dry	776652
			Loam, Colts Neck type, dry	877651
			Loam, Santa Barbara type, dry	765642

be done with the existing published laboratory data. It is possible that the soils mistaken for hematite and limonite contain those minerals, although the soil descriptions are not complete enough to draw such conclusions. Therefore, with spectral ratios calculated from laboratory data it is possible to approximate what other natural targets will be confused with a target of interest, when spectral ratios are used to produce either ratio images or automatic recognition maps.

Concerning the fourth characteristic, if only one small area in the scene is known, such that it can be used for ratio normalization, the possibility exists that spectral ratios calculated from laboratory data or even from previous multispectral scanner data can be used as a priori training sets to look for specific geochemical targets. In a LANDSAT experiment *(12)* an attempt to map hematite in the Wind River Basin of Wyoming was made. Ratio voltages were measured from LANDSAT data for a known limestone outcropping, and the only limestone (not from Wyoming) curve in the ERSIS data collection was used to calculate the six normalization factors between laboratory ratios and scanner ratios. Then hematite spectral ratios, calculated from the single laboratory spectrum for hematite, plus or minus an arbitrary 15 percent of each ratio value, were fed into the computer and the LANDSAT frame was searched for hematite. The only points recognized in the scene were known hematitic soils and rocks, although many other known hematite exposures were not recognized. Thus, the errors of omission were large, but there were no errors of commission (no false alarms). A later field trip confirmed that the limestone outcrop used for normalization was approximately 50 percent covered by coarse grass. To correct for this, the laboratory spectrum of limestone was mixed with a coarse grass spectrum, and normalization factors were recomputed. The resulting renormalized spectral ratios calculated from LANDSAT data were found to agree with ratios calculated from laboratory spectra within an average error of about 8.5 percent for four rock types. An experiment with aircraft data *(13)* collected over Halloran Springs, California, involved the use of a color composite of three different ratio images (from the visible, reflective infrared, and thermal infrared wavelength regions). Laboratory-measured reflectance data of rocks, minerals, and soils were used to interpret the ratio color composite. When this interpretation was compared with a geologic map, the average recognition accuracy over several rock types was approximately 78 percent over sparsely vegetated regions, with no a priori knowledge of the scene. These exercises have shown that the idea of using laboratory curves as training sets is plausible. Much more refinement is needed, however, before the experimental accuracies of such a procedure can be fully assessed.

The fifth desirable characteristic of a remote-sensing geochemical mapper can also be met by spectral ratio methods. As described earlier, the ratio voltages can be used to produce ratio images. Since ratio images have a continuous tone, they are similar to aerial photographs or single-channel images in appearance. Their use for this purpose will be discussed later. Ratio voltages can also be used as the basis for automatic recognition maps in at least three ways. If a given target has a unique range of values of a particular spectral ratio, compared to most or all other points in the scanned scene, that ratio can be gated (upper and lower limits are placed on the ratio to identify the target) to produce a recognition map in which the target appears bright and all other materials appear dark, or vice versa. A special case of single-ratio gating is thresholding, where the target of interest has a higher or lower value of a particular ratio than most or all other points in the scene. A second way is to gate more than one ratio and require that each ratio gate be satisfied before a point is identified as a target. The latter can be called multiple ratio gating logic, and it may involve any number of ratios. The third way is to use ratios as inputs to standard statistical decision logic, such as the maximum likelihood decision rule *(14)*. The latter is much more complicated than the other two and thus far has required known in-scene training sets for each target of interest. The other two methods are more amenable to the use of laboratory data as training sets.

The sixth characteristic can be met by the successful implementation of ratio gating logic, since nothing more complicated than dividers and AND-gates are required for the logic system. For geochemical targets, ratio gating logic appears to be a feasible candidate for on-board (satellite or aircraft) automatic recognition map data processors.

EXAMPLES OF RATIO IMAGES

At this point, it is instructive to review some examples of ratio images produced over the past 10 years. Figure 3 shows a thermal infrared ratio image from an early-morning, low-altitude (1 km) aircraft flight over Pisgah Center, in Southern California. These images were produced from the ERIM-NASA M-7 multispectral scanner, with a special Honeywell two-element Hg:Cd:Te thermal infrared detector *(15)*, which is capable of detecting radiances in two overlapping wavelength regions (channel 1, 8.2 to 10.9 μm, and channel 2, 9.4 to 12.1 μm) simultaneously, with the same instantaneous field of view. The top two strips are single-channel images of thermal channels 1 and 2, whereas the bottom strip is a ratio image of channel 1 divided by channel 2. Note that the single-channel images show very little contrast across the scene, owing to small temperature varia-

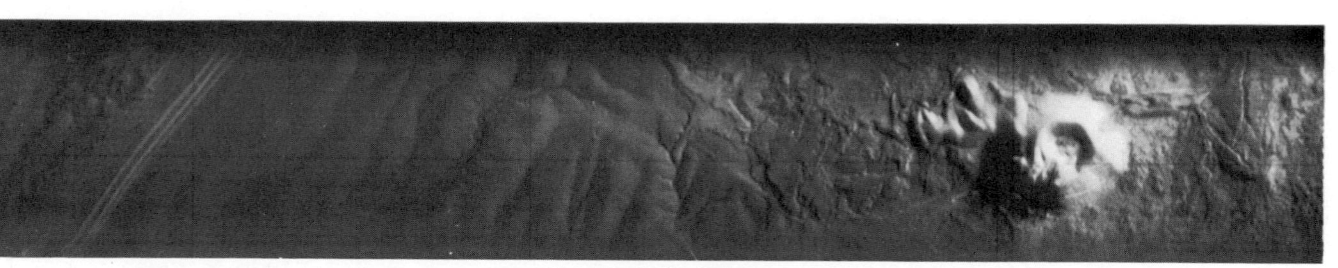

Channel 1: 8.2–10.9 μm

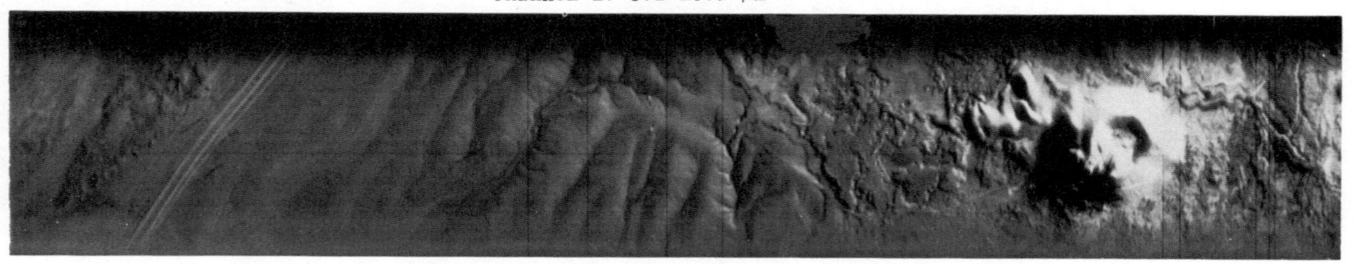

Channel 2: 9.4–12.1 μm

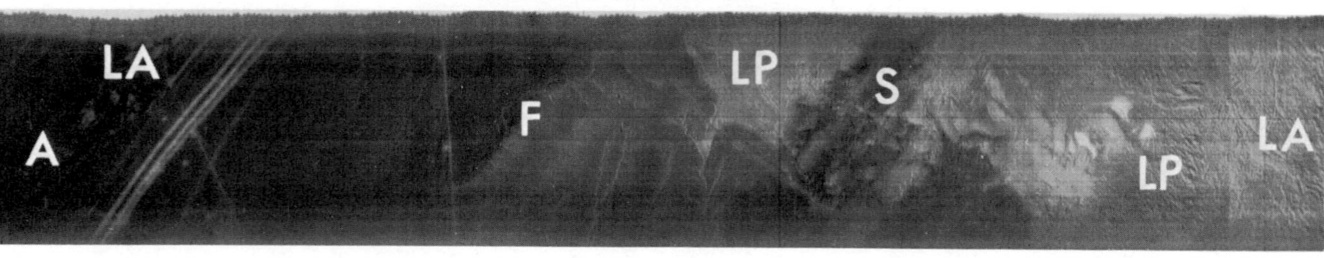

Ratio: $\frac{L_{8.2-10.9}}{L_{9.4-12.1}}$

FIG. 3

Analog infrared images of flight line 1, sec. 4, of a north–south flight over Pisgah Crater, Calif. North is toward the left; the image dimensions are approximately 1.5 km × 6 km. Left to right: alluvium (A); partially covered basaltic lava (LA) of phase 2; highway, fanglomerate, and gravel (F); Pisgah pahoehoe basaltic lava (LP) of phase 3; windblown sand and silt (S); and Pisgah as basaltic lava (LA) of phases 1 and 2.

tions, except for the sunward side of Pisgah Crater itself. The ratio, however, shows spectral emittance variations across the scene, which are produced in part by variations in the percentage of SiO_2 of the silicate rocks present *(16, 17).* Here the basaltic lava is brighter in the thermal ratio than the more felsic alluvium and wind-blown sand, which are dark.

Figure 4 shows the same thermal ratio compared with a color aerial photograph over an intermittent stream bed, approximately 5 miles southeast of Pisgah Crater. Volcanic ash flow outcrops (a rhyolitic tuff) are darker than the surrounding alluvium in the ratio image because the rhyolitic tuff has a higher percentage of SiO_2 than the surrounding alluvium. The aerial photo, taken simultaneously with the scanner data at a 1-km altitude above ground, fails to show these outcrops, because the greatest spectral discrimination between the tuff and alluvium occurs at thermal wavelengths, well beyond the spectral range of photographic film. In this particular case, malachite (an ore of copper) was associated with the volcanic tuff, but in insufficient quantity to be economical.

The thermal infrared region between 8 and 14 μm contains the best spectral information for discriminating among silicate rocks, but multiple channels in that wavelength region are required to take advantage of it. Unfortunately, no satellite has been launched with multiple thermal infrared scanner channels except for Nimbus E, which was short-lived with regard to its two-channel infrared scanner. It is important for geologic remote sensing, however, that multiple thermal channels be included in some earth resources satellite(s) in the future.

More can be expected from future multispectral infrared scanners in the way of classifying igneous rocks, however. In a recent thesis *(18)* a mineralogical parameter (called V_7) was defined, which correlates

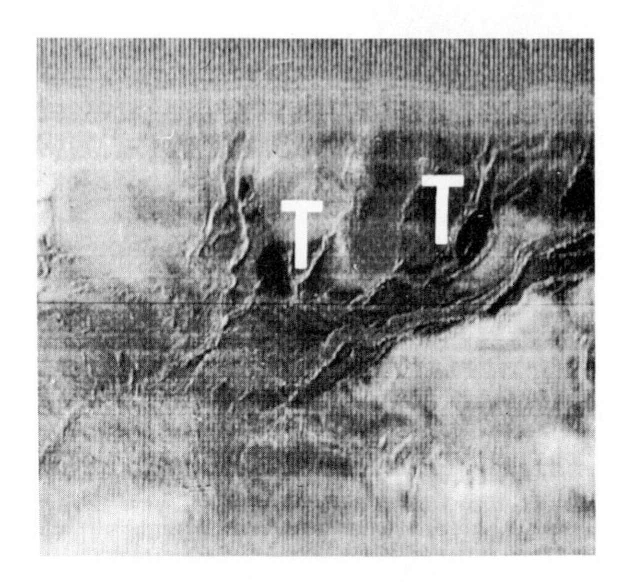

INFRARED RATIO IMAGE

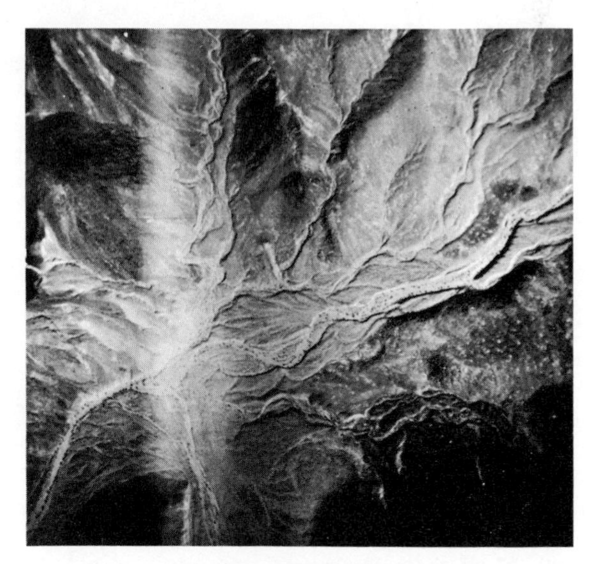

COLOR AERIAL PHOTO

FIG. 4
Comparison of thermal infrared ratio image with an aerial photo for a rhyolitic tuff (dark on ratio image) associated with malachite, near Pisgah Crater, Calif. North is toward the top; the images are approximately 1.5 km on a side.

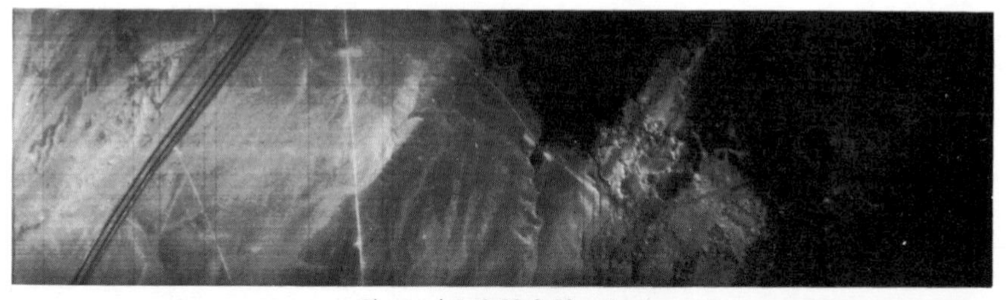

Channel 5 (0.50-0.52 μm)

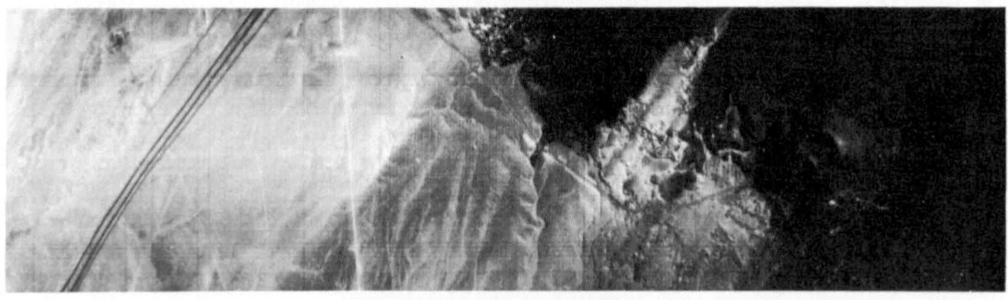

Channel 7 (0.74-0.85 μm)

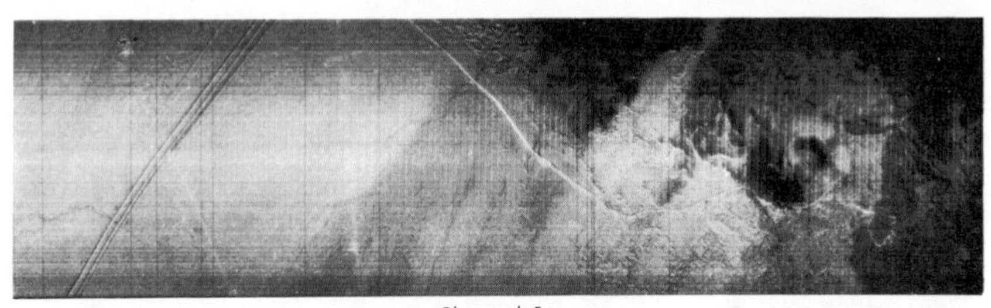

$$\text{Ratio: } \frac{\text{Channel 5}}{\text{Channel 7}} = R_{57}$$

FIG. 5
Analog visible and reflective infrared images of flight line 1, Sec. A, over Pisgah Crater, Calif. North is to the left; the images are approximately 1.5 km wide and 6 km long.

well with an accepted igneous rock chart by Travis (*19*). This parameter varies continuously between 100 and 0 as one moves from left to right across the rock chart. Table 2 shows the ranges of V_7 associated with 10 rock classifications on Travis's chart. The 8- to 14-μm wavelength region was arbitrarily divided into twelve 0.5-μm-wide "channels"; with laboratory spectra of 26 igneous rocks as a basis, a linear combination of 17 emittance ratios from 11 of these channels was found, which predicted V_7 within a standard error of 7.1 out of a possible range of 0 to 100. This procedure did not make use of wavelength regions outside 8 to 14 μm. Therefore, automatic mapping of exposed igneous

TABLE 2
Limits of V_7 for Rock Groups
Based on Igneous Rock Chart of Travis

Name	*Equigranular Member*	*V_7 Limits*
A	Granite	100–84
B	Syenite	84–81
C	Nepheline syenite	81–80
D	Quartz monzonite	80–67
E	Monzonite	67–77
F	Nepheline monzonite	66–65
G	Granodiorite	65–47
H	Quartz diorite and diorite	47–37
I	Gabbro, diabase, and theralite	37–18
J	Peridotite	18–0

silicates by future multispectral scanners may eventually be possible with a degree of accuracy comparable to that of accepted rock classification charts, as far as chemical composition is concerned. Textural classification may have to await the development of infrared laser or millimeter-wavelength radar techniques of the future.

However, as previously discussed, the visible and reflective infrared wavelength regions between 0.4 and 2.5 μm contain useful geological information also, but of a different nature. The first example *(20)* of a ratio image in these wavelength regions was a reflective infrared to visible green ratio of the same flight line over Pisgah Crater shown earlier (see Fig. 3). Figure 5 shows the single-channel images of a 0.50- to 0.52-μm channel and a 0.74- to 0.85-μm channel at top and middle, with the ratio image of the two at the bottom. Three eruptive phases of basaltic lava, which flowed from Pisgah Crater in the late Pleistocene or early Recent epochs, appear in the right half of these images. The lighter-toned region to the right of Pisgah Crater in the ratio image is phase 3 (youngest) lava. Whereas the phase 3 lava in the ratio image is distinct from phase 1 and 2 lava surrounding it, this distinction is not apparent in the single-channel images. Sparse chemical data from these flows indicate that this brighter appearance of phase 3 lava in the ratio image, which agrees *(6)* with laboratory spectra of rock samples from the area, may be caused by the presence of larger amounts of hematite in the youngest lava.

Other examples of ratio imagery can be taken from data collected by ERIM in May 1972 over a region in the Black Hills of South Dakota for the U.S. Bureau of Mines *(21)*. Figure 6 shows three ratio images of a flight line over Lead, South Dakota, where the Homestake gold mine is located. In these ratio images, a 0.62- to 0.70-μm channel was the common denominator, and the numerator channels were 0.67 to 0.94, 1.0 to 1.4, and 2.0 to 2.6 μm, respectively, from top to bottom in Fig. 6. In the top ratio image, called $R_{3,2}$, there is a marked contrast between vegetative (bright) and nonvegetative (dark) targets. This $R_{3,2}$ ratio has proved quite useful for detecting rock outcrops of all kinds in heavily vegetated regions. Even lichen-covered quartzites were easily distinguished from vegetative types of all kinds of this ratio image. This should be useful particularly in Alaska and Canada for locating exposed rock and soil outcrops. The 1.0- to 1.4-μm channel divided by the 0.62- to 0.70-μm channel ($R_{4,2}$ ratio) shows the conifers (primarily Ponderosa pine in this region) as dark, and vigorous grass as bright, whereas both conifers and grass are bright in the $R_{3,2}$ ratio image. The bottom ratio image in Fig. 6, 2.0 to 2.6 μm divided by 0.62 to 0.70 μm ($R_{6,2}$) shows good contrast between the slate, phyllite, and schist of the cut (to the left of the city) north of Lead and most other targets in the scene. The similarly composed circular mine dump on the other side of Lead and the Maitland tailings (seen at the upper left of the bottom ratio image) also appear darker than most other targets in the scene.

For comparison with aerial photography, Fig. 7 shows a mosaic of four aerial photos (taken at an approximate altitude of 1.5 km in mid-morning, simultaneously with the scanner data), showing the town of Lead on a different scale. Figure 8 shows a blowup of the top $R_{3,2}$ ratio image of Fig. 6 for the same region at approximately the same scale as the aerial photomosaic. The contrast between vegetative and nonvegetative targets is more pronounced in the ratio image. Furthermore, the mosaicking of the aerial photos is not necessary with the scanner imagery. With blowups of this $R_{3,2}$ ratio image, it is possible to detect road cuts on the order of 3 m in horizontal extent, which was not possible with the air photos.

In all, six ratio images were made from the Black Hills data. Besides the ones shown, images were also made of channels 1.0 to 1.4 μm divided by 1.5 to 1.8 μm ($R_{4,5}$), 2.0 to 2.6 μm divided by 1.5 to 1.8 μm ($R_{6,5}$), and 2.0 to 2.6 μm divided by 0.50 to 0.54 μm ($R_{4,1}$). Table 3 shows the channel number and ratio designations for the Black Hills data. (*Note:* channels 1 and 2 here are different from thermal channels 1 and 2 described earlier.) A gray scale was produced for each ratio image, which relates film density to ratio voltage. As a test to determine what targets in these ratio images could be discriminated, several areas of known rock outcrops and vegetative types were selected, and their ratio voltages were determined by comparing film densities from these areas

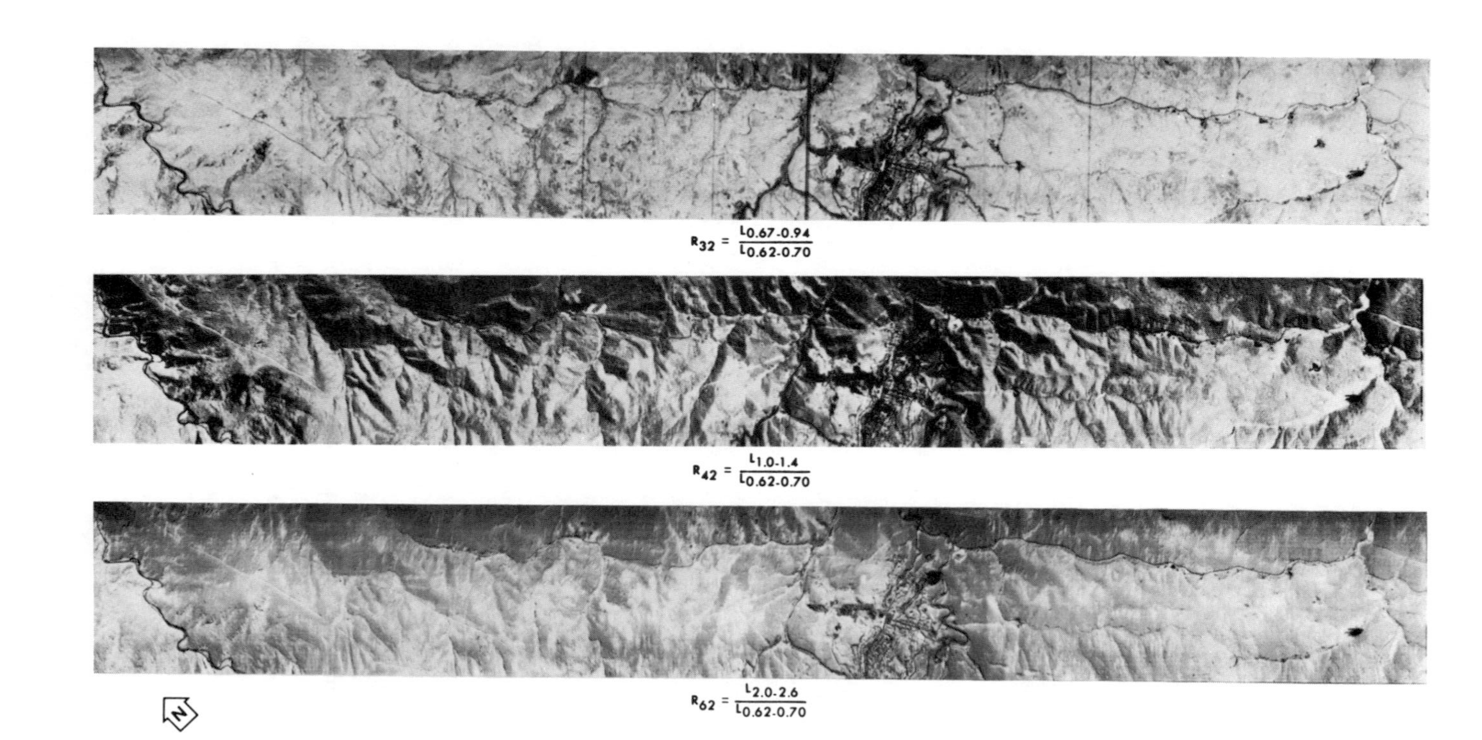

FIG. 6
Analog images of ratios $R_{3,2}$, $R_{4,2}$, and $R_{6,2}$ for flight line 2A near Lead, S.D. Dimensions of each image are approximately 3 km wide and 36 km long.

FIG. 7
Photomosaic of aerial photos showing the town of Lead, S.D. North is toward the upper-right corner; the mosaic represents an area approximately 1.8 km wide and 3.8 km long.

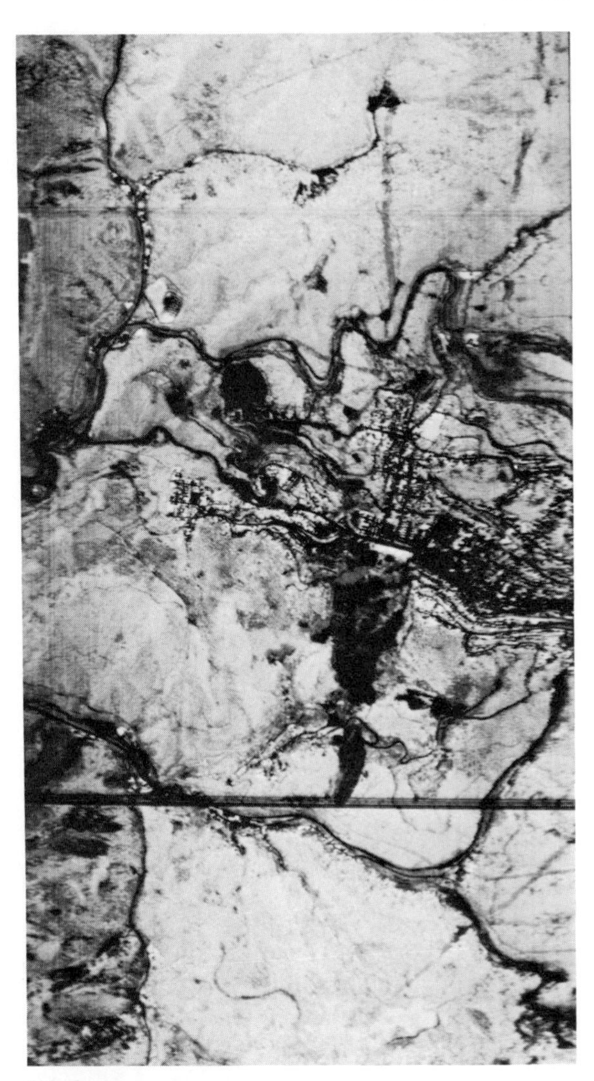

FIG. 8
Enlarged segment of the $R_{3,2}$ $L_{0.67-0.94}/L_{0.62-0.70}$ ratio image of Lead, S.D. North is toward the upper right corner; the dimensions of the image are approximately 3 km wide and 4 km long.

TABLE 3
List of Spectral Channels and Ratios Used with the Black Hills, South Dakota, Data

Single Channel No.	*Wavelength Region (μm)*	*Ratios Including That Channel*
1	0.50–0.54	$R_{4,1}$
2	0.62–0.70	$R_{2,3}$, $R_{4,2}$, $R_{6,2}$
3	0.67–0.94	$R_{3,2}$
4	1.0–1.4	$R_{4,1}$, $R_{4,2}$, $R_{4,5}$
5	1.5–1.8	$R_{4,5}$, $R_{6,5}$
6	2.0–2.6	$R_{6,2}$, $R_{6,5}$

with gray scales of the respective ratio images. Table 4 shows the resulting ranges of ratio voltage for each ratio and each target. Except for the all nonvegetation target, which encompasses targets 6 through 10, the targets shown in Table 4 are unique from one another in one or more ratios. Figure 9 is a schematic diagram summarizing the targets that could be separated from one another in this data set. If these ratio voltages had been reduced to reflectance ratios by normalizing to a known point in the scene, a table similar to Table 4 could have been made that was absolute, instead of relative. An example of what could be called relative ratio scannergrammetry is as follows.

A point of interest in the scene is located and its ratio voltages are determined by comparing its film density on each of the six ratio images with the respective gray scales. Consultation with Table 4 permits the classification of the point of interest as one of the target groups in this table. If, however, the ratio voltages were converted to reflectance ratios, the point of interest could be compared not only to the targets in Table 4, but also to targets for which laboratory spectra were available in the data bank. This latter method could be called absolute ratio scannergrammetry.

The most recent use of ratio images for geological remote sensing is with LANDSAT data. Figures 10, 11, and 12 show single-channel images of LANDSAT MSS channels 4 (0.5 to 0.6 μm), 5 (0.6 to 0.7 μm), and 7 (0.8 to 1.1 μm), respectively, as produced by ERIM's SPARC analog computer from LANDSAT computer-compatible tapes, for a small portion of

TABLE 4
Ratio Ranges of Black Hills Targets

Target No.	*Target Name*	*Ratio Voltage Limits (V)* $R_{3,2}$	$R_{4,2}$	$R_{6,2}$	$R_{4,5}$	$R_{6,5}$	$R_{4,1}$	*No. of Test Areas*
1	Vigorous range grass	6.6–7.0	4.9–5.3	3.7–4.3	4.1–4.7	2.9–3.5	5.3–5.5	5
2	Dormant and low vegetation	2.2–3.4	2.0–3.0	2.4–2.8	2.3–3.1	3.5–4.3	3.5–4.3	7
3	Conifer trees	4.3–5.9	2.7–3.7	2.6–3.6	3.5–4.5	2.7–3.5	3.6–4.4	3
4	Deciduous trees	3.0–4.2	2.6–3.6	2.9–3.9	2.3–3.1	3.4–4.2	3.7–4.5	1
5	All non-vegetation	0.6–2.2	–	–	–	–	–	–
6	Water	0.6–2.2	0.2–0.8	0.5–2.5	1.4–2.8	0.8–3.4	0.9–2.3	4
7	Quartzite	0.6–2.2	0.4–2.0	2.4–2.8	1.3–2.1	3.7–4.9	0.7–1.9	5
8	Slate, phyllite, schist	0.6–1.4	0.5–1.3	0.4–1.4	1.5–2.7	3.7–5.5	0.8–3.0	5
9	Igneous intrusives (rhyolite, quartz latite, and phonolite)	0.7–1.3	1.0–1.2	2.0–3.0	1.3–1.9	4.7–5.5	2.2–2.6	7
10	Deadwood and sandstone	0.6–2.2	1.7–1.8	2.2–2.7	2.0–2.5	4.3–5.5	3.0–3.2	2

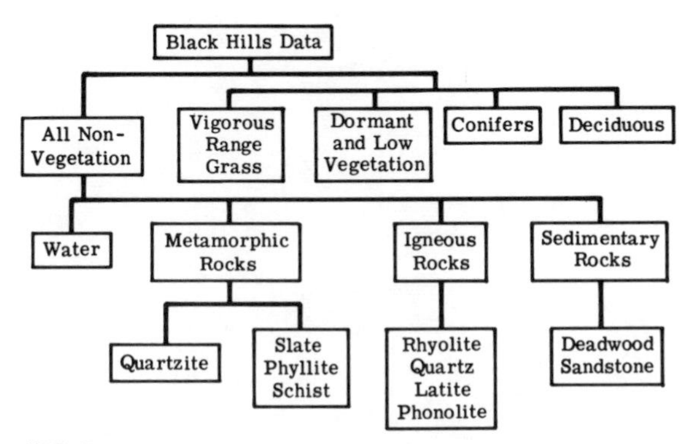

FIG. 9
Target categories for Black Hills, S.D., data set.

LANDSAT frame E-1013-17294, August 5, 1972, collected over the U.S. Steel iron mine near Atlantic City, Wyoming. (*Note:* the channel numbers of the LANDSAT MSS scanner should not be confused with the aircraft scanner channels defined above.) As an example of how laboratory data can be used to interpret ratio images, the cases of magnetite and hematite (two iron oxides) will be presented. The LANDSAT MSS ratio code *(7)* for a magnetite sample from Ishpeming, Michigan, is 200102, and for a hematite sample from Irontown, Minnesota, it is 888887. The ratio codes represent the ratios calculated from LANDSAT data, in order from left to right, $R_{5,4}$, $R_{6,4}$, $R_{6,5}$, $R_{7,4}$, $R_{7,5}$, and $R_{7,6}$. The $R_{7,4}$ ratio code digit for magnetite is 1, meaning that it should be darker in an $R_{7,4}$ ratio image than approximately 80 percent (brighter than 10 percent) of the materials represented in the laboratory data bank described earlier. Hematite has a ratio code digit of 8 for $R_{7,4}$, indicating that it should be brighter than approximately 80 percent of the materials in the laboratory data bank.

Figure 13 shows that the data bank must be fairly representative of the materials in this scene, because magnetite in the iron mine in the lower-right portion of the $R_{7,4}$ ratio image is darker than everything except water, and the hematitic Triassic formations in the upper-right corner appear fairly bright in the $R_{7,4}$ image. The R_{54}, ratio code digit for magnetite is 2, whereas it is 8 for hematite. Therefore, the magnetite should appear medium dark, and hematite should be brighter than about 80 percent of the materials represented in the laboratory data bank. Figure 14, an $R_{5,4}$ ratio image, shows that the iron mine is medium dark, and the Triassic red beds are the brightest objects in the scene. In fact, the $R_{5,4}$ ratio image of the whole LANDSAT frame, shown in Fig. 15, delineates all the hematitic Triassic formations that have been mapped in the area covered by the entire LANDSAT frame, as well as some other iron oxides. The only regions in the frame brighter than the Chugwater sandstone are agricultural fields along the river, where very fine grained hematite soils are exposed. Soils are practically the only materials in the LANDSAT data collection, besides 0- to 5-μm diameter hematite and limonite, with an $R_{5,4}$ ratio code of 9. Field checks have shown that even the slightly-above-average brightness areas in this ratio image are areas where surface exposures of iron oxides occur *(7)*.

For reference, Figs. 16 and 17 show the single-channel images of LANDSAT MSS channels 4 and 5, respectively. The iron oxides are neither brightest nor darkest in the scene in either single-channel image, but appear brightest in the $R_{5,4}$ ratio image because their spectral reflectance increases more from the visible green to red wavelength regions than any other common natural materials. The ratio codes for limonite (a field term for hematite or geothite that contains relatively large amounts of water) samples indicate that limonite will be similar to hematite in an $R_{5,4}$ ratio image, but will be much darker than hematite (in fact, darker than almost anything else) in an $R_{7,6}$ ratio image. The ability to separate these iron oxides from other natural materials and from each other is significant for mineral exploration.

IMPORTANCE OF SPECTRAL RATIOING TO GEOCHEMICAL MAPPING

The most well demonstrated accomplishment of spectral ratioing thus far is the mapping of iron oxides. An $R_{5,4}$ ratio image of LANDSAT data, for instance, can readily discriminate hematite, geothite, and limonite as a group from practically all other minerals. Laboratory data also indicate that ochreous limonite, brownish goethite, and reddish hematite can be discriminated from each other using additional

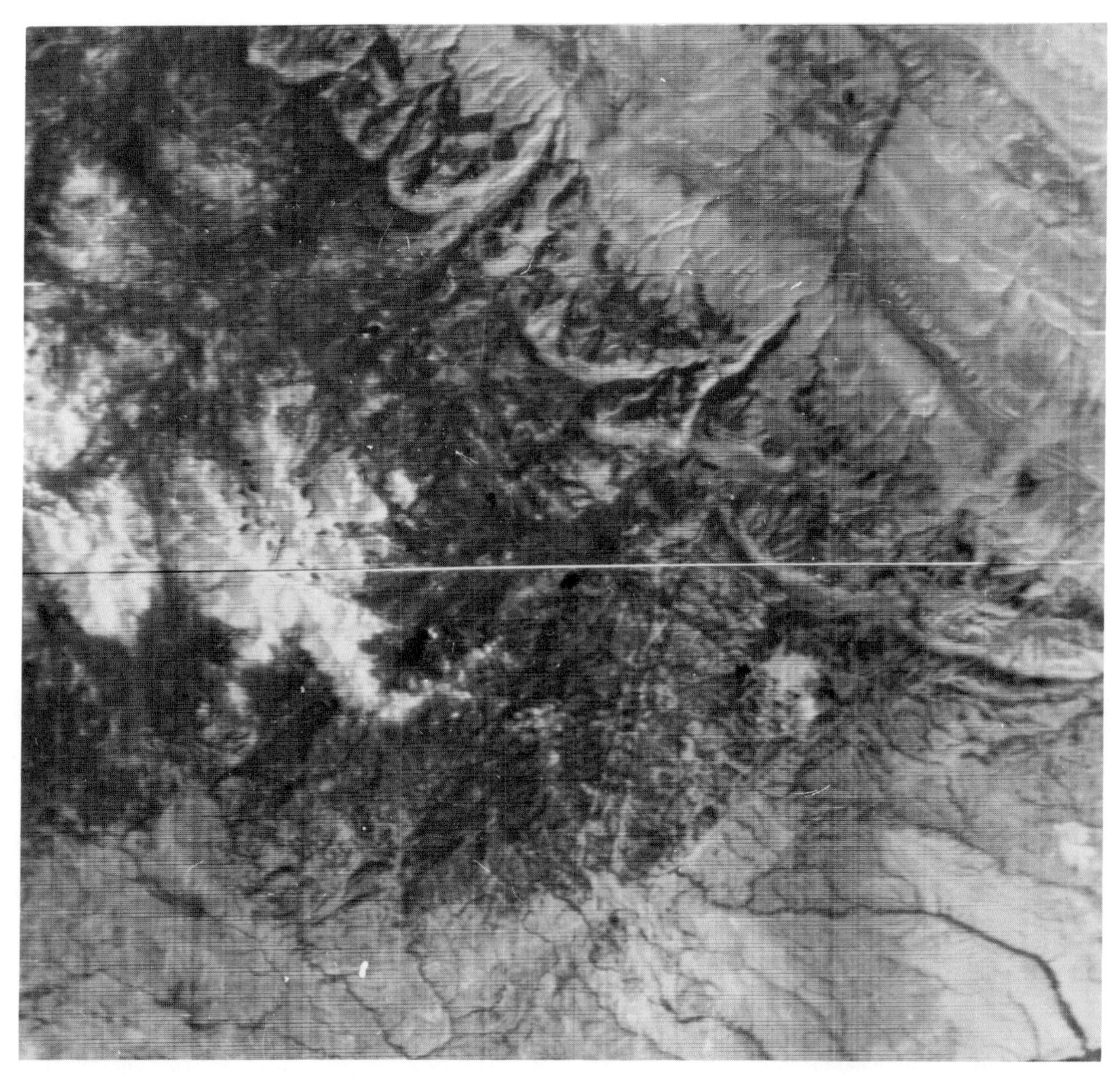

FIG. 10
Single-channel image of LANDSAT MSS channel 4 (0.5 to 0.6 μm) for a region near Atlantic City, Wyo. North is toward the top; the image dimensions are approximately 41 X 46 km.

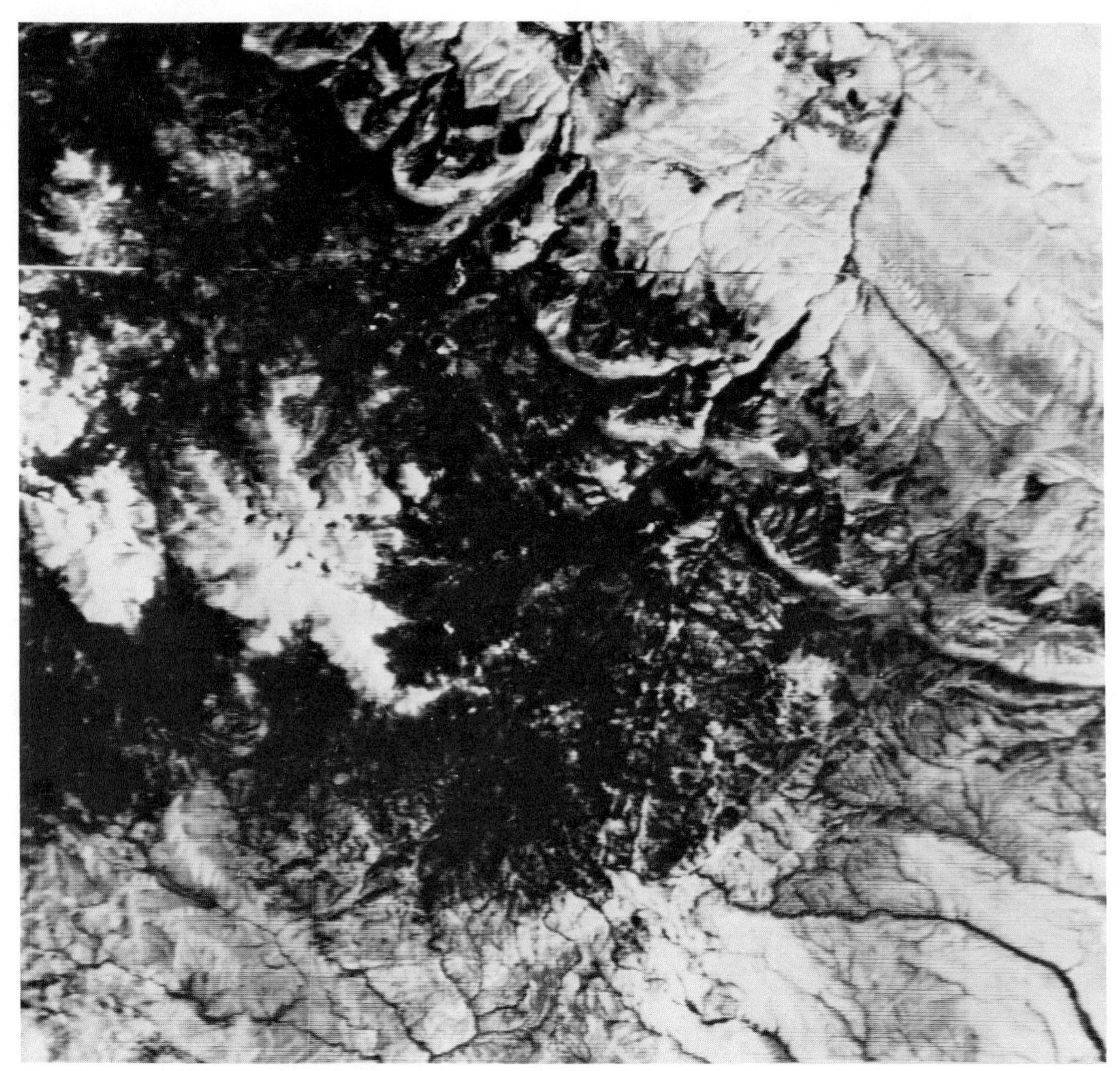

FIG. 11
Single-channel image of LANDSAT MSS channel 5 (0.6 to 0.7 μm) for a region near Atlantic City, Wyo. North is toward the top; the image dimensions are approximately 41 × 46 km.

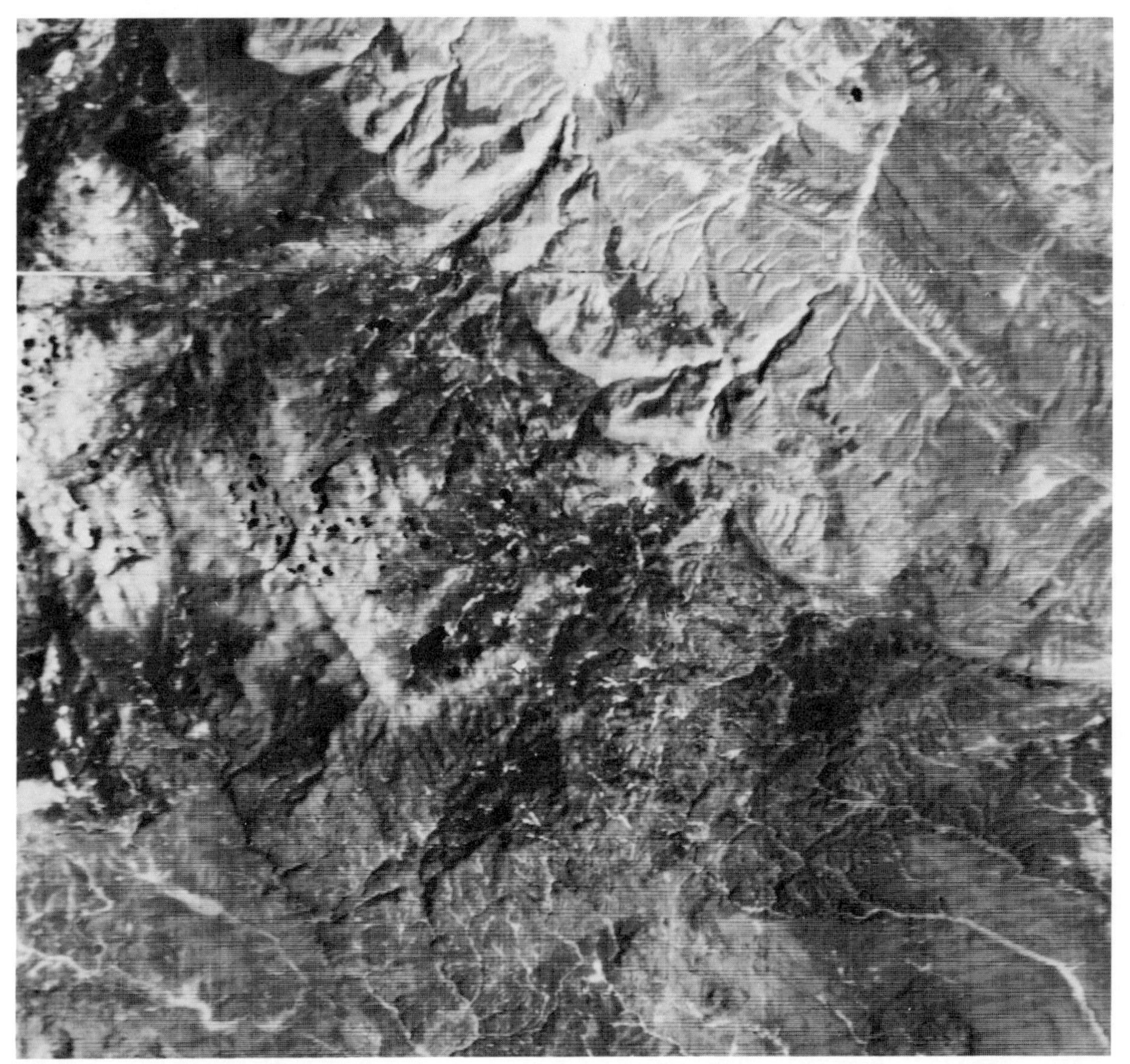

FIG. 12
Single-channel image of LANDSAT MSS channel 7 (0.8 to 1.1 μm) for a region near Atlantic City, Wyo. North is toward the top; the image dimensions are approximately 41 × 46 km.

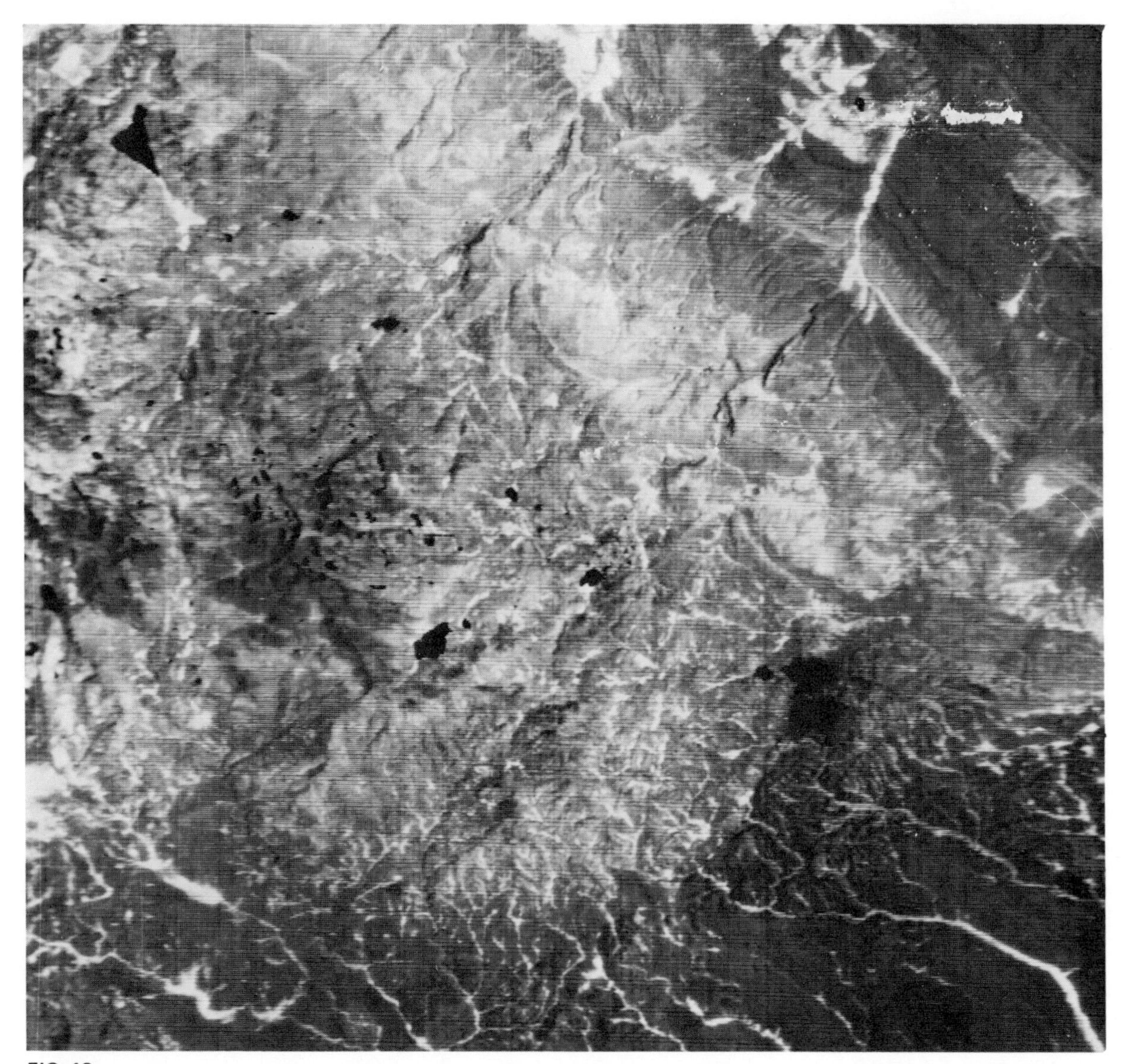

FIG. 13

LANDSAT $R_{7,4}$ $L_{0.8-1.1}/L_{0.5-0.6}$ ratio image for a region near Atlantic City, Wyo. Iron mine at lower right is dark and Triassic formations at upper right (trending northwest) are medium bright. North is toward the top; the image dimensions are approximately 41 × 46 km.

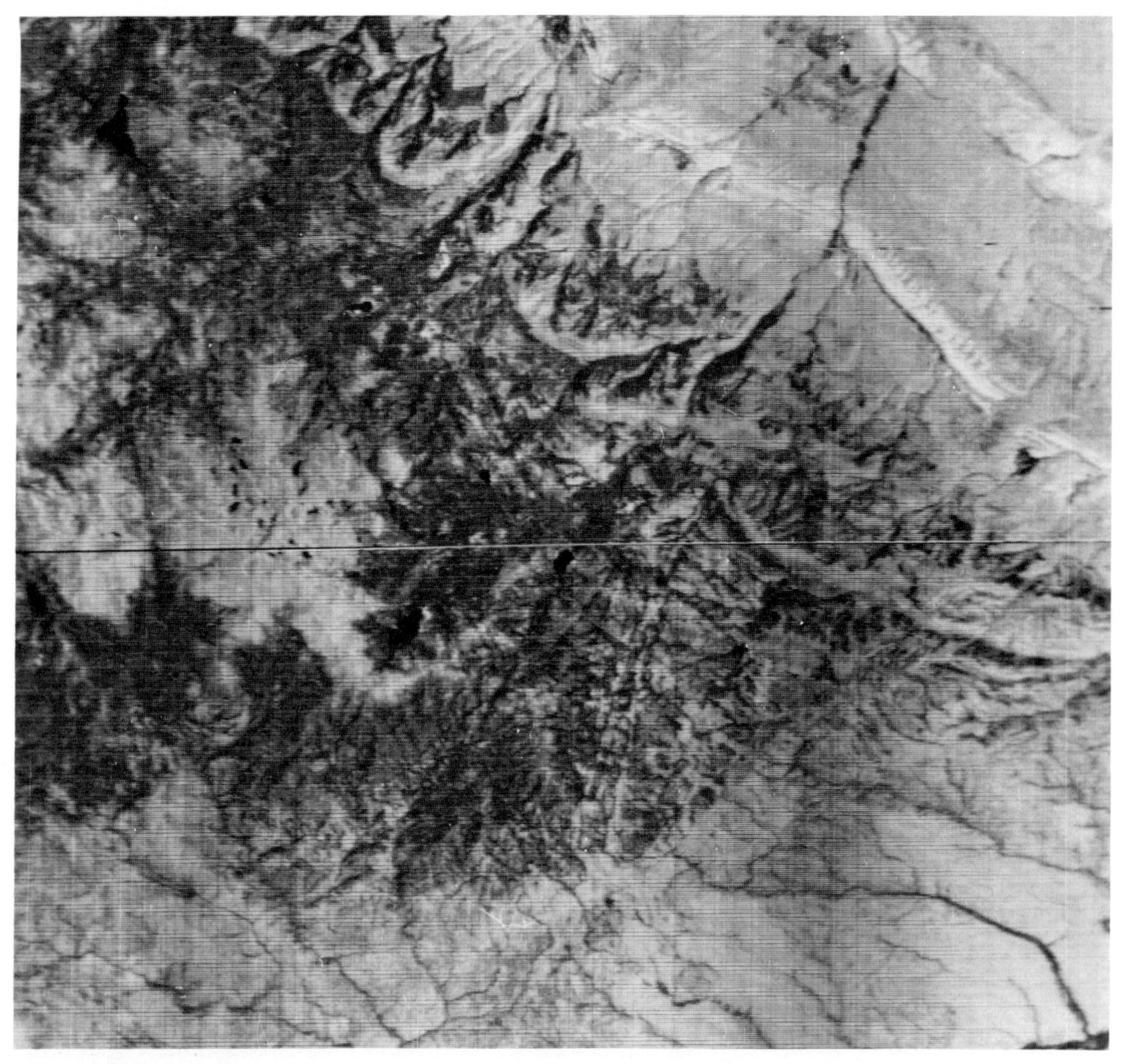

FIG. 14

LANDSAT $R_{5,4}$ $L_{0.6-0.7}/L_{0.5-0.6}$ ratio image for a region near Atlantic City, Wyo. Iron mine at lower right is medium toned and Triassic formations at upper right (trending northwest) are brightest in the scene. North is toward the top; the image dimensions are approximately 41 X 46 km.

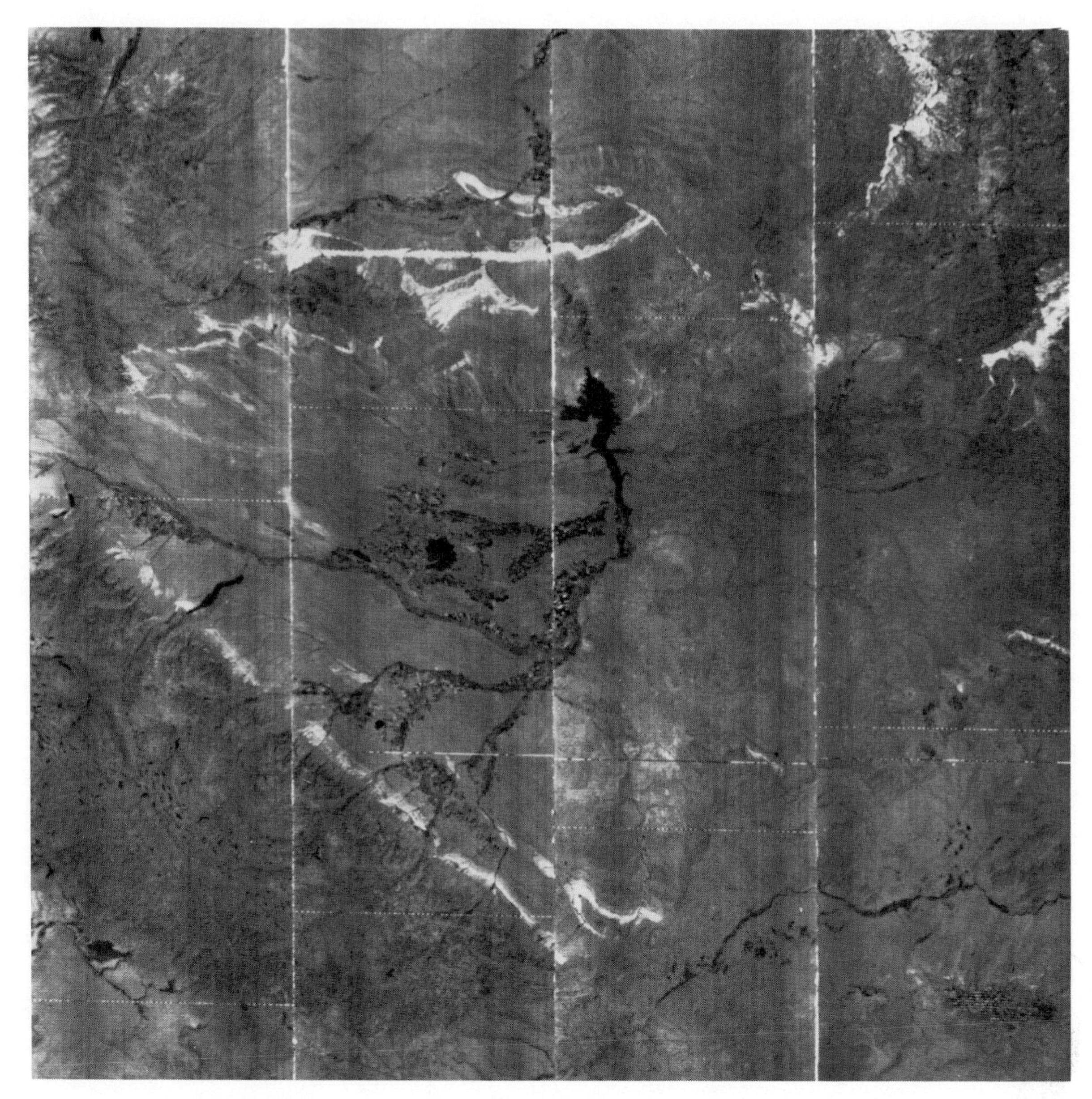

FIG. 15
LANDSAT $R_{5,4}$ analog ratio image of the Wind River Basin and Range, Wyo.

FIG. 16
LANDSAT MSS channel 4 image of the Wind River Basin and Range, Wyo.

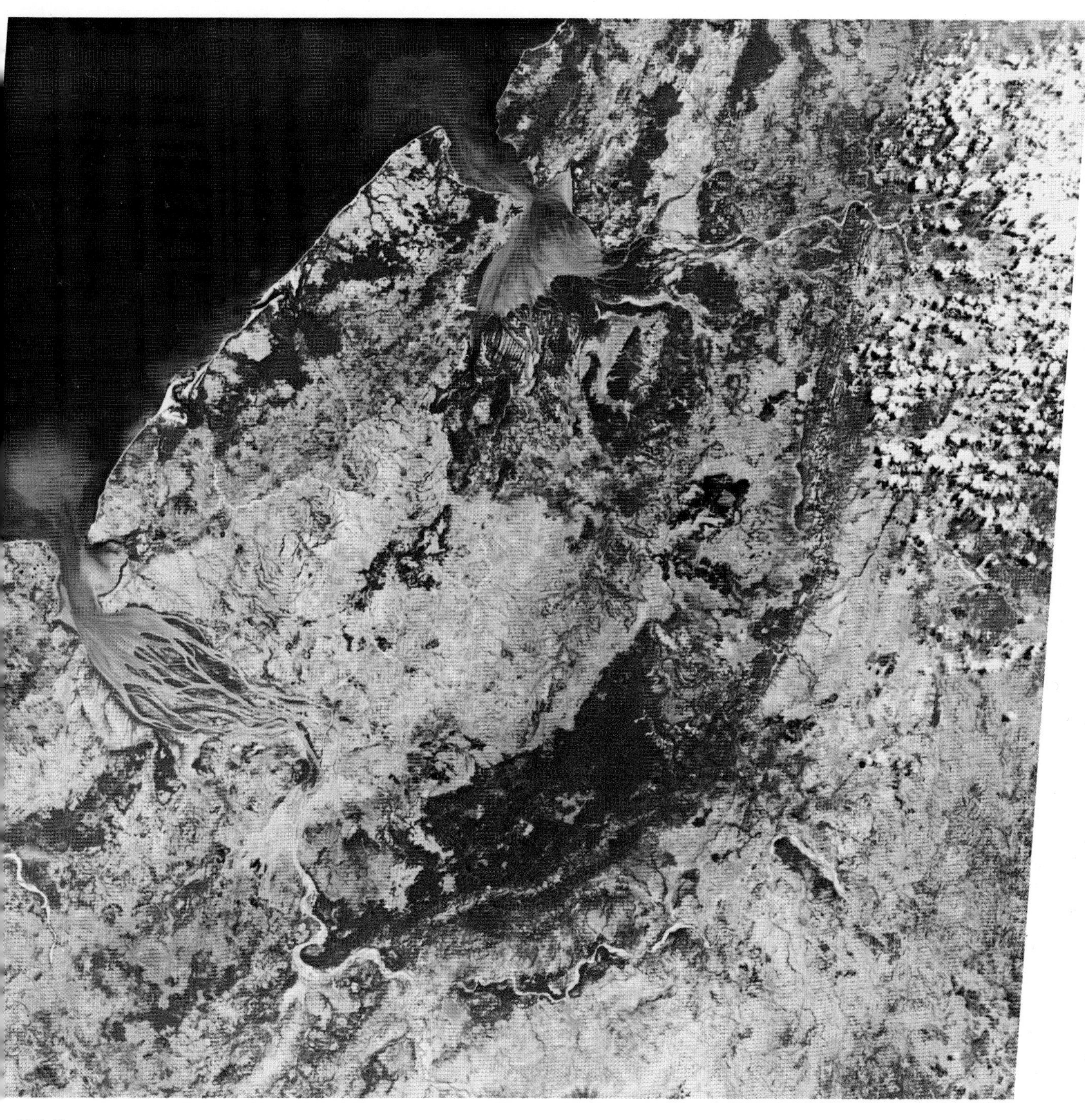

FIG. 2
Majunga, Madagascar.

FIG. 6
Peking, China.

FIG. 6
Computer-aided crop classification from LANDSAT-1 MSS data.

FIG. 18
Three LANDSAT scenes of Utah Lake showing variation in lake conditions.

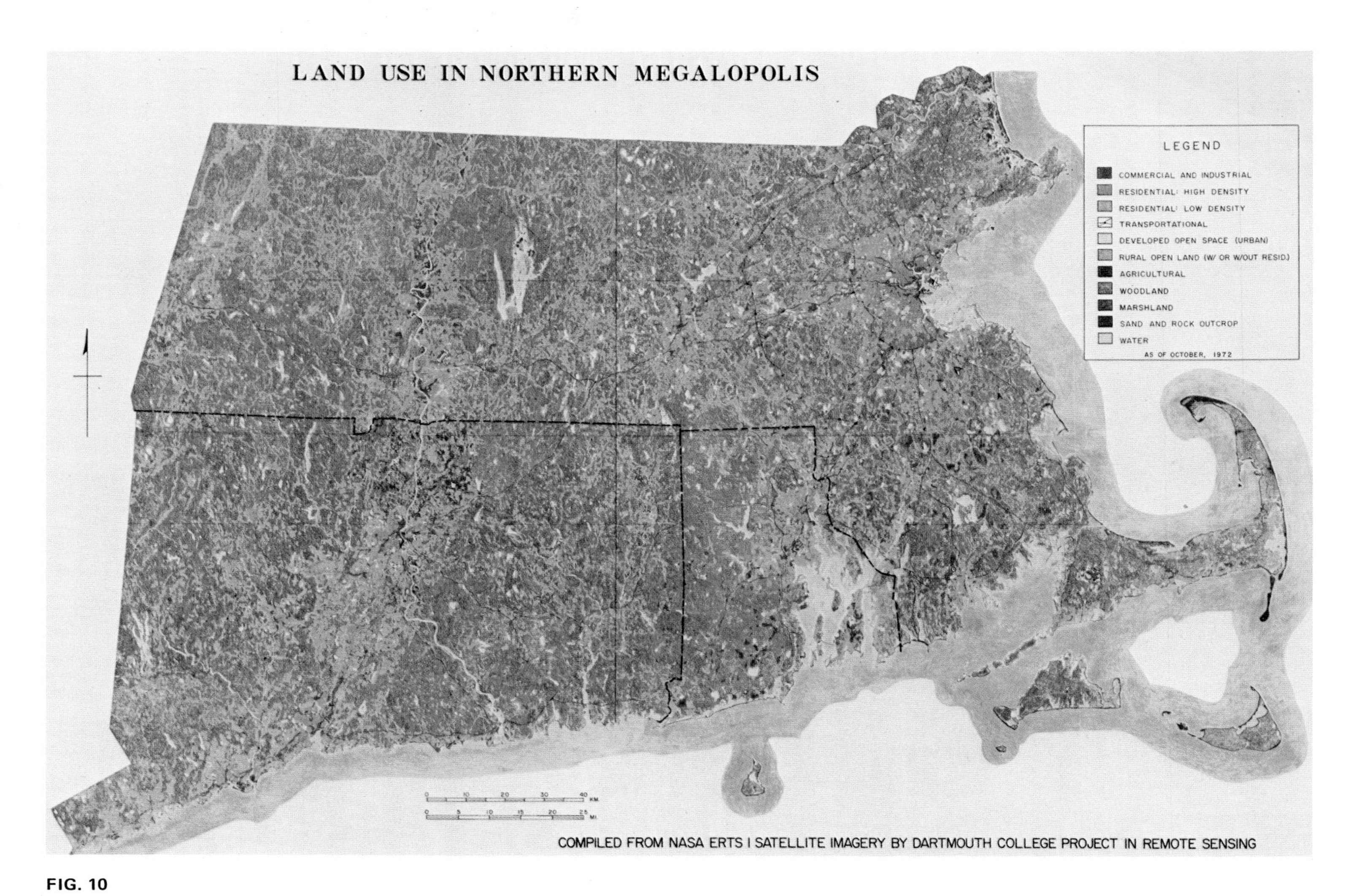

FIG. 10
Land-use map of three-state area in New England compiled from LANDSAT-1 imagery.

FIG. 8 (*above and on following page*)
Comparison of color composite LANDSAT-1 images 1050-07355 (September 11, 1972) and 1158-07363 (December 28, 1972) showing a portion of the Witwatersrand, South Africa, west of Johannesburg, in contrasting seasons.

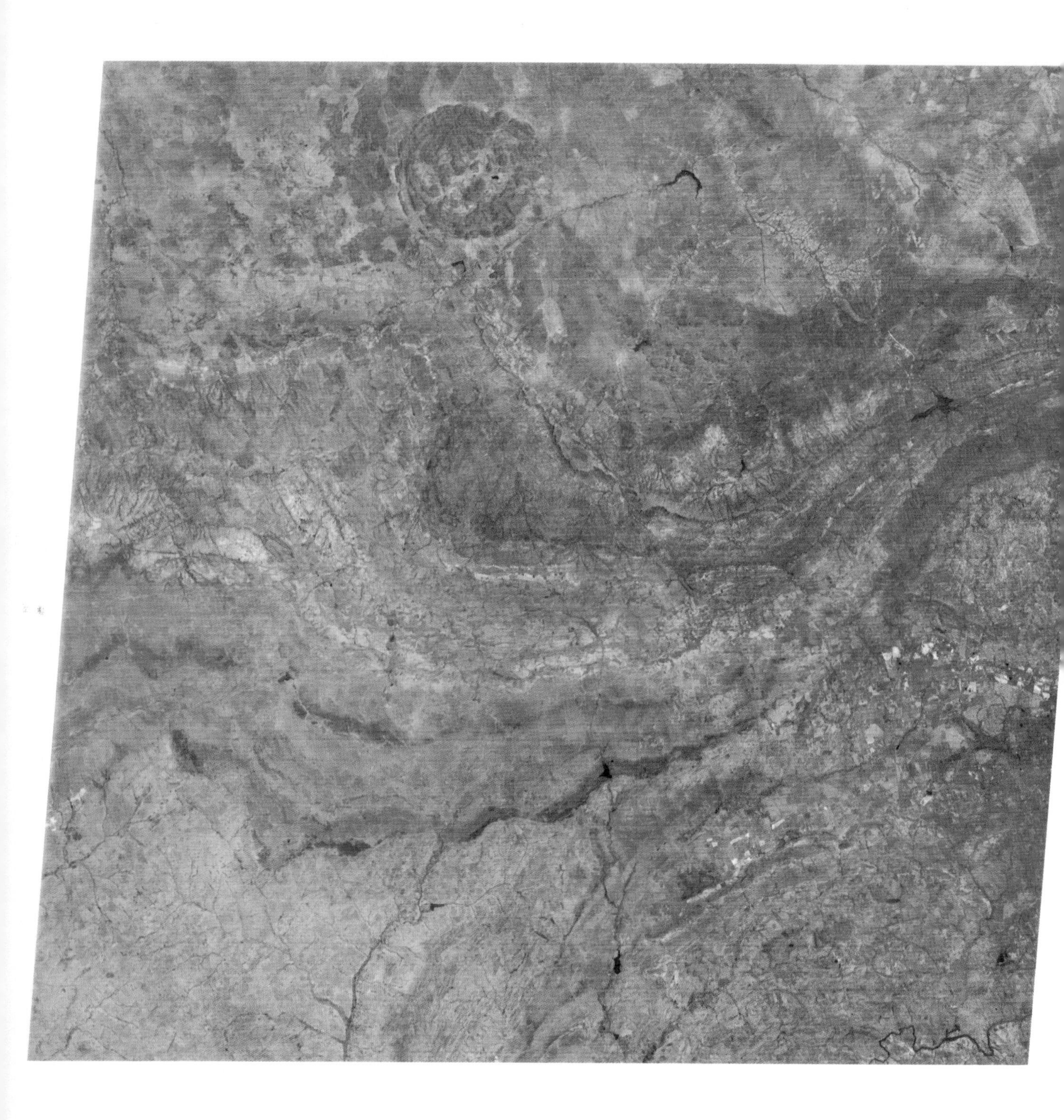

Chapter 6

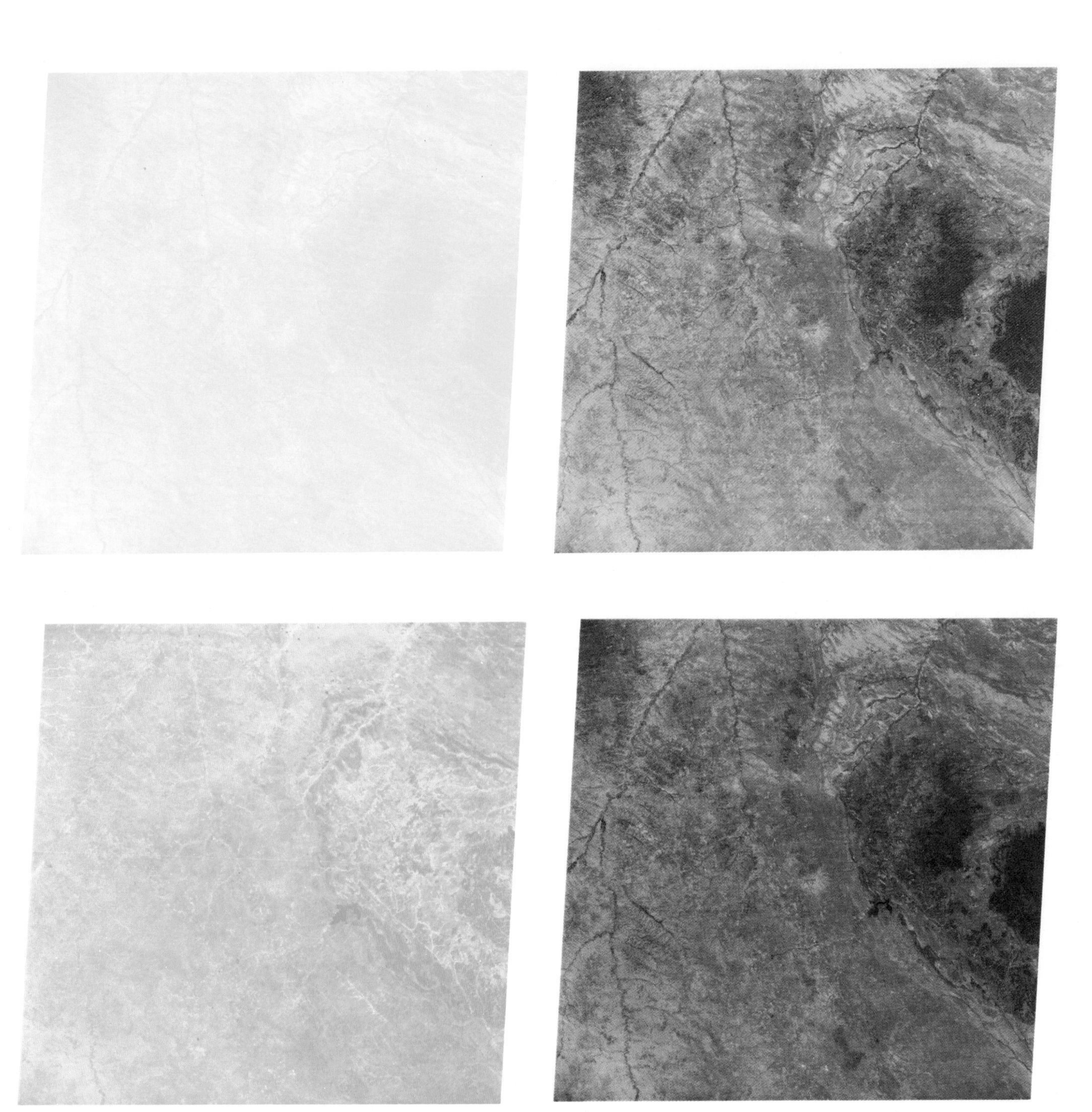

FIG. 13
False-color composite images are constructed from multispectral imagery by direct addition (superposition) of various bands presented in contrasting colors.

FIG. 17 *(above and on facing page)*
Comparison of a color and color-infrared photograph in a geologically complex region along the southern flank of the Owl Creek Mountains in central Wyoming. Note the striking contrast in colors presented in the color-infrared photograph versus those of the color photograph.

Chapter 6

FIG. 30(A)

LANDSAT-1 image of the Monterey Bay Area, Calif. Major structural and physiographic features are annotated (after Lowman, 1972). Mosaics of LANDSAT images have proved remarkably interesting and useful, more so than experience with mosaics of air photos would lead us to expect. This mosaic of California and adjacent Nevada illustrates their potential value in showing the geology of very large areas. The major faults actually visible on the mosaic (as well as a few, such as the Newport–Inglewood fault, added for completeness) have been labeled on the accompanying sketch map. Cities are also shown for orientation, although they may not be visible.

FIG. 45
Color composite image formed from band ratios of a LANDSAT image (1072-18001) of the Goldfield, Nev., area. Contrasts and colors seen on this composite are strikingly different from those on the original image (after Towan, 1973). In this image the altered and mineralized areas are expressed in shades of green and brown, clouds appear white, and vegetation orange.

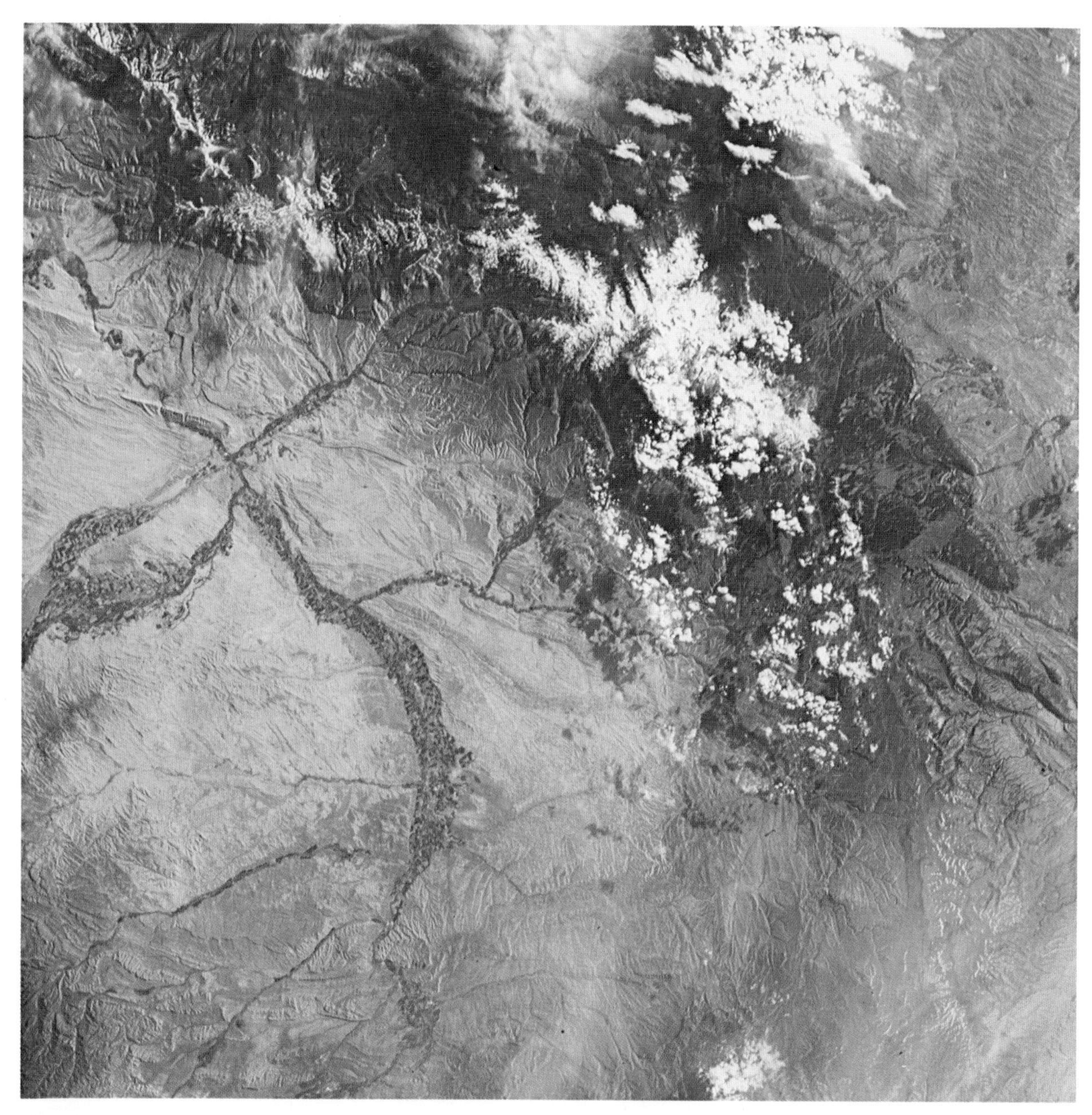

FIG. 43
Skylab S-190A color photograph of the Bighorn Mountains and Bighorn Basin, Wyo. Many rock and soil units appear in gray and buff shades that are not particularly distinctive. Haze or light cloud cover obscures portions of the scene.

FIG. 44
Skylab S-190A color-infrared photograph of the Bighorn Mountains and Bighorn Basin, Wyo. (compare with Fig. 43). The color-infrared presentation allows better distinction of some lithologic units and better haze penetration.

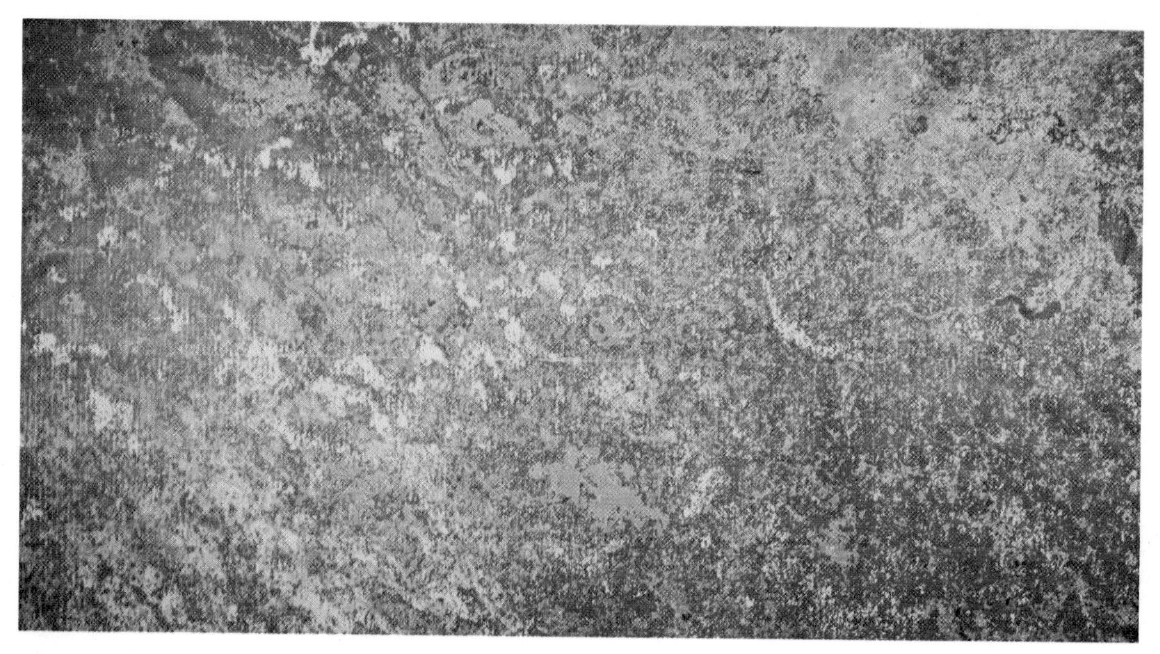

FIG. 14
Computerized land-use classification in eight major categories: saline dune areas, light red; bare field types 1 and 2, dark red; forest types 1 and 2, black and gray, respectively; water, blue; agriculture types 1 and 2, light and dark green, respectively. Four-band multispectral data utilized from LANDSAT, covering one eighth of the frame described in Figs. 12 and 13.

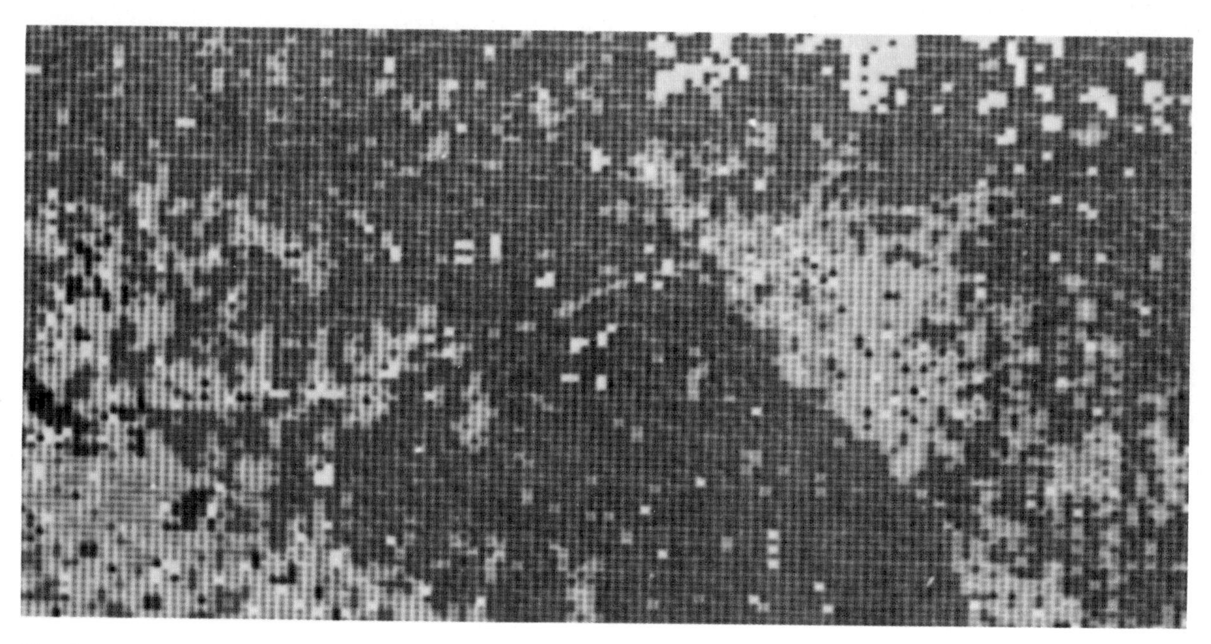

FIG. 15
Enlargement of a small portion of Fig. 14.

FIG. 8
Oil slicks in Baton Rouge Harbor. (Courtesy NERC—Las Vegas/EPA.)

FIG. 10
Evidence of surface pollution. (Courtesy NERC—Las Vegas/EPA.)

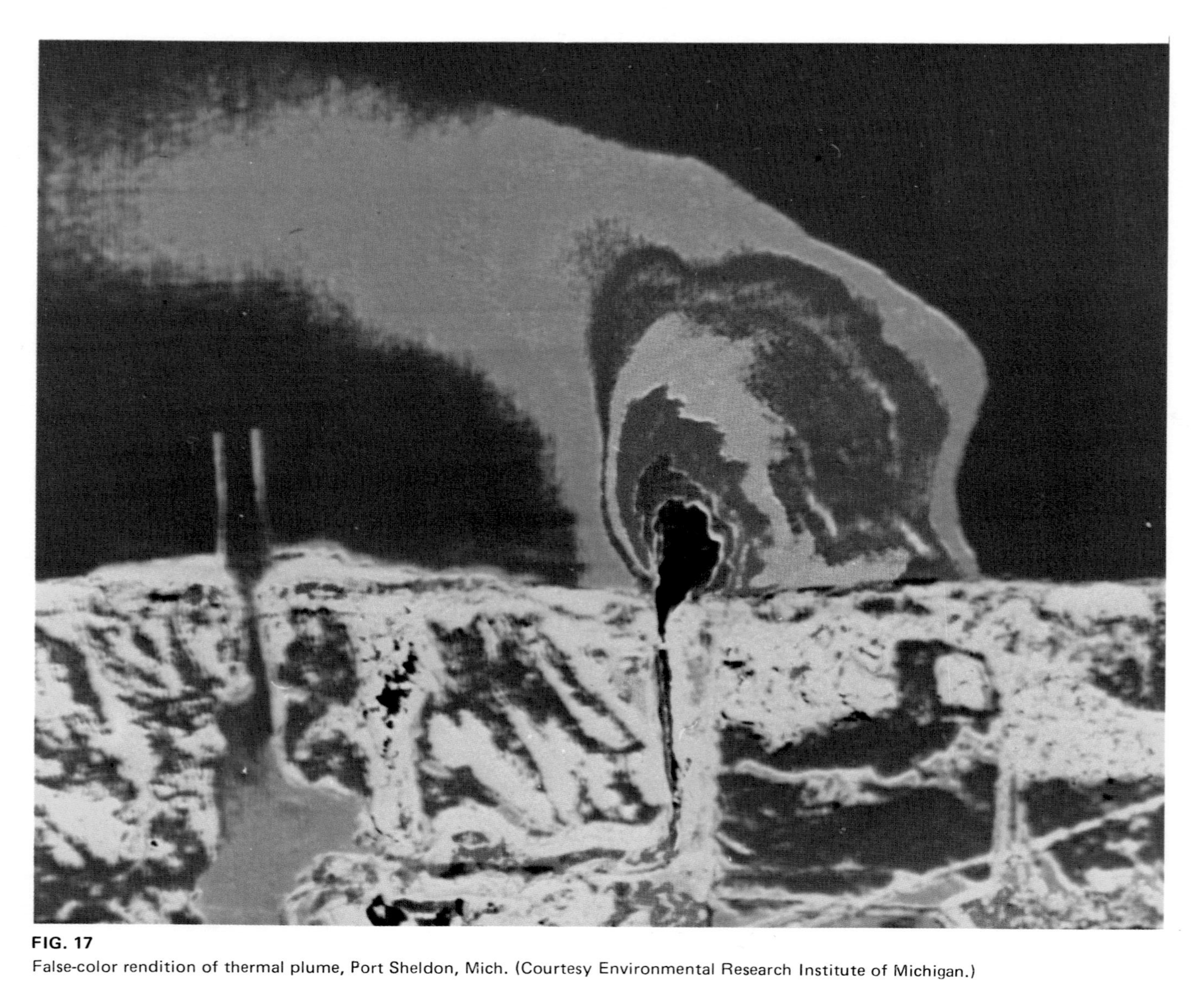

FIG. 17
False-color rendition of thermal plume, Port Sheldon, Mich. (Courtesy Environmental Research Institute of Michigan.)

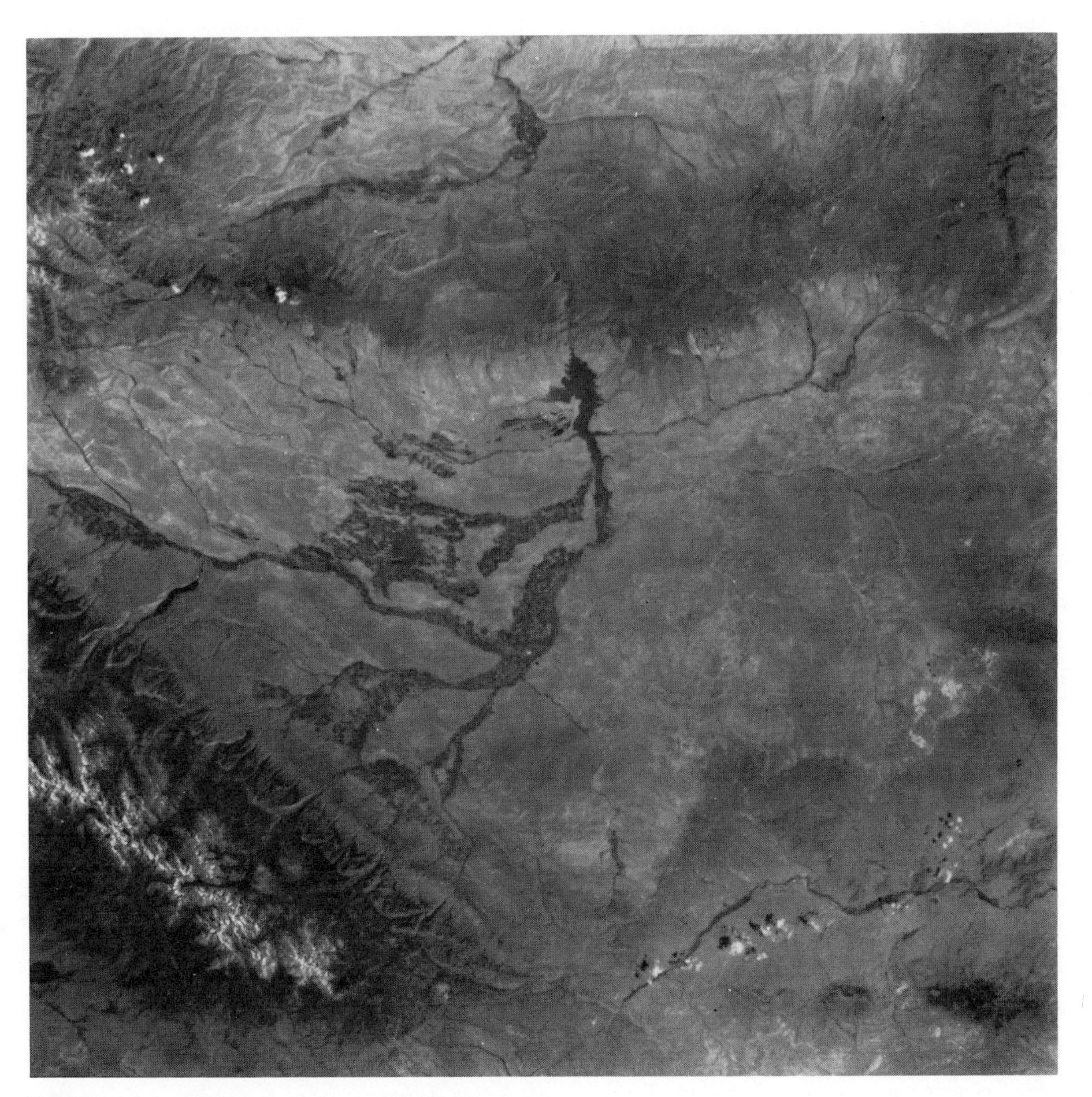

FIG. 17
LANDSAT MSS channel 5 image of the Wind River Basin and Range, Wyo.

LANDSAT ratios. Future laboratory and scanner data may show that other colors of iron oxides can be distinguished. What uses can be made of this capability?

Probably the most important use of iron oxide mapping will be for mineral exploration. Locke *(22)* was one of the first to notice correlations of secondary iron oxide colors to the sulfides from which the iron precipitate was derived. Weiser *(23)* showed that colors of synthetic hydrous ferric oxides reflect the chemical conditions under which colloidal iron particles were initially precipitated. Copper and iron orebodies were the first to receive much attention with regard to surface iron oxide exposures. Blanchard *(24)* has perhaps given the best guidelines for the use of iron oxide mapping related to copper and iron prospecting. He carefully limited its usefulness with the following conclusions:

1. Although reddish iron oxide colors most commonly point to hematite, orange–yellow–buff–ochreous colors to goethite and jarosite, no iron oxide color is a completely dependable representative of a specific ferric oxide hydrate or supergene mineral.
2. Although a particular iron oxide color may be characteristic of a specific sulfide in a given district, the same color may not represent the same parentage throughout that district.
3. Several iron oxides of contrasting colors may be associated with a single deposit.
4. Even when color is useful in determining the iron oxide parentage, it remains subordinate to other physical characteristics useful for this purpose.

The use of outcrop iron oxide colors as guides to lead and zinc ore has been investigated by Kelly *(25)*. Instead of simply describing the colors verbally, he related them to a standard rock color chart. In some of the areas Kelley visited, he found that the colors of iron precipitates in the outcrops bore a definite relation to original lead, zinc, and iron sulfides. He found dusky reds (5R 3/4 to 10R 2/2) associated with galena, light browns (5 YR 5/6) to yellowish oranges (10 YR 3/4) with sphalerite, and dusky browns (5 YR 4/4) to reddish browns (10R 3/4) for pyrite, except when oxidized in a reactive gangue, in which case yellowish-orange (10 YR 6/6) limonite can also occur. He also found that a color indicative of ore in one area may be insignificant or entirely absent in another. Kelley concludes that the iron oxide colors may be locally quite useful in certain districts, but should not be used indiscriminantly without allowance for local variations.

Gold is also sometimes associated with surface iron oxides. Rowan et al. *(26)* have used a color composite of three LANDSAT ratio images to map hydrothermally altered areas in south-central Nevada over a known gold deposit. The limonitic soil directly above the deposit has a unique color in the color ratio composite, but could not be discriminated from other areas in the scene by examining photos or single-channel images. The largest mined deposit of "invisible" gold, the Carlin gold deposit in western Nevada, is disseminated in a yellowish limonitic soil that should be similarly enhancible by LANDSAT ratio images.

Uranium is another metal that may often display an association between ore deposit and surface iron oxides. Shockey *(27)* used the term geochemical cell to describe the invasion of reduced, carbonaceous, and pyritiferous sandstone and uranium orebodies by oxygenated groundwater. A well-developed geochemical cell is shaped like a tongue, pointed down the hydrostatic gradient. The geochemical cell itself many times has colorful iron oxides in the interior of the cell, which can be exposed at the surface. The cell size is usually very large compared with the size of the mineralized "fronts," which occur along the edges of the cell and along irregularities on the upper and lower surface *(28)*. Primary uranium minerals, when oxidized to the hexavalent uranyl ion by oxygenated water, are mobile in weakly acidic solutions or in neutral to alkaline solutions if carbonate ions are present *(29)*. The uranium can be reduced by a number of reducing agents, including pyrite and organic matter, and deposited as uraninite or a hydrate. Iron oxide mapping plays an important role in uranium exploration, as it can with almost any ore associated with oxidation–reduction phenomena.

Along this line of thought, there may even be a role for iron oxide mapping in exploration for petroleum and natural gas. A recent paper by Donovan *(30)* on the Cement, Oklahoma, oil field has proved that striking bleaching of red beds at the surface can be caused by petroleum microseepage along structural zones of weakness. The typical Rush Springs Sandstone is red-brown (10R 4/6), but has been altered to pink (5 YR 8/1 to 5 YR 7/2), yellow (5Y 7/2 to 5Y 8/4), and almost white (5Y 8/1 to N8). He explains that hydrogen sulfide was probably liberated by the alteration of sulfates to carbonates (by microscopic seepage of hydrocarbons leaking along faults through a gypsum-rich stratum), and it then acted as a reducing agent, which bleached the surface red beds. One can only wonder if the reverse surface effect could occur, that is, reduction of subsurface red beds by hydrocarbon seepage and reoxidation of the vertically transported products to form iron oxides at the surface.

Although these applications of iron oxide mapping require similar spectral bands in multispectral scanners, they differ greatly in spatial resolution requirements. The strongly oxidized surface alterations of copper, iron, lead, and zinc deposits typically cover areas from a few to a few hundred meters in diameter (except for large porphyry deposits), which would suggest scanners with an instantaneous field of view on the order of 5 or 10 m. Aircraft scanners can accomplish this resolution and better, but satellite scanners may have to await the space-shuttle era to achieve these spatial resolutions with adequate spectral resolution. It has been this author's observation, however, that, although regions of strong alteration are typically small, the outcrop or soil surrounding them may be observable by enhancement techniques, thus raising the possibility that improved spectral techniques may reduce the restrictive spatial requirements for these orebodies.

The gold, uranium, and petroleum cases cited are quite different, in that alteration patterns can be 100 m to km in areal extent. For these applications, LANDSAT and Skylab resolution seem close to ideal. The approximately 80-m spatial resolution of LANDSAT permits the coverage of vast areas in a single frame of data, and also tends to "filter out" some of the smaller surface exposures of iron oxides that would seem to have lower probabilities of leading to large ore deposits. These properties lend themselves to rapid "high-grading" of large areas of land with inexpensive satellite data.

Besides iron oxide mapping, spectral ratioing techniques with MSS data have something to offer general geologic mapping, although the methods are much more complex and in early stages of development. The most obvious one is quick and dirty reconnaissance mapping over large unmapped terrain. Even though only a few lithologic or soil classes may be mappable for poorly known regions, the rapidity and economy of satellite mapping should be invaluable for this problem. Skylab 13-channel S-192 data *(31)* will no doubt be able to make many more distinctions among rock and soil types than can be made with the four channels of LANDSAT. However, one of the most important uses of multispectral techniques may be the improvement of existing geologic maps by what could be called intraformation mapping.

Consider the typical U.S. Geological Survey quadrangle geologic map. Formations of different ages are drawn in various colors, and the map key describes the compositional makeup of the members and facies of the formation. However, the actual locations of the members and facies are not given. With this amount of information, the mapping of member units and facies changes within formations seems to be a natural problem for multispectral recognition mapping, using a combination of spectral ratioing and in-scene training reference methods. The map shows where the formations lay. Two or three ratio images should show where the formation has compositional inhomogeneities. Training sets could be placed in each spectrally distinctive area and the average values of each spectral ratio extracted from each area. Knowing the composition of the formation members and facies, a comparison with laboratory data from an extensive collection of rock and soil spectra could probably be used to identify the composition of each training area. Finally, an automatic recognition map constructed on the basis of the signatures extracted from these in-scene training sets would then map those members and facies of the formation that are spectrally distinct. The important point to remember about this application is that any recognition errors that occur in the rest of the scene outside the formation of interest are incidental. It is probable that some members and facies of other formations may be spectrally similar to the training sets, but this application would be made one formation at a time. In this manner, practically every geologic quadrangle map could be improved. Possibly the greatest beneficiary of intraformation mapping would be mineral exploration, particularly for petroleum and natural gas. However, such mapping may greatly increase geologists' theoretical interests along intraformation lines of thought.

Although no one has yet tackled intraformation mapping with MSS data, it has been shown with LANDSAT data that the following lithologic groups can be separated *(7)*:

1. Triassic red beds, red pyroclastics.
2. Variegated shales, claystones, dolomitic silt.
3. Gray shales, gray sandstones, relatively fresh granite.
4. Buff-colored limestones and sandstones, weathered granite.
5. Magnetite, amphibole-rich mafic rocks.

These by no means represent the best that could be done for this problem from LANDSAT MSS data (some of these groups may be divisible), and certainly more could be done with a more suitable scanner. These groups are encouraging, however, in that even from four-channel LANDSAT data, all shales and sandstones do not look alike.

SUMMARY

Multispectral scanners are important for geochemical mapping because they are capable of collecting important compositional information in image form for wavelengths of radiation outside the spectral regions available to the camera and human eye. The 0.4- to 2.5-μm spectral region is most important for information pertaining to transition-metal ions (as in iron oxides), and the 8- to 14-μm region contains diagnostic compositional information concerning silicate rock type. Spectral ratio images from aircraft and satellite scanners can be used to map compositional information in image form. One of the most important capabilities of spectral ratios is iron oxide mapping. Iron oxides are sometimes associated with ores of copper, iron, lead, zinc, gold, uranium, and possibly even oil and gas. Exploration for these minerals from LANDSAT-1 computer-processed data is feasible. Future scanners with multiple channels in the 8- to 14-μm wavelength region, along with visible and reflective infrared channels, will probably be capable of mapping rock and soil types according to traditional classification schemes. Eventually, general geologic mapping should be aided by MSS techniques, especially in the form of intraformation mapping in sedimentary strata. Because of these scientific possibilities and the low cost and availability of satellite data on a global basis, spectral ratioing is becoming a revolutionary new tool for mineral exploration.

REFERENCES

1. Hunt, G. R., and J. W. Salisbury, Visible and Near-Infrared Spectra of Minerals and Rocks—I. Silicates, *Mod. Geol.,* v. 1, p. 283–300, 1970.
2. Hunt, G. R., and J. W. Salisbury, Visible and Near-Infrared Spectra of Minerals and Rocks—II. Carbonates, *Mod. Geol.,* v. 2, p. 23–30, 1970.
3. Hunt, G. R., J. W. Salisbury, and C. J. Lenhoff, Visible and Near-Infrared Spectra of Minerals and Rocks—III. Oxides and Hydroxides, *Mod. Geol.,* v. 3 (in press), 1971.
4. Hunt, G. R., J. W. Salisbury, and C. J. Lenhoff, Visible and Near-Infrared Spectra of Minerals and Rocks—IV. Sulphides and Sulphates, *Mod. Geol.,* v. 3, p. 1–14, 1971.
5. Logan, L. M., G. R. Hunt, J. W. Salisbury, and S. R. Balsamo, Compositional Implications of Christiansen Frequency Maximums for Infrared Remote Sensing Applications, *J. Geophys. Res.,* v. 78, p. 4983–5003, 1973.
6. Vincent, R. K., An ERTS Multispectral Scanner Experiment for Mapping Iron Compounds, *Proceedings of the 8th International Symposium on Remote Sensing of Environment,* Ann Arbor, Mich., p. 1239–1247, 1972.
7. Vincent, R. K., B. Salmon, W. Pillars, and J. Harris, Surface Compositional Mapping by Spectral Ratioing of ERTS-1 MSS Data in the Wind River Basin and Range, Wyoming, ERIM Technical Report, NASA Contract NAS5-21783, Environmental Research Institute of Michigan, Ann Arbor, Mich., in press.
8. Thomson, F. et al., ERIM Progress Report on

Use of ERTS-1 Data, Summary Report on Ten Tasks, Type II Progress Report, for Period 1 Jan.–30 June, 1973, ERIM, NASA Contract NAS5-21783, Report No. 193300-16-), Environmental Research Institute of Michigan, Ann Arbor, Mich., 1973.

9. Leeman, V., D. Earing, R. Vincent, and S. Ladd, The NASA Earth Resources Spectral Information System: A Compilation, NASA Rpt. NASA CR-31650-24-T, NASA Contract NAS9-9784, University of Michigan, Ann Arbor, Mich., 1971.
10. Leeman, V., The NASA Earth Resources Spectral Information System: A Compilation, First Supplement, NASA Report NASA CR-WRL-31650-69-T, NASA Contract NAS9-9784, University of Michigan, Ann Arbor, Mich., 1972.
11. Vincent, R. K., The NASA Earth Resources Spectral Information System: A Compilation, Second Supplement, NASA Report NAS CR-ERIM-31650-156-T, NASA Contract NAS9-9784, Environmental Research Institute of Michigan, Ann Arbor, Mich., 1973.
12. Salmon, B., and R. K. Vincent, Surface Compositional Mapping in the Wind River Range and Basin, Wyoming by Multispectral Techniques Applied to ERTS-1 Data, *Proceedings of the Ninth Symposium on Remote Sensing of Environment,* Ann Arbor, Mich., 1974, in press.
13. Dillman, R., and R. K. Vincent, Unsupervised Mapping of Geologic Features and Soils in California, *Proceedings of the Ninth Symposium on Remote Sensing of Environment,* Ann Arbor, Mich., 1974, in press.
14. Legault, R. R., R. Lyjak, J. R. Riodan, J. Penquite, W. Richardson, and J. C. Boyse, Studies in Spectral Discrimination, University of Michigan Report No. 5698-1-T, Report on Air Force Contract AF33(657)-10974, 42 p., 1964.
15. Halpert, H., and B. L. Musicant, N-Color (Hg,Cd)Te Photodetectors, *Appl. Optics,* v. 11, p. 2157–2161.
16. Vincent, R. K., and F. J. Thomson, Rock Type Discrimination from Ratioed Infrared Scanner Images of Pisgah Crater, California, *Science,* v. 175, p. 986–988, 1972.
17. Vincent, R. K., and F. J. Thomson, Spectral Compositional Imaging of Silicate Rocks, *J. Geophys. Res..,* v. 77, p. 2465–2471, 1972.
18. Vincent, R. K., A Thermal Infrared Ratio Imaging Method for Mapping Compositional Variations Among Silicate Rock Types, Ph.D. dissertation in the Department of Geology and Mineralogy, University of Michigan, Ann Arbor, Mich., 1973.
19. Travis, R. B., Classification of Rocks, *Quart. Colorado School of Mines,* v. 50, no. 1, p. 12, 1955.
20. Thomson, F., et al., ERIM Progress Report on Use of ERTS-1 Data, Summary Report on Ten Tasks, Type II Progress Report, for Period 1 Jan.–30 June 1973, NASA Contract NAS5-21783, Report 193300-16-P, Environmental Research Institute of Michigan, Ann Arbor, Mich., 1973.
21. Vincent, R. K., T. Wagner, B. Drake, and P. Jackson, Geologic Reconnaissance and Lithologic Identification by Remote Sensing, ERIM Technical Report 191700-8-F, ARPA-USBM Contract HO22064, Environmental Research Institute of Michigan, Ann Arbor, Mich., 1973.
22. Locke, Augustus, *Leached Outcrops as Guides to Copper Ore,* Williams & Wilkins Company, Baltimore, 1926.
23. Weiser, H. B., (1926) *The Hydrous Oxides,* McGraw-Hill, New York, 1926, p. 70–74.
24. Blanchard, R., *Interpretation of Leached Outcrops,* Nevada Bureau of Mines Bulletin 66, University of Nevada, Reno, Nev., 91 p., 1968.
25. Kelly, W. C., *Topical Study of Lead–Zinc Gossans,* Bulletin 46, State Bureau of Mines and Mineral Resources, New Mexico Institute of Mining and Technology, Socorro, N.M., p. 42–47, 1958.
26. Rowan, L. C., P. H. Wetlaufer, A. F. H. Goetz, Jr., F. C. Billingsley, and J. H. Stewart, Discrimination of Rock Types and Detection of Hydrothermally Altered Areas in South-Central Nevada by the Use of Computer-Enhanced ERTS Images, *U.S. Geol. Surv. Prof. Paper 883,* 1974, in press.
27. Shockey, P. N., R. I. Rackley, and M. P. Dahill, Source Beds and Solution Fronts: Remarks to Wyoming Metals Section, AIME, Feb. 27, 8 p., 1958.

28. Rackley, R. I., Environment of Wyoming Tertiary Uranium Deposits, *Am. Assoc. of Petrol. Geol. Bull.,* v. 56, p. 755–777, 1973.
29. Krauskopf, R. B., *Introduction to Geochemistry,* McGraw-Hill, New York, 721 p., 1967.
30. Donovan, T. J., Petroleum Microseepage at Cement Oklahoma: Evidence and Mechanism, *Am. Assoc. Petrol. Geol. Bull.,* v. 58, p. 429–446, 1974.
31. Vincent, R. K., and W. W. Pillars, SKYLAB S-192 Ratio Codes of Soil, Mineral, and Rock Spectra for Ratio Image Selection and Interpretation, *Proceedings of the Ninth Symposium on Remote Sensing of Environment,* Ann Arbor, Mich., 1974, in press.

11

LANDSAT APPLICATIONS IN THE LESS-DEVELOPED AREAS

William L. Smith

William L. Smith, formerly with System Planning Corporation, is currently at the Environmental Research Institute of Michigan, Arlington, Virginia.

THE ROLE OF REMOTE SENSING

In anticipation of the impact of LANDSAT-1 on the developing world, Charles Weiss, science advisor in the Economics Department of the World Bank, cited that imagery from satellites would offer

> great advantages to the less developed countries, which are in general poorly mapped and in great need of resource surveys. In many cases, satellite [products] provide small scale maps where none existed before. On the other hand, much of the value of satellite [imagery] is realized only after careful analysis by trained experts, often in conjunction with traditional methods of aerial and ground survey. . . . A satellite offers many advantages over ground and aerial surveys, possibly the most important being speed, the ability to obtain an overall view of a region, and repetitive coverage. The use of satellites should not, however, be thought of as replacing more traditional surveys; the various methods are complementary *(1)*.

The images are not the final product of remote sensing. The skills required to make the imagery useful include the selection and evaluation of data, photointerpretation by trained geologists, image enhancement, and the comparison of sensor data to the geology in the field.

In many developing countries the mineral industries are important foreign-exchange earners and a vital sector of the internal economy. This is particularly true in countries where there is geographical or climatic restraint upon extensive agriculture. In many other developing countries, the minerals sector could well become of greater economic significance were there better knowledge of regional geology, concerted efforts in systematic mineral search, and the capability of national and private resource management to make those critical decisions and to implement such actions as are essential to exploration planning, development, and exploitation. LANDSAT data can provide much of the information needed by resource management for initial decision making and planning.

The use of LANDSAT imagery in preparing base maps will be of foremost importance in those countries where current geological or topographic maps are inadequate. The multispectral scanner systems (MSS) on LANDSAT-1 and 2 have the geometric accuracy and registration required for excellent mapping fidelity, and the spectral range needed to enable the analysis of pertinent thematic content. Surface features of a geological nature 80 m or more in diameter may be detected, and the four different spectral bands can provide more lithologic and structural information than may be obtained on conventional aerial photographs. The LANDSAT system also provides computer compatible tapes with which geologists, cartographers, and other investigators may undertake signature analysis to identify and tabulate a wide range of information on surface phenomena, including variations in lithology and structural patterns. The multispectral information is recorded at specific combinations of densities, and different methods may be employed to enhance, extract, or isolate the image of a particular thematic content while retaining the geometry and resolution of the image. The different developing countries that are participating in the LANDSAT program employ different degrees of sophistication in their initial studies. These range from the inexpensive use of light tables and the production of diazochrome transparencies all the way to computerized processing and the use of color additive viewers, depending upon budget, national interest, and often upon the amount of enthusiasm that the investigators generate in the resource agencies.

But whether the national or private effort in a nation is maximum or minimal, the usefulness of the geological content of the LANDSAT products is only derived after study by skilled geologists, and preferably by those who are well acquainted with the regional geology and mineralization and who have had some experience in interpreting aerial photographs. It is essential that proper selection of data products be made and that the geologist be sufficiently familiar with the data-processing methods to be able to identify characteristic errors. The various skills required for interpreting multispectral imagery generally need to be acquired, usually through the short courses or workshops provided by NASA, the U.S. Geological Survey, and several universities. There are also practical minimum facilities for undertaking photographic reproduction, visual inspection of imagery, and production of color transparencies. Low-cost manual techniques are available and easily learned for enhancing imagery and preparing it in a format comprehensible to economic geologists and at planning levels. With this minimum requirement for equipment and training and with an aggressive program by industry or the resource agencies, LANDSAT imagery can provide the basis for new structural diagrams and updated geological and topographic maps of areas of difficult access, and provide much of the initial information required for resource management or commercial exploration decisions.

Oil and mineral revenues are often a major part of the economy of developing nations, particularly those with limited crops or manufacturing. Mineral and petroleum operations are often a major source of national wealth and local industry; ores, concentrates, or crude oil are important from the standpoint of hard currency exchange. Establishing a profitable minerals sector depends initially upon successful geological exploration and the development of discovered resources. In countries where exploration and mapping by current methods have been limited, a broad methodical search is usually desirable for obtaining the essential information as to the location and extent of sedimentary basins, structural patterns, the delineation of mineralized areas, and the identification of promising areas for closer exploration. These are essential for any underdeveloped country that is considering the development or expansion of an effective, operative mineral economy.

In the major industrial nations, when there is need of geological reconnaissance, there are detailed maps available from government and private sources, aerial surveys may be put into service rapidly, exploration companies or federal agencies have established procedures for sequencing exploration and the rapid interpretation of findings, and funds may be quickly available. In regard to the impact of LANDSAT products on exploration company ventures, an exploration geologist responsible for a major mineral investigation in the developing world was interviewed: Yes, he used LANDSAT data, and the most current interpretation methods. When inquired as to the economy of using LANDSAT data, he made it clear it was not the economy that was the primary factor, but the time element and the great information content of LANDSAT products not otherwise obtainable. His need was geological information, and his great advantage was a staff of skilled geologists and technicians who were able to extract the pertinent information expertly and quickly. But few developing countries have geological staff, facilities, or funding comparable to the major minerals or petroleum companies. In the underdeveloped countries, often a relatively few qualified geologists are responsible for an entire national effort in resource development. It behooves far-seeing geologists in these nations to convince their agencies of the potentialities of using LANDSAT data for the national interest.

The locating of mineral resources in relatively unexplored areas through remote-sensor data will require providing exploration geologists with information on lineaments, fault systems, igneous petrology, the loci of fossil river channels or detrital deposits with possibilities for placer operations, and various secondary clues of obscured structure. Although several developing countries are making remarkable progress in the use of LANDSAT products in mineral exploration, many more have either failed to grasp its importance and potentialities or remain convinced that the better route is resource development by foreign interests with accruing revenue benefits (or even of eventual Mexicanization?). On the other hand, the more progressive of these countries recognize the need for preinvestment surveys, and such may be assisted by the interpretation of multispectral imagery, particularly where requirements include large or essentially inaccessible regions. Such studies by conventional aircraft and ground surveys may often be costly, long-term efforts, but the insertion of LANDSAT data in many instances can localize and hence minimize conventional coverage. Such a use of LANDSAT data is proving to be a good investment of scarce capital funds, and permits resource agencies to have an earlier and more knowledgeable position when considering national ventures or when dealing with foreign development or exploitation of concessions.

Because existing maps and geological data on much of the interiors of the developing countries are inadequate, specifically local, or outright incorrect, it is difficult for planning levels in these nations to make decisions concerning funding exploration and the development of resources. The Agency for International Development (AID) and the United Nations Development Program (UNDP) have made significant efforts in this area, but the major incentive will have to be national if the efforts are to go beyond initial stages. It is quite possible that yet undiscovered reserves in such areas are adequate for providing raw material for the needs of world industry for a significant period, perhaps spanning the time when population pressures will otherwise exhaust many current sources. Major foreign aid may be needed from the industrial countries for a considerable period of time, and mining and exploration interests need to cooperate to the fullest with these countries. Because, although these areas may need our assistance for the present, the industrial nations will surely be a market for their resources in the foreseeable future. For political reasons, it is important that we provide all possible assistance in the application of LANDSAT technology to mineral exploration to consistently friendly sources of raw materials.

Although geological data are expensive to obtain by conventional methods, much of the initial exploration costs may be radically reduced through the interpretation of LANDSAT data for identifying the indicators of possible environments of ore and localizing the focus of exploration. On the other hand, using LANDSAT data entails certain skilled handling and interpretation before they become of practical value. Assuming the government or industry of a developing country has geologists experienced in interpreting aerial photography, and that the minimum equipment and facilities for enhancing LANDSAT data are available, the geological information derived will certainly prove beneficial for exploration decisions. In those areas where the geological agency or mining interests do not have such capabilities, the solution is either training or reliance upon foreign development or foreign research efforts. Clearly, training in such circumstances should be the desirable route for long-term national interests. In most developing areas the geological staffs are more than capable of identifying information pertinent to mapping and the surface expressions of structure. Training is only required in the use of specialized equipment for interpreting spectral band products and for relating the data to management. Workshops and other programs have taken LANDSAT data interpretation out of the realm of looking at pictures and into the area of assistance to national efforts. Developing a capability in interpreting LANDSAT data for geological purposes largely depends on the increased familiarity with the techniques by the investigator and the comparison of findings to field data.

IMPACT OF LANDSAT ON EXPLORATION IN REMOTE AREAS

In 1948, Hugh McKinstry advised his students to the effect that, where the broader structure of an area had not been defined, it was well worth hiring an airplane to obtain an aerial view of topography which could disclose geological relationships that might otherwise only be deciphered after extensive mapping *(2)*. Then, as now, aerial photographs were the best possible base for mapping surface geology, but their optimum application to geological exploration was limited to a relatively few far-seeing geologists.

A quarter-century later, proper aerial coverage is only poorly exploited. The advent of LANDSAT-1 made it possible for many countries to bridge this gap by providing synoptic pictures significantly better than existing maps for determining broad structural relationships, and at a resolution that is satisfactory for initial reconnaissance and for delimiting the areas where aerial coverage should be undertaken.

LANDSAT products have been found to have unique applications to geologic mapping in remote areas. In Brazil, for example, Amaral (*3*) reported to the COSPAR meeting at Sao Jose dos Campos, July 1974, a comparison of the use of LANDSAT-1 data and SLAR imagery for applications to geology and mineral resources in the Amazon region. He felt the repetitive capability of the LANDSAT system was of primary importance, and that through LANDSAT it had been possible to obtain an essentially cloud free coverage of 5 million km^2 of the region. He also felt that LANDSAT multispectral data were superior inasmuch as they provided as much structural information as the radar imagery, but provided a greater amount of lithological information. The LANDSAT imagery clearly shows the tin-bearing granites of the Xingu Valley and the iron deposits of the Serra dos Carajas. Using literature analysis as ground truth, it was possible to prepare a geological map of the region at 1:5,000,000 scale in about 3 months' time. Amaral concluded that the use of remote-sensing data from the initial phases of a systematic geological mapping program can result in an optimization of fieldwork as well as in a reduction of costs.

In Bolivia, Fernandez et al. *(4)*, with the purpose of extracting the maximum possible information from LANDSAT imagery for mapping natural resources, prepared 11 thematic maps from one LANDSAT color composite. The frame includes most of the western part of Oururo Province in the central sector of the Altiplano, the main feature being the Salar de Coipasa. Maps, overlays, and detailed interpretations were prepared for drainage, hydrology, volcanology, structure, physiography, surface lithology, and soil types. These results were compared with a parallel effort using black and white products obtained from band 7 of the MSS. Comparison indicated the color product contained some 50 percent more data than the black and white prints at 1:1,000,000 scale for most thematic products. LANDSAT-Bolivia is consequently processing color composites of the entire country to provide a photomap of Bolivia at 1:1,000,000 scale and making overlays of individual LANDSAT frames depicting the significant structural and morphological features. Such data would have taken years to acquire by conventional means at a major national expenditure.

In most developing countries a broad exploration program is required in the interior areas to locate possible economic mineral reserves. The exploration sequence will generally be integral with national-level decision making by its resource management agencies. The decision making could follow this order *(5–7)*:

1. Determine national need, consideration of mineral economy prospects, and market.
2. Whether to undertake reconnaissance.
3. Whether to mount a general exploration program.
4. Whether to undertake a surface survey.
5. Whether to test drill.
6. Whether to develop the resource.

In an exploration program, a parallel sequence of decisions would be addressed by geologists and exploration management and might include any of the following, depending on whether the resource involved is petroleum, metal ores, industrial minerals, or placer deposits *(5, 6)*:

1. Selection of promising area for exploration.
2. Review of available geological data.
3. Reconnaissance.
4. Regional exploration, including aerial methods.
5. Geologic mapping, sampling, geochemistry.
6. Test drilling, geophysical survey.
7. Systematic drilling, core sampling, evaluations.
8. Subsurface exploration, surveying, outlining ore, assessment of reserves.
9. Development.

When relevant LANDSAT data are inserted into the exploration decision-making sequence, planning-level decisions will be facilitated; the resulting benefits will include savings of exploration time, costs, and a possible accelerated rate of discovery and development. The models which are shown in Figs. 1 and 2 *(8)* follow the management decision-making points and the exploration sequence, presuming the area to be deficient in geological data.

When geological data are deficient or mapping is obsolete, LANDSAT data can play an important role in providing the regional information needed for investigating the nature and extent of resources. In many 5-year-plan-type countries, governments have relied upon private, often foreign-owned, petroleum and mining companies for much of their geological information. But no matter how thorough and accurate, the information is by necessity quite limited to the resource of interest. The decision to undertake a broad mineral resource inventory or even a specific-mineral exploration program depends foremost on national economic or industrial needs, and second on the probability that the preinvestment studies will prove to be practical and will identify more than one option for appraisal. Although LANDSAT data provide much geological information, they can only be used as a basis for exploration when carefully interpreted in terms of exploration objectives by skilled investigators. And although knowledgeable employment of LANDSAT data may significantly cut down the area to be explored and may even pinpoint areas where exploration is most apt to be successful, inept interpretation of LANDSAT data may lead costly efforts in the wrong direction. An inexperienced geologist–investigator may also unknowingly allow LANDSAT data to promote preconceived ideas.

Of the various steps in the conventional exploration sequence, LANDSAT will have the most impact on decisions related to initial reconnaissance, mapping, and the extent of aerial coverage needed. Any improvement in these areas will surely increase the cost effectiveness of exploration. Direct discoveries of ore are not a predicted or expected result of the use of LANDSAT data; yet, if any such discoveries are made, they may well be in the remote poorly mapped areas of the developing countries. The unique role of LANDSAT technology in the underdeveloped areas will not be in pinpointing locations for exploration but will be educational. LANDSAT will provide geologists with a better understanding of regional ore patterns and the possible extent of petroleum-bearing formations and structures.

POSSIBLE BENEFITS

The cost is only one of many concerns of exploration ventures by foreign mining and petroleum companies in the developing world, but it is often the determining factor for national efforts by these countries. Most previous attempts to assess cost savings or the future benefits to be derived from the use of LANDSAT or other remote-sensor data in mineral exploration has met with criticism or amusement, including the report of the National Academy of Sciences—National Research Council Geology Panel, 1969 Summer Study on Useful Applications of Earth Oriented Satellites *(9)*.

At this meeting, various approaches were tried to estimate the benefits that might accrue from a postulated program. Basing the inquiry on current annual exploration cost data, it was conservatively estimated that exploration costs for U.S. oil were $2.05 billion of which $345 million was attributable to geology and geophysics. Mining industry exploration costs were the most difficult to obtain, but were estimated at some $200 million for the United States and Canada, of which some $50 million related to geology and

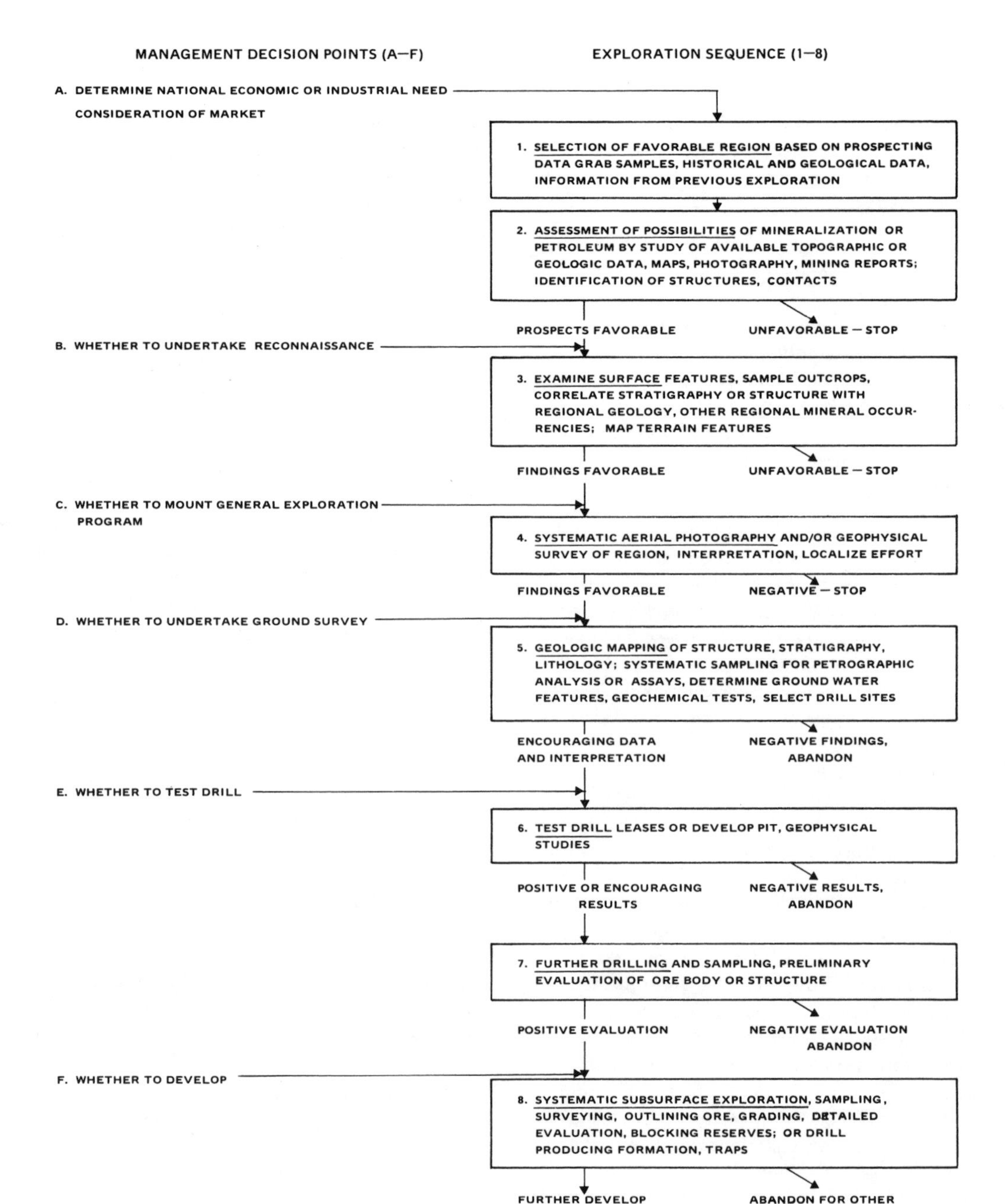

FIG. 1 *(above)*
Conventional exploration sequence in undeveloped area (presuming a deficiency of geologic data available).

FIG. 2 *(right)*
Exploration sequence in undeveloped area employing LANDSAT data (presuming a deficiency of geologic data available).

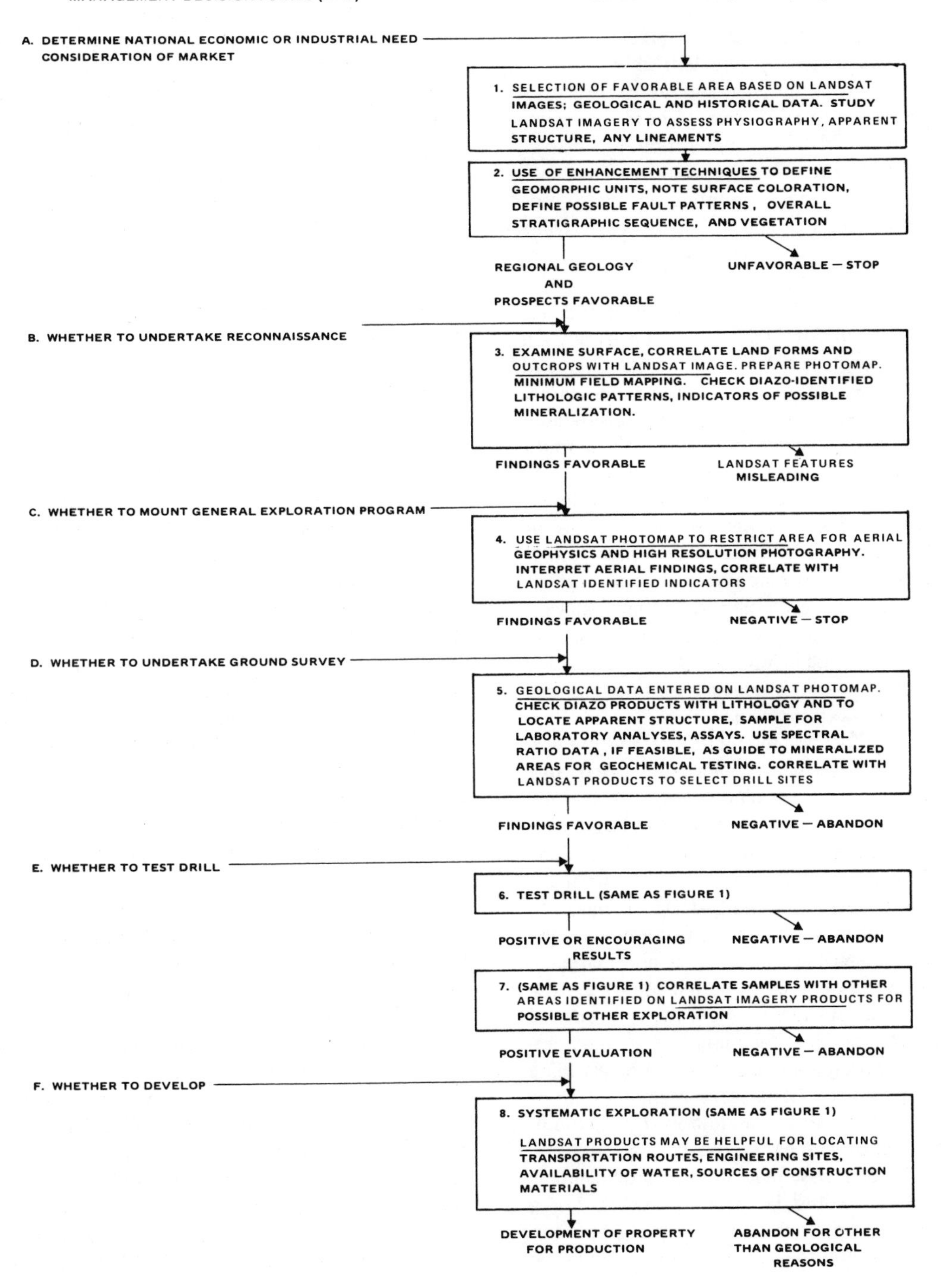
MANAGEMENT DECISION POINTS (A—F)
EXPLORATION SEQUENCE (1—8)
A. DETERMINE NATIONAL ECONOMIC OR INDUSTRIAL NEED CONSIDERATION OF MARKET
1. SELECTION OF FAVORABLE AREA BASED ON LANDSAT IMAGES; GEOLOGICAL AND HISTORICAL DATA. STUDY LANDSAT IMAGERY TO ASSESS PHYSIOGRAPHY, APPARENT STRUCTURE, ANY LINEAMENTS
2. USE OF ENHANCEMENT TECHNIQUES TO DEFINE GEOMORPHIC UNITS, NOTE SURFACE COLORATION, DEFINE POSSIBLE FAULT PATTERNS, OVERALL STRATIGRAPHIC SEQUENCE, AND VEGETATION
REGIONAL GEOLOGY AND PROSPECTS FAVORABLE
UNFAVORABLE — STOP
B. WHETHER TO UNDERTAKE RECONNAISSANCE
3. EXAMINE SURFACE, CORRELATE LAND FORMS AND OUTCROPS WITH LANDSAT IMAGE. PREPARE PHOTOMAP. MINIMUM FIELD MAPPING. CHECK DIAZO-IDENTIFIED LITHOLOGIC PATTERNS, INDICATORS OF POSSIBLE MINERALIZATION.
FINDINGS FAVORABLE
LANDSAT FEATURES MISLEADING
C. WHETHER TO MOUNT GENERAL EXPLORATION PROGRAM
4. USE LANDSAT PHOTOMAP TO RESTRICT AREA FOR AERIAL GEOPHYSICS AND HIGH RESOLUTION PHOTOGRAPHY. INTERPRET AERIAL FINDINGS, CORRELATE WITH LANDSAT IDENTIFIED INDICATORS
FINDINGS FAVORABLE
NEGATIVE — STOP
D. WHETHER TO UNDERTAKE GROUND SURVEY
5. GEOLOGICAL DATA ENTERED ON LANDSAT PHOTOMAP. CHECK DIAZO PRODUCTS WITH LITHOLOGY AND TO LOCATE APPARENT STRUCTURE, SAMPLE FOR LABORATORY ANALYSES, ASSAYS. USE SPECTRAL RATIO DATA, IF FEASIBLE, AS GUIDE TO MINERALIZED AREAS FOR GEOCHEMICAL TESTING. CORRELATE WITH LANDSAT PRODUCTS TO SELECT DRILL SITES
FINDINGS FAVORABLE
NEGATIVE — ABANDON
E. WHETHER TO TEST DRILL
6. TEST DRILL (SAME AS FIGURE 1)
POSITIVE OR ENCOURAGING RESULTS
NEGATIVE — ABANDON
7. (SAME AS FIGURE 1) CORRELATE SAMPLES WITH OTHER AREAS IDENTIFIED ON LANDSAT IMAGERY PRODUCTS FOR POSSIBLE OTHER EXPLORATION
POSITIVE EVALUATION
NEGATIVE — ABANDON
F. WHETHER TO DEVELOP
8. SYSTEMATIC EXPLORATION (SAME AS FIGURE 1)
LANDSAT PRODUCTS MAY BE HELPFUL FOR LOCATING TRANSPORTATION ROUTES, ENGINEERING SITES, AVAILABILITY OF WATER, SOURCES OF CONSTRUCTION MATERIALS
DEVELOPMENT OF PROPERTY FOR PRODUCTION
ABANDON FOR OTHER THAN GEOLOGICAL REASONS

geophysics. This provides a basis for assuming which exploration costs would be affected: about 17 percent of oil exploration costs and about 25 percent of mining industry exploration costs. These figures are surely too conservative for exploration in undeveloped areas where there is a deficiency of available geological data and where mounting an exploration would be more costly. The following cost-savings discussion applies mostly to these 17 and 25 percent sectors of total exploration costs.

As noted with regard to the exploration models, the impact of LANDSAT data would be largely on initial reconnaissance, in aiding geologic mapping, and in reducing the extent of aerial coverage required. The specific cost savings would differ considerably from country to country depending upon the amount of geological information available, whether mineral activity is already a significant factor in the economy, and the nature and occurrence of the minerals of interest. Although detailed cost-effectiveness studies are not yet available on mapping and geological reconnaissance in the developing countries, relevant time and cost data have been compiled by several U.S. investigators, some involving remote areas with a lack of geological data. These are pertinent and a reasonable base for extrapolation.

In terms of geological reconnaissance and the production of photo-base metric maps from LANDSAT imagery, the value to developing countries is beyond estimate if considered in regard to the accelerated discovery of economic deposits. In terms of geological exploration costs alone, it is possible to make rough assessments, assuming the LANDSAT images to be superior to any existing maps. In summing up the applications of LANDSAT to land use and mapping, Lindgren and Simpson *(10)* show the use of LANDSAT imagery to be cheaper by more than an order of magnitude over similar mapping using medium-altitude aerial photography as a base. They estimate that preparing a 10- to 12-category thematic map of an area the size of Iowa or Illinois by medium-altitude aerial photography would cost $1 million. The preparation of a base map and thematic overlays for geological use would be of similar cost. For a similar mapping from high-altitude aerial photography, roughly assuming the cost of mapping would be proportional to the scale of the map, the same job would cost only 20 percent as much, or about $200,000. Extrapolation to the use of LANDSAT imagery would put the mapping at about $80,000 or one twelfth the cost of a conventional job.

Insofar as saving time is concerned, Miller and Belon *(11)* report the preparation of a resource survey of a 13,000-square-mile area of a remote section of the Brooks Range of Alaska. Using color and black-and-white prints of the essentially inaccessible area, a multidisciplinary team with a minimum of orientation spent some 25 man-hours to make a survey, including maps of gross geologic features, previously unknown structures, tectonic zones pertinent to economic minerals, vegetation types, watershed drainage, and potential land use.

Lauer and Krumpe *(12)* present a detailed cost-effectiveness study related to using LANDSAT products for a wildland resource inventory of the Feather River watershed in northern California as compared to a conventional black-and-white aerial photography coverage, which took three years and was completed before LANDSAT, and in comparison to a CIR high-flight coverage. The LANDSAT and high-flight products were superior to the conventional coverage, but the LANDSAT study was clearly superior and showed much better delineation density. Their itemized analysis showed favorable cost ratios of 1:20 for LANDSAT compared to conventional coverage and 1:9 for LANDSAT compared to high-flight coverage.

In terms of geological mapping of structures, there is the often-referenced case of the University of Wyoming geologist who had mapped the major fracture systems of the Wind River Mountains for 5 years by conventional methods, completing less than 15 percent of the range. He completed the remaining part of the map with comparable accuracy in 3 hours after he received the LANDSAT imagery of the range *(13)*. A lineament map of a 34,000-km^2 area of the Canadian Shield was produced in 45 min because IR imagery clearly delineated water-filled fractures *(13)*. In such an area of dense vegetational cover, similar to some of the developing countries, this degree of structural detail is most difficult to obtain either by aerial or surface methods. Amaral *(3)* reported that, for purposes of mapping at 1:5,000,000 scale in the

Amazon basin, not only were LANDSAT images more effective than SLAR mosaics, but that 1 km^2 of LANDSAT imagery costs about U.S. $0.15 (including the cost of acquisition and maintenance of the Brazilian receiving and processing system), whereas the costs of SLAR imagery acquisition were about U.S. $5.00 per km^2.

In comments on the applications of LANDSAT to Department of Interior programs, DeNoyer *(14)* noted in effect that although efficiency factors of 6 to 20 and higher had been reported at the second LANDSAT symposium, the real benefits will come from analysis of the data and taking timely actions based on the efficiently obtained information.

In summary, 17 to 25 percent of total exploration costs are conservatively estimated to be affected by the use of LANDSAT data. It is estimated that aerial geophysical coverage (hence costs) would be affected somewhat similarly to aerial photography owing to the reduction of the area of study. The efficiency ratio for geological search and mapping is currently estimated to range up to 1:20, and more in some instances, depending upon the type of deposit or structure involved, the availability of geological data, and the amount of ground that would be reduced by the study of LANDSAT images.

Insofar as ultimate benefits are concerned, it is of course impossible to estimate the kind, number, or value of discoveries, if any, that might accrue from the use of LANDSAT data. The best approach to such a question is probably the assumption that, although total reserves would not be increased by the use of satellite data, there would be an accelerated rate of discovery of existing deposits. The NAS/NRC study *(9)* suggested, for example, that a possible 5 percent of ultimate reserves might be brought into accelerated production over a 20-year period following the then-postulated GEROS program. On the other hand, the same study noted that in the United States the cost of geological search for oil is about 3 percent of oil production values, and the cost of geological search by U.S. and Canadian mining industries is equivalent to 4 to 5 percent of mining industry production values. This does not in any sense predict a return on investment, but it illustrates the possible orders of magnitude between the relatively inexpensive application of LANDSAT data and possibly large or accelerated benefits.

Short and Lowman *(15)*, considering the cost effectiveness of using LANDSAT data in mineral exploration, state

> The outlook is especially promising for certain parts of the world where LANDSAT images represent the first detailed surface coverage of regions that have never been surveyed or mapped beyond a reconnaissance level. This applies particularly to inaccessible high mountain areas, broad deserts, tundra and steppes, and polar regions as well as sections of underdeveloped or low population countries.

The ability to update maps or to acquire new, good-quality photo-base maps from LANDSAT imagery with a minimum of expense and fieldwork is in itself a major step for many developing countries. United States government agencies spend some $450 million annually on mapping. Even at this rate, the United States is inadequately covered by topographic maps, with an estimate (pre-LANDSAT) of completing continental U.S. topographic coverage over the next 20 years. The map production cycle in the United States for updating topographic maps is 3 years between acquisition of aerial photography and the revised product. Each map only has a 5-year useful life for urban or developing areas, and somewhat greater for remote areas. In general, map revision is between 5 and 10 years for those areas most highly in demand and 20 to 30 years for other areas. A considerable part of the United States is at best mapped at 1:63,000 scale and some of the maps were published over 60 years ago *(6, 16)*. As for geologic maps, only a minor part of the United States and most other developed countries is covered by quadrangle maps showing geologic structure or surface geology adequate for engineering use or for exploration for minerals. In the developing countries most of the intensive geologic mapping has been local and restricted to areas of active extractive industry. Whereas the observation techniques available before LANDSAT required a sectional approach to mapping, LANDSAT data will now make it possible to undertake broad programs in topographic and geologic mapping, at practical scales for reconnaissance and exploration, particularly in the more remote areas of the developing countries.

DISCUSSION

The early returns of mineral resource studies employing LANDSAT data in remote or developing areas has been most encouraging. Computer analysis of tapes and the use of a color-additive viewer are often desirable for obtaining the maximum benefits from LANDSAT products, but this degree of sophistication is by no means essential. Successful results have been obtained by investigators in several developing countries using light tables, manual plotting equipment, and low-cost techniques for enhancing LANDSAT imagery. National programs have been undertaken at low budgets by skilled geologists with experience in aerial photograph interpretation or photogrammetry.

Various "workshops" have been of exceptional value to investigators and their staffs. In some of the developing countries, there is a great need for training personnel in LANDSAT technology, the use of data products, and the use of the newly developed enhancement methods. This need, however, is far less among geologists than among users from other disciplines, inasmuch as the study of maps, landforms, etc., is an essential part of their thinking. As the information gained from the interpretation of LANDSAT data becomes usable in economic planning, it will be important that other elements of the national economy become better informed as to the potentialities of LANDSAT, including economists, conservations, and the domestic extractive industries, as well as personnel in government planning, budgeting, and resource offices.

The use of a diazo machine is an example of a low-cost technique for enhancing LANDSAT imagery and preparing it for natural resource interpretation as a product compatible with current information systems. A procedure developed at the New York State College of Agriculture *(17)* employs photographic techniques and standardization of 70-mm NASA film clips. A subtractive color process is used producing 1:3,300,000 images to scales up to 1:66,000. Diazo transparencies are produced in magenta, cyan, and yellow for each MSS band. Investigators state that data retrieval is possible from a great number of diazo color combinations. Map transfer is easily done at 1:250,000 scale, and some enlargements are feasible up to 1:10,000 using overhead projectors. The approximate cost of preparing photographically enhanced enlarged negatives and positives and the diazo materials is about $0.01 per square mile. The specific cost of making and mapping a land-use classification of 12 use types at a 1:250,000 scale is about $1 per square mile. This requires no expensive hardware. The wide range in color selection allows adaptation for a wide range of applications. Malan *(18)*, discussing LANDSAT imagery use in South Africa, noted the production of diazo color composites to be the technique most generally applicable and available to LANDSAT investigators; he described how diazo exposures can be made of black and white ERTS transparencies through a plate of glass in sunshine, which in turn may be developed in a container with ammonia liquor at the bottom. Such a process also permits the investigator to experiment with color and contrast to fit the investigation, as opposed to relying on LANDSAT products produced by a standard method.

At the 1970 AID Symposium on the Utility of Remote Sensing as an Aid to Developing Countries, Washington, D.C., it was noted that there would be a need to develop an indigenous analytical capability in a developing country so that practical problems could be addressed. Also, training and technical assistance in earth resources survey systems and techniques is a first order of business in the developing countries *(19)*. Data alone is of little value to the minerals sector of a developing country. Only in the hands of skilled geologists trained in the use of LANDSAT data do the products become information for aiding management decisions.

Training at the workshops and various short courses that have been made available to geologist investigators has emphasized analysis of imagery and interpretation, enhancement techniques, information extraction, and the insertion of LANDSAT data into decision-oriented models. The ultimate value of this training will be agency capabilities in extracting information (e.g., through models) pertinent to the requirements of technical and policy management levels. An oversimplified statement of the route to assist national project planning in mineral related problems is as follows:

1. Evaluate potential utility of resource inventory or exploration programs.
2. Compile photomaps and thematic overlays of areas of interest.
3. Interpret thematic maps of surface characteristics and resource-related features.
4. Insert information into management decision points of the exploration sequence model (Fig. 2).

The rapidly advancing technology, however, has an attendent problem: at the Ninth International Remote Sensing Symposium, Brown *(20)* observed that, although remote sensing might in time afford all nations with practical and economic means for locating new natural resources, the technology remained well ahead of interpretive knowledge. Henderson *(21)* has identified a gap between the rapid progress in developing remote-sensing techniques and their application, and Grossling and Johnston *(5)* noted a gap existing between raw remote-sensor data and the resource information needed by several units of the nation's economy. Particular insight into the problem of relating remote-sensing research to ultimate users in developing countries was shown by Bond *(22)* at the 1974 COSPAR meeting in Brazil. She emphasized that many of these countries have a problem of incorrect or inadequate resource data, which limits their abilities to make plans and hence their economic development. If LANDSAT data are to become an important source of information and effectively used, better communication has to be established between researchers and users. It may be added that the same problem existed among U.S. government agencies and still does to some degree. Major efforts had to be made by NASA to determine user agency needs in order to select appropriate sensors and orbital parameters.

The immediate problem that many undecided developing countries are concerned with is whether to appropriate the necessary funds to undertake LANDSAT-related resource investigations. To help answer this question, System Planning Corporation, Mathematica Inc., and the Environment Research Institute of Michigan undertook a cost–benefit study for the AID Office of Science and Technology. Summers stated the situation as follows at the 1974 COSPAR meeting *(23)*:

> No aerospace hardware would need to be acquired by developing countries since it is expected that projected U.S. satellite programs would provide adequate international coverage. However, the imagery produced will require processing and interpretation by trained personnel, and to be useful the information must be disseminated to user groups, all of which is true for aircraft and ground surveys as well. Thus, there are additional costs that an operational natural resource information system would incur. The problem confronting the developing countries, then, is whether this improved natural resource information system will justify the required investment in trained personnel, equipment, and information dissemination costs.

Perhaps the answer to this problem so far as geology is concerned may be found in these examples from the early "Reports Containing Author-Identified Significant Results" section of the National Technical Information Service Abstracts of the NASA Earth Resources Survey Program (NTIS report numbers in parentheses):

Bolivia:

> It has been possible using ERTS data to define obscured structures and lineaments, to differentiate the extent of volcanic formations and to correlate regional fracture zones with volcanism . . . (E73-10296).

Botswana:

> The imagery appears to confirm a new theory that Archean greenstone belts in northeast Botswana and southwest Rhodesia are co-extensive . . . (E73-10753).

Guatemala:

> There are promising indications that Altiplano pumice basins important to our nation's agriculture can be delineated and identified on ERTS imagery . . . (E73-10609).

Mali:

> The potential capacity to map laterite deposits in the Inland Delta from space was established with the help of aerial and ground surveys . . . (E73-10436).

Pakistan:

> Geomorphic features related to known porphyry despoits . . . are easily distinguished on ERTS images . . . (E73-10401).

Thailand:

A geologic map of Thailand has been constructed from ERTS imagery . . . (E73-10746).

These representative, encouraging reports by the principal investigators attest to the value of the program and provide us with the hope that follow-on exploration in these areas may yet provide industry with the essential raw materials needed to hold the line against shortages, while other branches of technology address the ultimate problems of substitute materials and new methods for utilizing the vast reserves of submarginal ores.

REFERENCES

1. Weiss, C., Satellites and International Resource Development, *Finance and Development,* Vol. 9, No. 2, International Monetary Fund and the World Bank Group, Washington, D.C., June 1972.
2. McKinstry, H. E., *Mining Geology,* Prentice-Hall, Englewood Cliffs, N.J., 1948.
3. Amaral, Gilberto, Remote Sensing Applications for Geology and Mineral Resources in the Brazilian Amazon Region, Paper B.4.1, COSPAR, Sao Jose dos Campos, Brazil, July 1974.
4. Fernandez, A. C., C. Brockmann, S. Kussmaul, and O. Unzueta, Applicacion y Evaluacion de Imagenes ERTS de Composicion de Color al Inventario de Recursos Naturales, *Rev. Tech. Yacimientos Petrolifros Friscales Bolivianos,* v. 3, no. 1, Apr. 1974.
5. Grossling, B. F., and J. E. Johnston, On Bridging the Gap Between Raw Material Remote-Sensor Data and Resource and Environment Information, Technical Symposium of Soc. Photo-Optical Instrumentation Engineers, Anaheim, Calif., Sept. 1970.
6. Smith, W. L., *Remote Sensing for Mineral Deposits,* Bellcomm, Inc., Washington, D.C., TM-69-10-15-5, 1969.
7. Zissis, G. H., K. P. Heiss, and R. A. Summers, Design of a Study to Evaluate Benefits and Cost Data from the First Earth Resources Technology Satellite (ERTS-A), Final Report, Willow Run Laboratories, Mathematica, System Planning Corporation, 1972.
8. Lowe, D. S., R. A. Summers, and E. J. Greenblat, An Economic Evaluation of the Utility of ERTS Data for Developing Countries for AID, by Environmental Research Institute of Michigan, System Planning Corporation, and Mathematica, 1974.
9. Lyon, R. J. P., et al., Geology, Panel 2 Report, *Useful Applications of Earth Oriented Satellites,* NAS–NRC, to NASA, Washington, D.C., 1969.
10. Lindgren, D. T., and R. B. Simpson, Land Use and Mapping, *Symposium on Significant Results Obtained from Earth Resources Technology Satellite-1,* Vol. II, Summary of Results, NASA/Goddard Space Flight Center, Greenbelt, Md., Mar. 1973.
11. Miller, M. M., and A. E. Belon, A Multidisciplinary Survey for the Management of Alaskan Resources Utilizing ERTS Imagery, *Symposium on Significant Results Obtained from Earth Resources Technology Satellite-1,* Vol. II, Summary of Results, NASA/Goddard Space Flight Center, Greenbelt, Md., Mar. 1973.
12. Lauer, D. T., and P. F. Krumpe, Testing the Usefulness of ERTS-1 Imagery for Inventorying Wildland Resources in Northern California, *Symposium on Significant Results Obtained from Earth Resources Technology Satellite-1,* v. I, NASA/Goddard Space Flight Center, Greenbelt, Md., Mar. 1973.
13. Short, N. M., *Mineral Resources, Geological Structure and Landform Surveys,* Vol. III, Second ERTS Symposium, New Carrollton, Md., 1973.
14. DeNoyer, J., Applications of ERTS-1 Results to U.S. Department of Interior Programs, *Symposium on Significant Results Obtained from Earth Resources Technology Satellite-1,* v. II, Summary of Results NASA/Goddard Space Flight Center, Greenbelt, Md., Mar. 1973.
15. Short, N. M., and P. D. Lowman, Jr., Earth Observations from Space: Outlook for the Geological Sciences, NASA/GSFC X-650-73-316, Oct 1973.
16. Doyle, F. J., et al., *Geodesy-Cartography, Vol.*

13, Useful Applications of Earth-Oriented Satellites, prepared by Panel 13 of the Summer Study on Space Applications, NAS–NRC, for NASA, 1969.

17. Hardy, E. E., J. E. Skaley, and E. S. Phillips, Evaluation of ERTS-1 Imagery for Land Use/ Resource Inventory Information, N.Y. State College of Agriculture and Life Sciences, Third ERTS Symposium, Washington, D.C., Dec. 1973.
18. Malan, O. G., The Application of ERTS Results in the Republic of South Africa, Paper B.1.3, COSPAR, Sao Jose dos Campos. Brazil, 1974.
19. Summers, R. A., Institutional and Training Requirements for Exploiting Remote Sensing in Developing Countries, AID Symposium on the Utility of Remote Sensing as an Aid to Developing Countries, Washington, D.C., Nov. 1970.
20. Brown, G., The Importance of Remote Sensing Technology to the International Community and in Particular to the Third World, Keynote Address, Ninth International Remote Sensing Symposium, Ann Arbor, Mich., 1974.
21. Henderson, F. M., Research versus Applications, a Perspective on Remote Sensing, *Photogrammetric Eng.* v. 40, no. 8, Aug. 1974.
22. Bond, T. S., Linking End-Users and Remote Sensing Research in Developing Countries, Paper B.4.6, COSPAR, Sao Jose dos Campos, Brazil, 1974.
23. Summers, R. A., E. J. Greenblat, and D. S. Lowe, An Economic Evaluation of ERTS Data Utilization in Developing Countries, Paper B.4.5, COSPAR, Sao Jose dos Campos, Brazil; System Planning Corporation, Mathematica Inc., Environmental Research Institute of Michigan, for AID Office of Science and Technology, 1974 (abs.)

12

ANALYSIS OF GEOLOGICAL STRUCTURES BASED ON LANDSAT-1 IMAGES

Carlos E. Brockmann
Alvaro Fernandez
Raúl Ballón
Hernán Claure

Carlos E. Brockmann is Director of and Alvaro Fernandez, Raúl Ballón, and Hernán Claure are members of the Programa del Satelite Technologico de Recursos Naturales, Servicio Geologico de Bolivia (GEOBOL), La Paz, Bolivia.

Experience has shown that images from the LANDSAT satellite, particularly those from the multispectral scanner system (MSS), provide a great wealth of information about geologic structure in comparison to other areas of geology. The obvious advantages that this unique imagery has to offer, as compared to conventional products, are due specifically to the peculiar characteristics derived from its broad coverage, the altitude of acquisition, the repeat coverage, and the range within the electromagnetic spectrum, which we may summarize as follows:

1. The broad coverage of LANDSAT imagery permits the main structural elements to be recorded on a single frame, often with considerable detail, as compared to other methods, which only furnish local data. This synoptic view assists in comprehending the regional structural framework and its bearing on other factors or related features.
2. It is possible to extrapolate regionally those structures which appear to be truncated and/or which have been obscured by soils or vegetation.
3. The ability to view the topography of a broad area in a single image permits a better interpretation of the physiography and provides an indirect approach for defining structure hidden by vegetation.
4. Because the images were obtained from a great altitude, they show structural features that cannot be detected by other means or which cannot be discerned on the surface owing to their great size or because other natural features obscure them.
5. The high-altitude images show ancient faults, impossible to identify on the surface. In many cases the fault traces have been modified by subsequent geological processes and show only as lineaments.
6. The sequential character of LANDSAT imagery permits the monitoring of phenological changes and provides seasonal data on soil moisture and vegetation, which aids in making assessments as to the possible presence of fracture zones.
7. The LANDSAT MSS system records the topography in green and red wavelengths within the visible spectrum (bands 4 and 5) and in the near infrared (bands 6 and 7). This makes it possible to composite spectral band images to enhance structural features. Inasmuch as most structural fea-

tures show up as topographic alignments or tonal variations related to vegetation or moisture, bands 5 and 7 are the more appropriate for structural interpretation.

The objective of this chapter is the analysis of geological structures through interpretation of their surface expressions, and the formulation of an interpretive methodology, based on classical methods, for their use as guides in exploration for mineral deposits.

ANALYSIS OF STRUCTURAL ELEMENTS

Folds

The possibility of identifying either anticlinal or synclinal folds increases with the scale of the LANDSAT images. At a scale of 1:1,000,000 only large regional folds may be identified. At scales of 1:500,000 and 1:250,000, the maximum limit of enlargement without loss of resolution, it is not only possible to define folds, but to obtain other detailed information. Figures 1A, 1B, and 1C show a comparison of data obtained at the different scales.

Inasmuch as LANDSAT images do not provide for stereoscopic viewing outside of areas of overlap, in many cases it is difficult to determine the direction of dip. This reduces the possibility of a rapid identification and classification of folds, as compared to using other methods such as conventional aerial photography. Nevertheless, experience has shown that by using other criteria, such as topographic expression and patterns of drainage, it is possible to obtain indirect evidence with which to identify and classify folds successfully. In general, we can show that these criteria are as applicable for flat tropical areas without outcropping rock as they are for desert areas, void of soil cover and vegetation.

1. Synclines generally show lobate forms, allowing the viewer to see cuestas or basins with a larger regional development than in the case of anticlines.
2. Patterns of centripetal drainage permit determining the directions of structural slopes associated with synclinal folds.
3. Anomalous drainage, showing radial patterns on remnants of ancient erosion surfaces, aids in locating synclines by means of inversion of relief.
4. Repeated sequences of rocks of different resistance, produced by differential erosion, may be grouped together to assist in defining structural slopes and for identifying synclinal structures.
5. Repeated sequences of rocks of different chemical composition and age may furnish tonal and morphological patterns that help to define these features. However, it is most difficult to identify folds in homogeneous rocks of the same age.

Illustrations of the manner for applying these recognition criteria are shown in Figs. 2A, 2B, and 2C.

The best criterion for identifying anticlinal folds is geomorphological expression. Elliptical to elongated structures with positive topographic expressions in the majority of cases indicate anticlines. Simple deflections of stream courses may indicate the nose of a plunging anticline. Patterns of annular drainage may also identify anticlines with double plunges or structural domes. Figures 3A and 3B demonstrate these recognition criteria. These examples show the possibilities that space imagery offers in defining folds; however, the system has its limitations because of the factor of scale.

Minor folds and drag folds are impossible to identify. On the other hand, the methodical use of the listed criteria increases the information that may be obtained from the analysis, often eliminating the need for more absolute data. Figure 4 shows a structural interpretation of an LANDSAT frame. The density and quality of the data obtained can serve as a classic example of the application of LANDSAT imagery in this area of analysis.

Unconformities

By their nature, unconformities are not easily identified structural features. Work in the field or the interpretation of small-scale aerial photographs often does not permit the locating of unconformities inasmuch as they seldom display any outstanding charac-

FIG. 1 (A and B)
Structures of Incapampa–Cantar Gallo. (Scale: A, 1:1,000,000; B, 1:500,000.)

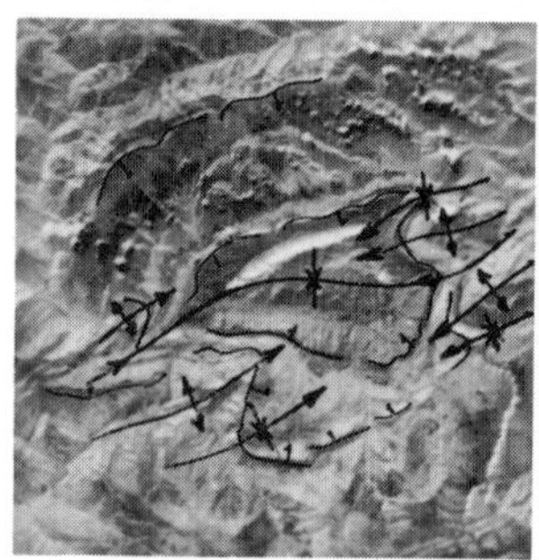

A

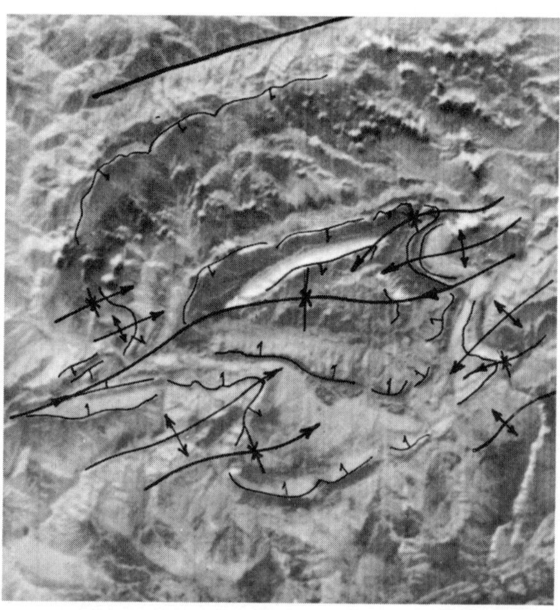

B

teristics. Nevertheless, some applicable criteria exist to make up for this.

By applying geomorphological and tonal guides, it is possible to delimit with relative accuracy lithological unconformities between volcanic rocks (lava fields and ignimbrites) and adjacent, strongly folded Paleozoic units (Fig. 5). Lithological, structural, and tonal guides permit defining different trends in the orientation of groups of rocks with adequate precision for defining the contours of the unconformity (Fig. 6). These criteria should be used rather cautiously as the success of the application depends largely upon the experience of the interpreters.

Lineaments, Fractures, and Faults

A lineament is any line that appears to be controlled by structural factors on aerial photographs or space imagery. In this category one includes all structural alignments, topographic alignments, vegetational linears, shapes, and lithological boundaries on contacts between physiographic units that appear to reflect subjacent geology, in other words, those which may be surface expressions of buried structure.

Generally, lineaments are seen as straight lines; however, they may also be expressed as curvilinear lines. On the western side of Fig. 7 a straight lineament is seen trending in a north–south direction, plus two others trending in an east–west direction, which cut the image by a length of some 80 to 100 km. As one can see, the trace is quite obvious and easy to identify, standing out sharply along its entire length.

Figure 8 clearly shows a curved lineament, which is the surface expression of a large fissure, across which may be seen projections of ignimbrite deposited by *nuees ardentes.*

These lineaments clearly demonstrate the quality of registration of LANDSAT images and the possibilities they offer for structural analysis. However, for a correct interpretation, it is necessary to propose a genetic classification that will permit defining the geological significance.

Taking into account the manner in which the lineaments are displayed on the LANDSAT images, the spatial relationships, frequency, magnitude, continuity, and relationship to other regional geologic features, the lineaments are divided into five categories: (1) lineaments along lithostructural contacts,

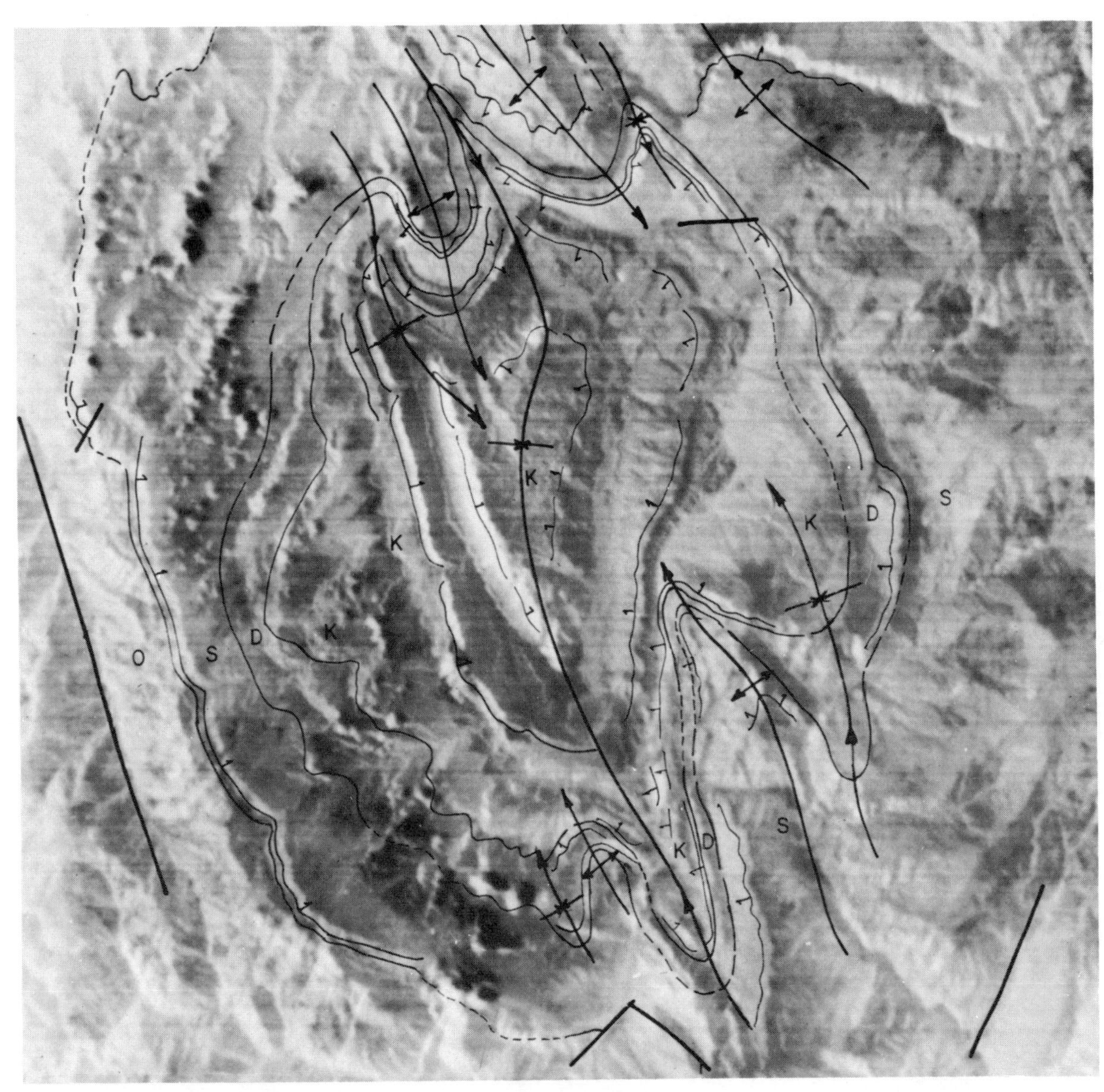

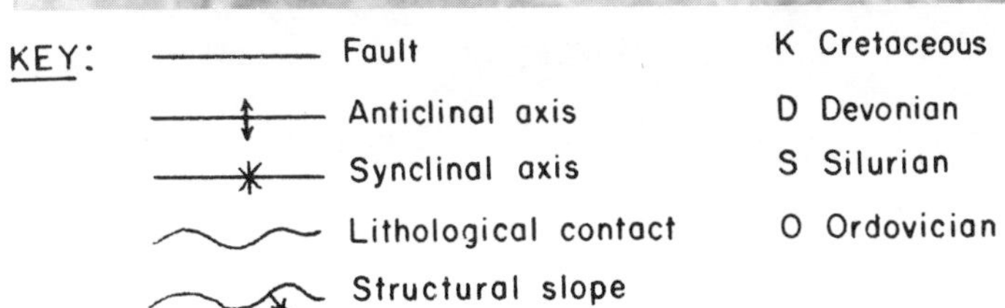

FIG. 1(C)
Structures of Incapampa–Cantar Gallo. (Scale: 1:250,000.)

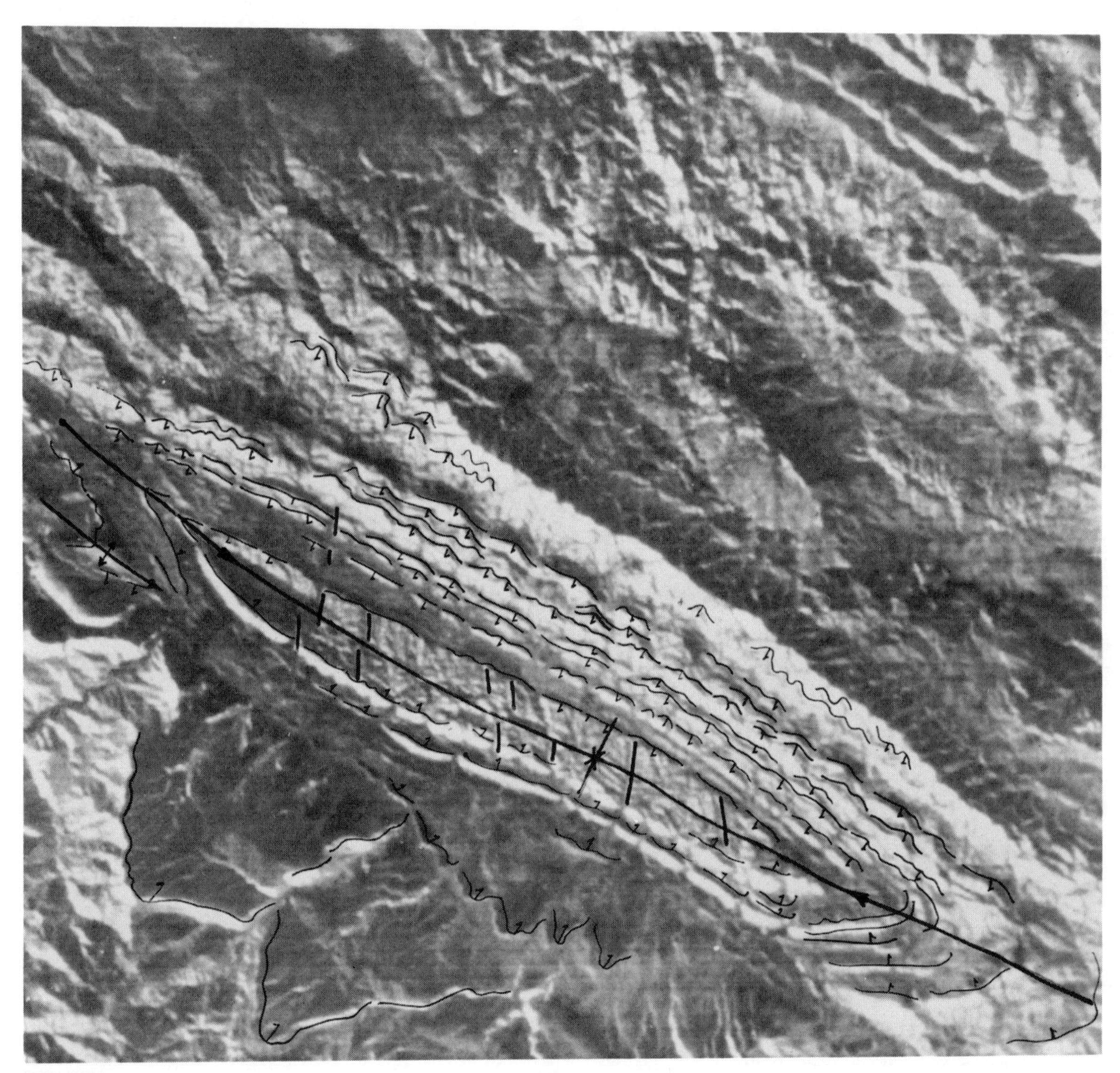

FIG. 2(A)
Rio Caine syncline. (Scale: 1:250,000.)

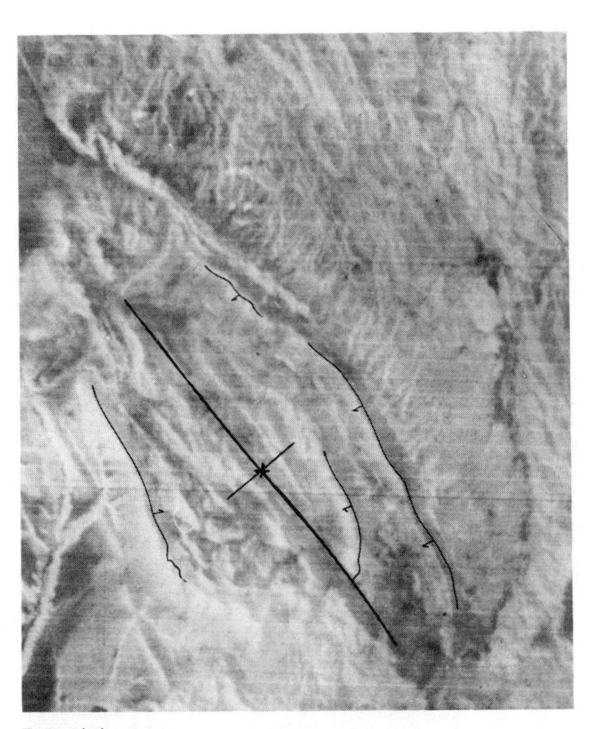

FIG. 2(B)
San Lucas syncline. (Scale: 1:250,000.)

FIG. 2(C)
Maragua syncline. (Scale: 1:250,000.)

(2) topographic lineaments, (3) fracture zone lineaments, (4) lineaments along fault traces, and (5) lineaments related to major faults or "geofaults."

Lineaments along lithostructural contacts. The best criteria for the recognition of these lineaments are the changes in drainage patterns on either side of the linear. Generally, one may observe a correlation of this type of lineament with contacts between different regional stratigraphic units. Sometimes they may be related to regional high-angle faults. In Fig. 9 such a lineament may be seen in the central sector of the image corresponding to a line between Paleozoic rocks and those of the Altiplano-fossa (Quarternary). In the south-southeast sector of the same image another lineament can be seen (not a straight line, but somewhat irregular) coincident with the contact between volcanic rocks and Paleozoic sediments.

Topographic lineaments. This cateogory includes those lineaments which reflect structural control of the lineal arrangement of surface features. These may be expressed as alignments of lakes, volcanic cones, alluvial fans, talus cones, intrusive bodies, and peneplain surfaces. Most are related to faults or zones of weakness.

In the western sector of Fig. 9 may be seen a lineament along which lakes are aligned toward the south. In Fig. 10 one may identify an east–west trending lineament along which are six volcanic cones. A typical example of an alignment of alluvial fans is seen in Fig. 11, where they are arranged along a curvilinear lineament. Figure 12 illustrates the emplacement of igneous bodies along a north–south trending regional lineament.

Fracture zone lineaments. These are characterized by linear outlines connecting zones of weakness or fracture zones, principally generated by diaclase; that is, the lines of weakness or fractures in the surface rocks are roughly perpendicular to one another. Because of the similarity of these linears to those related to faults, it is often difficult to tell them apart on LANDSAT images. Nevertheless, distinct criteria exist, these being lesser magnitude, greater discontinuity, greater density of areal distribution, and arrangements in several modes of orientation.

In Fig. 13 one may notice a series of geometrically oriented lakes whose borders are shaped and controlled by fractures in the substratum. Figure 14 shows evidence of movement in three different directions along lineaments in sedimentary rock.

Lineaments along fault traces. The criteria for identifying this category of lineaments include coincidence with zones of block displacements, abrupt changes in the orientation of folds along a line, truncation of structures, a noticeable reduction of stratified units, and anomalous curving in the general orientation of folds. The first characteristic can be seen in Fig. 14, where displacement of the northern block is evident in relation to the southern block of a synclinal fold. Using the criteria of tone, an appreciable reduction in the thickness of the sediments in the central sector of the same image may be observed until it essentially disappears along a conspicuous lineament. Changes in the general trend of folds may be observed in Fig. 15. The structures to the west of the lineament run in a general northwest–southeast direction; those seen to the east trend nearly north–south.

Lineaments related to geofaults. These lineaments have characteristics similar to those along fault traces, but are of greater geographic continuity. This category includes lineaments greater than 100 km in length, which generally correspond to geocrustal fractures. A lineament of this type over 200 km long is clearly shown in Fig. 11.

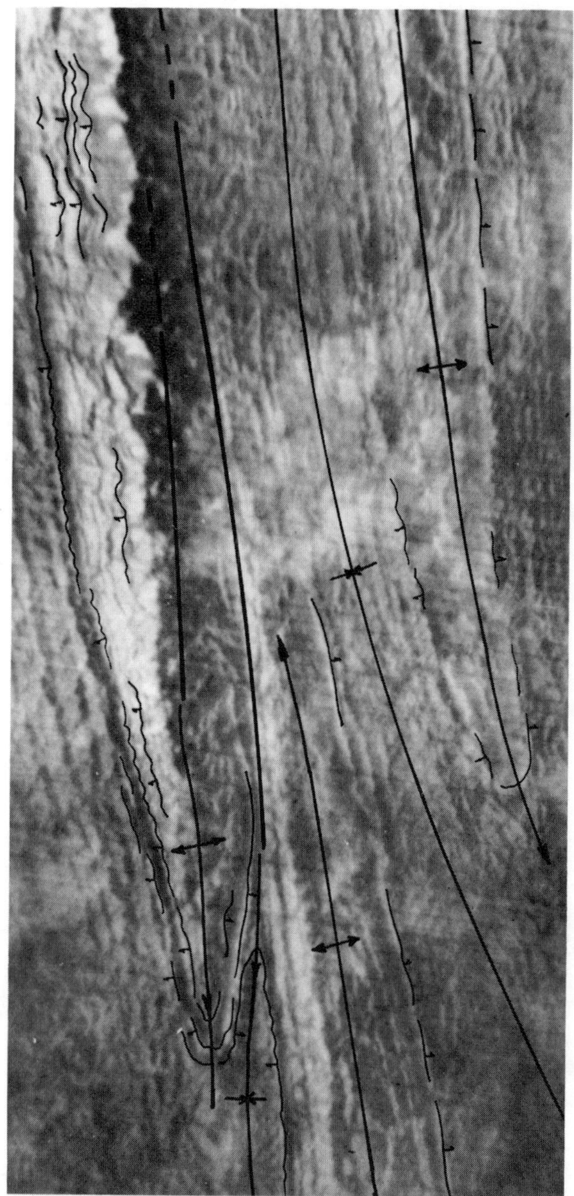

FIG. 3(A)
Anticlines of sub-Andean zone. (Scale: 1:250,000.)

APPLICATIONS OF STRUCTURAL DATA IN MINERAL EXPLORATION

In recent years there has been much speculation as to the possible relationship between mineral deposits and the circulation of magmatic mineralizing fluids along geocrustal fractures. The correlation of mineral deposits with structural lineaments and major fault systems has been verified. Also, emphasis has been placed on the role played specifically by the intersection of regional lineaments in the emplacement of mineral deposits. The role of intersecting north–south and east–west lineaments has been adequately presented in Brockmann and Fernandez (1973).

The N–S lineaments are generally well known Andean tectonic features in Bolivia; on the other hand, the E–W linea-

FIG. 3(B)
Elipsoidal and double plunging anticlines. (Scale: 1:250,000.) Northwest sector of the image of Tupiza-Image LANDSAT 1008-13531.

ments are not mentioned. The fact that the latter do not cut the Hercynian folding indicates that they are older than this orogeny. If we consider that the mineralization of the Bolivian Cordillera starts in the Jurassic, it becomes evident that the mineralizing fluids availed themselves of available fractures as avenues for upward movement, and that these fluids were checked in their movement by N–S faults of the Andean orogeny of Tertiary age. It then becomes obvious that those areas which offer the best possibilities for finding mineral deposits are those where the two fracture systems intersect.

It is also evident that fold structures control mineralization. According to Herness, there is a frequent occurrence of mineralization in anticlinal arches and similar tectonic zones in the Bolivian Andes. Although Herness's concept is not universally accepted, it is a known fact that anticlinal folds are seen to be accumulators of mineralization in the majority of the mineral deposits of the Antiplano and Andean Cordillera.

The concepts of lithologic control and of selective mineralization of host rocks are universally recognized as guides in the search for mineral deposits. Rocks that have been bleached by hydrothermal alteration are seen on space imagery as linear or circular tonal anomalies. Topographic guides such as anomalies in drainage may identify subjacent igneous bodies, which are important to the selection of areas for exploration. In summary, the usefulness of space imagery in the search for areas favorable for mineralization depends upon the application of structural, lithological, and physiographic guides.

With the purpose of applying these criteria to verify that the proposed correlations are valid in the Altiplano and Eastern Cordillera of Boliva, let us consider the area situated between the geographic coordinates of 16° to 20°S and 66° to 72°W, using as a base the 1:1,000,000 scale photomap compiled from LANDSAT imagery by the EROS program of the U.S. Geological Survey (Fig. 16).

Figure 17, showing the location of lineaments, geomorphologic anomalies, structure, tonal anomalies and the distribution of known mineral deposits has

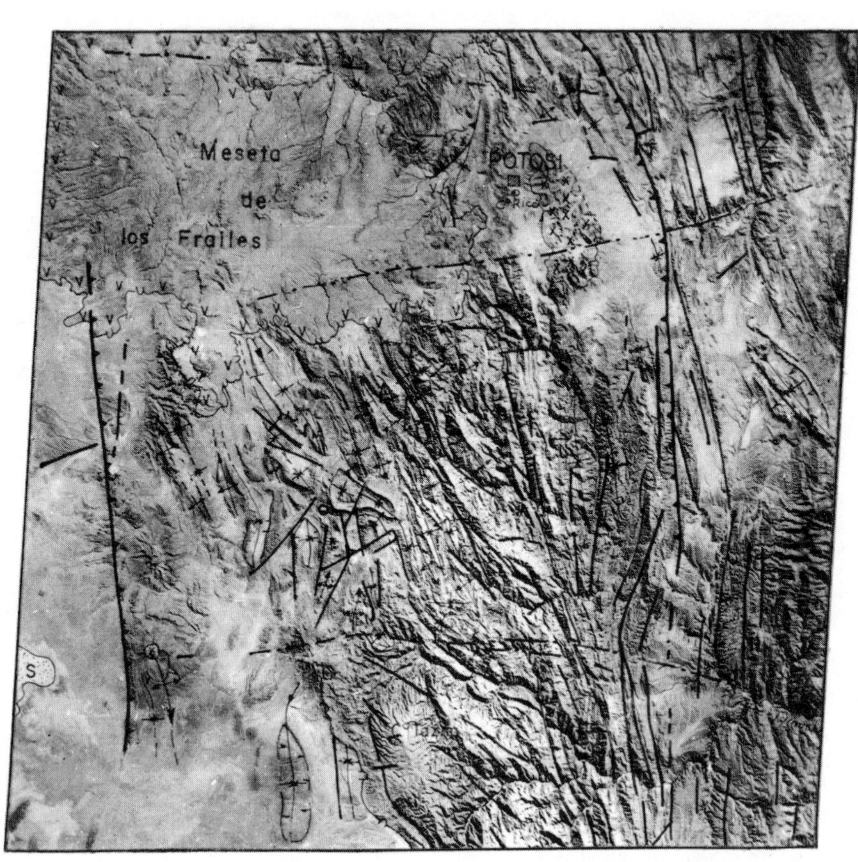
Meseta
de
los Frailes
POTOSI
S

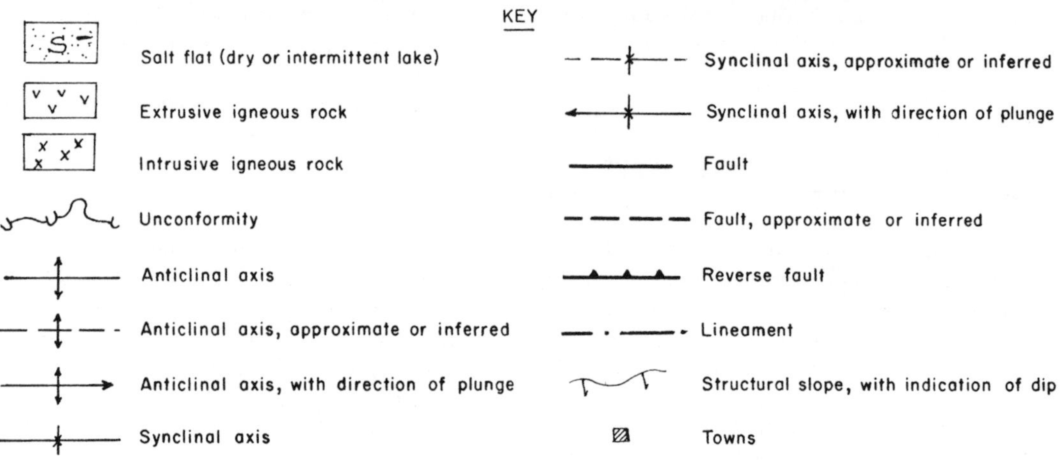
KEY
S
Salt flat (dry or intermittent lake)
Extrusive igneous rock
Intrusive igneous rock
Unconformity
Anticlinal axis
Anticlinal axis, approximate or inferred
Anticlinal axis, with direction of plunge
Synclinal axis
Synclinal axis, approximate or inferred
Synclinal axis, with direction of plunge
Fault
Fault, approximate or inferred
Reverse fault
Lineament
Structural slope, with indication of dip
Towns

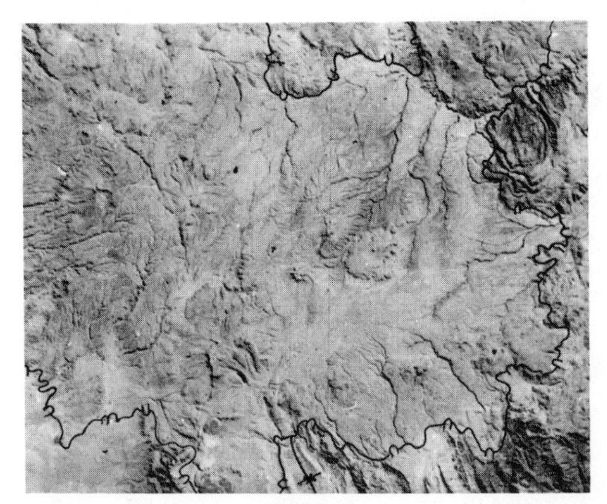

FIG. 5
Unconformity between volcanic and paleozoic rocks.

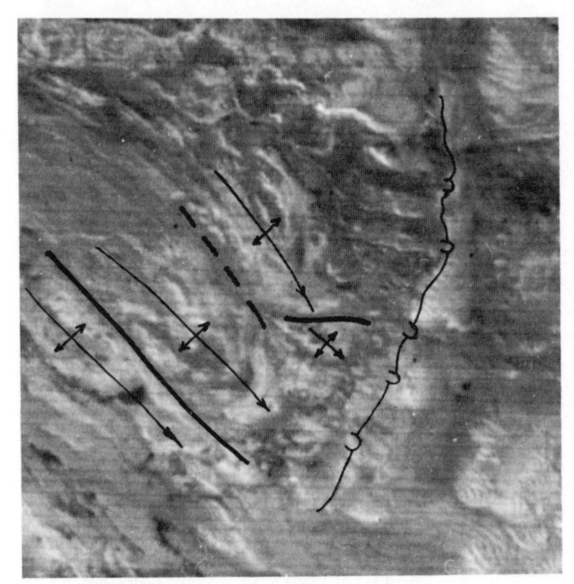

FIG. 6
Unconformity in sedimentary rocks.

FIG. 4 *(left)*
Structural interpretation of LANDSAT image 1008-13524 (Potosi).

been compiled at the same scale as the photomap. Study of this map permits the identification of the following relationships:

1. Correlation between occurrences of mineral deposits and major parallel or subparallel lineaments and the general trend of fold structures. In area A of Fig. 17 one may see an alignment of copper deposits along a large northwest–southeast lineament. A similar relationship may be seen along a lineament in area B, except in this case the deposits are tin and wolfram. Also, in area B it is interesting to note the major concentration of tin and wolfram deposits at areas where differently trending linear systems intersect. A high concentration of such occurrences may be seen in area B. Area C has characteristics similar to area B. Mineral deposits can be seen to correlate with igneous bodies and lineaments.

2. Correspondence between tonal anomalies, probably related to hydrothermal alteration, lineaments, and mineral deposits. In area D these relationships may be seen very clearly. In area E similar characteristics are associated with a strip of known antimony mineralization. A similar relationship may be seen in area F. In this area, however, the mineralization is silver.

3. Areas G and H show correlations of all four previously noted characteristics. Area H provides an exceptionally good example of the correlation of lineaments, intrusive bodies, tonal anomalies, and occurrences of thermal springs with deposits of tin, silver, bismuth, and gold.

To prove the validity of these guides as indicators of patterns of mineralization, and with the objective of locating areas where similar characteristics might justify mounting an exploration, a test area was chosen for study and for the application of the described methodology on a larger scale. For this purpose LANDSAT image 1008-13524 (band 7) was selected (scale 1:1,000,000), the area of Potosi-Kari Kari, Porco-Canutillos-Pulacayo.

Following the proposed method, maps were compiled at the same scale showing lineaments (Fig. 18), known mineral deposits (Fig. 19), geomorphology and tonal anomalies, igneous bodies, and the location of thermal springs (Fig. 20). Similarly, aerogeophys-

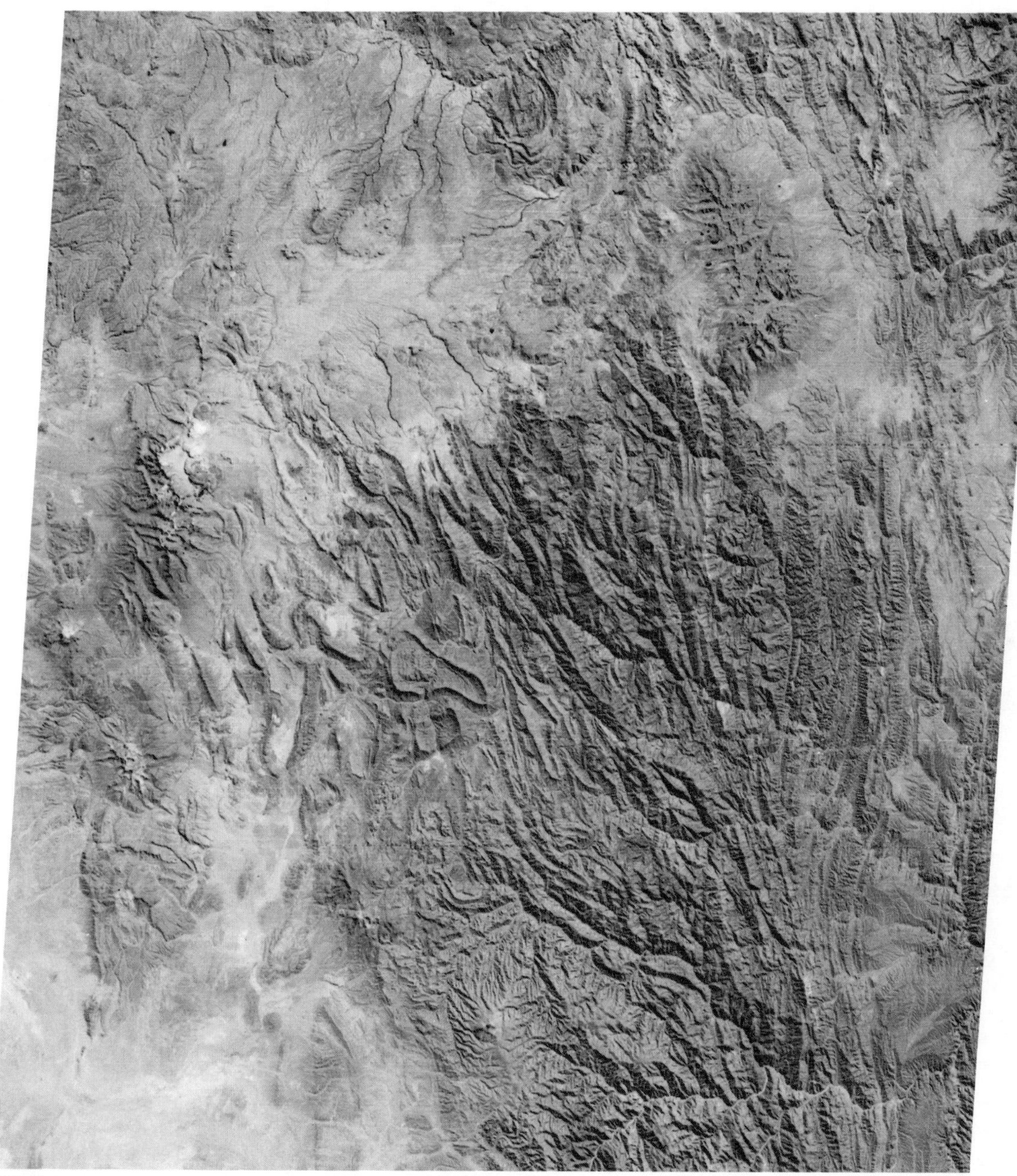

FIG. 7
Straight lineaments trending N–S and E–W with lengths of about 100 km (LANDSAT-1 1008-13524, band 5, Potosi-Ubina, Bolivia).

FIG. 8
Curved lineament related to a large fissure across which can be seen deposits of ignimbrite (LANDSAT 1243-14001, band 5, Laguna Colorada, Alota, Bolivia).

FIG. 9

In the larger eastern sector there can be seen a topographic lineament that corresponds to the line between the Paleozoic rocks and Quaternary sediments associated with a reverse fault. To the S–SE is a lineament of equal proportions that follows the contact between volcanic rocks and Paleozoic sediments. In the central zone is an alignment of seven elongate lakes (LANDSAT 1243-13592, band 7, Lago Poopó, Bolivia).

FIG. 10
Topographic lineament showing an alignment of volcanic cones (LANDSAT 1010-14035, band 5, Salar de Coipasa, Bolivia).

FIG. 11
In the eastern sector of the image there is a curvilinear lineament corresponding with a reverse fault. Cones and alluvial fans are aligned along the structure. Elsewhere, two prominent lineaments are related to two major faults (geofaults) (LANDSAT 1153-13583, band 5, Cochabamba, Bolivia).

FIG. 12
Alignment of igneous bodies along a zone of weakness seen as a lineament. In the eastern section two lineaments may be seen that correlate with tectonic zones known from work in the field (LANDSAT 1008-13531, band 5, San Vincente–San Antonio de Lípez, Bolivia).

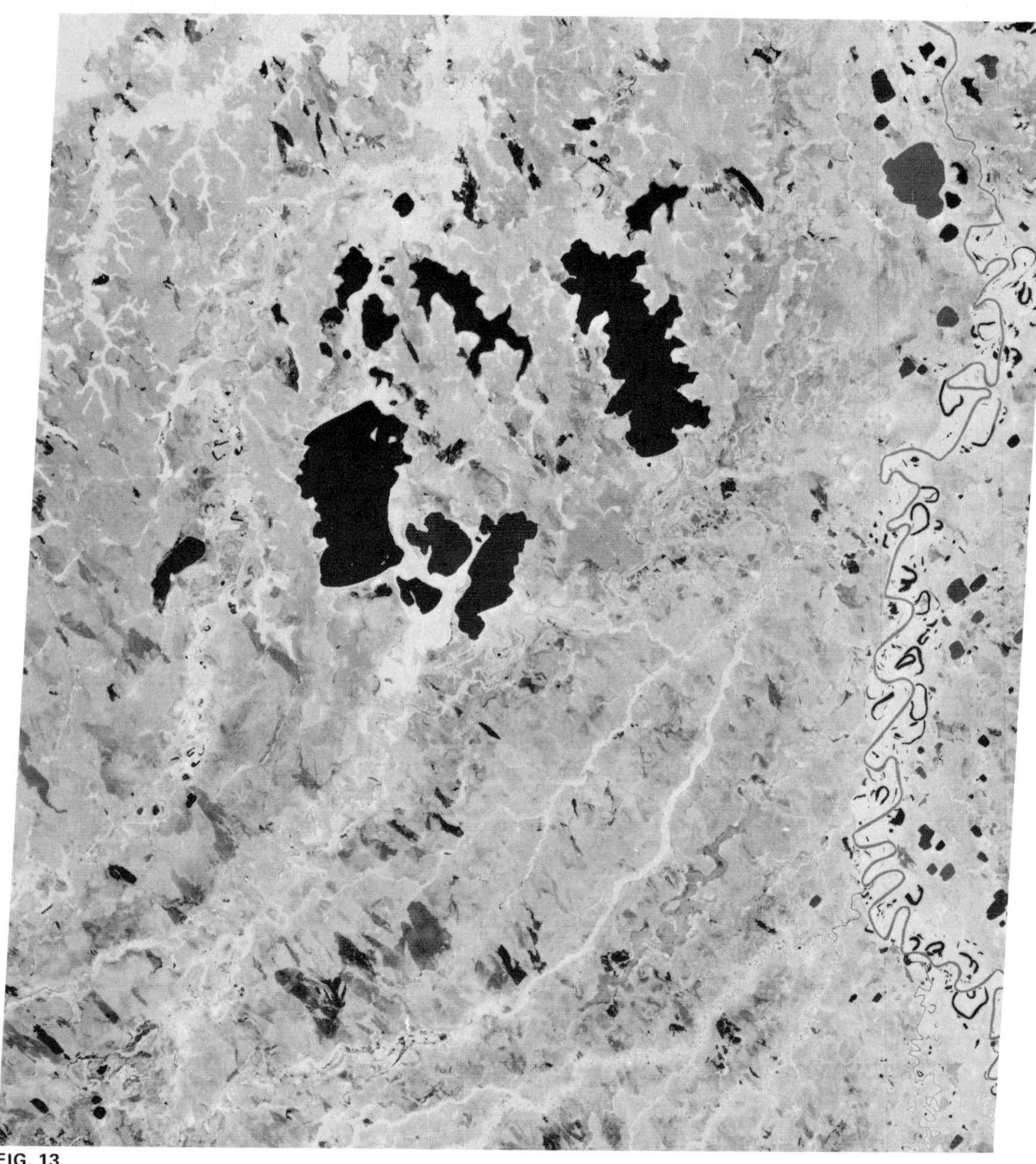

FIG. 13
Lineaments of limited regional development, great density of spatial distribution, and various trends of orientation, corresponding to fractures that control the geometrical shapes of lakes along the border of the Brazilian shield (LANDSAT 1045-13563, band 7, Lago Rogaguado–Rio Mamoré, Bolivia).

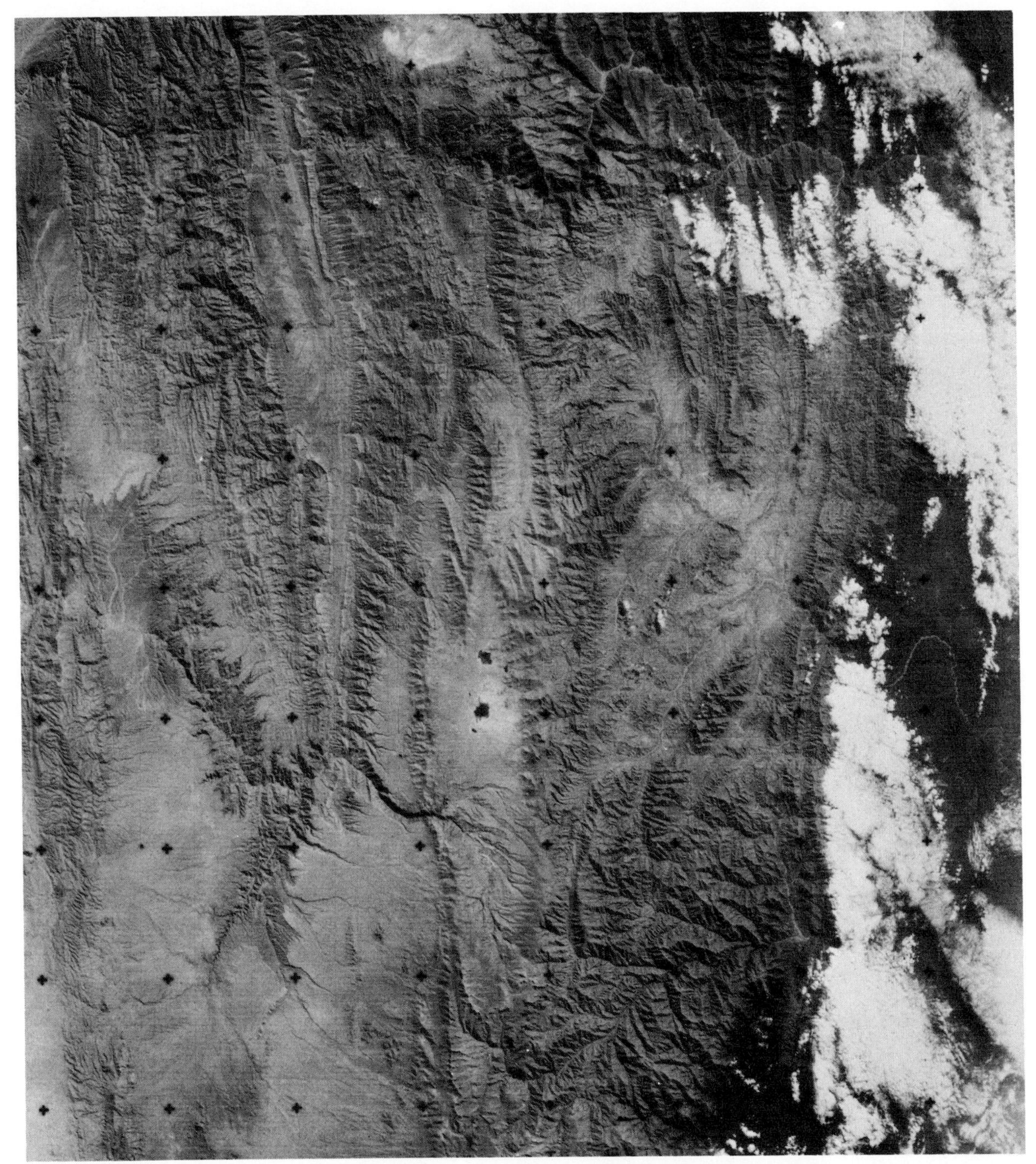

FIG. 14

In this image there are various lineaments associated with fracture and fault zones. In the south-central region are small lineaments associated with diaclase. In the same area a SE trending lineament can be seen along which are several displacements. In the extreme west is a lineament corresponding to a reverse fault, which cuts through the thick terrain of tertiary sediments (LANDSAT 1007-13472, band 2, Tarija-Villazón, Bolivia).

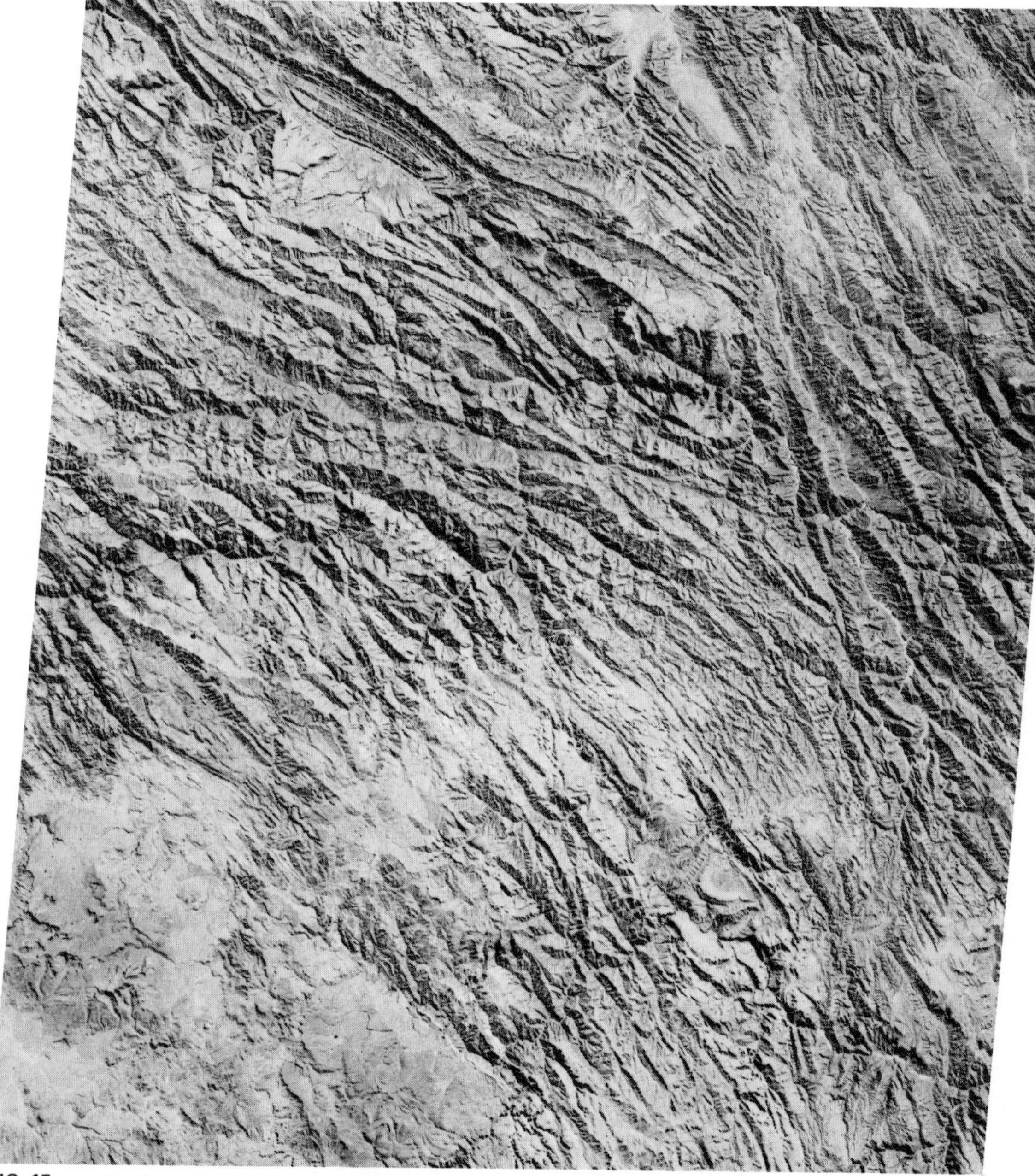

FIG. 15

Regional lineament associated with a large fault. Note the NW–SE trend of structure to the west of the lineament and the nearly N–S trend of structure to the east (LANDSAT 1008-13522, band 7, Torotoro-Sucre, Bolivia).

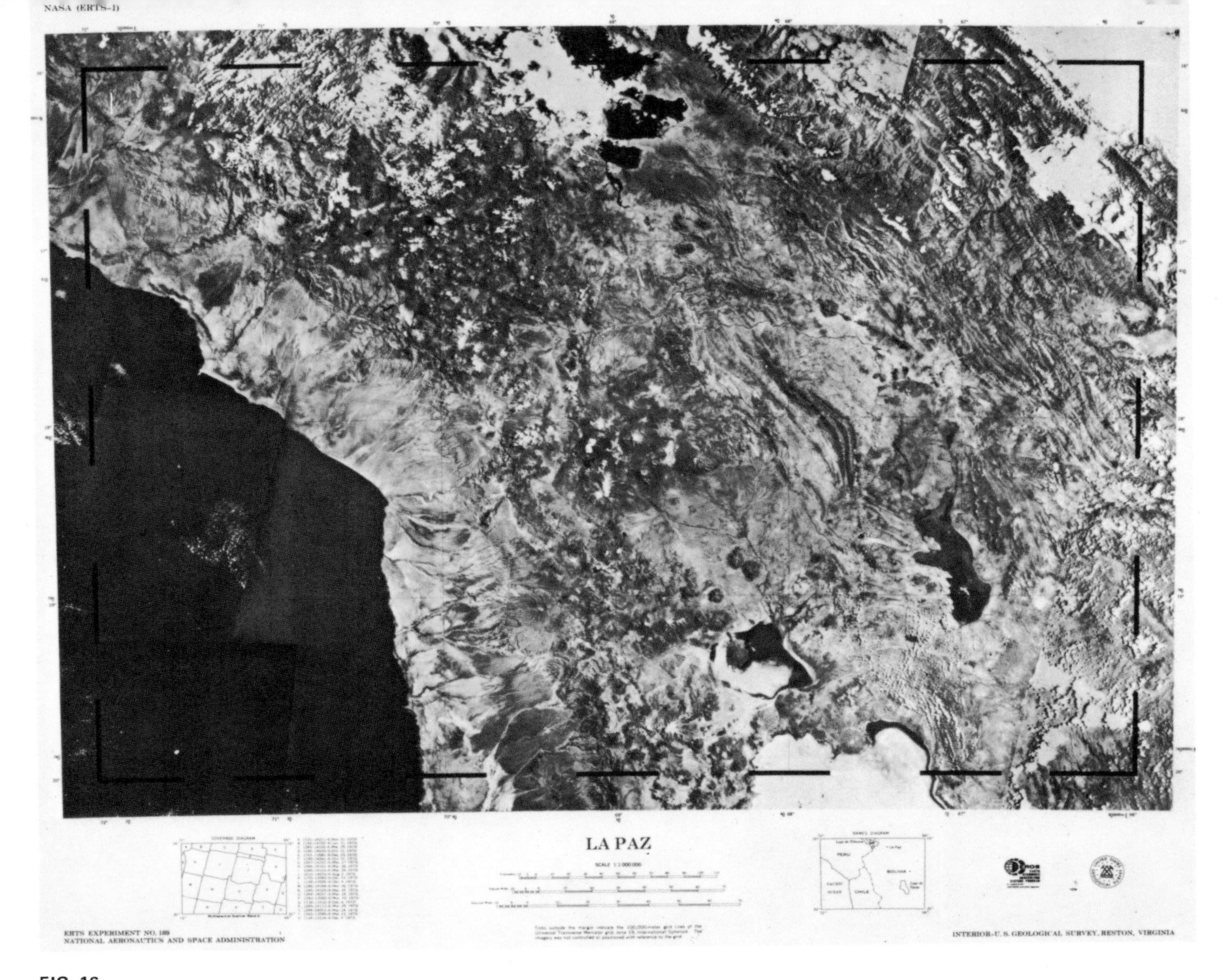

FIG. 16
Photomap of Bolivia, originally compiled from LANDSAT images at 1:1,000,000 scale by the EROS Program of the U.S. Geological Survey.

FIG. 17
Map of lineaments and anomalies interpreted from LANDSAT imagery.

FIG. 18
Lineament map. (Scale: 1:250,000.)

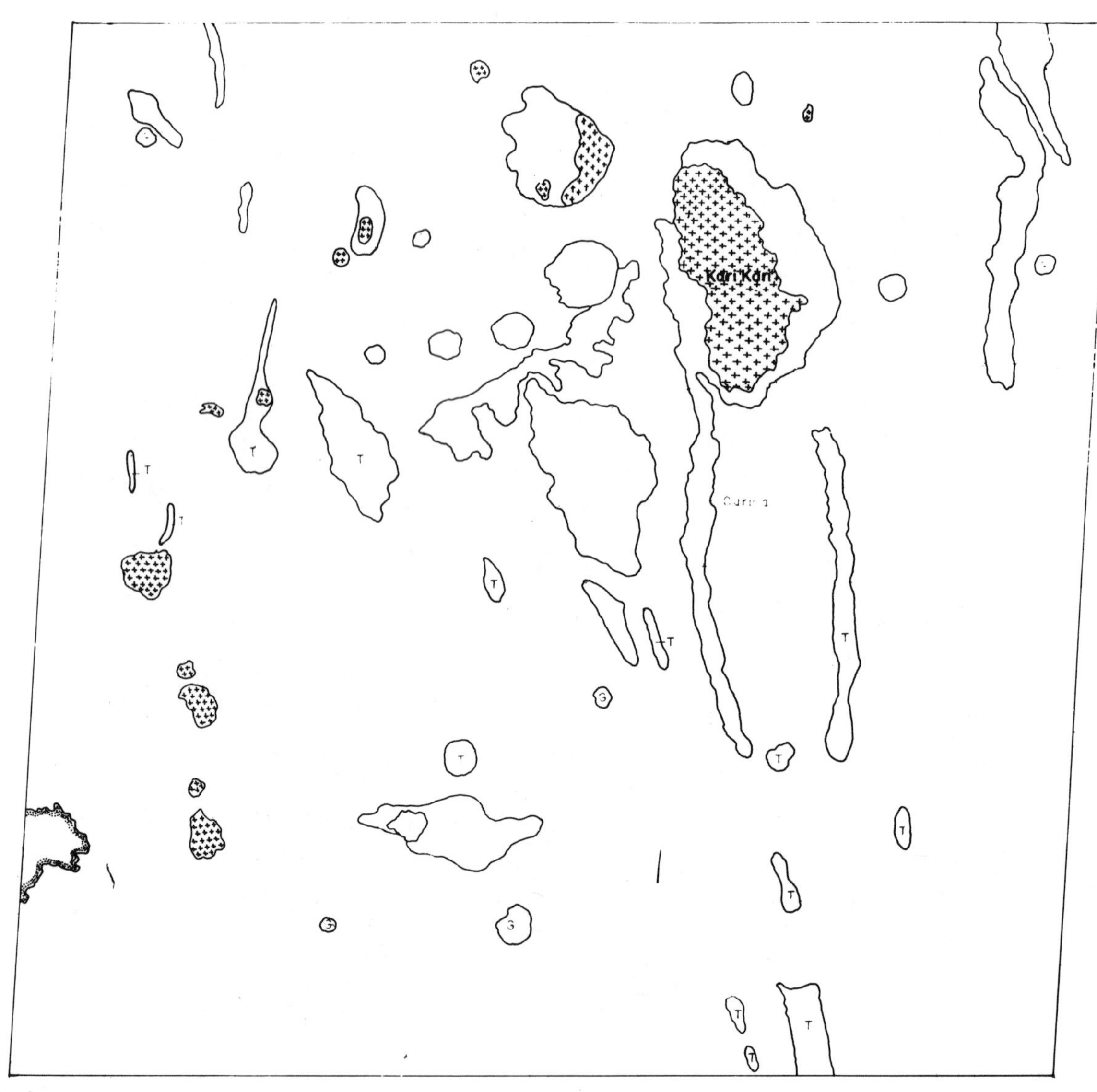

FIG. 19
Map of mineral deposits. (Scale: 1:250,000.)

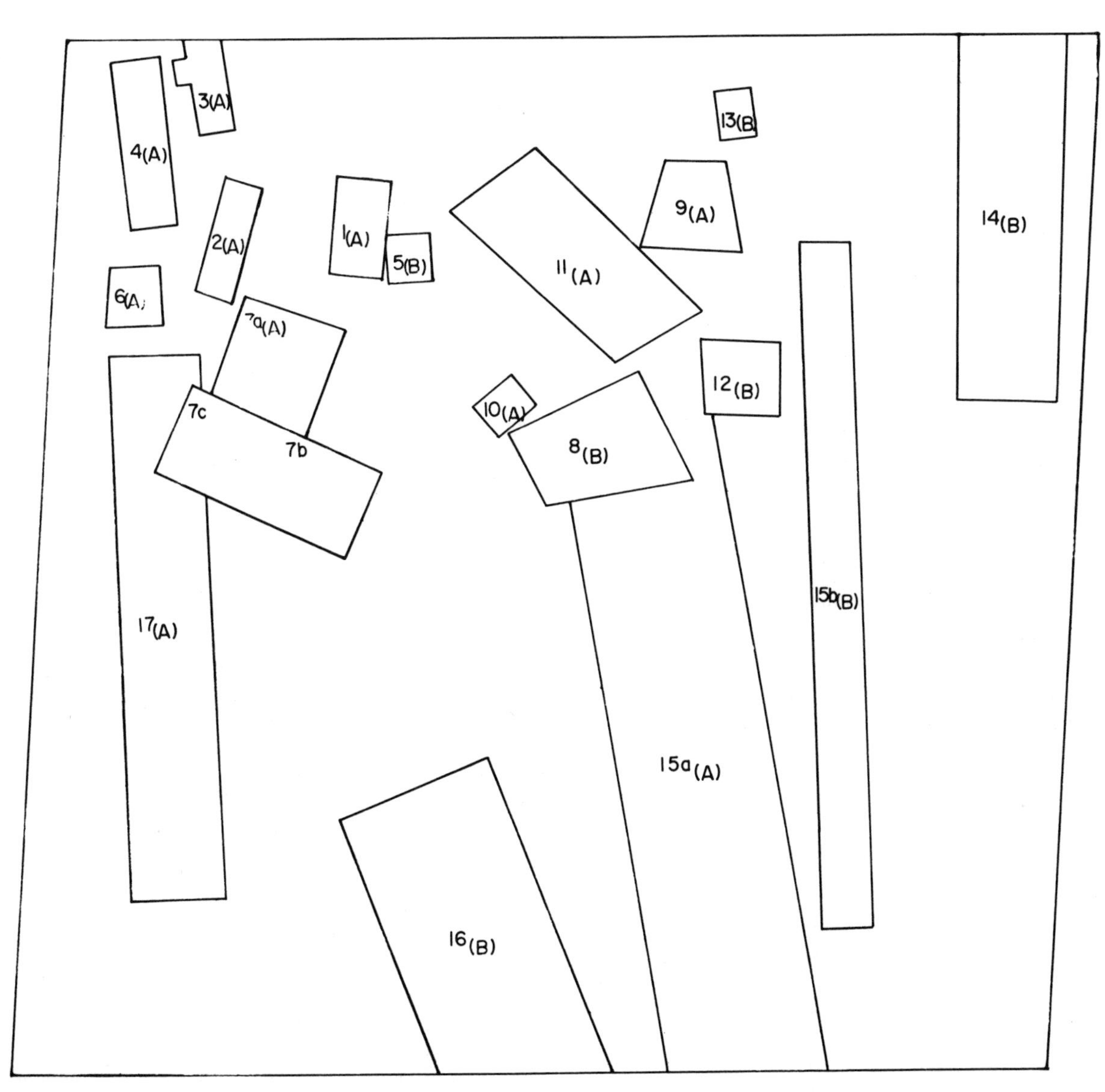

FIG. 20
Map of anomalies. (Scale: 1:250,000.)

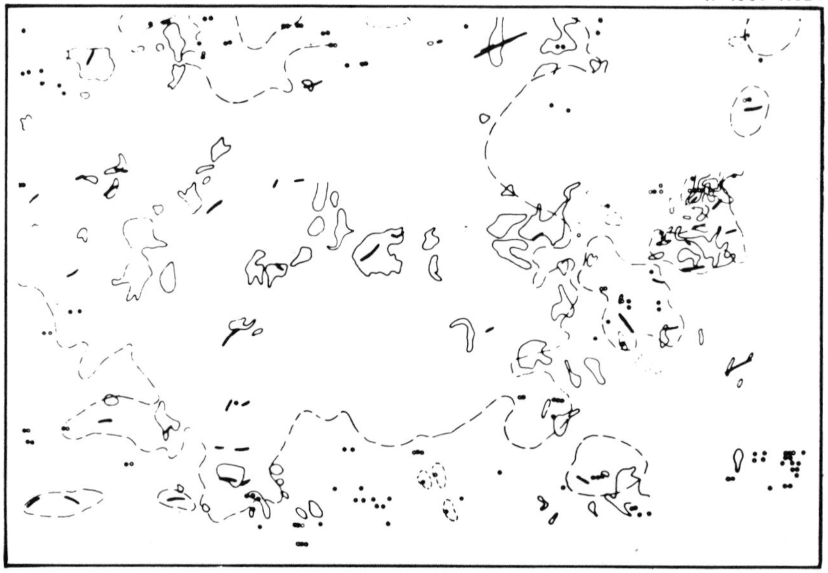

POTOSI N° 1008 13524

FIG. 21
Map of geophysical anomalies. *Solid dots,* location of electromagnetic anomaly of first order; *open dots,* location of electromagnetic anomaly of second order; *dashed lines,* boundary of magnetic anomalies area; *short, heavy lines,* location of cause of magnetic anomaly; *solid lines,* boundary of radioactive areas. (Scale: 1:250,000.)

ical (magnetometry) and geochemical data were compiled at a scale of 1:250,000 as additional input of value to the objectives of the study (Fig. 21).

Detailed study of the parameters permitted the preparation of a map indicating likely areas of economic interest. Fifteen of the areas were selected for follow-on exploration by conventional methods.*

CONCLUSIONS

It is possible to draw the following conclusions from the above study:

1. LANDSAT imagery and other space imagery make it possible to locate, delineate, and classify geologic structures by a method that has advantages over conventional techniques.
2. Such data permit us to better and more clearly define the tectonic framework of an area and its control over the emplacement of mineral deposits.
3. The data permit relating structural controls to the patterns and concentrations of mineralization.
4. Based on the relationships of lineaments to other guides to mineralization (thermal springs, regions of hydrothermal alteration, geophysical data, the mode of occurrence of igneous bodies), it is possible to define those parameters which control mineralization in a determined area.
5. With such data one may compile maps to indicate target areas of different priorities or likelihood of mineral discovery.

It must be emphasized, however, that the application of such a methodology differs from the conventional exploration route by orders of magnitude of scale. Whereas ground prospecting is highly detailed work, the interpretation of space imagery is on a regional scale. However, only at the latter scale is it possible to locate and delineate areas favorable for further exploration.

In summary, one should realize that the use of space imagery as a tool in the search for mineral deposits provides such obvious advantages that it

*Fieldwork subsequent to this writing proved the presence of mineralization in several of the selected areas.

would be inconceivable at the present time to consider an exploration program without it.

REFERENCES

Ballón, Raúl, Interpretación geólogica de imágenes ERTS, región occidental de Bolivia, internal report of the LANDSAT program.

Brockmann, Carlos, ERTS-B, Sensor Data for Mineral Resource Sector Development and Regional Land Use Survey, internal report of the LANDSAT program.

_____, and Alvaro Fernandez, Aplicación de imágenes ERTS, a la cartografía y geología, *Proc. 1st Panam. Symp. Remote Sensing,* 1973, p. 40–52.

_____, et al., Informe técnico del proyecto cartográfico temático del Dpto. de Potosi, internal report of the LANDSAT program.

Claure, Hernán, Evaluación de imágenes ERTS compuesta a color, para su aplicación a problema minero, internal report of the LANDSAT program.

_____, Imágenes ERTS en investigaciones para exploración minera, internal report of the LANDSAT program.

Fernandez, Alvaro, and Carlos Brockmann, Aplicación y evaluación de imágenes ERTS de composición a color al inventario de recursos naturales, *Rev. Tec. YPFB,* vol. 3, no. 1, Apr. 1974.

_____, et al., Aplicación de imagenes ERTS, al mapeo geológico, *Soc. Geol. Boliviana Bol. 20,* 1973, p. 23–29.

_____, et al., Lineamientos en imágenes ERTS, su significado geológico y su aplicación en la exploración de minerales, *Rev. Tec. YPFB,* vol. 3, no. 3, Dec. 1974.

Herness, Kermit, *Zonificación mineral en los Andes Bolivianos,* Vol. 2, Bolivian Institute of Petroleum, 1961.

Kusmaul, Sigfried, Interpretación vulcanológica de la parte Septentrional de la Cordillera Occidental de Bolivia, utilizando imágenes ERTS, internal report of the LANDSAT program.

Martinez, Claude, Geología estructural del altiplano Norte, unpublished.

_____, and Pierre Tomasi, Historia estructural del Alplano de Bolivia, unpublished.

Tomasi, Pierre, and Raúl Ballón, La Microtectónica, un método de verificación en el campo de lineamientos mayores observados en una imagen del satélite ERTS-1 (Bolivia: 17°20′–18°00′ latitud sud), *Rev. Tec. YPFB,* vol. 4, no. 3, Dec. 1975.

Vargas, Carlos, Result of Geomorphological–Geological Study of Two ERTS Images of the zones San Borja–Mamoré River, Rogaguado Lake East, Bolivia, internal report of the LANDSAT program.

_____, Interpretación de drenaje de imágenes ERTS del área Cobija-Puerto Heath-Rio Madre de Dios-Ixiamas, internal report of the LANDSAT program.

13

THE GEOLOGICAL APPLICATION OF LANDSAT IMAGERY IN BRAZIL

Aderbal C. Corrêa
Fernando de Mendonça
Chan C. Liu

Aderbal C. Corrêa, Fernando de Mendonca, and Chan C. Liu are members of the Instituto de Pesquisas Espaciais, Brazil.

Brazil is a country with more than 8.5 million km^2, which is only partly known with respect to its natural resources potential. Regional evaluation studies of such a large area can be carried out much more efficiently and in less time with the application of remote-sensing mapping techniques. Data obtained from spacecrafts especially designed to acquire information for studies of earth's natural resources are particularly useful because of their ability to provide a synoptic view, which is necessary in regional mapping programs.

The satellite LANDSAT-1 (previously known as ERTS-1) has provided multispectral imagery of Brazil that has been available since 1972. This imagery has allowed thematic mapping, even in areas where more conventional mapping material (i.e., aerial photography) is difficult to obtain because of prevailing poor weather conditions. The Amazon Region in northern Brazil is a typical example of one of these problematic areas; for most of the year the weather does not allow photogrammetric flight (Girardi, 1973). The periodic coverage provided by LANDSAT-1, which systematically obtains imagery over the same area, is a relatively inexpensive way to obtain data over such regions.

METHODOLOGY FOR INTERPRETATION

The basic data used in geological interpretation were LANDSAT's multispectral scanner (MSS) imagery, especially bands 5 (0.6 to 0.7 μm) and 7 (0.8 to 1.1 μm), and color composite images. Conventional photointerpretation techniques were employed in the analysis without the benefit of stereoscopic viewing, except for the area of side lap, which is approximately 15 percent of the total area.

The results obtained from the study of LANDSAT imagery therefore are highly dependent on the geological knowledge of the photointerpreter and his experience in extracting useful information from the images. The photointerpreter is required to take into consideration the influence of topography, color (tone in the imagery), distribution of natural vegetation and cultivated land, sun illumination (sun azimuth and elevation above the horizon), and drainage patterns when extracting geological information from the satellite imagery. The relationships between these factors and maximum use of space images has been discussed by Lee et al. (1974) and was carefully considered during analysis of LANDSAT imagery.

CASE STUDIES

Three areas in Brazil were selected to serve as examples of the application of LANDSAT imagery in geological studies. These areas (Fig. 1) are located in different geological settings, have been studied already with different degrees of detail, are covered by different types of vegetation, and are within different climatic zones.

As a result of geological mapping in Brazil in recent years, valuable information has been available about previously poorly known regions, mainly the central and northern parts of the country. Based on this geological knowledge and radiometric dating available today, it is possible to outline the tectonic framework of Brazil which will serve as a reference to parts of the discussion in the following sections (Fig. 2).

Brazil and the entire Atlantic coast of South America are part of a geotectonic unit known as the South American Platform (Almeida, 1971). Precambrian rocks, locally overlain by a thin sedimentary cover, are exposed in the Guyana, Central Brazilian, and Atlantic shield areas (Fig. 2). The Guyana and Guaporé cratons (which include the Central Brazilian shield area) may have been a single stable block during late Precambrian time (Amaral, 1970). The large Amazon, Parnaiba, and Paraná sedimentary basins are located also between the shield areas mentioned above (Fig. 1).

The three areas selected as case studies for presentation in this paper are the following:

1. São Domingos Range area.
2. Poços de Caldas area.
3. Area of the Middle Araguaia and Tocantins rivers.

São Domingos Range Area

The São Domingos Range area (Fig. 1) includes a representative part of the Brasilia fold belt and of the sedimentary cover of the São Francisco craton, besides a few outcrops of crystalline basement rocks. The different types of sedimentary rocks present in this area and the sharp contrast between two tectonic domains are some of the interesting observations that can be made in LANDSAT images over this area (Fig. 3).

This area is characterized by a tropical climate with a dry winter (rainy summer) corresponding to Köppen's Aw type (Azevedo, 1964). The vegetation predominant in the area is a cerrado (savanna-like vegetation) with a small occurrence of tropical deciduous forest along the northwestern side of the area. The images in Fig. 3 were obtained in August 1973 during the dry season (winter) and allow a very good discrimination of lithologic types in the area. Imagery taken during the rainy season (not shown here) gives essentially the same information for most of the area as observed in dry-season imagery (Fig. 3). Little geological information is available, however, in the rainy-season imagery from the area covered by a tropical deciduous forest which is well developed at that time of the year. A stratigraphic column of the São Domingos Range as based on previous geological mapping is given in Table 1.

A careful analysis of drainage patterns, landforms, tone, vegetation, and soil-use patterns in the area led to the identification of "remote-sensing rock units." A tentative correlation of remote-sensing rock units with stratigraphic units is presented in the legend of Fig. 4, but it is emphasized that in some cases there is not a unique or complete correlation between these two types of mapping units.

A comparison between the geological interpretation (Fig. 4) and maps available at a larger scale for this area (DNPM, 1965–1968a, b; 1968a, b) shows that stratigraphic units defined by a single or few rock types and cropping out in large areas can be well defined in the imagery. The best results in discrimination between stratigraphic units was attained for alluvium (unit Qa) and the sedimentary deposits designated as Tq, both of them include unconsolidated sediments. The Tertiary lateritic zone (unit Tql) found throughout the area is characterized by a uniform image tone and smooth topography and has

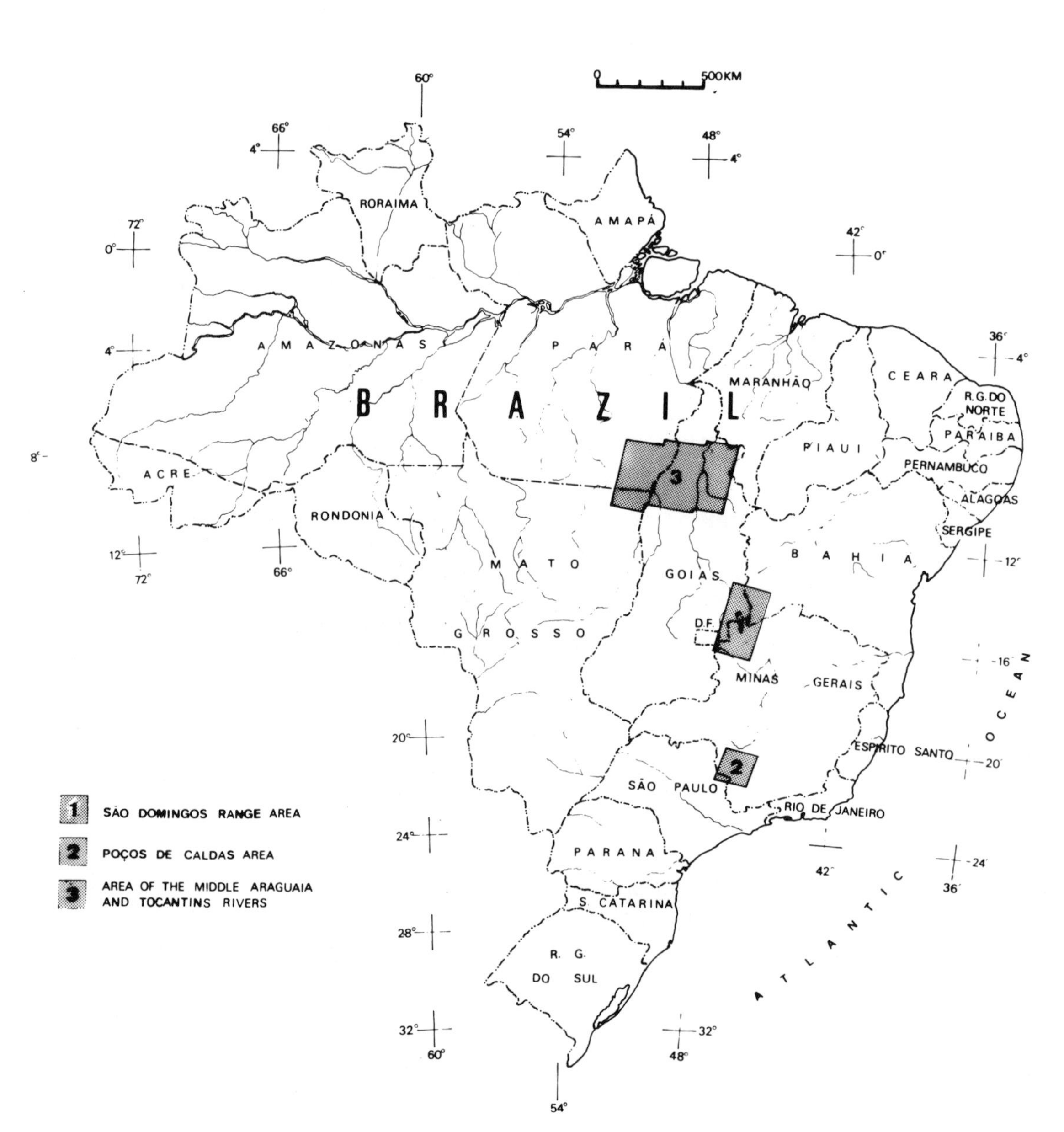

FIG. 1
Location of case study areas.

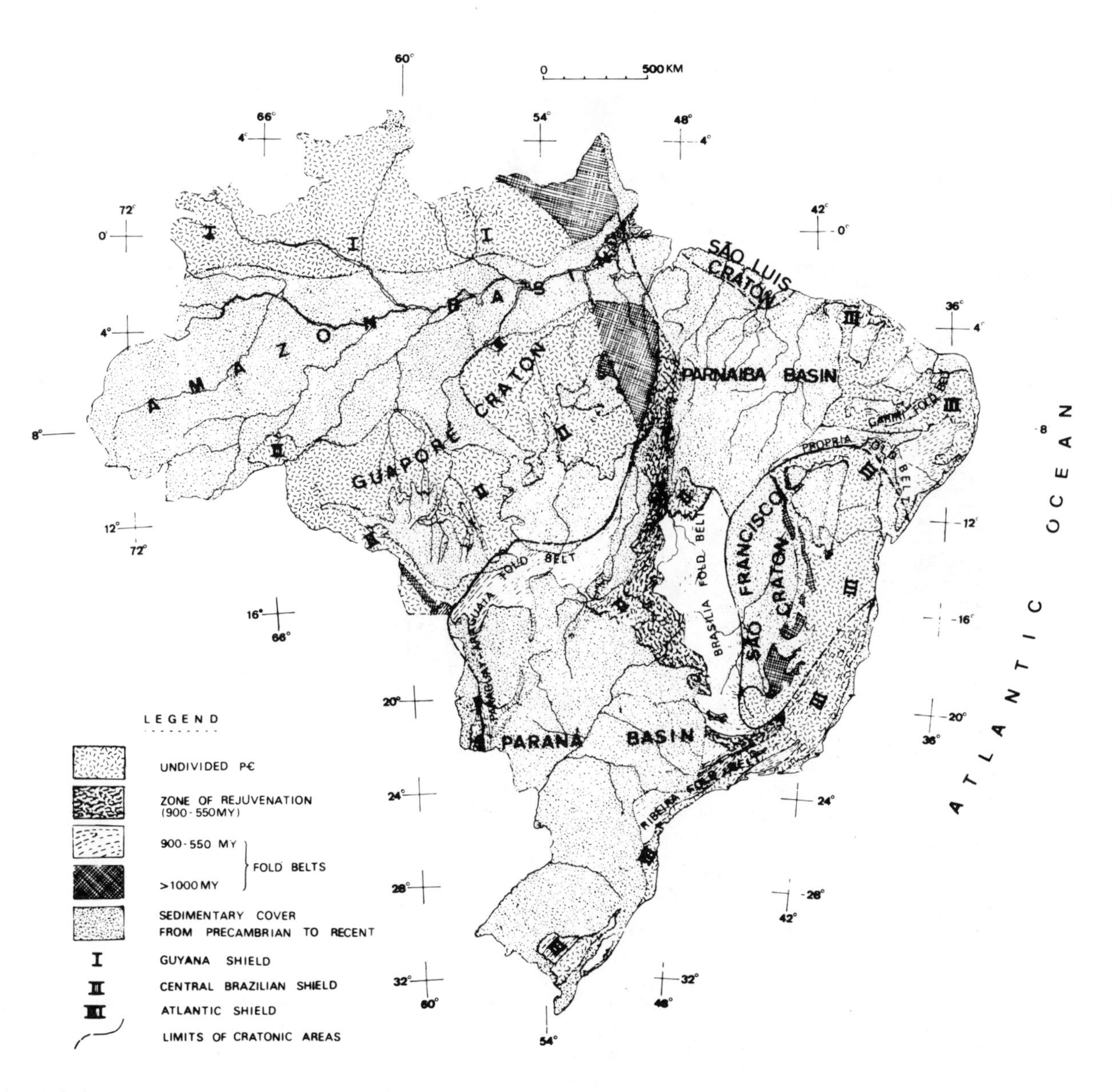

FIG. 2
Tectonic framework of Brazil (after Cordani et al., 1968; Ferreira et al., 1970; Amaral, 1970; Almeida et al., 1973).

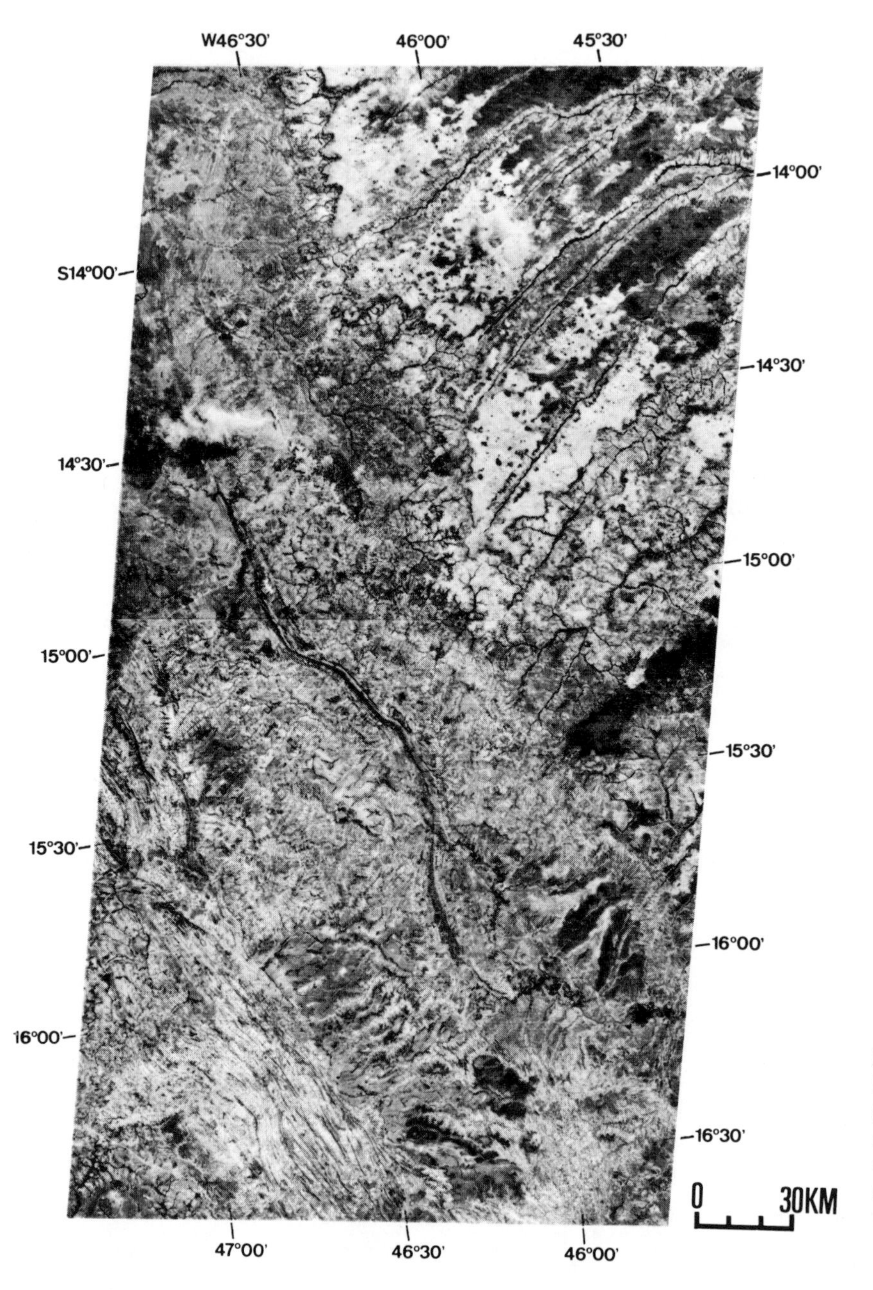

FIG. 3
Photomosaic of the São Domingos range area from LANDSAT's multispectral scanner imagery (band 5: 0.6–0.7 μm). Imagery obtained on Aug. 18, 1973 (top: E-1391-12362; bottom: E-1391-12365).

TABLE 1
Generalized Stratigraphic Column for the São Domingos Range Area
(after Braun, 1968; DNPM, 1965–1968a, b, 1968a, b; Ladeira et al., 1971)

Age	*Lithologic Units*		*Summary Description*
Quaternary	Alluvium deposits		Sand, clay, and gravel
Tertiary (?)			Sandstones, conglomerates, and argillaceous sand deposits
			Lateritic cover
Cretaceous	Urucuia Formation		Red sandstones, argillites, and a basal conglomerate
Upper Precambrian–Eocambrian (?)	Bambuí Group	Três Marias Fm.	Arkose, micaceous, siltstone, and arkosic sandstone
		Paraopeba Fm.	Carbonatic and pellitic rocks
		Paranoá Fm.	Quartzites interbedded with fillites and metasiltstones
Precambrian (undifferentiated)	Crystalline basement		Granite and gneiss complex

been identified with good precision from LANDSAT imagery.

The sedimentary sequence of the Urucuia Formation, composed predominantly of sandstones, overlies unconformably the Bambuí Group and forms the large plateau in the northeastern side of the area where a typical parallel drainage pattern can be observed (Fig. 3). The areal extent of the three remote-sensing rock units identified in the São Domingos Range area (Kus, Kum, and Kui) is correlated in part with the distribution of the Urucuia Formation shown in published maps and is discussed in more detail later.

From the geological analysis of satellite imagery it is possible to identify units Kus, Kum, and Kui, each with its own spectral and textural characteristics (Fig. 3). The uppermost unit, Kus in Fig. 3, is separated from the other units by a very steep erosional scarp (Fig. 3), which is clearly visible in the images (Fig. 2). This remote-sensing rock unit indicates the areal extent of the sandstone upper member of the Urucuia Formation (Oliveira, 1967). The other units, Kum and Kui in Fig. 3, enclose areas where previous maps based on photointerpretation of 1:60,000 aerial photographs and some fieldwork (DNPM, 1965–1968a) indicate the presence of scattered lateritic cover. At the 1:1,000,000 scale, however, these patches of laterite cannot be detected, and it is believed that units Kum and Kui correspond to the sequence of shale, siltstone, and silty sandstone of the lower member of the Urucuia Formation (Oliveira, 1967).

The other stratigraphic unit present in the area, the Upper Precambrian sequence of the Bambuí Group, has a more diversified lithology, and its outcrops in a structurally complex region are relatively small. The Bambuí Group has been divided into three formations, but it is not possible to identify them accurately in LANDSAT-1 imagery.

The crystalline basement rocks in the northern part of the area, because of their relatively small

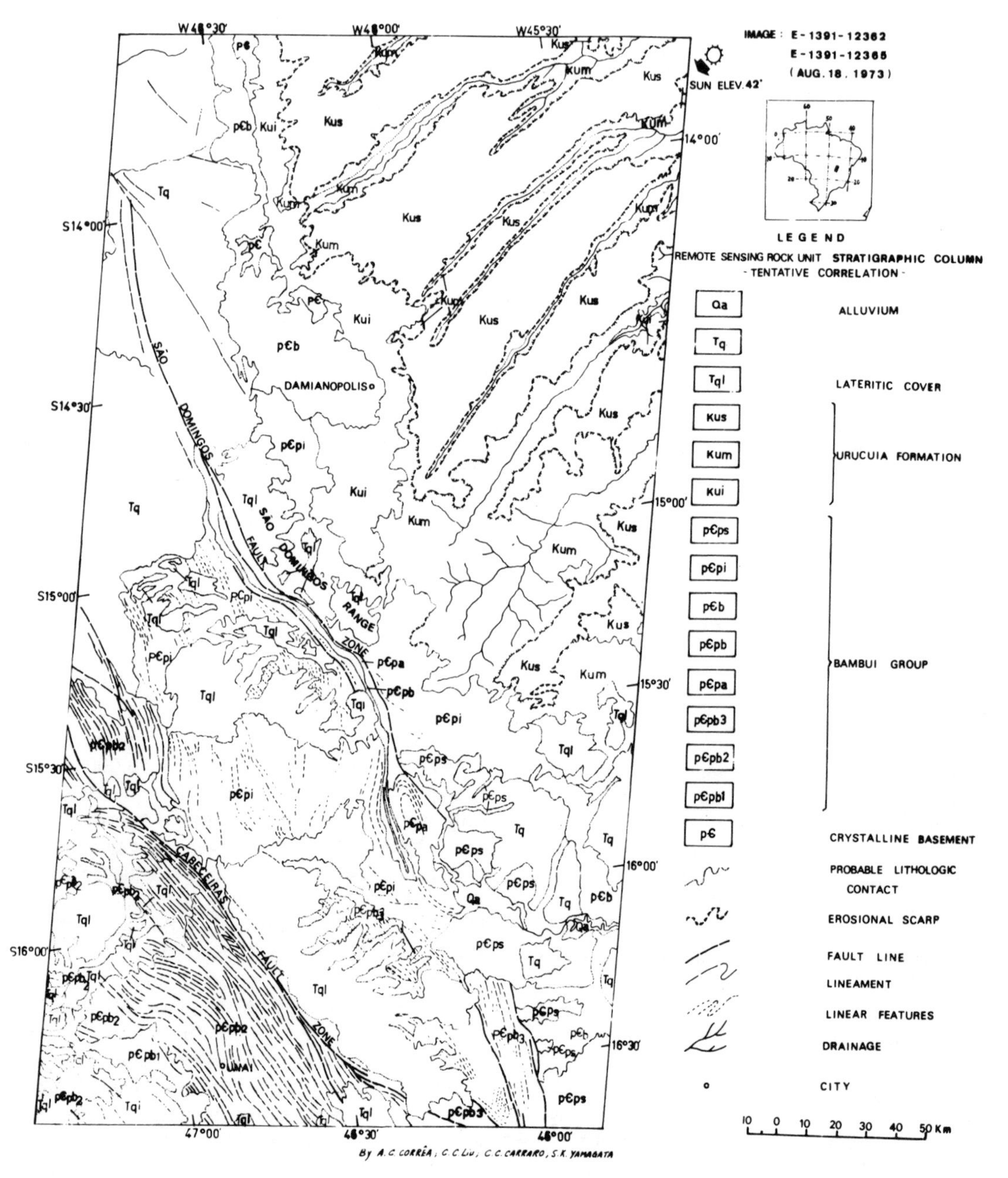

FIG. 4
Geological interpretation of the São Domingos range area based on LANDSAT-1 imagery.

outcrops, are very difficult to identify in LANDSAT-1 images.

The São Domingos Range area comprises the contact between the ancient São Francisco craton, which is covered by Precambrian and younger sediments, and the Brasilia fold belt (Fig. 2). The change in structural features between these two tectonic units is well defined in the images.

The Brasilia fold belt is characterized in the area under discussion by a series of northwest trending, elongated anticlines and synclines. These structures cannot be individually identified in the space imagery because of their relatively small lateral extension, which makes the determination of the attitude of their flanks extremely difficult. The linearity shown by fold axes, however, permits a precise indication of the dominant structural trend in the area. Lineaments shown in the geological interpretation map include both folds and probable fractures in the area of the Brasilia fold belt (Fig. 4). The term "lineament" is used in this chapter to indicate those lines observed in the imagery which are structurally controlled.

The region between the Cabeceiras and São Domingos faults can be considered as the contact zone between the Brasilia fold belt and the São Francisco craton. Structures detected in this zone have been called "linear features," a term used to indicate features with uncertain origin, some of which are certainly structurally controlled. Previous maps of the area at the 1:250,000 scale (DNPM, 1965–1968 a, b) show a few folds in this zone, which do not account for all the linear features detected in the imagery. Braun (1968) observed also that folding in this general area decreases toward the northeast, and that there are no indications of structures east of the São Domingos fault zone, where bedding is almost horizontal. Probably, the linear features observed in space imagery with a trend parallel to the Brasilia fold belt indicate that this zone could have been affected to a small degree by the tectonic activity in the fold belt.

Poços de Caldas Area

The area considered for analysis in this section lies on the border of the states of Minas Gerais and São Paulo in southeastern Brazil (Fig. 1). The larger feature just to the southwest of the center of the LANDSAT-1 image, which outlines the Poços de Caldas alkaline complex, has attracted the interest of geologists for some time (Fig. 5). Derby (1887) wrote the first geological report on the area indicating the presence of this alkaline intrusive and was followed by others, who described with more detail the geology and mineral resources.

The Poços de Caldas alkaline complex is one of the several alkaline intrusives located along the northeastern boundary of the Paraná Basin and scattered along the Atlantic coast (Fig. 2).

The Poços de Caldas alkaline intrusive is emplaced in granite and gneiss of Precambrian Basement Complex, which occupies most of the area in the imagery. The contact between the Precambrian igneous and metamorphic terrain and the sedimentary sequence of the Paraná Basin can be outlined with very good precision using bands 5 and 7 (Fig. 6). The alluvial deposits (unit Qa) surrounding the Furnas Dam area are identified also using these two bands. The alkaline intrusive rocks in the Poços de Caldas district do not present in band 5 any difference from the surrounding igneous and metamorphic terrain. The medium gray tone and smooth texture observed in band 7 for the area enclosed by the circular structure delineates very well the extent of the intrusive rocks (unit K). A reconnaissance mapping of the circular structure immediately south of the Poço Fundo Dam, which has approximately the same gray tone in band 7 as that of the Poços de Caldas district, has not revealed the presence of outcrops of alkaline rocks (Almeida Filho and Paradella, 1975).

The geological interpretation of this LANDSAT-1 frame identified a large number of previously unknown fault zones, lineaments, and linear and circular features. The Poços de Caldas area is located in Precambrian terrain that was rejuvenated during the Upper Precambrian and Cambrian (Fig. 2). The large approximately east–west trending fractures in the Furnas Dam area and south of the Poços de Caldas district may be related to the deformation associated with the Brasilia and Ribeira fold belts, respectively (Fig. 5). The Mesozoic activation of the South American Platform, which caused an increase in the sub-

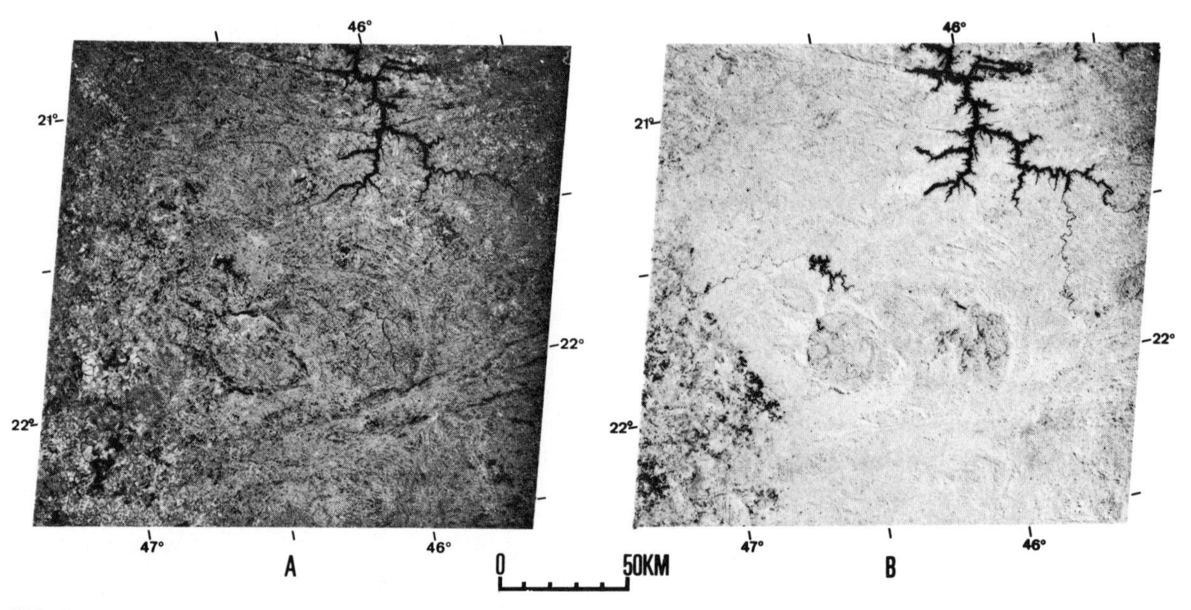

FIG. 5

LANDSAT-1 images of the Poços de Caldas area: (A) band 5 (0.6–0.7 μm); (B) band 7 (0.8–1.1 μm). Imagery obtained on Sept. 9, 1972 (E-1048-12330).

sidence rate of the Paraná Basin and intense magmatic activity (Almeida, 1972), was responsible for important fissure volcanism, which covered most of the basin with basaltic lava. The fissure volcanism during the Jurassic and extending until the Albian (Amaral et al., 1966) was simultaneous with the intrusion of alkaline centers in southern Brazil. The volcano-plutonic magmatism that is present in the Poços de Caldas district occurred during the Upper Cretaceous (Senonian) and is associated with trending fractures in a zone of monoclinal flexure between the Paraná Basin and the São Francisco craton (Almeida, 1972). The alkaline intrusive bodies are also considered to represent the rejuvenation of zones of rifting or fracturing that were active at the time of the initial separation of the South American and African continental plates (Neill, 1973).

The alkaline intrusive centers associated with the Mesozoic activation of the South American Platform are of great importance because of the economic mineralization. The underlying rocks of the Poços de Caldas district are mainly of nepheline syenite (referred to also as foyaite) and phonolite (minor amounts of volcanic and sedimentary rocks are present also), which are deeply weathered in most of the plateau area. In the northern part of the district, aluminous laterite or bauxite is the product of weathering forming commercial deposits (Teixeira, 1937; Weber, 1959) that are being mined today.

Uraniferous zirconium deposits are associated with the Poços de Caldas alkaline intrusive, and a preliminary delineation of radioactive anomalies was accomplished with data acquired by airborne and ground radiometric surveys. The deposits are located in three main areas within the circular structure: the east-central part of the Poços de Caldas plateau (Pocinhos area), the west-central part (Cascata area), and the south-central area (Fig. 6). The primary uraniferous zirconium desposits are believed to originate from the action of hydrothermal solutions on the zirconiferous syenite and subsequent deposition of the stable oxide (baddeleyite, ZrO_2) and silicate (zircon, $ZrSiO_4$) in fissures (Franco and Loewenstein, 1948; Tolbert, 1966). Uraniferous molybdenite mineralization of economic value is found in Campo do Agostinho at the center of the Poços de Caldas plateau. Uranium and molybdenum ore minerals with variable amounts of pyrite and fluorite are found in veins located in fracture or breccia zones in hydrothermally altered tinguaite (Gorsky and Gorsky, 1970).

In the south-central area of uraniferous zirconium mineralization, at Morro do Ferro, the alkalic igneous rocks were percolated by mineralizing solutions rich in thorium and rare-earth elements, which followed the existing fracture system (Wedow, 1967). Secondary enrichment of rare-earth elements and thorium in the upper part of the stockwork was caused by the deep weathering typical of the Poços de Caldas area (Wedow, 1967).

The geological interpretation of the LANDSAT-1 image suggests that within the large circular structure of the Poços de Caldas district, smaller circular structures and linear features probably indicate secondary intrusive centers and fractures, which may have controlled to a great extent the hydrothermal mineraliza-

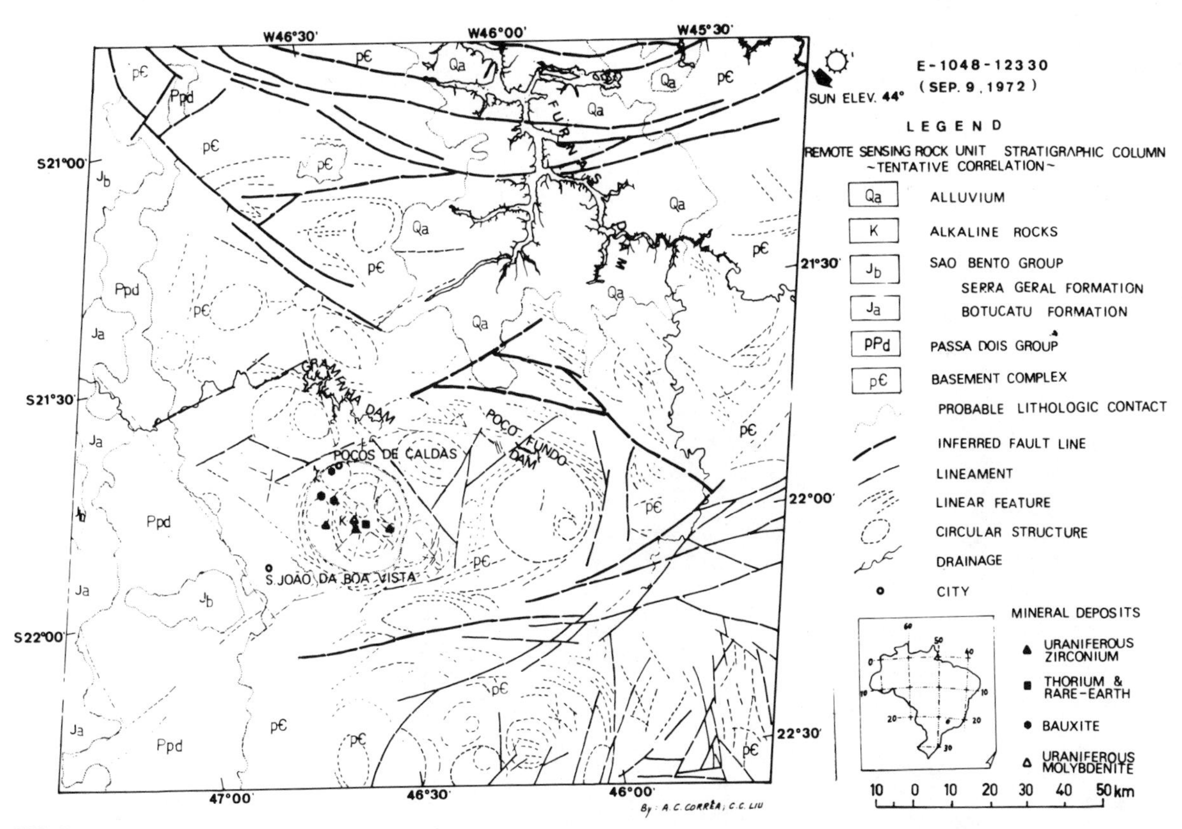

FIG. 6
Geological interpretation of the Poços de Caldas area based on LANDSAT-1 imagery.

tion. These structural features are not all shown in the most recent geological map available for the district (Ellert, 1959). Other previously unknown circular features identified in the Poços de Caldas area can be considered as likely mineral exploration targets and should be examined in the field. One of the authors (A. C. Corréa) visited the circular features in the Graminha Dam area (Fig. 6), but has not found at the surface any indication of alkaline igneous intrusive rocks; this does not exclude the possibility of their occurrence subsurface or at other nearby areas.

Area of the Middle Araguaia and Tocantins Rivers

An area in north-central Brazil was chosen as the third case study for the application of LANDSAT-1 imagery in geological research (Fig. 1). The Araguaia and Tocantins rivers, flowing toward the Amazon river to the north, cross the region from south to north (Fig. 7). Two climates and vegetation types prevail in the area with a boundary zone located approximately along the Araguaia river. To the west of the river, a hot and humid equatorial climate (Köppen's Am type) and a broad-leaf equatorial forest are dominant; to the east a tropical climate with distinct dry and rainy seasons (Köppen's Aw type) and cerrado (savanna-like) vegetation are present (Azevedo, 1964).

LANDSAT-1 imagery for this area provided structural data that were interpreted in order to outline structural systems which can define regions with similar tectonic style. The small-scale satellite images, which provide a synoptic view of large areas, are extremely useful in the identification of the regional characteristics that are not evident in low-altitude aerial photography.

The purpose of this structural interpretation is not to discuss specific linear features but to identify those sets of structures that determine regional trends and to point out their relationship with other structural systems.

Two basic assumptions are made concerning the

FIG. 7
Photomosaic of the area of the middle Araguaia and Tocantins rivers from LANDSAT's multispectral scanner imagery (band 7: 0.8–1.1 μm). Identification of LANDSAT-1 frames given in Fig. 8.

discrimination of structural systems (Rich and Steele, 1974). The first is that any system in a given region originates as a result of the action of the stresses present in that region at a certain period of geologic time. A structural system may have been rejuvenated as a consequence of younger tectonic activity. The second assumption is that the original structural trend has been preserved, and that any later changes which could have affected a given trend must have been recorded also in the whole system. Even in the case where a younger system is developed along a trend intersecting an older system, indications of the older system can still be detected, subtly in some cases, in the geomorphic characteristics of the terrain. Recent alluvial deposits or thick sedimentary sequences may also obliterate to some extent trends of old linear systems, but erosion and tectonic reactivation, which may reinforce preexisting trends, allow their identification.

Some of the structural systems detected have been recognized by geologists working in the region, but others have not been identified or described in the available literature and will possibly encourage a reexamination of present ideas.

The preliminary structural map of the area of the middle Araguaia and Tocantins rivers is shown in Fig. 8. The structural features mapped were detected using traditional photogeologic interpretation, not only in the images that were used to make the photomosaic, but also in other images available. The terms lineaments and linear features were used in the same way as defined in a previous section.

STRUCTURAL SYSTEMS

The structural systems identified in the study area are shown schematically in Fig. 9 and described briefly. The western side of the study area is characterized by an orthogonal system of short linears consisting of a north–south or north–northeast trending set and another set that trends east–west or east–southeast. This structural system is called Guaporé, a reference to the Guaporé craton which is outlined by it. The eastern border of the Guaporé craton, which is present in our study area, according to age determinations may be as old as 3000 m.y.; however, most ages fall into the 2600- to 1800-m.y. range (Amaral, 1970). This crystalline basement consists mainly of migmatites, gneiss, granite, and amphibolite (Amaral, 1974); however, in the area under consideration it is indicated by highly fractured granitic bodies. A Tertiary sedimentary cover is widespread and obliterates to a great extent the basement rocks.

In the northwest corner of the area shown in Fig. 8 there is a predominance of northwest trending linear structures overprinted on the Guaporé system. This structural system, referred to as the Carajás system, is related to the sequence of fractured and folded metasediments, including iron formations, intrusive granites, quartzites, phyllites, and mafic rocks, defined as Serra dos Carajás Group (SUDAM/HIDROSERVICE, 1973) and is considered to be more than 200 m.y. (Amaral, 1974).

The central area of the structural map (Fig. 8) is characterized by an approximately north–south set of linears designated as the Tocantins–Araguaia structural system. This system encloses the fold belt along the eastern border of the Guaporé craton, where metasediments of the Araxá and Tocantins groups are present. According to age determinations, these two groups are probably 2000 m.y. (Amaral, 1974) and were rejuvenated during the Braziliano tectonic event (900 to 550 m.y.) (Fig. 2) (Ferreira, 1971). These two groups also include iron formations associated with metabasites to a lesser extent than the Serra dos Carajás Group. Small basic and ultrabasic bodies, which intruded along the deep fractures formed during the separation of the Guaporé and São Francisco cratons (Almeida, 1967), are part of the serpentine belt located within this structural system.

The eastern side of the study area corresponds to a border of the Parnáiba sedimentary basin (Fig. 2). This region is characterized by a predominantly north–northwest to north–northeast set of linears defining the Parnaíba structural system (Fig. 9). This set of linears observed in the Paleozoic sedimentary sequence may have been present at the basement of this basin. During the activation of the South American Platform, during the Mesozoic, the Parnaíba basin underwent an increase in subsidence rate (Almeida, 1972), and this old fracture system, partly due to its

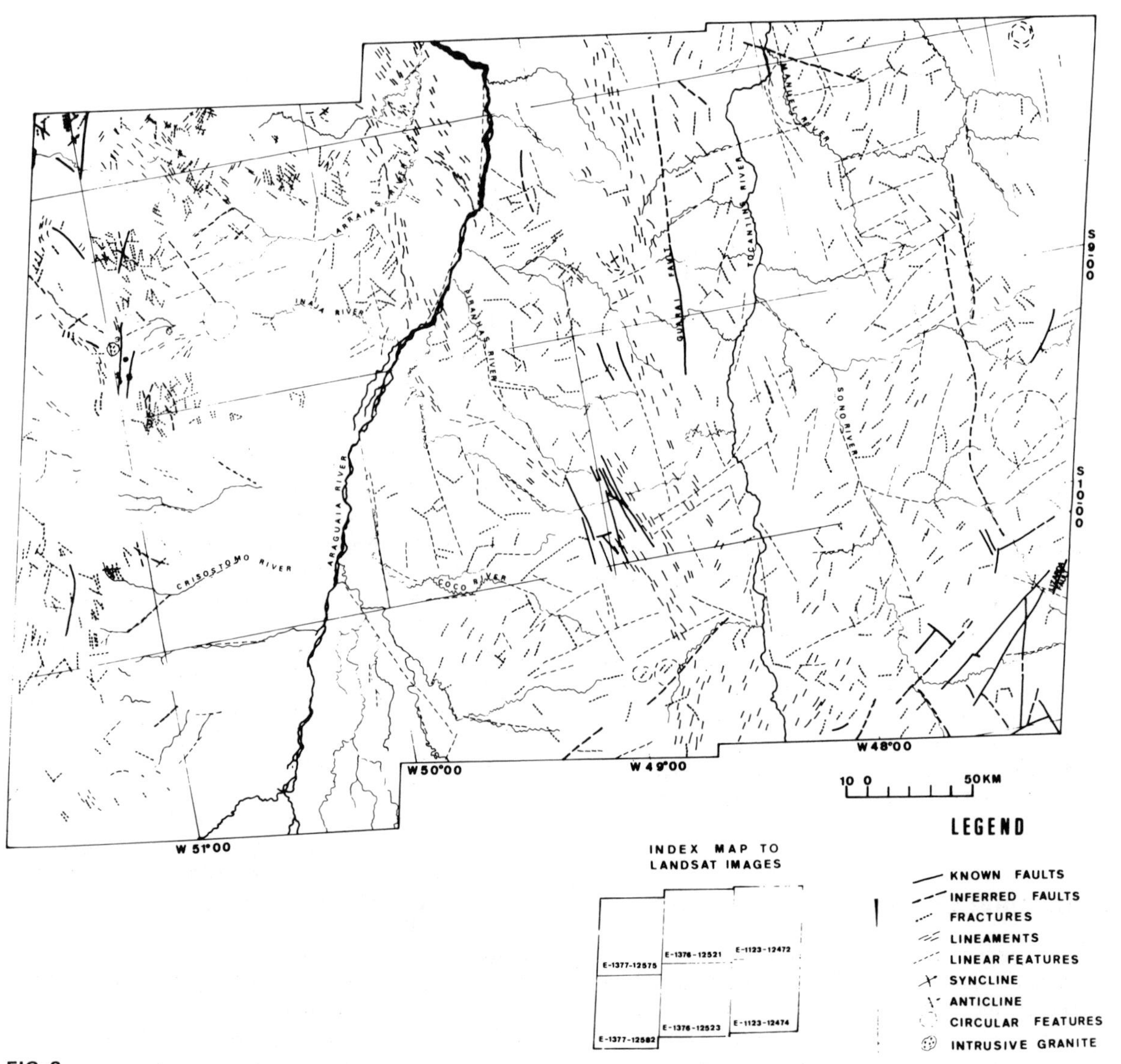

FIG. 8
Structural interpretation of the area of the middle Araguaia and Tocantins rivers based on LANDSAT-1 imagery.

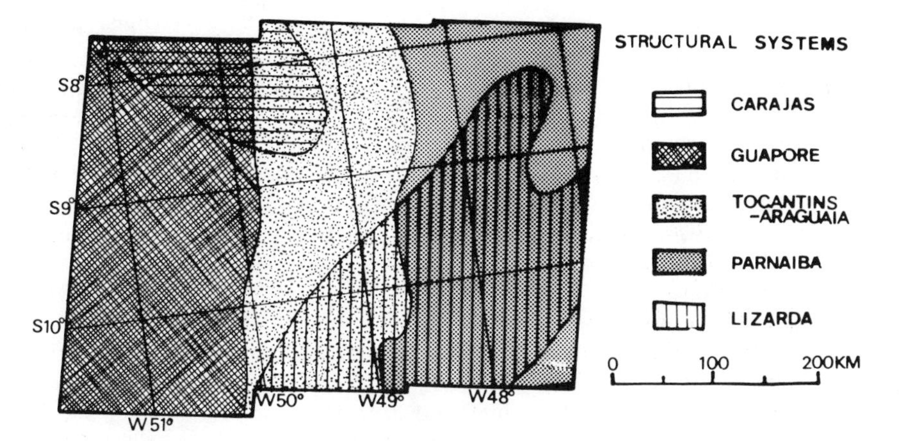

FIG. 9
Structural systems defined for the area of the middle Araguaia and Tocantins rivers. Boundaries shown are approximate only; refer to Fig. 8 for their geographic extent and to the text for their description.

positions at the border of the basin, may have been reactivated.

The Lizarda structural system, overprinted on the Tocantins–Araguaia and Parnaíba systems in the study area, is characterized by a predominant northeast trending set of linears. This system may have formed during the separation of the Guaporé or São Francisco cratons; its trend suggests that it may be associated also with the northern boundary of the São Francisco craton. Large northeast trending faults such as the Lizarda fault, which extends to the east of the study area (Fig. 8), are present in this structural system.

CONCLUSIONS

Structural information extracted from LANDSAT-1 imagery allowed the identification of several structural systems that may be important for the interpretation of the geologic evolution of north-central Brazil. The structural systems defined mainly by the trends of linears, their size and spacing, seem to be more directly related to the physical properties of the lithosphere and the deformational evolution of a region, as pointed out by Corréa and Lyon (1974); however, a chronologic sequence may be suggested.

The Mesozoic activation of the South American Platform appears to have enhanced the trends of the Tocantins–Araguaia and Lizarda structural systems, which are reflected in the younger Parnaíba system. The Carajás structural system is an older system than those mentioned above and was superimposed on the Guaporé system, which may contain the fundamental structures of this part of the crust.

ACKNOWLEDGMENTS

The authors are thankful to Gilberto Amaral for critically reading the manuscript and providing helpful suggestions. S. Yamagata, P. Menezes, U. Santos, C. Carraro, geologists with the Instituto de Pesquisas Espaciais, helped in the interpretation of imagery and their contribution is acknowledged. Maria do Carmo S. Soares typed the manuscript.

REFERENCES

Almeida, F. F. M., 1967, Observacões sobre o Pré-cambriano da regiaõ central de Goiás, *Bol. Paranaense Geociências*, v. 26, p. 19–22.

____, 1971, Geochronological Division of the Precambrian of South America, *Rev. Brasil. Geociências,* v. 1, no. 1, p. 13–21.

____, 1972, Tectono-Magmatic Activation of the South American Platform and Associated Mineralization, 24th Intern. Geol. Congr., Sect. 3, Montreal, p. 339–346.

____, G. Amaral, U. G. Cordani, and K. Kawashita, 1973, The Precambrian Evolution of the South America Cratonic Margin South of the Amazon

River, in A. E. M. Nairn, ed., *The Oceans' Basins and Margins,* vol. 1, Plenum, New York, p. 411–466.

Almeida Filho, R., and W. R. Paradella, 1975, Relatório preliminar das atividades do Projeto Alumínio, durante o ano de 1974, Instituto de Pesquisas Espaciais, INPE-601-RI/286, 45 p.

Amaral, G., 1970, Precambrian Evolution of the Brazilian Amazon Region (abs.), 1st Latin Am. Geol. Congr., Lima, Peru.

——, 1974, Geologia Pré-cambriana da região Amazônica, unpublished Ph.D. thesis at Universidade de São Paulo, Brazil, 212 p.

——, U. G. Cordani, K. Kawashita, and J. H. Reynolds, 1966, Potassium–Argon Dates of Basaltic Rocks from Southern Brazil, *Geochim. Cosmochim. Acta,* v. 30, p. 159–189.

Azevedo, A., 1964, Brasil—A Terra e o Homem, Cia. Editora Nacional, v. 1, S. Paulo, Brazil, 571 p.

Braun, O. P. G., 1968, Contribuição à estratigrafia do Grupo Bambuí, Soc. Brasileira de Geologia, Anais do XXII Congresso, p. 155–166.

Cordani, U. G., G. C. Melcher, and F. F. M. Almeida, 1968, Outline of the Precambrian Geochronology of South America, *Can. J. Earth Sci.,* v. 5, p. 629–632.

Corrêa, A. C., and R. J. P. Lyon, 1974, An Application of Optical Fourier Analysis to the Study of Geological Linear Features, *Stanford Remote Sensing Laboratory Tech. Rept. 74-9,* 34 p.; also *Proc. First International Conference on the New Basement Tectonics,* Utah Geol. Assoc. (in press).

Derby, O. A., 1887, On Nepheline Rocks in Brazil, with Special Reference to the Association of Phonolite and Foyaite, *Geol. Soc. London Quart. J.,* v. 43, p. 457–473.

DNPM, 1965–1968a, Projeto Brasília—Geologia da região central de Goiás, Dept. Nacl. Produção Mineral, Folha Posse S.D. 23 N, 1:250,000 scale.

——, 1965–1968b, Projeto Brasília—Geologia da região central de Goiás, Dept. Nacl. Produção Mineral, Folha Alto Paraíso de Goiás, S.D. 23 M 1:250,000.

——, 1968a, Projeto Goiânia—Geologia da região sul de Goiás: Dept. Nacl. Produção Mineral, Folha Brasília S.D. 23 S, 1:250,000 scale.

——, 1968b, Projeto Goiânia—Geologia da região sul de Goiás, Dept. Nacl. Produção Mineral, Folha Buritis, S.D. 23 T, 1:250,000 scale.

Ellert, Reinholt, 1959, Contribuição à Geologia do maciço alcalino de Poços de Caldas, Fac. Filosof. Ciênc. Letras, Univ. São Paulo, Bol. 237, Geologia no. 18, p. 1–63.

Franco, R. R., and Walter Loewenstein, 1948, Zirconium from the Region of Poços de Caldas, *Am. Mineralogist,* v. 33, no. 3–4, p. 142–151.

Ferreira, E. O., 1971, Carta tectônica do Brasil—notícia explicativa (Tectonic map of Brazil—explanatory note), Dept. Nacl. Produção Mineral, 19 p.

——, F. F. M. Almeida, E. F. Suszcynski, and G. R. Derze, 1970, Mapa tectônico do Brasil, Dept. Nacl. Produção Mineral, 1:5,000,000 scale.

Girardi, Carlos, 1973, Áreas e épocas favoráveis aos vôos aerofotogramétricos, Centro Técnico Aeroespacial, Rel. IAE-M-03/73, 22 p.

Gorsky, V. A., and E. Gorsky, 1970, Diferentes tipos de mineralização no planalto de Poços de Caldas, M.G., Soc. Bras. Geol., Resumo das Conferências e Comunicações, Brasília, Bol. Especial 1, p. 135–137.

Ladeira, E. A., O. P. G. Braun, R. N. Cardoso, and Y. Hasui, 1971, O Cretáceo em Minas Gerais, Soc. Bras. Geol., Anais do XXV Congresso, v. 1, p. 15–31.

Lee, K., D. H. Knepper, and D. L. Sawatzky, 1974, Geologic Information from Satellite Images: Colorado School of Mines, Remote Sensing Rept. 74-3, 37 p.

Neill, W., 1973, Possible Continental Rifting in Brazil and Angola Related to the Opening of the South Atlantic, *Nature Phys. Sci.,* v. 245, no. 146, p. 104–107.

Oliveira, M. A. M., 1967, Contribuição à geologia da parte sul da Bacia do São Francisco e áreas adjacentes, Cienc. Tecn. Petróleo, Publ. 3, Coletânea de Relatórios de Exploração, v. 1, p. 71–105.

Rich, E. I., and W. C. Steele, 1974, Speculations on Geologic Structures in Northern California as Detected from ERTS-1 Satellite Imagery, *Geology,* v. 2, no. 4, p. 165–170.

SUDAM/HIDROSERVICE, 1973, Geologia e recursos minerais, *in Plano de deservolvimento integrado da áreas da bacia do Rio Tocantins,* Superintendência do Desenvolvimento da Amazônia/HIDROSERVICE, v. 4, p. 29–260.

Teixeira, E. A., 1937, Bauxita no Planalto de Poços de Caldas, Estados de São Paulo e Minas Gerais (Brasil), Brasil, Serviço Fomento Produção Mineral Avulso 15, 19 p.

Tolbert, G. E., 1966, The Uraniferous Zirconium Deposits of the Poços de Caldas Plateau, *U.S. Geol. Surv. Bull. 1185-C,* 28 p.

Weber, B. N., 1959, Bauxitização do distrito de Poços de Caldas, Minas Gerais, Brasil, *Bol. Soc. Bras. Geol.,* v. 8, no. 1, p. 17–30.

Wedow, Helmuth, Jr., 1967, The Morro de Ferro Thorium and Rare-Earth Ore Deposit, Poços de Caldas District, Brazil, *U.S. Geol. Surv. Bull. 1185-D,* 34 p.

14

STUDIES UTILIZING ORBITAL IMAGERY OF INDIA FOR GEOLOGY AND LAND USE

Ravi D. Sharma
B. N. Raina
Mukhtar S. Dhanju

Ravi D. Sharma is Scientific Secretary of the Indian Space Research Organization (ISRO), Bangalore, India. B. N. Raina is with the Geology Division of the Indian Photo Interpretation Institute (IPI), Dehra Dun, India. Mukhtar S. Dhanju is with the Remote Sensing and Meteorology Applications Division of the Space Applications Center (ISRO), Ahmedabad, India.

Apollo photographs and LANDSAT-1 imagery of selected parts of India have been studied with a view toward applications in geology, drainage patterns, and land-use classification. Interpretations of multiband imagery in photographic form and quantitative analyses with the help of computer-compatible tapes (CCT) have been carried out.

Specifically, an Apollo photograph of the Himalayan region has been interpreted for regional geology and new features have been identified. A brief discussion of the principles of photointerpretation is presented. Geological maps have been made through the interpretation of LANDSAT-multiband imagery of the northern Terai area of Uttar Pradesh and of areas in Himachal Pradesh, Uttar Pradesh, and the Punjab. Similarly, geological maps have been made of areas in the southern parts of Tamil Nader and Andhra Pradesh states, showing various linear features and some lithology (Raina, 1972a).

A detailed study was made of the drainage pattern in the Sawrashtra area using LANDSAT-1 imagery. Four images were used to compile a map of the drainage basin. The drainage pattern is most conspicuous in band 7 (0.8- to 1.1-μm range). In some instances the information in band 6 (0.6 to 0.7 μm) was also usefully employed. A watershed drainage map was prepared showing the different regions, which have separate stream patterns. Calculations were made for the drainage density of the watershed basins and were compared with the climatological information of the area.

An agricultural area of approximately 25 X 45 nautical miles in the Nabha–Patiala region of the Punjab was selected from a LANDSAT-1 frame of northwest India for detailed quantitative studies utilizing multispectral classification techniques for resource-inventory and land-use categorization. Twenty training sets were selected from available maps. These included known forest, water, and saline dune areas and assumed agricultural and fallow fields. Statistics (i.e., mean, variance, and covariance) for each training set were generated assuming multivariate Gaussian distributions and utilizing data from all four bands of the LANDSAT-1 multispectral scanner (MSS). Closer examination revealed that the 20 training sets could be combined into eight separable classes, perhaps signifying distinguishable categories. New statistics on the combined sets were generated, and these signatures were used for creating a recognition map, which was color coded. Water, saline sand-dune areas, and two types each of bare-field, agriculture, and forest areas could be classified and mapped. Acreage for each recognized class is available. These results signify the capability of land-use classifications from LANDSAT data. For a large country such as India, such data are of economic importance since accurate and repetitive quantitative information is not so easily available from other sources (Sharma, 1972).

The following sections provide examples of the earlier analyses of orbital imagery of India. More recently, subsequent studies have been undertaken on the automatic classification of various geological and land-use parameters.

PRINCIPLES OF PHOTOINTERPRETATION

Aerial photointerpretation, the interpretation of orbital imagery, and remote-sensing techniques permit the observation and measurement of geological features using both direct and indirect methods. The subjective judgment of facts and observations based on objective reasoning and personal experience are in part being replaced by statistical treatment and computerization.

The interpretation of aerial and orbital imagery for geological studies, although still largely a subjective science, is a distinct improvement over conventional methods of reconnaissance mapping. Examples of the advantages of this technique are well known and, therefore, will be mentioned only briefly here.

Aerial Photographs

The various types of aerial photographs available are vertical, high and low obliques, trimetrogon, and multiple lens; of these the first variety is commonly used. Photographs can be black and white, colored, infrared, or multispectral photographs. Commonly, vertical aerial photography is used for geological interpretation; its advantages are the following:

1. The stereo model gives a better idea of terrain geomorphology, forest cover, and accessibility, and hence is useful in the planning of a survey.
2. It reduces fieldwork, as only a few critical sections may need checking.
3. It permits greater coverage and more accuracy in less time, resulting in greater economy.
4. Major geological features, linear features, fold patterns, and general overall structural details can be better interpreted.
5. Its greatest use is for inaccessible areas like the Himalayas, where the season for fieldwork is very limited and many sections are not accessible for field checking. Such areas can be viewed easily in three dimensions in stereo models, often permitting an accurate tracing of geological boundaries and the observation of contact details, dips, strike, continuity, etc.
6. It is possible to observe the area repeatedly to clarify doubts and problems. This possibility of taking repeat observations is a great asset of photogeology.

Orbital Photography

Gemini- and Apollo-mission astronauts took several photographs of the earth by hand-held cameras. Many of these photographs were remarkable for their clarity and were used for geological interpretation (Lowman, 1971). They permitted certain observations and inferences that were not possible before. The synoptic coverage available on a single orbital photograph permits observation and interpretation of regional geomorphological characteristics and large-scale geological features. Taking sequential photography at desired intervals is a boon for the observation of such short-lived phenomena as catastrophies and other changes. However, these photographs do not have the geometrical fidelity of aerial photographs nor can they be used satisfactorily in stereopairs to obtain usable three-dimensional views.

Interpretation Techniques

The geological interpretation of aerial photographs is a two-step process—observation and deduction (Ray, 1960). The deduction process involves either matching photographic images with similar reference images or reasoning by analogy. The methodology can generally be characterized by four stages: surveying, analysis, synthesis, and hypothesis. The synthesis attempted and the hypothesis propounded should be based on a critical examination of the data. The interpretation is based on the identification of surface features of geological significance.

This identification is carried out by the (1) qualitative analysis of some recognition elements of the photograph, and (2) the geotechnical analysis of the photograph.

Recognition elements

1. *Tonal variations* of a photograph.
2. *Colors* seen on colored photographs.
3. *Texture* of the photograph, which is a measure of the frequency of tonal changes within the image.
4. *Patterns* seen on the photograph, which may be due to some orderly spatial arrangement of geological, topographical, or vegetation features.
5. *Shape of objects* as seen on the photograph, permitting direct identifications.
6. *Size of observed features,* as compared to some known objects. This is useful in the classification of the different lineaments and fault offsets seen on the photograph.

Geotechnical analysis. This is based on study of the following:

1. *Landform* analysis.
2. *Drainage* analysis.
3. *Erosion* pattern of the rocks.
4. *Vegetation* cover.
5. *Land-use* pattern in the area.

All these observations give direct or indirect clues as to the geology of the area; in the end the interpretation is based on the convergence of evidence, that is, generalization covering the largest amount of data.

Scale and vertical exaggeration are also significant factors that aid in dip estimation and exaggerate minor topographic differences, which may reflect underlying geological structure.

GEOLOGICAL INTERPRETATION OF ORBITAL PHOTOGRAPHY AND IMAGERY

The interpretation of orbital imagery is somewhat different inasmuch as the imagery is not generally available in stereopairs. Even where stereo is possible, the vertical exaggeration observed under a stereoscope is too small and is of limited use. Hence, as Carter and Stone (1974) have pointed out, the interpretation of hyperaltitude imagery is normally carried out by using a magnifying lens and a light table. They have further pointed out that no single recognition parameter is most useful or dominant. In their view, a combination of several recognition elements, usually tone, color, and texture, are most useful; shape may be used for those features which can be recognized on the small scales of hyperaltitude photographs. Such may provide direct information on sediment circulation in the sea, or a drainage pattern may provide an idea as to the nature of the underlying geology.

The photographic texture has been defined (Ray, 1960) as the frequency of tonal change within the image produced by an aggregate of unit features too small to be clearly discernible on the photograph. It is a function of the photographic scale. The photographic pattern is defined as an orderly spatial arrangement of geological, topographical, and vegetation features. When the features making up a pattern become too small to be identified individually, as on small-scale photographs, they may then form a photographic texture. Thus, we see that the scale of a photograph is of considerable importance in the interpretation technique.

At the very small scale of hyperaltitude photographs (e.g., 1:1,000,000), most patterns are recorded as photo textures only; large-scale features, such as landforms, drainage, and vegetation, are reduced to pattern scale. Carter and Stone (1974) have mentioned drainage patterns as useful elements in the interpretation of hyperaltitude photographs. We believe that landform patterns and vegetational patterns can be as effective. Landform patterns are dependent on the geomorphological evaluation of an area. At the small scale of hyperaltitude photographs, the individual details of terrain features are not visible, but the overall geomorphological response of major lithological units or structure is shown as a pattern that can be interpreted to identify the physiography.

The interpretation of a Gemini 5 photograph (NASA 65-HC-821) of the Changchenmo region (Raina, 1972b) is such an example. The landform pattern of the Pangong slates, which shows a subdued topography of irregular hills, is in considerable contrast to the pattern exhibited by the surrounding Ladakh granite, which permits separation of the two geological units. Similarly, a close scrutiny reveals a distinct and different landform pattern for the Permian limestone outliers as compared to the slates and granites that surround them. It is possible to differentiate these units on a photomap with full confidence as to their being different from the surrounding rocks.

The Siwalik Group rocks from the southern or outer ranges of the Himalayas are densely forested, whereas the rest of the Himalayas are barren. On the Gemini 6 photograph (NASA 65-HC-2089) interpreted by Raina (1972b), the Siwaliks show a much darker tone because of the vegetation and present a distinct pattern on the photograph. Thus, it is indicated that landform patterns and vegetation patterns at orbital photographic scales can be usefully employed for geological interpretations.

It should also be noted that only major lineaments, fold patterns, and those geological features which can be recognized on the photos from their shape alone can be directly identified on these pictures. Dips in general are not distinct, but trend lines are often clearly inferred. Tonal variations and landform or drainage patterns give indications of rock types, but their interpretation in terms of lithological details and the geological significance of their contacts is only possible through additional study of conventional aerial photography and available geological literature, including ground survey data.

Interpretation of Orbital Imagery of India*

Apollo-7 photograph, NASA AS 7-7-1748 (Figs. 1A and 1B). This was first studied by Abdel Gawad (1971), who interpreted two major lineaments, the Kali Gandaki lineament (GG) and one to the west of Everest (EE). Further detailed study of this photograph has helped in interpreting other linear structures (Raina, 1974). Two sets of lineaments are seen; one set of lineaments is trending north and south (these are in the nature of tear faults); the other set runs east–west and may represent block faulting.

The salient geological features that can be observed on this photograph are the following:

1. The Kali Gandaki valley appears to be a graben running in a north–south direction across the trend of the Himalayas.
2. There are a number of tear faults running nearly parallel in roughly north–south or northeast to southwest directions.
3. The tear fault west of Everest (EE) is a major geosuture. It is probably a sinisteral shear and has dragged and cut off a major north–south fold, called here the Arun Khola Structure.
4. To the east another parallel geosuture, also in the nature of a sinisteral tear fault (KK) separates the Kanchenjunga Structure from the Arun Khola structure.
5. The Brahmaputra Valley appears to be a graben bounded on the north and south by lineaments. The fault marking the northern border of this graben appears to have affected the physiographic development of the Nyenchen Tang Lha Range and the Tibetan Plateau.
6. It is also seen that the physiography of the Kunlun, Kailash, and Karakoram ranges has been affected by major shears.
7. Another interesting feature is the development of the Peiku Tso and Talung basins, which may be structural depressions, the former plunging to the south and the latter to the north. These were first observed in this Apollo orbital photograph; their existence was not previously known.

It is difficult to explain the full geological significance of the lineaments and geosutures seen on this photograph. Some details need to be checked out in the field. The geological significance of others may only be explained when the geology and structure of the area are studied in detail on conventional aerial photographs. However, the presence of these geosutures indicates that the present concept of the Himalayas having been formed owing to northeast–southwest compression and crustal shortening by the

*Interpretation of the orbital photographs received through the courtesy of NASA, ISRO, and ITC (the Netherlands) was carried out by the staff members of IPI.

FIG. 1(A)

Apollo 7 earth-sky view (NASA AS-7-7-1748). The world's dozen peaks that reach a height of greater than 5 miles above sea level are seen in this remarkable photograph from the Apollo 7 spacecraft at an altitude of 130 nautical miles. The 29,028-ft-high Mount Everest is at lower center. On the central horizon can be seen 28,250-ft-high Mount Godwin-Austen (K-2), some 800 miles northwest of Mount Everest. In the lower right, Mount Kanchenjunga rises 28,208 ft to separate Nepal from Sikkim. The snow line on the peaks was at 17,500 ft. In the upper right, the lake-studded highlands of Tibet are visible.

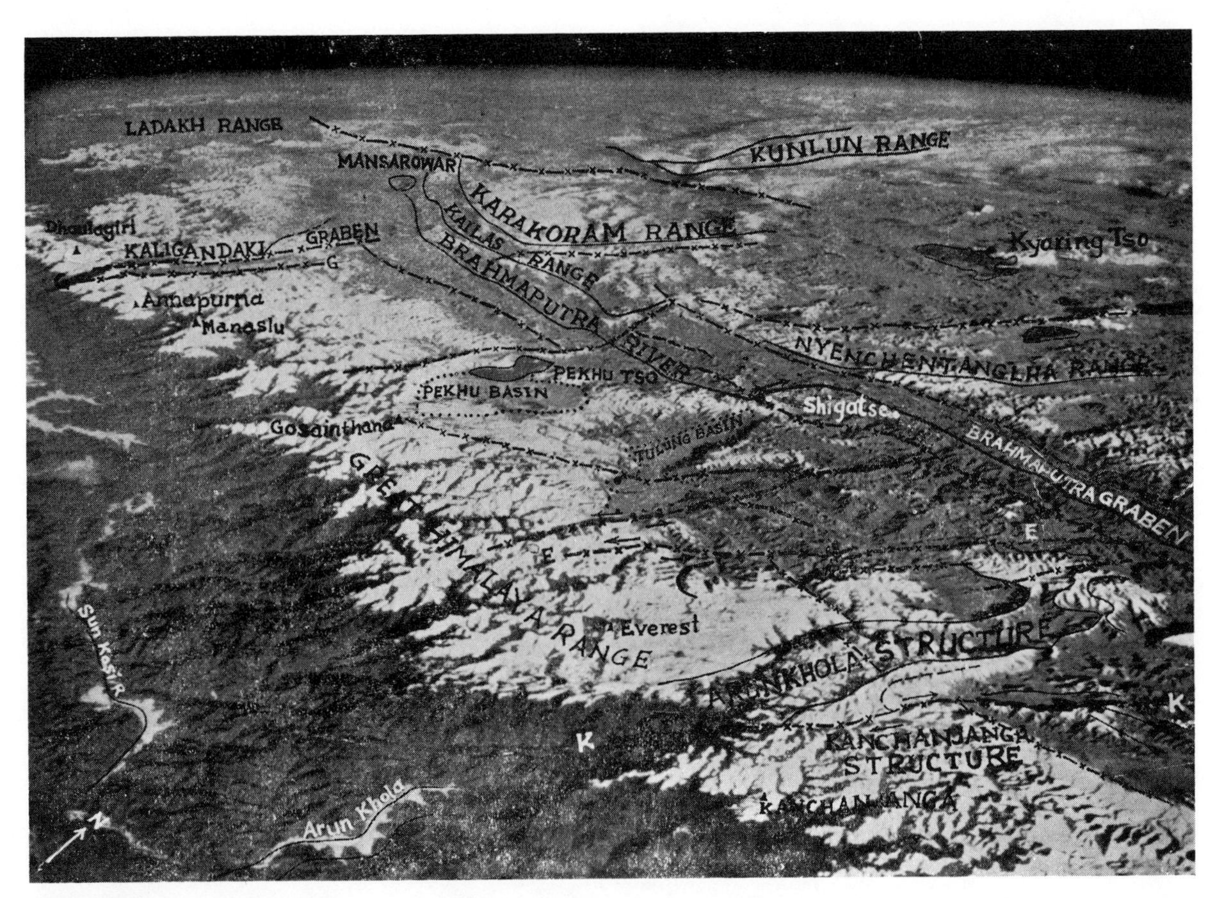

FIG. 1(B)
Interpreted Apollo 7 photograph of the Himalayas (NASA AS-7-7-1748).

northward drift of India against the Central Asian mass needs modification, especially the concept that the major Himalayan tectonic features are aligned in a northwest–southeast direction. The north–south trending tear faults, like the ones affecting the Arun Khola and the Kanchenjunga structures seem to divide the northwest–southeast trending Himalayan folds into a number of tectonic blocks and suggest major crustal shearing in that direction.

LANDSAT-1, imagery of the Terai region: NASA E1095-04454-7 (Figs. 2A and 2B). This picture is remarkable for its clarity. Its greatest advantage is the sharp delineation of land and water surface boundaries. The hill ranges in the northern part of the picture comprise the Siwaliks in the south and pre-Tertiaries in the north, separated by the Main Boundary Fault. This fracture plane can be clearly identified on the picture as a long sinuous boundary. The rocks to the north of it (pre-Tertiary quartzites, limestones, etc.) show a markedly different landform pattern than the Siwaliks lying to the south.

The Siwaliks in the south are bounded by a zone of subrecent conglomerates, representing coalesced fanglomerates brought down by the numerous streams and deposited on this piedmont zone owing to the change in their velocity. Farther south, ad-

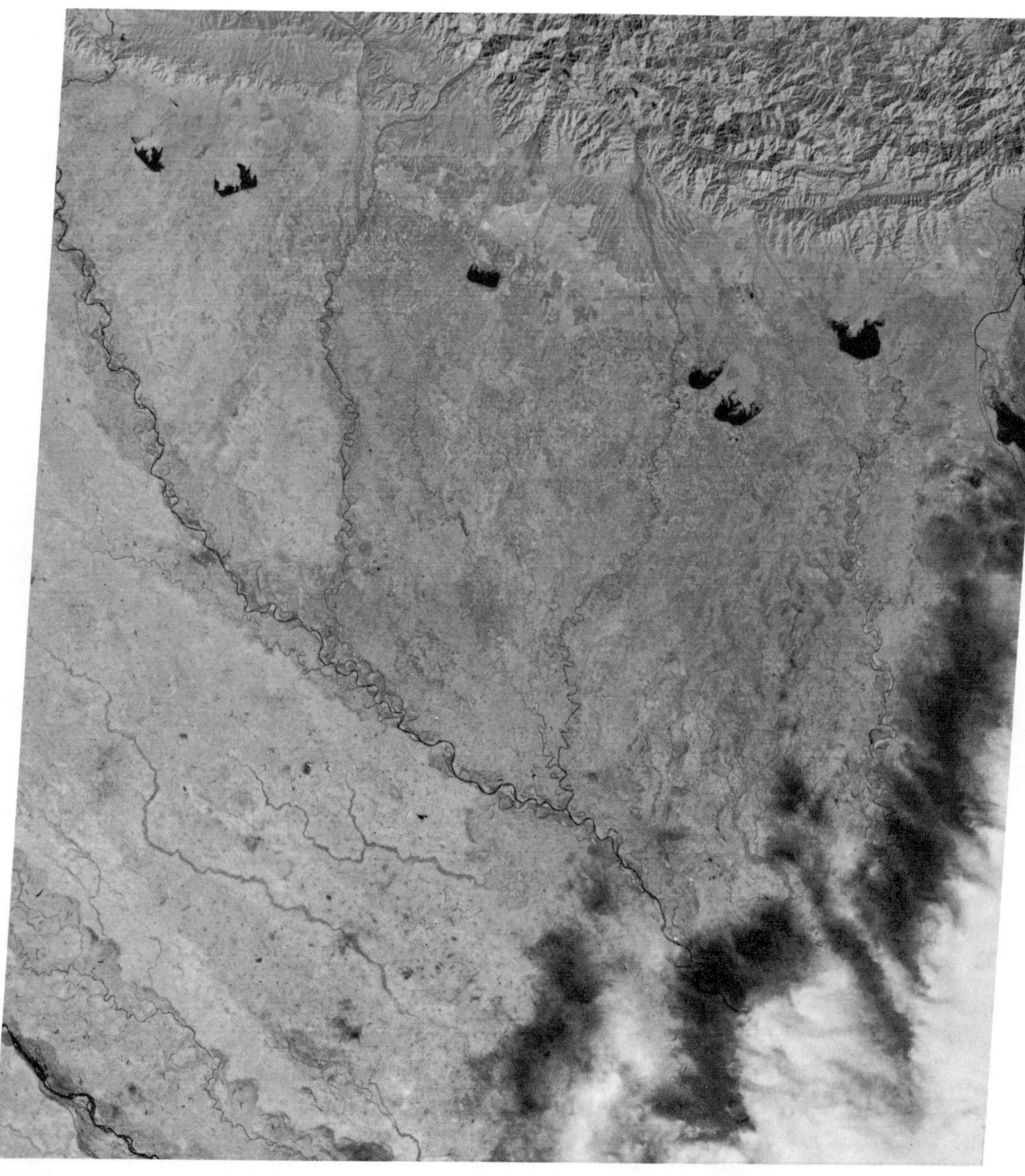

FIG. 2(A)
LANDSAT-1 image of Uttar Pradesh Terai region (NASA E-1095-04454-7).

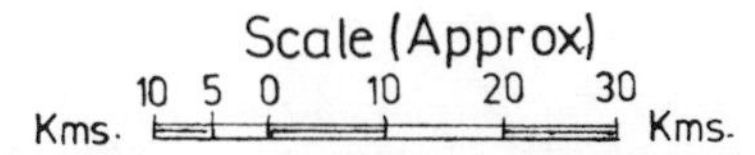

FIG. 2(B)
Interpreted LANDSAT-1 picture of Uttar Pradesh Terai region (NASA E-1095-04454-7).

jacent to this fanglomerate zone called Bhabar, is the alluvial plain called the Terai. Previously, the exact contact of the Siwaliks with the Bhabar and that of the Bhabar with the Terai were not clearly defined owing to the dense vegetation occurring in the area. However, infrared radiation has differentiated this vegetation growth and, for the first time, with the help of the band 7 image, it is possible to delineate these contacts (Raina, 1974). On the image, the nature of the Bhabar coalesced fanglomerates is clearly seen and some of the fans are evident. This type of study is of great help for determining the geohydrological regimen of the area and for locating suitable tube-well sites. The picture also demonstrates the advantage of infrared imagery in geohydrological studies.

LANDSAT-1, imagery of Mettur-Coimbatore area: NASA E-1201-4393-5 (Figs. 3A and 3B). A part of the Indian peninsula is shown on this image, including two broad categories of terrain features. These are the high-relief outcrop area and the low-relief pediment and buried pediment areas. Based on tonal variations and landform patterns, and with the aid of available geological literature, the high-relief outcrop areas may be subdivided into four litho-units: Dharwars Peninsular gneiss, charnokite, and dunite. The charnokites

FIG. 3(A)
LANDSAT-1 image of Mettur–Coimbatore area (NASA E-1201-4393-5).

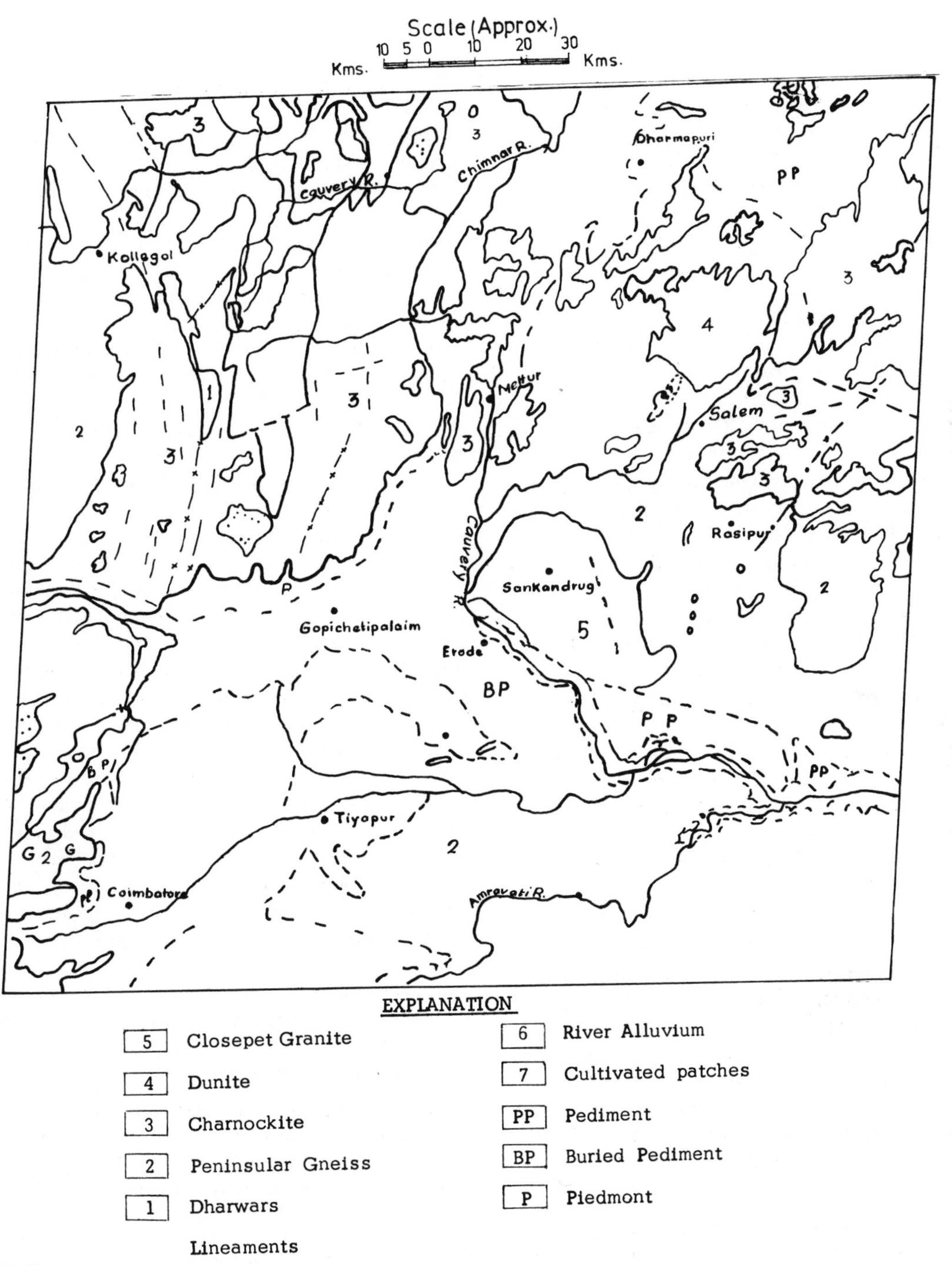

FIG. 3(B)
Geological map of Mettur–Coimbatore area interpreted from LANDSAT-1 picture (NASA E-1201-4393-5).

form positive relief and dark tonal expressions. They are characterized by north–south structural trends and a dissected upland. The Closepet granite has a lighter tone, and the dissection of the upland is finer compared to the charnokites; that is, the landform patterns of the two show subtle variations. The Dharwars are rather inconspicuous, light toned, and have been delineated from preexisting geological data only. The dunites have a dark tone, positive relief, and a landform pattern different from the charnokites. They also lack any distinct structural trend. Their identification as dunite is based on existing literature and is also tentative. Dip or foliation planes are not distinguishable, but trend directions are easily observed.

The low-lying areas have been interpreted as the Peninsular gneiss. The light-toned low-lying areas are interpreted as exposed pediment of this gneiss, whereas the darker-toned area is interpreted as buried pediment.

At the foot of the charnokite and granite escarpments, a wide belt or piedmont zone is distinguishable by its whiter tone, which possibly consists of coarse detritus derived from the adjacent hills by gravity and water channels.

River alluvium, characterized by a darker tone and a distinctive landform (bordering rivers), is mainly restricted to the major river courses like the Cauvery and the Amravati. The high moisture content and phreatophytic vegetation are responsible for the dark tone of these areas.

Light-toned patches on the uplands and other outcrop areas are interpreted as cultivated ground.

LANDSAT-1, imagery of Chandigarh–Chor area: NASA E-1079-04562-5 (Figs. 4A and 4B).* Some lineaments, the contact of the Siwaliks with the Bhabar, and the extent of the Siwaliks are well shown. Several tonal variations are seen, which are expressions of litological units. Their interpretation in terms of geological features and lithological details have been worked out on the basis of the familiarity of the interpreter with the terrain and the available geological literature.

Some new lineaments have been indicated in the Tertiary rocks, which need confirmation through field checking and detailed study of aerial photographs.

The water bodies and the drainage networks are clearly shown in the picture. Different tonal contrasts are also seen in the Ganga alluvium, perhaps due to salinity variations.

LANDSAT-1, imagery of Madras–Chittoor area: NASA E-1182-01330-5 (Figs. 5A and 5B).† The more important geological features that may be delineated are the outcrop boundaries and certain broad structural elements. Outcrops with low relief, such as the Gondwanas and Neogene (Cuddalores) sediments, have been interpreted from tonal variations and comparison with existing literature.

The Dharwars have low relief but show clear northwest–southeast structural trends. The basement gneisses and granites have prominent east-northeast, west-southwest trending fractures and have been mapped in the western part. Their interpretation becomes difficult when they are covered with alluvium. The charnokites and Closepet granite are definable because of their distinctive trends and relief.

The Cuddapahs with their high relief and sedimentary characteristics can easily be identified. Several bedding dips are visible, and the less resistant units (Cumbums) have a different landform pattern.

The Neogene sediments that occupy the coastal areas show low relief and a slightly darker tone. These are distinguishable from the other lighter-toned alluvium.

In Pulicat Lake, the shallow regions with their sandy beds are clearly visible and help in indicating water depth. Other interesting features that are identified include the following:

1. A northwest–southeast trending major fault (F′F′) running from west of Pulicat lake up to the Palar River may represent a boundary fault. It seems to divide the high-relief basement and the Neogene

*The image covers parts of Himachal Pradesh, Haryana, and Uttar Pradesh. It was interpreted by R. P. Sharma, Research Professor at IPI.

†A very clear image of part of the eastern coast of India. It has been interpreted by D. P. Rao, Professor of Geomorphology at IPI.

sediments in the east. Toward the south this fault has been offset by another northwest–southeast fault.

2. Another north-northeast, south-southwest trending lineament is seen running from the south of Pulicat Lake to the east of Arcot.
3. South of the Palar River a geological feature is marked by tonal and drainage anomalies. Its significance is not clear and needs more detailed study. It may represent a structural high in the basement rock.
4. A buried channel is seen east of Arcot, running in a northeast direction. It probably represents a former course of the Palar River, which is now flowing in a southeast direction.
5. The remnants of two or three highly dissected pediplaned surfaces may be seen.
6. The low basement surface east of the fault F′F′ appears to represent a down-faulted block where later sedimentation has taken place. This view is supported by the fact that the drainage to the west of this fault is largely fracture controlled, whereas in the east it is mainly of an insequent type.
7. Pulicat Lake, which lies east of the fault (F′F′), is a possible remnant of the Neogene sea and is still undergoing deposition.
8. The high beach ridges–strand lines indicate a retreating sea and an emergent coastline.

DRAINAGE PATTERN OF SAURASHTRA, GUJERAT, AS DERIVED FROM LANDSAT-1 IMAGERY

Four images (Figs. 6, 7, 8 and 9) having MSS information in the spectral band of 0.8 to 1.1 μm were used to compile the drainage pattern (Table 1). One picture covers an area of about 186 × 186 km. It is difficult to assign proper dimensions to the drainage patterns in terms of their width because these have been recorded by the process of scanning; their width is shown in relation to their contrast with the background. Since the basic resolution is about 60 m, it is not possible to observe the finer drainage patterns that pertain to the lower order of drainage density.

Drainage Pattern as Observed in LANDSAT Imagery

Figure 10 shows the map of the Saurashtra area showing third- and fourth-order drainage patterns prepared from an overlay of the LANDSAT picture. The latitude and longitude lines were taken from the annotation marks of the latitude and longitude provided on the image itself.

Since adjoining images have some areas that overlap, it was possible to have continuity of the drainage pattern from one picture to the next. No geometrical corrections were applied to the pictures, and the drainage patterns have been drawn as they were observed on the images in their respective locations. A drainage pattern can have coarse, medium, or fine texture. Certain materials will yield a fine texture, whereas one may find a medium texture in silty or clay soils and coarse textures in sandstone belts.

Figure 11 shows the different drainage basins based upon the information in Fig. 10. The area is composed of six distinct watershed regions. Region I drains into the Gulf of Kutch and region II drains into the Arabian Sea. Regions III, IV, and V drain into the Gulf of Cambay, and region VI drains into the Little Rann.

If we consider that the differences between rainfall and runoff is due to infiltration, it should be possible to determine the discharge due to surface runoff from drainage basins by considering the drainage density and climatic conditions. Thus, the drainage density becomes an important parameter in determining the hydrology of any basin. The rate of infiltration is inversly proportional to the drainage density in a watershed. Thus, the water that cannot filter into the ground must run off. Runoff conditions will in turn depend upon the precipitation history, slope, plant cover, and various other factors.

To investigate the characteristics of any drainage basin, one should have sufficient discharge and precipitation data, which can define the relationship between drainage densities and discharge characteristics.

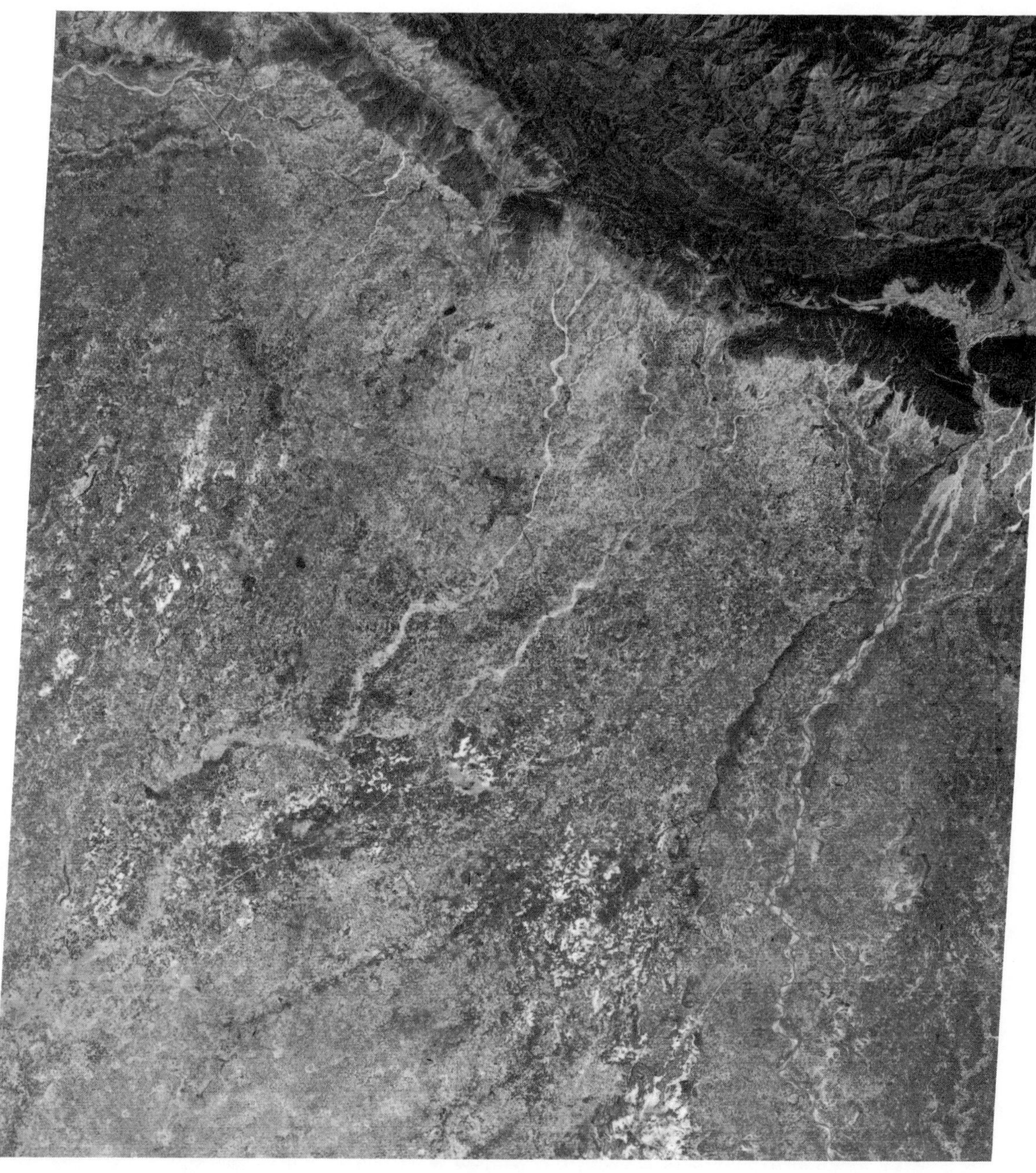

FIG. 4(A)
LANDSAT image of Chandegarh–Chaur area (NASA E-1079-04562-5).

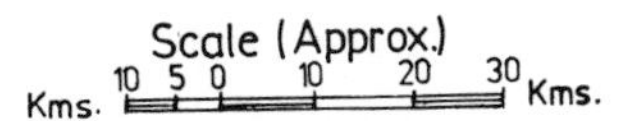

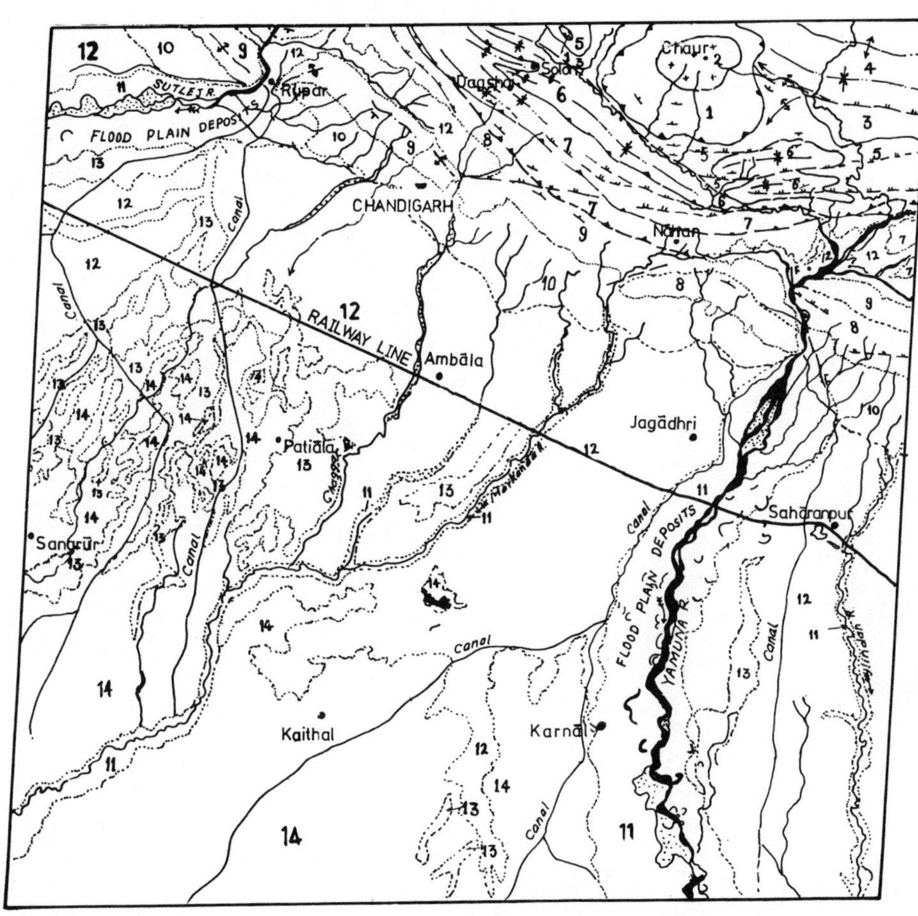

EXPLANATION

14 Back swamps	7 Tertiaries, Sandstones, Shales.
13 Natural levees	6 Krols:- Limestone, Shales, Phyllites
12 Alluvium	5 Jaunsars:- Slates, Phyllites, Quartzites
11 Flood plain deposits	4 Deobans:- Limestone, Slates, Quartzites
10 Piedmont deposits	3 Simlas:- Slates, Phyllites, Quartzites
9 Upper Siwaliks:- Conglomerates, Clay	2 Chaur Granite
8 Siwaliks:- Clay, Sandstone, Shale; Conglomerates	1 Jutoghs:- Phyllite, Schists, Gneisses

Thrust — Fault — Dip — Anticlinal axis — Synclinal axis

FIG. 4(B)

Geological map of Chandigarh–Chaur area interpreted from LANDSAT-1 picture (NASA E-1079-04562-5). Interpreted by R. P. Sharma, Research Professor, IPI.

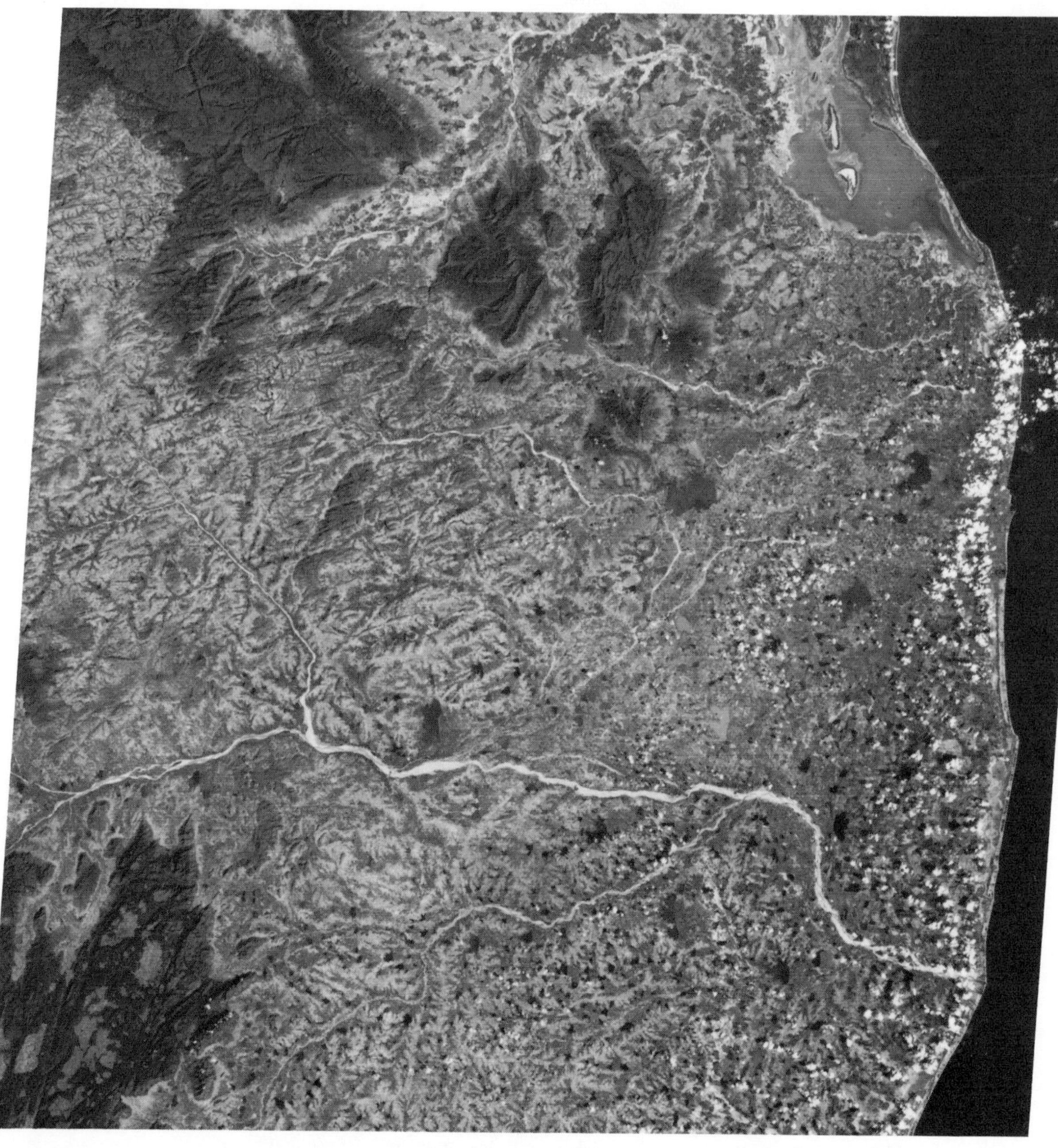

FIG. 5(A)
LANDSAT image of Madras–Chittoor area (NASA E-1182-04330-5).

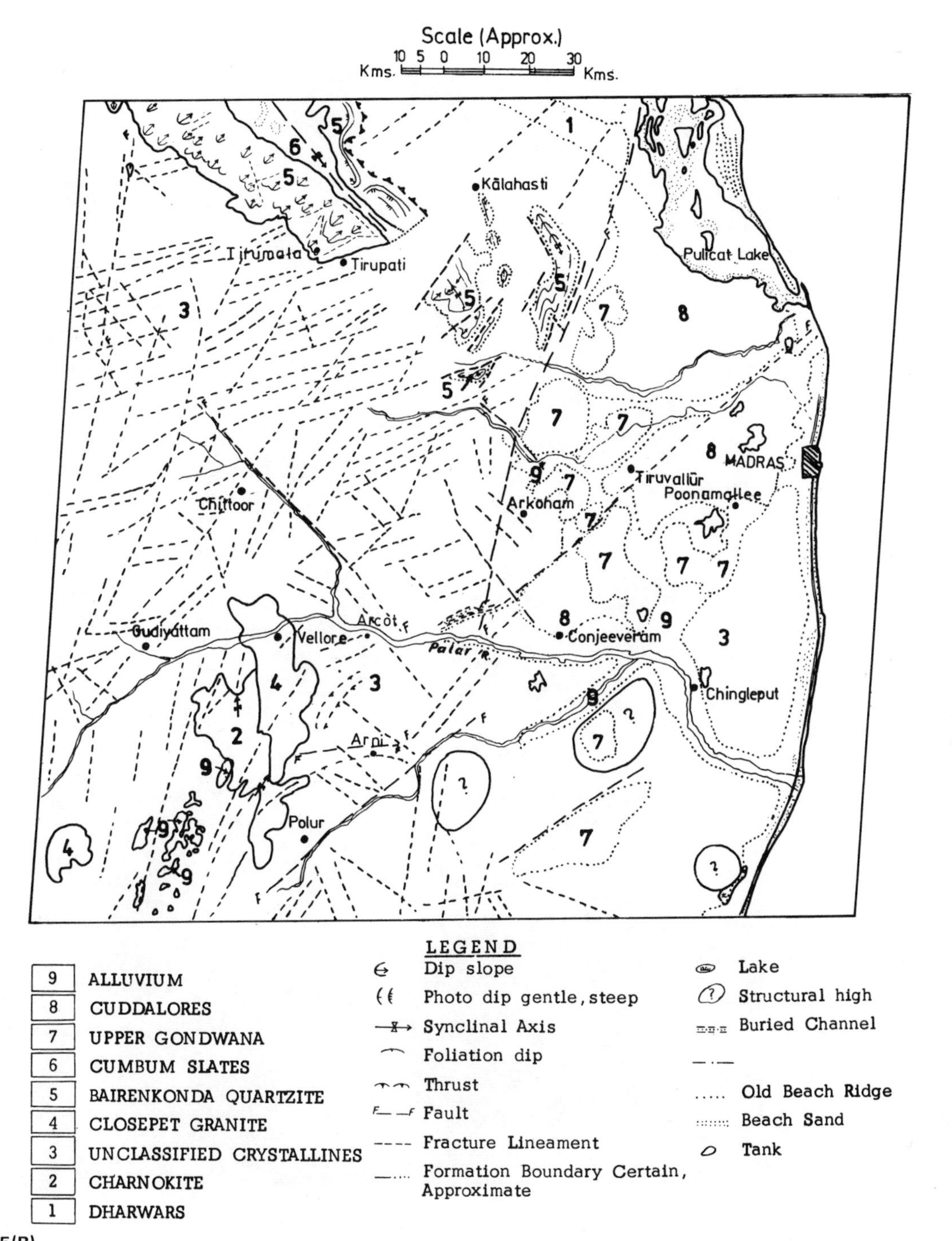

FIG. 5(B)

Geological map of Madras–Chittoor area interpreted from LANDSAT-1 picture (NASA E-1182-04330-5). Interpreted by D. P. Rao, Geomorphology Professor, IPI.

FIG. 6
LANDSAT image of Kutch (Gujerat) area (NASA E-1117-05103-7).

FIG. 7
LANDSAT image of Kutch (Gujerat) area (NASA E-1117-05110-7).

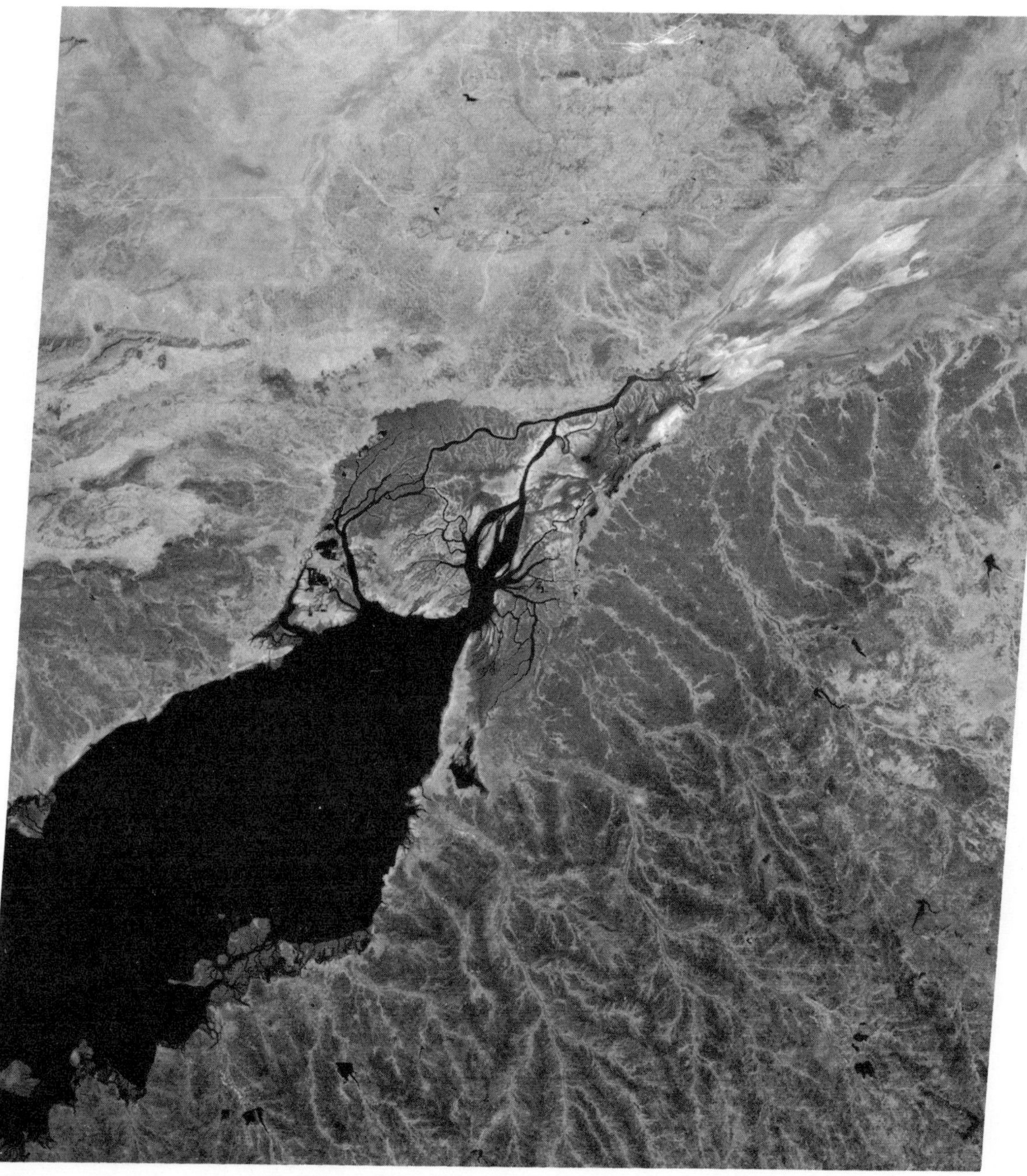

FIG. 8
LANDSAT image of Kutch (Gujerat) area (NASA E-1118-05162-7).

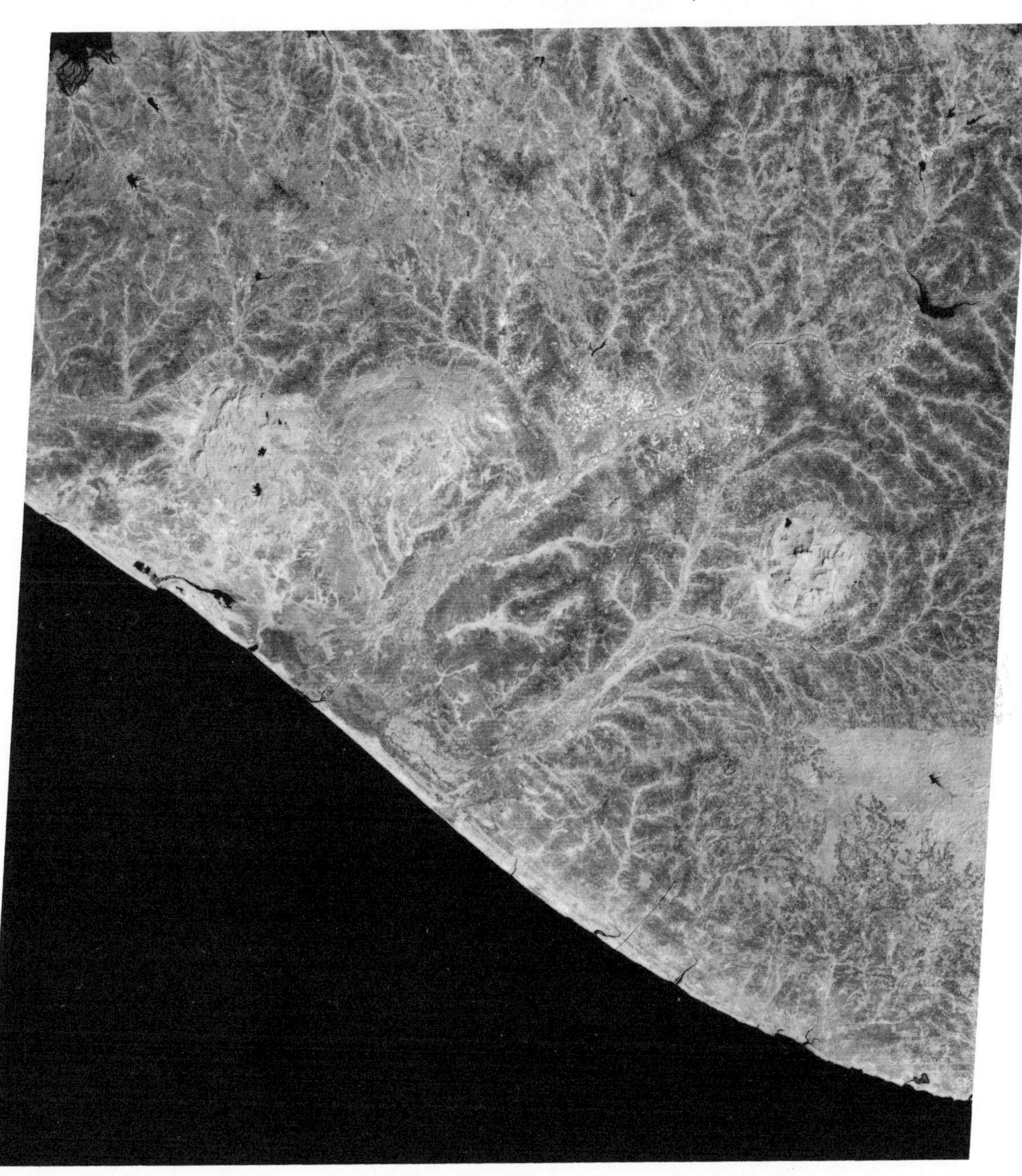

FIG. 9
LANDSAT image of Kutch (Gujerat) area (NASA E-1118-05164-7).

TABLE 1
Details of the LANDSAT Imagery Used for the Kutch (Gujerat) Analysis

Fig. No.	*I.D. No.*	*Date*	*Image Format Center*	*Position of the Sun*	
6	11177-05103	Nov. 17, 1972	Lat.: 23°03′N	Elev.	40°
			Long.: 71°55′E	Az.	146°
7	1117-05110	Nov. 17, 1972	Lat.: 21°36′N	Elev.	41°
			Long.: 71°33′E	Az.	145°
8	1118-05162	Nov. 18, 1972	Lat.: 23°00′N	Elev.	40°
			Long.: 70°28′E	Az.	146°
9	1118-05164	Nov. 18, 1972	Lat.: 21°34′N	Elev.	41°
			Long.: 70°06′E	Az.	145°

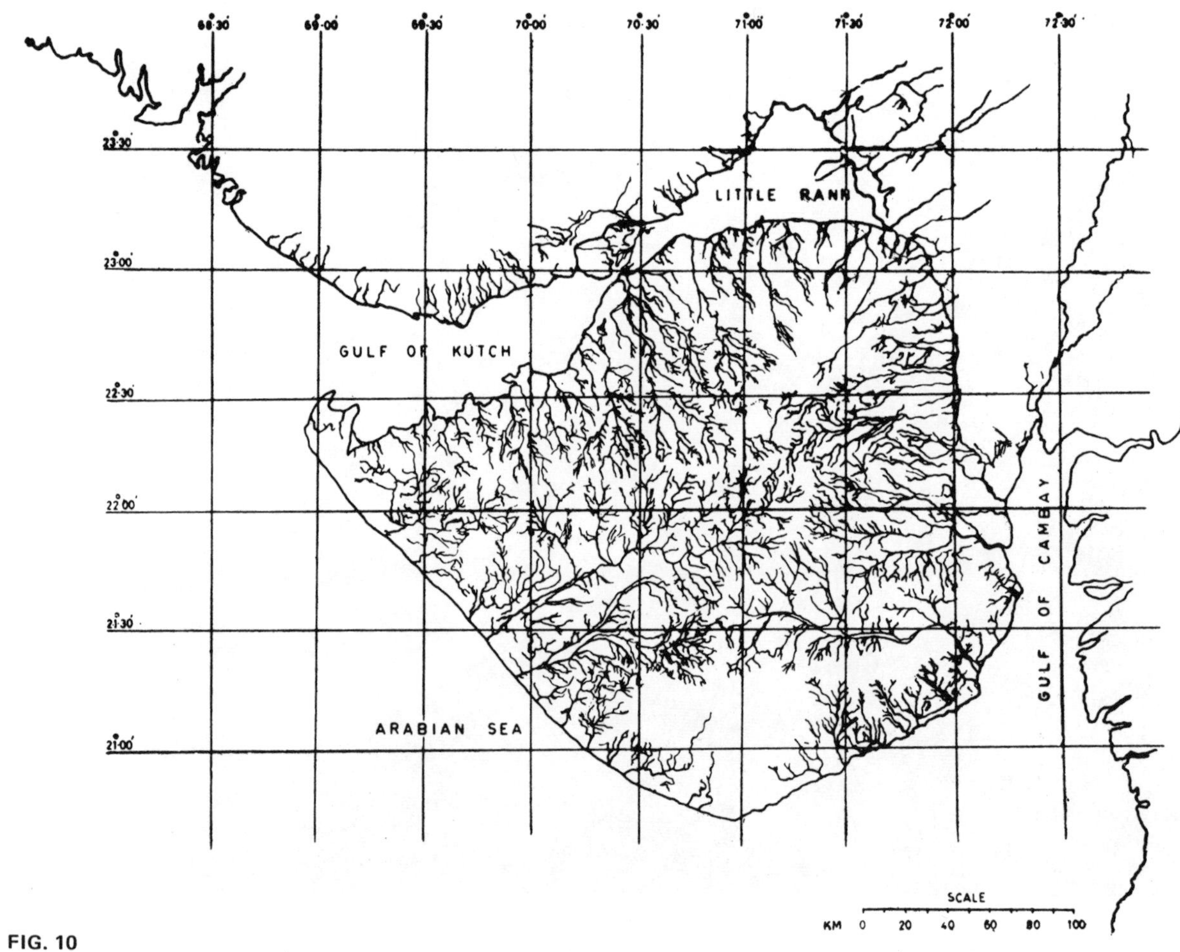

FIG. 10
Drainage pattern as observed from LANDSAT pictures of Saurashtra area.

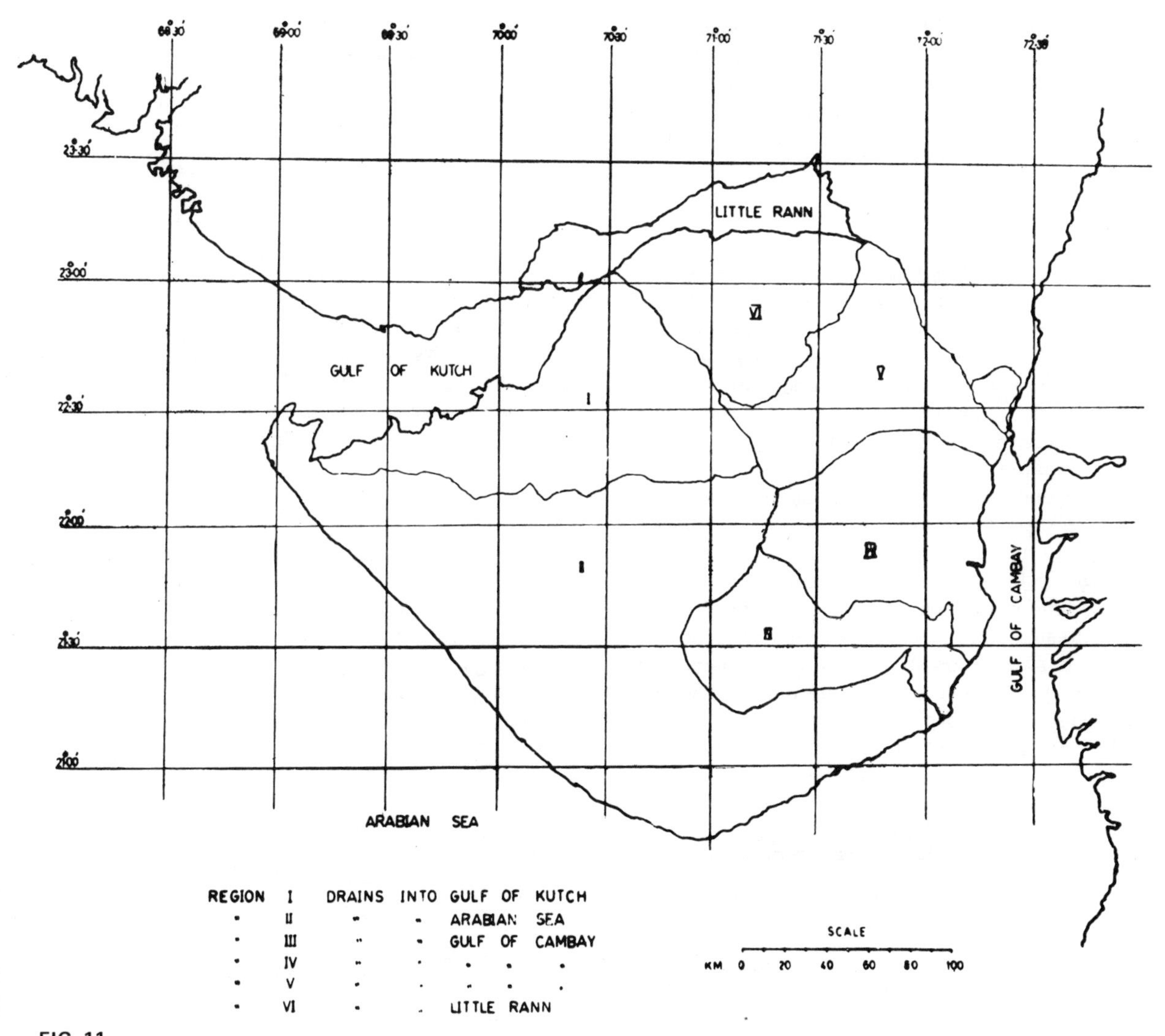

FIG. 11
Watershed drainage basins prepared on the basis of information in Fig. 10.

Rainfall and Water Bodies

As has been shown, to calculate the runoff for different watersheds, certain rainfall information is required. Standing water bodies form naturally, depending on the topography of a basin. Artificial water bodies also form as the result of damming stream channels. Together these comprise an important unit for purposes of irrigation and public water supply. Table 2 describes the average rainfall characteristics and the number of water bodies for each of the watersheds shown in Fig. 11. The precipitation data were obtained at locations typical of each of the drainage basins.

The climatic classifications of the six watersheds depend on the annual rainfall. Region VI may be classified as semiarid, since, according to Blair's classification system, it has an annual rainfall of less than 500 mm. All the other regions shown in Table 2 belong to the subhumid class.

Calculation of Drainage Density

In sketching the different streams seen on the LANDSAT imagery, it was found that certain limitations need to be placed in recognizing all orders of streams. This is important, because for making calculations of drainage density all orders of drainage streams must be considered. There are several limitations because of the scale size and various other factors required for recognizing streams. Thus, it is important that certain conventions be followed which will allow systematization of the process of computing drainage density. In the case at hand, it is difficult to recognize first- and second-order streams and one must be limited to the higher orders. Inasmuch as the major streams and some of the tributaries do not reflect the average long-term infiltration characteristics, it is not possible to compute drainage density from this scale of information.

As there is no standard method of classifying the drainage system, it is difficult to compare the results of the calculations with any standardized values. Table 3 shows the drainage density calculated for watershed regions II, IV, and V of Fig. 11.

TABLE 3
Drainage Density for Kutch (Gujerat)

Watershed Region	*Drainage Density (km/km²)*
II	0.174
V	0.330
IV	0.415

The drainage density of the drainage basin is calculated by applying the following formula:

$$D = \frac{L}{A},$$

where D = drainage density (km per km^2)
L = total length of channels in the drainage basin (km)
A = area of the drainage basin (km^2)

Considering the "law of stream lengths," it is possible to calculate the drainage length for the lower orders when information is available only for higher orders. The law of stream lengths states that the

TABLE 2[a]
Rainfall Characteristics in Kutch (Gujerat)

Region as Indicated in Fig. 11	*Place Taken as Typical for the region*	*Normal Rainfall (mm, annual)*	*Drainage Runs Into*	*No. of Water Bodies, Natural and Artificial (as seen in LANDSAT pictures)*
I	Rajkot	594.3	Gulf of Kutch	12
II	Junagadh	843.7	Arabian Sea	18
III	Amreli	515.0	Gulf of Cambay	4
IV	Bhavnagar	620.1	Gulf of Cambay	9
V	Ahmedabad	782.2	Gulf of Cambay	5
VI	Surendranagar	487.0	Little Rann	5

[a]For region V, rainfall data have been taken as represented by Ahmedabad, even though geographically it is not situated in the region. The maximum size of a water body is found to be about 4 km^2.

length ratio R_L tends to be constant throughout the successive orders of a watershed (Chow, 1964):

$$\frac{R_{Lu}}{R_{Lu-1}} = \frac{Lu}{Lu-1},$$

where R_{Lu-1} = length ratio for order $u-1$
Lu = stream length for the order u
$Lu-1$ = stream length for the previous order

Applying this formula, the values calculated for the drainage density of the three basins given in Table 3 can be modified. The resultant values are given in Table 4.

Thus, according to the classification given by Belcher and Majtenyi (1967), region II has coarse drainage density, region IV is classified as fine, and region V as medium.

In using the LANDSAT pictures for studies of drainage density, we have seen that it is possible to sketch the drainage pattern and the watersheds. This information in conjunction with relevant data on the basins can lead to very useful information about the hydrology of the area.

The information provided is limited by the 60-m resolution of the image; however, it is possible to extrapolate this information and calculate the drainage density for the entire watershed. Confirmation can be derived from high-resolution aerial photography where it is possible to detect first- and second-order channels also. The use of climatological data in conjunction with the information derived from the satellite imagery can lead to detailed information on the hydrological cycle of a region.

TABLE 4
Modified Drainage Density for Kutch (Gujerat)

Region	*Drainage Density (modified)*
II	1.74
V	8.25
IV	26.56

PRELIMINARY QUANTITATIVE LAND-USE CLASSIFICATION IN THE PUNJAB*

The NASA LANDSAT-1 data, that is, single-band (black and white) transparencies and prints, and the magnetic tapes containing digitized radiance values for each resolution element from the scanner, can be employed by a variety of users, for example, photo-interpreters and computer-oriented natural resource scientists. The four-band scanner obtains information that is quantitative (*Data User's Handbook,* 1972) and which can be analyzed statistically. In addition, the LANDSAT-1 provides orthogonal coverage of large regions under uniform and repeatable lighting conditions; preliminary evaluations of scanner imagery have suggested that it is essentially of cartographic quality. These attributes have special significance in relation to a developing country such as India, where a large part of the natural resource evaluation process has to depend upon less sophisticated techniques and yet, where required, quantitative analyses should be possible.

The LANDSAT coverage of India until September 1972 was examined and, as expected, most of the country was found to be cloud covered due to the monsoons. A relatively clear frame in the north-western area of the country, including portions of the Punjab, Uttar Pradesh, and Himachal Pradesh states, was chosen for investigation. This area of the country is agriculturally productive, especially in the state of Punjab. Black and white positive photographs of this frame made from the transmitted data are shown in Figs. 12 and 13 and correspond to the scene in bands 5 and 7, that is, the red-infrared and infrared bands of the multispectral scanner (MSS 5 and 7).

Various features on the two frames can be correlated with information on standard maps. Ground truth for this season and area is not available; hence, certain features and the various agricultural categories will have to be verified further.

*The computer-based analysis was carried out in 1972 when one of the authors (Ravi D. Sharma) was a Visiting Research Physicist at the Environmental Research Institute of Michigan.

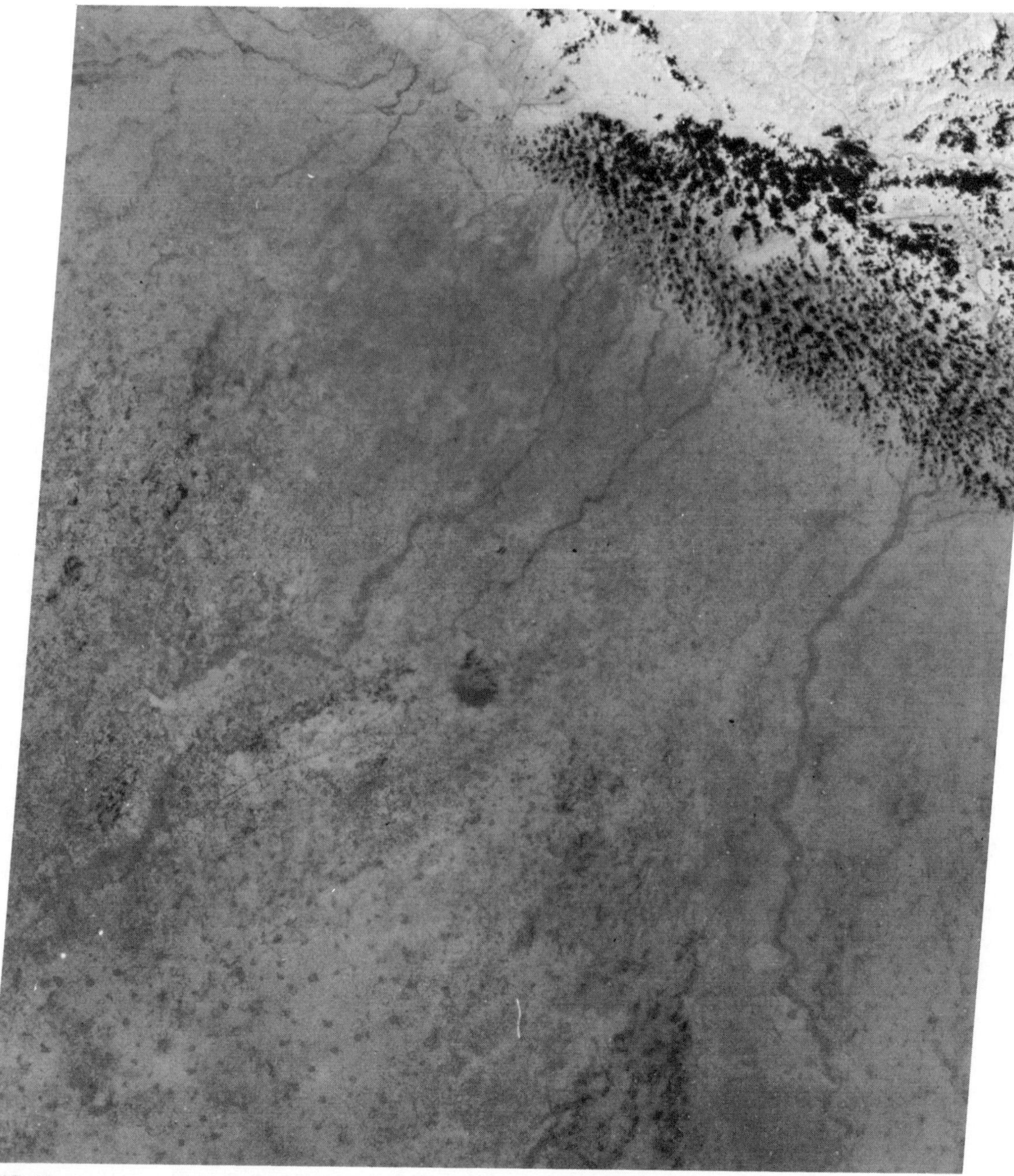

FIG. 12
LANDSAT image of the Punjab, Uttar Pradesh, and Himachal Pradesh areas of India (NASA E-1043-04562-5).

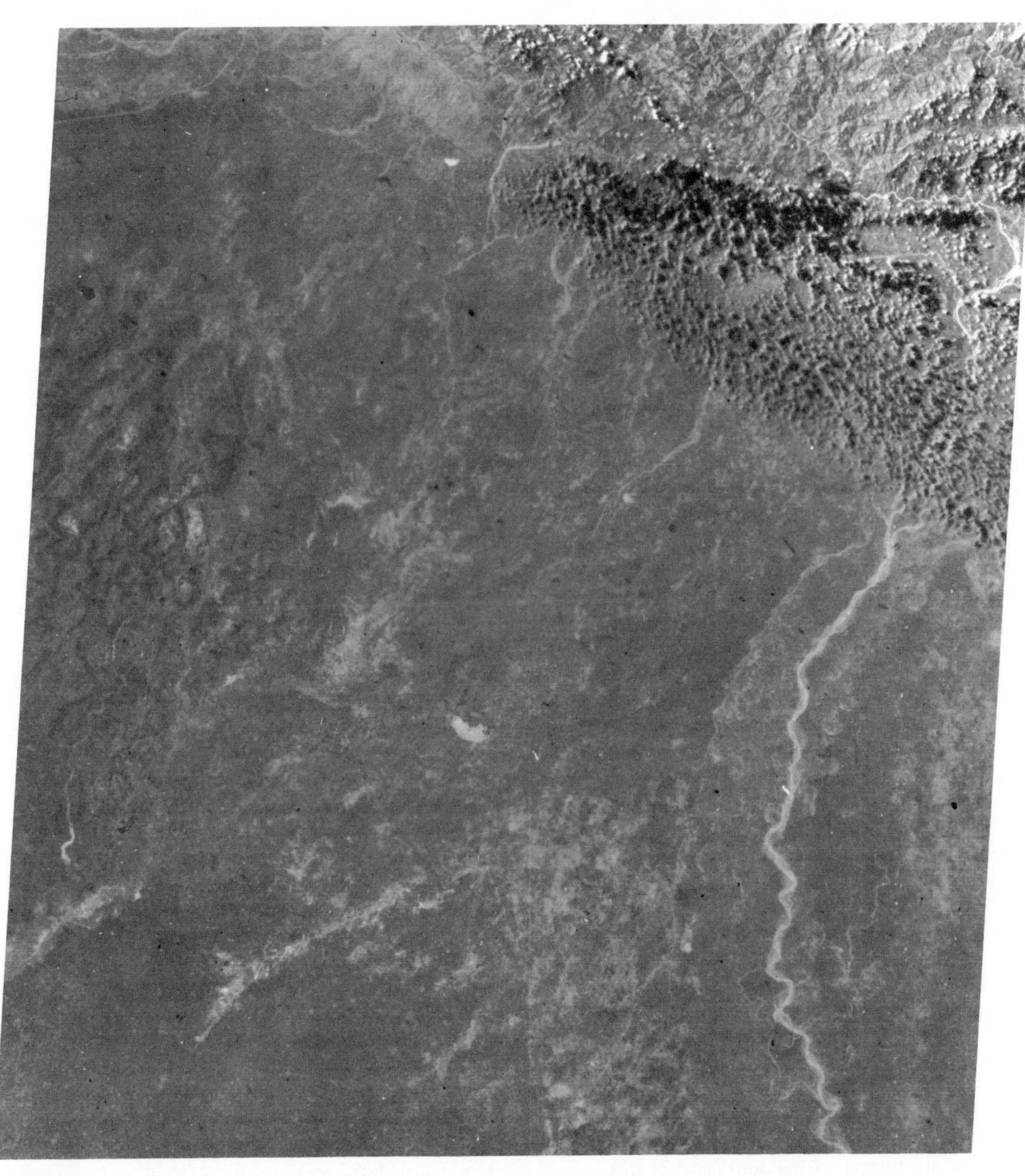

FIG. 13
LANDSAT image of the Punjab, Uttar Pradesh, and Himachal areas of India (NASA E-1043-04562-7).

Single-Band LANDSAT Images

A striking characteristic of Fig. 12, particularly of MMS-4, is the presence of atmospheric haze over the Indo-Gangetic plain while the Himalayan peaks and other high areas have very clear definition. Thus, the haze is restricted to the lower altitudes. The matter of correcting for the atmospheric effects in LANDSAT imagery in general has been described elsewhere (Sharma, 1972). The clouds are confined to the northeast corner of the frame, essentially along the mountain–plain boundary. The major river formed by the three rivers in the Himalayan region and trending north–south (Fig. 2) is the Jumna (Yamuna) River, which is second only to the Ganges and which joins it in the eastern part of India. The average height of the plains is about 150 m above mean sea level; the peaks in this LANDSAT-1 frame rise from 1000 m to more than 3000 m. The valleys are heavily forested and have Reserve Forests. Figure 13 shows major water bodies and wet soils as bright areas and also some old and new runoff features running south-southwest from the mountains. Toward the west-central portion is the inverted-question-mark-shaped feature called Choa Nadi (Choa River). A clear print also shows the major canals in this area. The most striking among them is the Sirhind canal in the west at the northernmost tip of the frame. It comes from the Sutlej River in the north just at the western edge of the frame and runs westward and again southeast. The Kotla branch goes out to the southwest, and on a clear negative one can also see the Chaggar branch again toward the southwest. The Choa branch can also be seen in a clear negative and appears nearly parallel to the Choa Nadi.

Figure 12 shows the agriculture and vegetation aspects of the frame. The bright areas are vegetated areas, and the dark areas correspond to bare fields and saline sand dunes. The sandy bare soil and water features also appear dark. Thus, the vegetated areas are concentrated in the western portion of the frame, excepting the forests in the mountain regions. The western region has a large saline sand dune at the mid-point of the frame and west of the Nabha-Patiala area. There are forests to the southeast of this area running diagonally from the southwest to northeast, and pehaps along the Ghaggar River. With magnification, further details can be obtained, but single-band visual interpretations from photographic data are difficult to quantify.

Quantitative Analysis Using Electronic Image Data

Our attempt was to obtain the maximum detail from data as telemetered from the satellite and reformatted. The magnetic tapes contain quantitative information, and the spatial resolution obtainable depends upon the contrast between the objects that one attempts to distinguish. However, the instantaneous resolution element corresponds to approximately 1 acre of ground. The magnetic tapes contain the four-band radiance values as output voltages from the on-board detectors, which in the case of the scanner are digitized and transmitted. A gray-level slicing (i.e., voltage step chopping) of single-band imagery and display of these gray levels in the form of a computer printout is a conceptually simple operation but requires a large area for display if every scan line and every resolution element is to be printed on a gray map. For example, the LANDSAT frame under discussion would lead to printing about 3000 $\times$ 3000 (i.e., 10^7) characters on computer paper, which would be more than 240 m^2 in area.* For simplicity of display, therefore, a smaller area approximately 25 $\times$ 45 nautical miles (i.e., more than 1600 km^2) in the Nabha-Patiala area covering the central half of the westernmost 40-km-wide strip was chosen. This area has agricultural, forest, and irrigation aspects, and merges in the south with desert areas. The symbols for each gray-level map were assigned after 300 lines were sampled to determine their distribution in various voltage values. These could also be manually set if it were so desired.

The specific software for processing the digital multiband data had been developed at the Environmental Research Institute of Michigan (ERIM) over the past several years. The same was modified to accept LANDSAT four-band data from computer compatible tapes. The computer used was the IBM-

*Now visual displays and a squirt-gun system are available but are still an expensive hardware.

7094, the selected area corresponding to the first quarter-frame of LANDSAT. Points 1 through 500 and lines 500 to 1700 were chosen to create the gray map, which amounted to approximately one eighth of the total LANDSAT frame. Band 5 was mapped using the auto level set feature of the program by which the first 300 lines were sampled to determine the distribution of points with various voltage values, and gray-level symbols were assigned according to this distribution. Since MSS-5 does not show water areas very clearly, MSS-7 was chosen with the auto level set to gray map the known water areas. Twenty "training sets" were initially selected, based on knowledge of known features on maps and from the gray-level printouts, which included several assumed agricultural areas. These agricultural areas were derived primarily from field patterns noticed on the gray maps of MSS-5 and were concentrated primarily in the northeastern portion of the selected frame. Thus, three types of forested areas, three saline sand dune areas, three bare soil areas, two water areas, and the remaining nine presumed agricultural varieties were delineated on the gray map in terms of several points corresponding to each of the above type areas as samples.

The program STAT was used to calculate statistics (i.e., the mean signal, variance, covariance, etc.) for each of the 20 training sets. The signatures were plotted and were found to overlap considerably. Also, the probability of misclassification for each pair of signatures was calculated using the program DIST. This probability was found to be high, especially in the agricultural categories. Therefore, the several signatures that looked alike were combined, the number of classes was reduced to eight object classes (training sets), and the program DIST gave low pairwise probability of misclassification for each pair of signatures. The average probability of misclassification thus reached was 0.0518. For the combined classes, the statistics were again computed. The initial classes that emerged were water, saline dune areas, bare fields 1 and 2, forests 1 and 2, and agriculture 1 and 2. The eight training set signatures were used to create a recognition map. However, all points for which the probability of the point belonging to the winning signature was less than 0.01 percent were edited out, and the remaining points were mapped into the recognition map utilizing all four channels of data. The resulting recognition map was color coded with the following code:

1. Saline dune area: light red.
2. bare field types 1 and 2: dark red.
3. Forest types 1 and 2: black and gray, respectively.
4. Water: blue.
5. Agriculture types 1 and 2: light and dark green, respectively.

The acreage for each class is available. The recognition map is shown in Fig. 14*, and a blowup of a portion of it is shown in Fig. 15. The vegetation along the canal and the water features can be identified easily, together with the fact that the agricultural areas are primarily concentrated in the northeast and the desert–bare soil area increases toward the southwest portion of the selected area. The average field size may be derived by looking at the field patterns on the original computer printout, which correspond to 6- to 12-ha field sizes. However, at this stage of classification, crop identification is difficult; only the agricultural areas can be identified. Since Fig. 14 is a photographic reduction of a computer printout measuring 3.5×2.5 m^2, it is not possible to identify many of the details that may be seen normally in the original printout.

There has been no detailed ground correlation of the features observed on the LANDSAT data products, and a variety of conjectures may be drawn from the data alone. Was a large portion of the Ghaggar basin inundated? Is the clear definition of the basin due to high soil moisture? Had it rained just before the September 4, 1972, orbit of LANDSAT-1? Without further ground truth, the interpretation remains conjectural.

In conclusion it may be observed that physiographic and geological features are well defined. Broad land-use classifications may be mapped and seasonal land-use changes may be monitored using LANDSAT imagery. On the other hand, detailed crop classifications, which have proved to be successful for areas of the midwestern United States, appear to raise

*Figures 14 and 15 will be found in the color insert.

formidable difficulty in the context of the small field sizes and mixed crops of India. It is possible that convex-mixture techniques will be useful when more LANDSAT data are obtained and correlated with more detailed ground truth.

ACKNOWLEDGMENTS

Ravi D. Sharma would like to acknowledge Mr. Frederick J. Thomson and Mr. James Morganstern for their contributions and participation in the computer analysis of land-use classification studies relating to Punjab, India, using LANDSAT imagery. (*See* All India Workshop and Symposium on Digital Image Processing, Nov. 30, 1973, Indian Institute of Science, Bangalore, India.)

REFERENCES

Abdel Gawad, Monem, 1971, Wrench Movement in the Baluchistan Arc and Relation to Himalayan–Indian Ocean Tectonics, *Geol. Soc. Am. Bull.*, v. 82, p. 1235–1250.

Belcher, D. J., and S. Majtenyi, 1967, A Preliminary Investigation of Discharge Characteristics of Drainage Basins Derived from Airphoto Analysis and Climatic Data, *Cornell University Water Resources Center Tech. Rept. 4,* Ithaca, N.Y.

Carter, L. D., and R. O. Stone, 1974 nterpretation of Orbital Photographs, *Photogrammetric Eng.*, v. 40, no. 2, p. 193–197.

Chow, V. T. (ed.), 1964, *Handbook of Applied Hydrology,* McGraw-Hill Book Company, New York.

Data Users' Handbook, 1972, NASA Earth Resources Technology Satellite, Goddard Space Flight Center, Published by General Electric Co., Space Division, 5030 Harzel Place, Beltsville, Md.,

Lowman, P. D., Jr., 1971, Geologic Uses of Earth Orbital Photography, x-644-71-359, NASA/Goddard Space Flight Center, Greenbelt, Md.

Raina, B. N., 1972a, The Use of Satellite Photography in Interpretation of Regional Geology, VIII Remote Sensing Seminar, Center of Remote Sensing, University of Michigan, Ann Arbor, Mich.

____, 1972b, Photogeology and Himalayan Geology, *Himalayan Geol.*, v. 2, p. 527–536.

____, 1974, Structure of Himalaya–New Dimension from Space Imagery, International Seminar on Tectonics and Metallogeny of Southeast Asia and Far East, Geological Survey of India, Calcutta.

Ray, R. G., 1960, Aerial Photographs in Geologic Interpretation and Mapping, *U.S. Geol. Surv. Prof. Paper 373.*

Sharma, R. D., 1972, Enhancement of Earth Resources Technology Satellite (ERTS) and Aircraft Imagery Using Atmospheric Corrections, *VIII International Symposium on Remote Sensing of the Environment,* Environmental Research Institute of Michigan, Ann Arbor, Mich.

15

ENVIRONMENTAL MONITORING OF MINERAL-RELATED INDUSTRIES

S. Sidney Verner

Sidney S. Verner is in the Office of Monitoring and Technical Support, U.S. Environmental Protection Agency, Washington, D.C.

The enterprises associated with industrial development (e.g., mining, smelting, drilling) are activities that tend to degrade environmental quality unless adequate precautions are instituted. Scarring of terrain due to strip mining, acid runoffs from mines, smokestack emissions from smelting furnaces, and spills from drilling and transporting of oil are all too familiar examples of insults to the environment from industrial processes. Because the quality of the environment was becoming sufficiently degraded to threaten the health and well-being of populations, the U.S. Congress passed a series of laws, notably the Federal Water Pollution Control Act Amendments of 1972, Public Law 92-500, and the Clean Air Act of 1970, Public Law 91-604, to control the discharge of pollutants into the environment by industry and other institutions (e.g., sewage treatment plants and incinerators). The most notable feature of these two acts is the ultimate goal of zero pollution discharge into waters and streams and zero visible emissions from stationary sources. Although industry is expected to control its own operations in limiting discharges of pollutants, the states and federal government will monitor source emissions for compliance with statutes.

Historically, routine monitoring of sources of air and water pollution involved analysis of selected samples collected at or near the point of discharge. Such monitoring techniques are not only time consuming, but, more important, it is often difficult for responsible monitoring agencies to gain access to these sources or to collect samples when active discharges are occurring. A further limitation to the effective use of these monitoring techniques for comprehensive coverage is the near-ubiquitous dispersion of pollution sources throughout communities. Clearly, the routine monitoring of effluents into air and water where many sources exist over large geographic areas is best suited to remote-sensing instruments—automated in situ or noncontact.* Here we limit our discussion to noncontact instruments mounted on airborne platforms.

*A noncontact instrument is one in which the sensor element is not in contact with the medium being examined (e.g., instruments depending on light absorption, reflection, or scattering).

Generally speaking, any pollutant that absorbs, reflects, or scatters light can be detected by photosensitive surfaces such as film or by light-analyzing systems such as spectrometers, radiometers, and monochrometers. The type of instrument used will be dependent on a number of considerations, for example, the matrix in which the pollutant is contained, the spectral region of interest, and whether measurement is to be qualitative, quantitative, or both.

Visible pollutants in air are limited to particulate matter, which may be detected by qualitative instruments (e.g., cameras). Most noxious gases are invisible, but detection is possible with spectral analyzing instruments that detect the characteristic light scattering or absorbing signature of the gas; these instruments frequently are quantitative as well as qualitative.

The detection of pollutants in water is more complex because the light attenuation characteristics of water limit detection of below-surface pollutants to the visible and near-visible portions of the spectrum. Even for surface pollutants, detection is often difficult, because the characteristic scattering or reflection of sunlight by pollutants is a function of the state of surface roughness as well as the angle of incident and reflected sunlight. Also, many dissolved chemicals have no spectral signature detectable through remote analysis. On the other hand, there are classes of pollutants that may be detected when water surface conditions and sun angle permit. These are particulates, algae, petroleum products, and thermal anomalies.

Several classes of instruments lend themselves to remote sensing from aircraft. A few are operational; many more are still in development. The major concern of this chapter will be to examine these instruments and assess their potential usefulness for detection of pollutants in air, in water, and on land. In pursuing this objective, the primary emphasis will be on the operational and practical information extraction concepts rather than theoretical principles of instrument design and performance. A slight exception will be made for imaging systems, because at this time these are the most universal sensors for remote acquisition of environmental information; and principles governing their operation are basic to understanding their potential for remote environmental monitoring of mineral-related industries.

IMAGING SYSTEMS

Cameras

Among all classes of instruments proposed for remote sensing of environmental parameters, the camera is the most advanced operationally, having undergone the longest period of development. And although the technology for producing high-quality, diffraction-limited lenses, with large apertures and long focal lengths, as well as high-resolution, high-speed panchromatic and color film is very sophisticated, the camera remains the simplest of all instruments, both in concept and use. Furthermore, information extraction from imagery is by far the simplest and least costly of all forms of data analysis. However, one must recognize that, although the camera has many desirable qualities, it is not a quantitative instrument for environmental analysis and therefore has very serious limitations. Nevertheless, the simplicity, ease of use, and ease of photographic interpretation will ensure the continued popularity of this instrument in remote-sensing applications.

Basically, a camera consists of a light-refracting element (the objective lens), a shutter for controlling the amount of light entering the chamber, a light-tight box for rigidly maintaining a fixed spacing between lens and film, and a light-sensing element (film or electrooptic element). An aerial camera is a highly refined electro-optic development of the basic camera; refinement is necessitated by several factors, most notably the desire for high spatial resolution. Regardless of the degree of sophistication of a particular camera, all instruments utilizing acquisition optics are governed by a common set of optical principles. This requires that the refracting element or lens be highly corrected for a number of aberrations, including curvature of field, coma, astigmatism, distortion, and chromatic and spherical aberrations. In a matched system, the degree of lens correction is determined by the limiting resolution of the photo detector or film. In practice, computer-designed optics using modulation transfer function analysis and advanced fabrication techniques are today almost universally employed in producing high-quality lenses so that matching lens and film is no longer necessary. As a result, camera, lens, and film selection is governed by such practical considerations as size of image format, total angular field of view, required exposure time per frame, spectral requirements, ground resolution requirements, and need for image motion compensation (IMC) to correct for camera movement during frame exposure.

Many of the factors that one must consider in selecting an imaging system (e.g., image format, field of view, focal length, film resolution) are related and not independent parameters at all. In addition, the altitude at which flights are made will also affect ground (spatial) resolution and thus may necessitate changes in some or all of the other system parameters. These relationships may be seen in the discussion that follows.

The basic camera relation is illustrated in Fig. 1 and given in equation (1):

$$\theta \leqslant \frac{W}{FL} \tag{1}$$

where θ = field of view, radians
W = film width (format), centimeters
FL = focal length, centimeters

Equation (1) expresses the built-in limits to a camera's field of view (FOV) based on the film dimension and the focal length of the lens. Similar expressions, based on the limiting resolution of the film and the map scale may be obtained from (1). Thus,

$$\theta \leqslant \frac{1}{r \times FL} rW \tag{2}$$

where r = film resolution expressed in resolution elements per mm.* In this expression, $10^{-1}/(r \times FL)$ represents the angle a resolution element makes at the

*Historically, there has been a difference of opinion concerning the units for expressing film or photo surface resolution. Common terms are line pairs per mm, line elements or resolution elements per mm, or cycles per mm. Because the term resolution elements per mm most clearly expresses the information-packing potential, this unit will be used in this discussion. For conversion, a line pair or cycle contains two resolution elements.

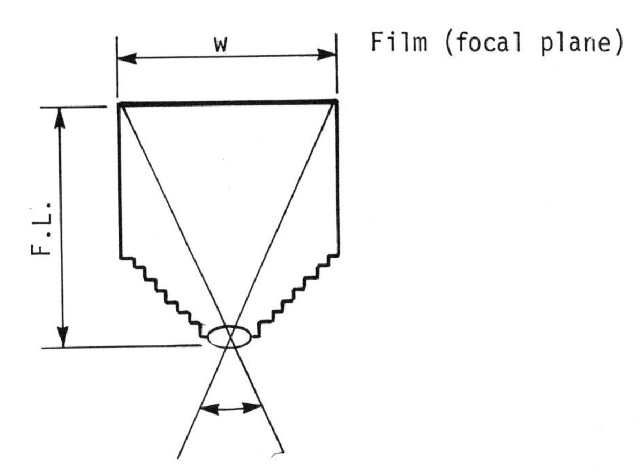

FIG. 1
Basic camera relationship.

focal plane of the camera lens and $(10rW)$ is the total number of resolution elements. Similarly,

$$\theta \leqslant \frac{W\ h}{h\ FL} \tag{3}$$

where h = altitude of platform in meters. By definition, $(h/FL) \times 10^2$ is the map scale.†

The inequalities expressed in these equations are necessary and must be satisfied; they indicate that (1) the film must not limit the system performance, rather, limitations should result from mission parameters, and (2) the map scale must be chosen to give meaningful imagery. Two examples are given to illustrate these points. Suppose that the resolution element angle is 0.1 milliradian (e.g., 200 resolution elements per mm); the camera lens has a 5-cm focal length and the film is 70 mm ($2\frac{1}{4}$ in. format or 57 mm). Then, by equation (2),

$$\theta \leqslant 1.14 \text{ rad} = 65.5^\circ$$

On the other hand, assume a flight altitude of 5000 m and a map scale of 1:50,000. Then, by equation (3),

†This is the acquisition map scale; useful map scale can vary from this through enlargement. Through geometric relationships, map scale may also be thought of as the ratio of the distance between two points on the ground to the distance between the corresponding points on the film, assuming a flat terrain.

$$\theta \leqslant 0.57 \text{ rad} = 32.8^\circ$$

In this illustration the limitation in FOV is imposed by the map scale. That is, one cannot fly a camera with a greater FOV than 32.5° at 5000 m and obtain a map scale of 1:50,000.

In addition to limitations in FOV, camera lenses impose certain other restrictions on imaging systems performance. As an aid to understanding these constraints, Figs. 2 through 4 are included. Figure 2 illustrates how the map scale is related to flight altitude for several typical lens focal lengths. Thus, one can acquire 1:20,000 imagery, for example, by flying at 1000 m with a 5-cm lens or at 4000 m with a 20-cm lens. If, for example, the choice is a 20-cm lens, Fig. 3, which illustrates the FOVs required for various film sizes (a square format is assumed throughout), indicates that a camera employing 70-mm film would require a lens with a 30° FOV. Figure 4 illustrates how the ground swath varies as a function of flight altitude for representative FOVs. Continuing the use of the above example, we see that at an altitude of 4000 m and with a 30° FOV each photographic frame would encompass an area approximately 1500 m on a side.

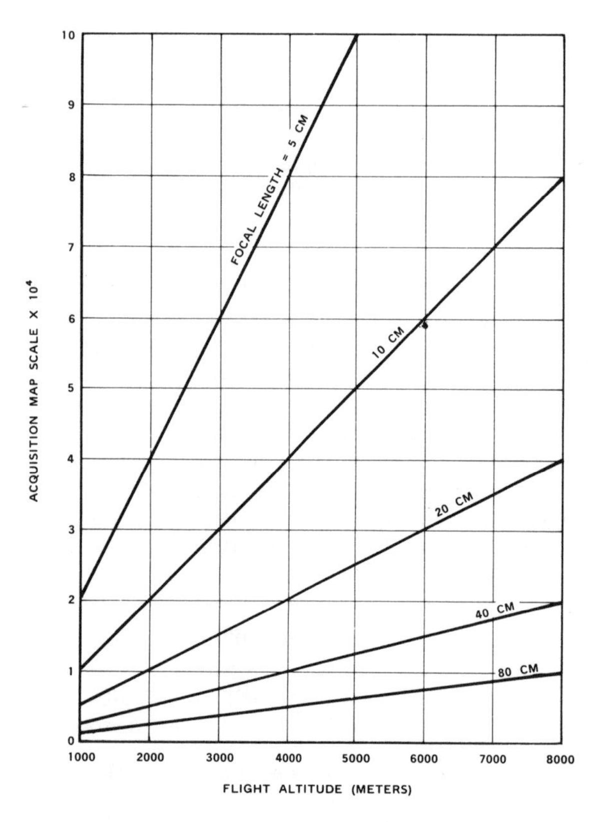

FIG. 2
Map scale vs. flight altitude for typical lens focal lengths.

When planning a flight covering a specified geographic location, the data presented in Figs. 2 through 4 are useful in allowing one to balance parameters and options in arriving at an optimum system. A system for a photographic mission would include camera, lens, film, flight altitude, and number of passes required for complete coverage, always allowing for minimum specified overlap of frames.

This discussion of imaging systems focuses primarily on frame-type cameras because these are simple and versatile; any frame camera will accept almost every type of film available for aerial photography. Also, frame cameras vary in complexity and can be as simple as a hand-held 35-mm instrument flown aboard a light aircraft to very intricate and expensive systems, such as are used in mapping services. A different class of imaging system is the scanner, or multispectral scanner (MSS), so called because it records several spectral regions simultaneously. Although the principles of photography are identical in both types of cameras, the scanner camera is much more sophisticated. This camera receives its name from the nature of its image recording, which is a continuous record as the optics "scan" across the terrain (i.e., transverse to the flight path). Because simultaneous images are recorded in different wavelength regions, the MSS allows images in any combination of spectral regions to be combined to form composits, which often enhance details not visible otherwise. Multispectral scanners represent the most advanced type of imaging system; unfortunately, they are expensive, and the processing of the film is also very expensive since it requires careful computer analysis to arrive at the best spectral combination for maximum enhance-

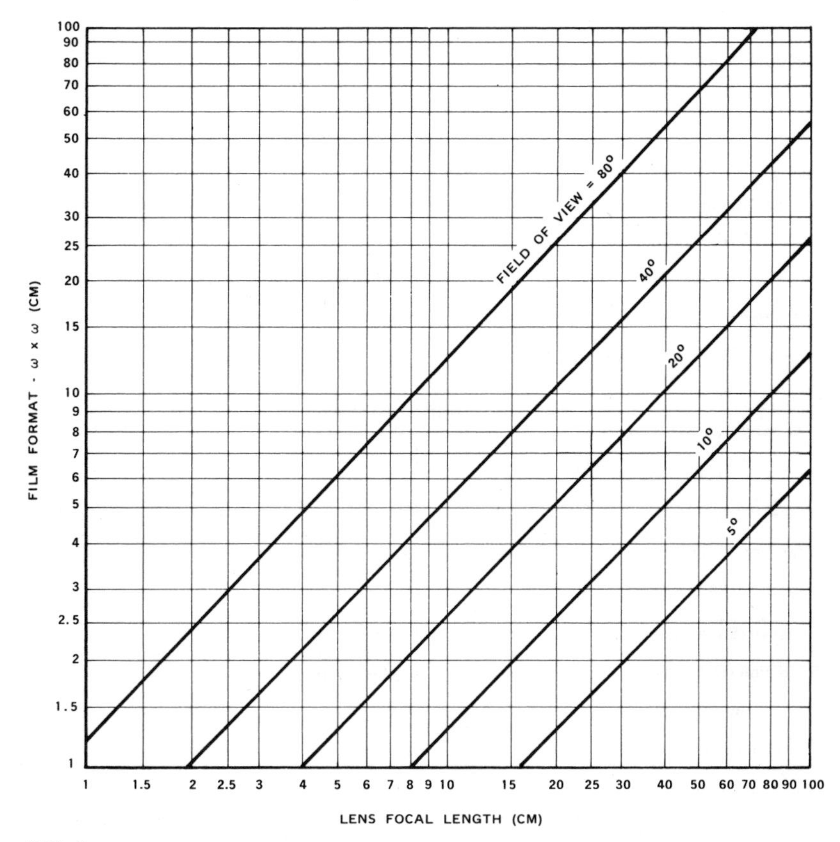

FIG. 3
Film format vs. lens focal length for typical fields of view.

ment of details and information. The inherent high costs of MSS imagery generally place it beyond the means of most institutions. Multispectral line scanners generally operate in spectral bands between 0.35 and 15 μm. For detection above about 1 μm, radiometric techniques are used, and the medium of record is generally tapes, which may later be converted to film if a visual record is desired. This class of instrument is most useful in detecting heated effluents from power plants and industrial processes, monitoring surface temperatures of rivers, lakes, and streams, and has been highly effective in detecting oil slicks in coastal waters and lakes. These instruments have the great advantage of operating during the day or night since the thermal IR region ($>$ 3 μm) is an emissive region and does not depend on the sun's reflected energy.

Occupying a position somewhere between the simple frame camera and the multispectral line scanner is the multiband camera. This camera combines the taking of a photograph of a large scene with the ability to discriminate among objects in that scene. These analyses are made possible because spectral differences in reflection have been shown to correlate closely with compositional and textural properties of material types, thus providing a means for discrimination. The synoptic multiband camera system has uses in many applications for remotely monitoring water

TABLE 1
Characteristics of Kodak Aerial Films

Film Type	KODAK Film[1]	Film Number[2]	Sensitivity	Description and Applications
Camera Acquisition Films	PLUS-X AEROGRAPHIC (ESTAR Base)	2402	Panchromatic (with extended red)	Medium-speed, high dimensional stability for aerial mapping and reconnaissance
	TRI-X AEROGRAPHIC (ESTAR Base)	2403		High-speed, high dimensional stability for aerial mapping and reconnaissance under low levels of illumination
	DOUBLE-X AEROGRAPHIC (ESTAR Base)	2405		Medium- to high-speed, standard film for mapping and charting; high dimensional stability
	PANATOMIC-X Aerial (ESTAR Thin Base)	3400		Intermediate-speed, high-contrast, medium- to high-altitude reconnaissance film; suitable for small-negative format
	PLUS-X Aerial (ESTAR Thin Base)	3401		Medium-speed, high-contrast, fine-grain, medium- to high-altitude reconnaissance film
	High Definition Aerial (ESTAR Thin Base)	3414		Thin base, slow-speed, high-definition film for high-altitude reconnaissance
	High Definition Aerial (ESTAR Ultra-Thin Base)	1414		Similar to 3414; ultra-thin base for maximum spool capacity; slow-speed, high-definition film for high-altitude reconnaissance
	Infrared AEROGRAPHIC (ESTAR Base)	2424	IR	Black-and-white film for reduction of haze effects, water location, vegetation surveys, and multispectral aerial photography
	AEROCHROME Infrared (ESTAR Base)	2443		False-color reversal film, high dimensional stability for vegetation surveys and camouflage detection
	AEROCHROME Infrared (ESTAR Thin Base)	3443		Similar to 2443 Film; thin base for high spool capacity and minimum storage space
	AEROCOLOR Negative (ESTAR Base)	2445	Color	High-speed, **color-negative** film without integral masking for mapping and reconnaissance
	AEROCHROME MS Film (ESTAR Base)	2448		**Color-reversal** film for low- to medium-altitude aerial mapping and reconnaissance
	Aerial Color (ESTAR Thin Base)	SO-242		Slow-speed, high-resolution color-reversal film for high-altitude reconnaissance
	Aerial Color (ESTAR Ultra-Thin Base)	SO-255		Similar to SO-242; ultra-thin base for maximum spool capacity; high-resolution film for high-altitude reconnaissance
	EKTACHROME EF AEROGRAPHIC (ESTAR Base)	SO-397		High-speed, color-reversal film for aerial mapping and reconnaissance
Duplicating Films	AEROGRAPHIC Duplicating (ESTAR Base)	2420	Blue	Extremely fine-grain film for duplicating from medium-grain aerial negatives; high dimensional stability
	Aerial Duplicating (ESTAR Thin Base)	SO-122		Similar to 2420; thin base for maximum spool capacity and minimum storage space
	AEROGRAPHIC Duplicating (ESTAR Thick Base)	4427		Similar to 2420; thick base for maximum dimensional stability (Cut sheets for aerial diapositives)
	AEROGRAPHIC Direct Duplicating (ESTAR Base)	2422		Extremely fine grain film for **one-step duplication** of high-definition aerial negatives or positives
	Fine Grain Aerial Duplicating (ESTAR Base)	2430		Extremely fine grain film for duplicating high-definition aerial negatives; high dimensional stability
	Low Contrast Fine Grain AEROGRAPHIC Duplicating (ESTAR Base)	SO-355		Microfine-grain film for duplicating high-definition aerial negatives
	AEROCHROME Duplicating (ESTAR Base)	2447	Color	Low-contrast, color-reversal film for making duplicate transparencies from Ektachrome and Aerochrome film originals; good color balance; high resolution and high dimensional stability

quality and land-use conditions. Functional examples include multispectral photographic studies to identify coastal water radiance, water color, algal blooms, sedimentation, and strip mining.

Generally, the user of imagery obtained from aircraft is interested in four quantities: spatial resolution, spectral content, map scale, and ground swath width. Spatial resolution presents no problem, since almost any desired resolution, that is, ground detail, may be obtained with available aerial film, and spectral content is possible with the many color films in addition to the false-color enhancement techniques that are available.

Knowledge of these four quantities enables one to

Factory Stocked[3]	Nominal Base Thickness (mils)	Nominal Total Thickness (mils)	Backing	Aerial Exposure Index[4]	Aerial Film Speed[5]	Effective Aerial Film Speed[5]	Resolving Power (lines/mm) T.O.C. 1000:1	Resolving Power (lines/mm) T.O.C. 1.6:1	Diffuse RMS Granularity[6]	KODAK Safelight Filter	KODAK Developers[7]	Processes and KODAK VERSAMAT Chemicals[8]	Kodak Literature References[9]
√	4	4.3	Fast-Drying	80	—	250	100	50	19	Total Darkness	**D-19**; DK-50	**885**; 641; A	**M-45**; M-29
√	4	4.5	Fast-Drying	250	—	640	80	25	33	Total Darkness	**D-19**; DK-50	**885**; 641; A	**M-24**; M-29
√	4	4.4	Fast-Drying	125	—	320	100	50	26	Total Darkness	**DK-50**	**885**; 641; A	**M-29**
	2.5	2.9	Dyed-Gel	20	64	—	200	80	16	Total Darkness	**D-19**	**885**; 641; A	**M-29**
	2.5	3.1	Dyed-Gel	64	200	—	125	40	30	Total Darkness	**D-19**	**885**; 641; A	**M-29**
	2.5	3.0	Dyed-Gel	2.5	8	—	630	250	9	Total Darkness	**D-19**	**885**[10]; 641[10]	**M-73**; M-29
	1.5[11]	2.0	Dyed-Gel	2.5	8	—	630	250	9	Total Darkness	**D-19**	**885**[10]; 641[10]	**M-73**
√	4	4.3	Fast-Drying	100[12]	—	200[12]	80	40	33	Total Darkness	**D-19**; DK-50	**885**; 641; A	**M-58**; M-29
√	4	4.8	Fast-Drying[13]	10[14]	—	40[14]	63	32	17	Total Darkness	—	**EA-5**	**M-69**; M-29
	2.5	3.6	Clear-Gel	10[14]	—	40[14]	63	32	17	Total Darkness	—	**EA-5**	**M-69**; M-29
√	4	4.9	Fast-Drying	32[15]	—	100[15]	80	40	13	Total Darkness	—	**AERO-NEG Color**[16]	**M-70**; M-29
√	4	4.8	Fast-Drying	6[15]	—	32[15]	80	40	12	Total Darkness	—	**EA-5**	**M-29**
	2.5	3.7	Clear-Gel	2	—	6	200	100	11	Total Darkness	—	**Process ME-4 (modified)**[17]; EA-5	**M-74**
	1.5[11]	2.7	Clear-Gel	2	—	6	200	100	11	Total Darkness	—	**Process ME-4 (modified)**[17]; EA-5	**M-74**
	4	4.9	Fast-Drying	12[15]	—	64[15]	80	40	13	Total Darkness	—	**EA-5**[18]	**M-78**
	4	4.2	Fast-Drying	RMPI With and Without UV Filter[20]	**160**	*640*	160	63	9	1A (light red)	—	**885**; 641; A; B[19]	**M-29**
	2.5	2.9	Clear-Gel		**160**	*640*	160	63	9	1A (light red)	—	**885**; 641; A; B[19]	**M-29**
	7	7.4	Clear-Gel		**160**	*640*	160	63	9	1A (light red)	—	**885**; 641; A; B[19]	**M-29**
(√)	4	4.2	Fast-Drying		*8*	**12**	500	160	6	**1 (red)**	—	**885**; 641	**M-41**
	4	4.2	Fast-Drying		*25*	**50**	250	125	5	1A (light red)	—	**885**; 641	**M-61**
	4	4.2	Fast-Drying		*32*	**125**	250	125	7	1A (light red)	—	**885**; 641; A	**M-29**
	4	4.7	Fast-Drying	—	—	—	100	50	9	Total Darkness	—	**EA-5**	**M-72**

Reprinted with permission from a copyrighted Kodak publication (No. M-57, Oct. 1973).

make a quick estimate of the system's overall performance. If these "performance figures" satisfy requirements, further evaluation is indicated, involving such things as system reliability, costs, film- and data-processing requirements, and image motion compensation requirements to correct for the effect of aircraft movement during frame exposure.

In selecting a film for aerial photography several factors must be considered. Choices will be based on the end use of the image, the subject, and atmospheric conditions. For example, is the subject smoke from an industrial site? If so, black and white film will probably suffice, particularly if the atmosphere is hazy; black and white infrared photography is espe-

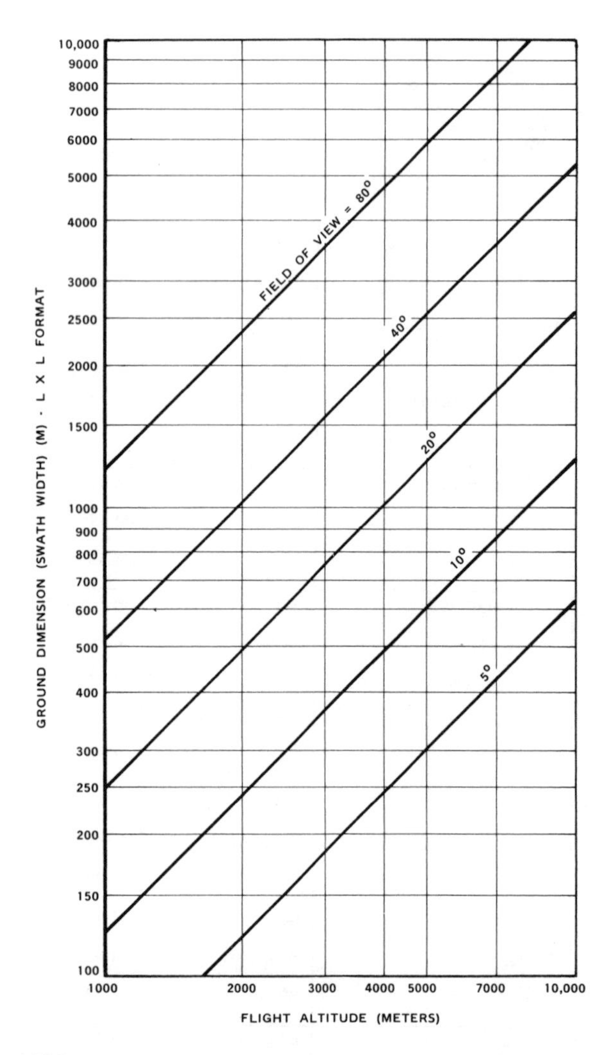

FIG. 4
Swath width vs. flight altitude for typical fields of view.

cially good at penetrating haze to yield photographs with high contrast. However, for general photography, the film of choice will be some type of color film because the information content is inherently greater than for black and white or IR film. Other factors that influence the selection of film include weather, altitude, light levels, and format.

Table 1 gives some general characteristics and applications of several specialized Kodak aerial films. For example, Tri-X pan film is fast, even when using haze-control filters, has good exposure latitude (shutter speeds and *f*-stop settings are not critical), and its relatively fine grain will allow substantial enlargements. Black and white IR film has several advantages not found in panchromatic film—higher contrast and an ability to see through haze. Infrared color films were originally designed for camouflage detection and emphasize differences in IR reflectance between live, healthy vegetation and visually similar areas; consequently, these films may be used in certain instances of pollution detection such as overfertilization of streams and lakes due to introduction of nutrients (phosphates, nitrates, etc.) from outfalls of processing plants.

The remote surveillance of land and water bodies has a history almost as old as the camera itself. Practically speaking, however, it has the technical advancements in cameras, and particularly heat-sensitive or IR film during and following World War II, that placed remote sensing on a sound basis. Infrared film can readily distinguish between vegetation, bare land surfaces, and water as can other conventional color and black and white films; in addition IR film lends itself to evaluating thermal pollution and, to some extent, the biological productivity of lakes and streams. Photographic techniques have successfully been used for detecting underwater outfalls, plumes of light-colored effluents resulting from municipal waste treatment discharges, downstream eutrophication from waste discharges, and drainage wastes from processing facilities whose biological reduction by natural processes may induce deoxygenation. In addition, algal and other biological activity can be imaged by IR aerographic film.

When preparing a flight it is essential that proper annotation of imagery be included in the flight plan. Each frame or exposure should be accompanied by at least the following information: date, flight path, time, altitude, type of film, weather conditions, exposure time, aircraft ground speed, and what development procedure was used. Many aerial cameras automatically record some of these parameters on the film during each exposure. In addition to annotations,

TABLE 2
Criteria for Detecting and Interpreting Several Types of Pollution

Type of Pollutant	*Detection Criteria*
Oil leaks and seepage	Dark blue or black or rainbow coloration; difference in reflectivity from surrounding area
Water outfalls	Characteristic outfall plume; color contrast; turbulence
Thermal pollution	Characteristic red color on IR color film; light shading on black and white IR film
Eutrophication	Characteristic green color; inability to see beneath surface
General water quality	Color, shading, or tone contrast; water opaqueness
Stack effluent aerosols	Characteristic stack plumes
Haze and smog	General decrease in ground detail

additional collateral information derived from ground inspection at the time of overhead surveillance may sometimes be necessary for proper interpretation of the imagery. The degree of ground truth needed will depend on user requirements, the nature of the imagery, including target characteristics, and the skill of the interpreter. Criteria for detecting several common types of pollution are given in Table 2. It should be noted that these criteria only serve as general guides, and that detection of many types of pollutants will depend on skills developed after considerable experience with these techniques has been acquired.

Radiometers

Many types of remote sensing are made possible only because all substances above absolute zero emit electromagnetic energy, or photons. This emission is dependent on the absolute temperature of the substance and the emissivity of its surface. The distribution of the energy throughout the electromagnetic spectrum and the position of the intensity peak are strictly temperature dependent, however. In the 1870s Stephan, experimentally, and Boltzmann, theoretically, conceived the physical relationship which has come to bear their name, that radiant emittance of an object was directly proportional to its absolute temperature raised to the fourth power. The Stephan–Boltzmann law may be written as

$$W_B = \sigma T^4 \tag{4}$$

where W_B is the radiance emitted by a perfect radiator, or blackbody; σ is the Stephan–Boltzmann constant, and has a value of 5.67×10^{-8} W m^{-2} ($^\circ$K)$^{-4}$; and T is the absolute temperature. Equation (4) has limited usefulness, however, since very few (if any) objects are ideal radiators. A more useful expression for computing the radiant emittance of a real body is given by

$$W = \epsilon \sigma T^4 \tag{5}$$

where ϵ is the emissivity of a real body and is defined as the ratio of radiation emitted by a given body to that emitted by a blackbody at the same temperature. For real bodies, ϵ always has a value less than 1.

Although equation (5) is a valid expression of radiant emittance, it is still not a very useful one, because it describes the total emission from a warm body, whereas detectors of electromagnetic radiation, which include all classes of remote sensors, are wavelength selective and do not, actually cannot, detect across the entire spectrum of electromagnetic radiation. The wavelength dependence of an ideal radiator or blackbody is given by Planck's radiation law, which is a modification of the Stephan–Boltzmann law, and may be written as

$$W_\lambda = 2\pi C^2 h \lambda^{-5} \left(\frac{hc}{e^{\lambda k T} - 1} \right)^{-1} \tag{6}$$

where W_λ = radiant emission in the wavelength interval λ
C = speed of light
e = natural logorithm base
h = Planck's constant (6.62×10^{-34} J-s)
k = Boltzmann's constant (1.38×10^{-23} J $^\circ$K^{-1})
λ = wavelength
T = absolute temperature

As before, it is necessary to adapt expression (6) to

real bodies by introducing the emissivity factor. Thus, the radiant emittance for a gray body (i.e., nonideal radiator) becomes

$$W_\lambda^2 = 2\pi C^2 h\epsilon_\lambda \lambda^{-5}\left(\frac{hc}{e^{\lambda kT}-1}\right)^{-1} \tag{7}$$

Integrating Planck's law over all wavelengths should give the Stephan–Boltzmann equation. The Planck radiation equation shows that the peak of the radiation spectrum shifts toward shorter wavelengths as the temperature of the radiating body increases. This is explicitly defined by the Wien displacement formula, which can be derived directly from Planck's equation. Wien's formula states that

$$\lambda_n T = 2897.9 \ \mu\text{m} \ ^\circ\text{K} \tag{8}$$

Thus, the peak radiation of the sun (assumed surface temperature of 6000°K) occurs at approximately 0.5 μm, and for the earth's surface (assumed surface temperature of 300°K), the radiation peak occurs at 10 μm. It should be noted that the emissivity term in equation (7) is wavelength dependent and may be written as

$$\epsilon_\lambda = \frac{R_r(\lambda_n, T_r)}{R_B(\lambda_n, T_B)} \tag{9}$$

where ϵ_λ = spectral emissivity
R_r = real surface radiance
R_B = radiance of a blackbody at the same temperature
λ_n = nth wavelength interval
T_r = transmittivity of real surface
T_B = transmissivity of blackbody

For opaque materials, where transmissivity (T_r, T_B) are zero, equation (3) takes the form

$$\epsilon_\lambda = 1 - \rho_\lambda \tag{10}$$

where ρ_λ is the spectral reflectivity.

However, most substances are not opaque for all wavelengths. An example would be the relative opaqueness of water in the 8- to 14-μm region and its transparency in the visible.

In the microwave region, where wavelengths are on the order of centimeters, emissivity variations have great effect on thermal emission of radiation as can be seen from equation (11), which is the microwave power measured from an observed surface element extended over the entire receiving beamwidth:

$$\rho_\lambda = \frac{2k(\epsilon_\lambda T)B}{\lambda^2} \tag{11}$$

where k = Boltzmann's constant
T = absolute temperature
ϵ_λ = wavelength-dependent emissivity
B = receiving beam width in steradians
λ = radiation wavelength

In practice, equation (11) is modified by background contributions resulting from atmospheric radiations, stray emissions from other sources, cosmic noise, reflectance, etc.

What the foregoing discussion tells us is that substances can be detected other than visually because of thermal emission of radiation, as distinguished from reflected radiation, the underlying principle in photographic techniques. In thermal radiation the distinguishing characteristic is the emissivity, so although two different objects may have the same real temperature, their apparent temperatures will be different because their emissivities will make them appear to be at different temperatures. The analogous situation in photography is reflectivity. Different objects have different spectral reflectivities and thus are differentiated visually and photographically. However, photography is generally employed as a nonquantitative detection device and thus does not require being related to laws of radiation or reflection.

Examples of the potential of imagery obtained from aircraft-mounted instruments are shown in Figs. 5 through 20. Figure 5 is a panchromatic photograph of a leaking oil platform off the Santa Barbara coast. The photograph was taken in early morning light at approximately 8:15 A.M. at an altitude of 610 m with a K-2 haze filter, which rejects blue light. Figure 6 is a multispectral comparison of a steel plant "pickling liquor" effluent into Lake Michigan. The bands shown are blue, green, and red, respectively, reading from top to bottom. Figure 7 is another multispectral illustration demonstrating the selective enhancement of pollutants in various spectral regions. The fourth band, 8.0 to 13.5 μm, is the thermal IR region.

In Figs. 8* to 11 are photographs (0.40 to 0.72 μm) of selected areas of the Mississippi River Delta taken with a framing camera. The camera had a 7.62-cm focal length lens with a 74° angular field of view. The photographs were taken at altitudes ranging from approximately 900 to 2100 m. Figure 8 shows oil slicks spreading out from a loading platform. Figure 9 also reveals oil slicks in the vicinity of docking platforms. Figures 10 and 11 are illustrative of anomalies on water surfaces, which, if occurring near industrial sites, would require more intensive investigation, including ground inspection.

Shown in Figs. 12 and 13 are images of a power plant recorded in color and the thermal IR region, respectively, from an altitude of about 1525 m. Although the temperature range in this scene is only 2.5°C, the heated-water distribution is easily observed in the IR image. However, the thermal image does not differentiate ground details such as the power plant, roads, and other cultural landmarks.

Radiometers are usually employed in a scanning mode with data recorded on magnetic tape. Pictorial conversions of the IR brightness data stored on the tapes allow photograph-like images to be produced, often containing details not discernable in conventional photographs. Examples of imagery obtained from IR radiometers (8 to 14 μm) are illustrated in Figs. 14 through 17. Figure 14 is the radiometric image of an oil platform fire in the Gulf of Mexico taken almost 2 hours apart. The extent of the oil slick is easily seen; the heated smoke rising from the platform is black but, having a high brightness, is recorded as white. In Fig. 15 a major industrial outfall into the Gulf of Mexico is seen; the mixing patterns occurring are readily discernable. The data for Fig. 16 were calibrated against a blackbody reference, which made possible the generation of the thermal contour map of Big Lake at Biloxi, Mississippi. The data for this image was collected by NASA and processed by the Environmental Protection Agency at their National Environmental Research Center—Las Vegas facility. The detector was a cooled mercury–cadmium–telluride crystal. Figure 17 is a false-color rendition of radiometric data in which colors have been arbitrarily assigned to brightness (temperature) intervals. In this instance, temperatures varied from 22.5 to 26°C. This figure shows a thermal discharge plume from a power plant into Lake Michigan and vividly illustrates the dynamic nature and variability of mixing patterns in a large body of water. False-color rendition is a data-enhancement technique allowing both qualitative and quantitative information to be presented simultaneously in a photograph-like format with dramatic effect.

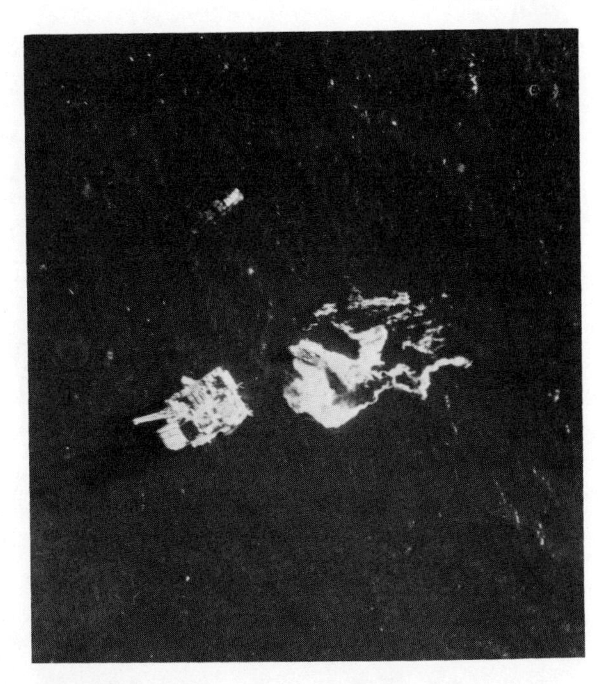

FIG. 5
Santa Barbara oil slick. (Courtesy Environmental Research Institute of Michigan.)

Uncontrolled strip mining presents severe esthetic and ecological problems. In addition to disrupting the topography to the point where the land is no longer productive, strip mining causes land erosion and drainage of highly mineralized water, which is destructive to vegetation. Figure 18 illustrates a large strip-mine operation in Ohio; the scarred earth is in

*Figures 8, 10, and 17 will be found in the color insert.

0.40 to 0.44μ

0.55 to 0.58μ

0.62 to 0.66μ

FIG. 6

Multispectral comparison of effluent contrast. (Courtesy Environmental Research Institute of Michigan.)

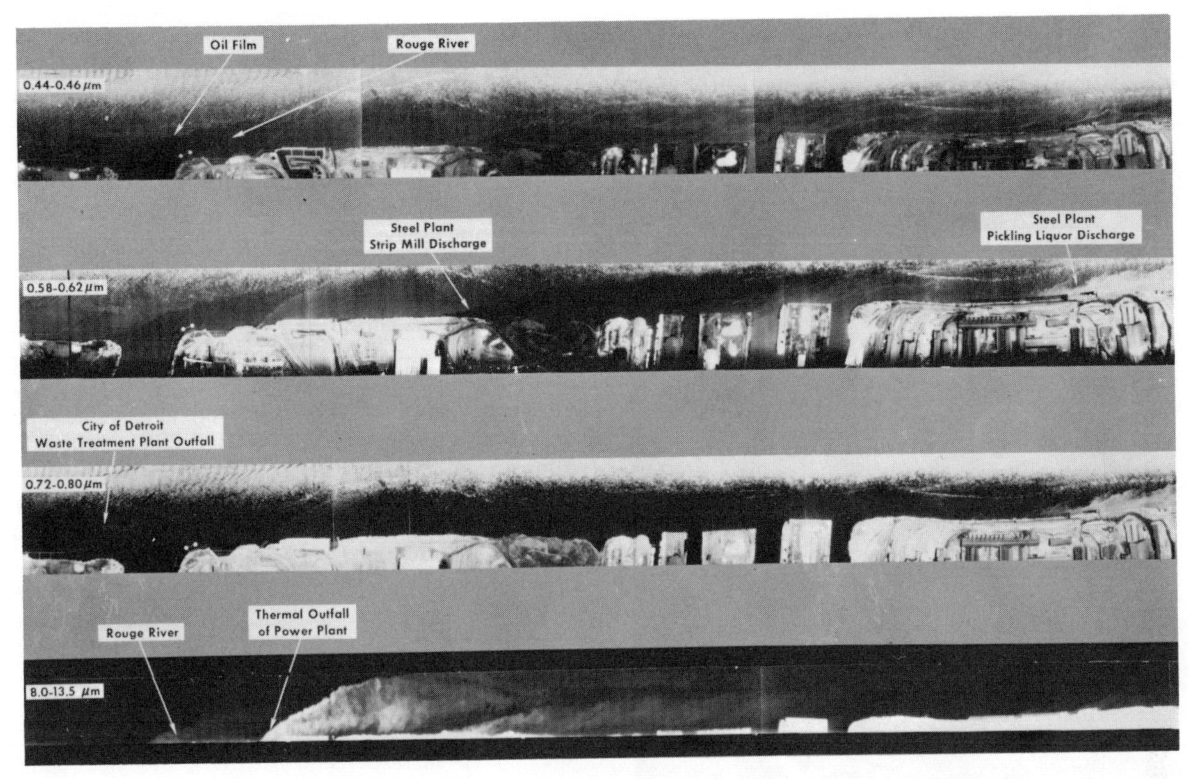

FIG. 7
Detection and identification of industrial pollutants by spectral analysis. (Courtesy Environmental Research Institute of Michigan.)

stark contrast to the surrounding vegetation. Figure 19 is a photograph in blue light of an adjacent stripping operation; some ground detail is lost in this image but the contrast is enhanced over the black and white rendition of Fig. 18. Figure 20 is interesting by illustrating how oblique photography can highlight terrain features, which tend to get flattened in overhead viewing. This photograph illustrates a spoil pile from a nearby strip mine before it has been graded and revegetated as required by recent environmental statutes.

ALTERNATIVES TO IMAGING

Environmental pollutants occur in air, land, and water. Since remote-sensing techniques are generally specific as to pollutant and the matrix in which it occurs, the discussion that follows will be broken down into air quality monitoring and land and water quality monitoring.

Remote Sensing of Air Quality

At present nearly all commercially available instruments for measuring concentrations of gaseous pollutants in stack effluents require extraction of a gas sample from the stack into a separate container for subsequent analysis. Such a sampling procedure is of necessity an averaging technique, which tends to wipe out any changes in concentration that may occur along the sampling train and fails to detect variations in pollution concentration which may be present in the stack. These difficulties can be largely overcome by utilizing *in situ* absorption spectroscopy, in which radiant energy from a thermal source is transmitted through the stack gases and detected on the other side after passing through a dispersive element or narrow-band filter. By comparing the amount of transmitted energy at characteristic wavelengths, it is possible to determine the instantaneous average concentration in the sampling field. Although simple in concept, this technique suffers from the fact that narrow-band filters usually do not provide the high spectral resolution required for good specificity (discrimination against other gases having absorption characteristics in the same spectral interval), whereas instruments employing dispersing elements are complex, expensive, and subject to misalignment.

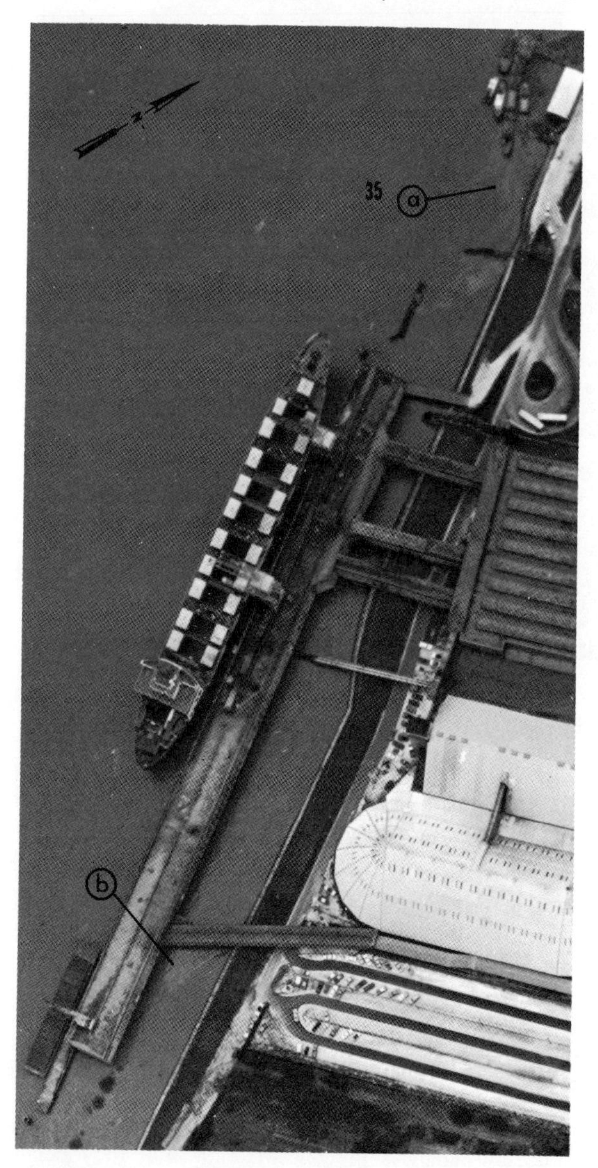

FIG. 9
Oil seepage in port facilities on Mississippi River. (Courtesy NERC—Las Vegas/EPA.)

Progress in remotely monitoring air quality has been impeded in the past by a lack of sufficiently sensitive real-time techniques for identifying and quantifying pollutants with sufficient resolution in time and space. However, in recent years a wide range of applications has been explored with increasingly sophisticated techniques. The advent of the pulsed laser coupled with radar techniques (LIDAR) marked the beginning of major progress in quantitative remote sensing of the atmosphere. Lasers are used as atmospheric illuminators in preference to other light sources because the monochromaticity of the laser allows discrimination against background noise through the use of narrow-band filters. In addition, the high peak power and narrow pulses yield good signal-to-noise ratios and range resolution, even for low-cross-section scatterers. Operational laser systems range in peak pulse power to several hundred megawatts. Pulse widths are as short as 10^{-9} s, and pulse repetition rates may be as high as several thousand per second. However, not all the desired characteristics of frequency, power, pulse width, pulse repetition rate, coherence, beam divergence, efficiency, and compact size are available in the same laser system.

Because of specific molecular interaction, backscattering, and absorption, laser techniques have the unique capability of identification of atmospheric gases. Generally, the very weak IR absorptions produced by air pollutants are detected by resonance absorption of the gas rather than the attenuation of the beam as it passed through the sample, since the attenuation will be very small for reasonable path lengths and therefore hard to detect. Other techniques employed for detecting air pollutants include Mie and Rayleigh scattering, Raman backscattering,

FIG. 11
Surface anomaly (probably oil). (Courtesy NERC—Las Vegas/EPA.)

resonance backscattering, and angular scattering phenomena involving polarization effects. In addition to studying the properties of the gaseous components of the atmosphere, the distribution and dynamics of aerosols are also potentially available for investigation.

The techniques cited above using lasers as atmospheric illuminators are examples of active remote-sensing methodologies. In contrast to this approach, other techniques concentrate on passive methods utilizing emission principles combined with sophisticated correlation techniques for detecting atmospheric pollutants. Until relatively recently, the passive approach centered on dispersive elements such as gratings and prisms for separating the information content of incoming signals. This method had the disadvantage of requiring long integration times since incoming signals were usually faint. More recent techniques are based on interferometric principles, which carry out correlation in real time. These devices correlate against the Fourier transform of the spectrum rather than the spectrum itself. Correlation interferometry has particular application in the infrared due to its large light throughput in this region and the

FIG. 12
Color aerial photograph of the harbor at Michigan City, Ind. (Courtesy Environmental Research Institute of Michigan.)

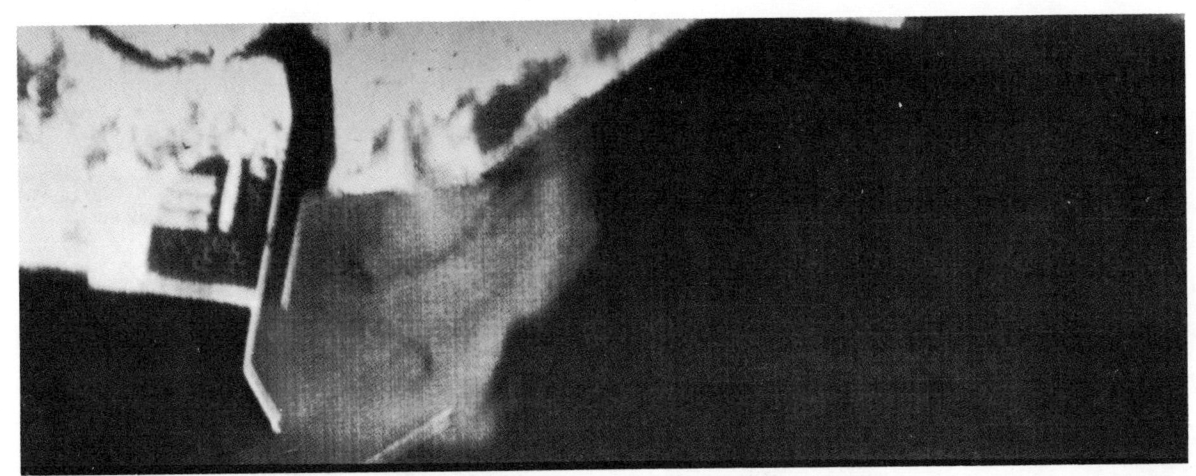

FIG. 13
Thermal image (8.0 to 13.5μ m) of discharge plume at Michigan City, Ind. (Courtesy Environmental Research Institute of Michigan.)

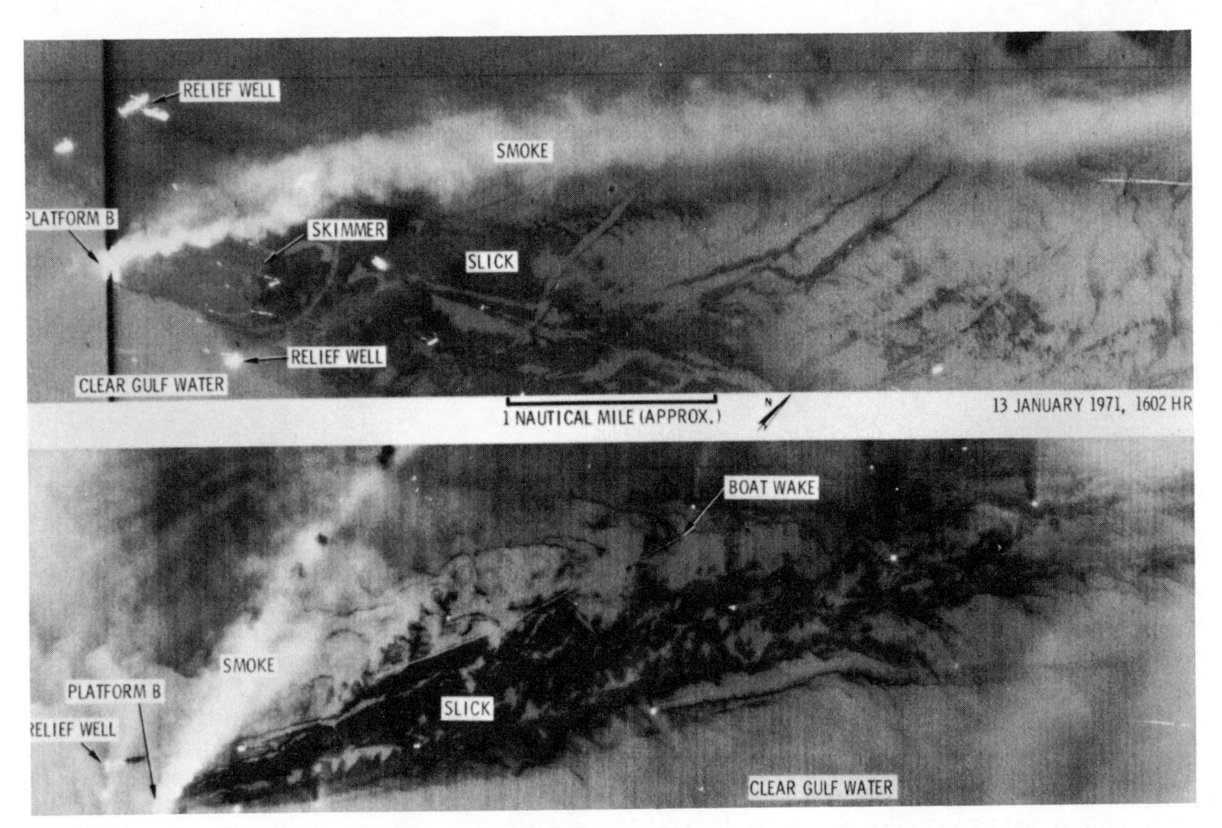

FIG. 14
Infrared imagery, Shell Oil Platform-B fire, Gulf of Mexico, January 13, 1971. (Courtesy EPA.)

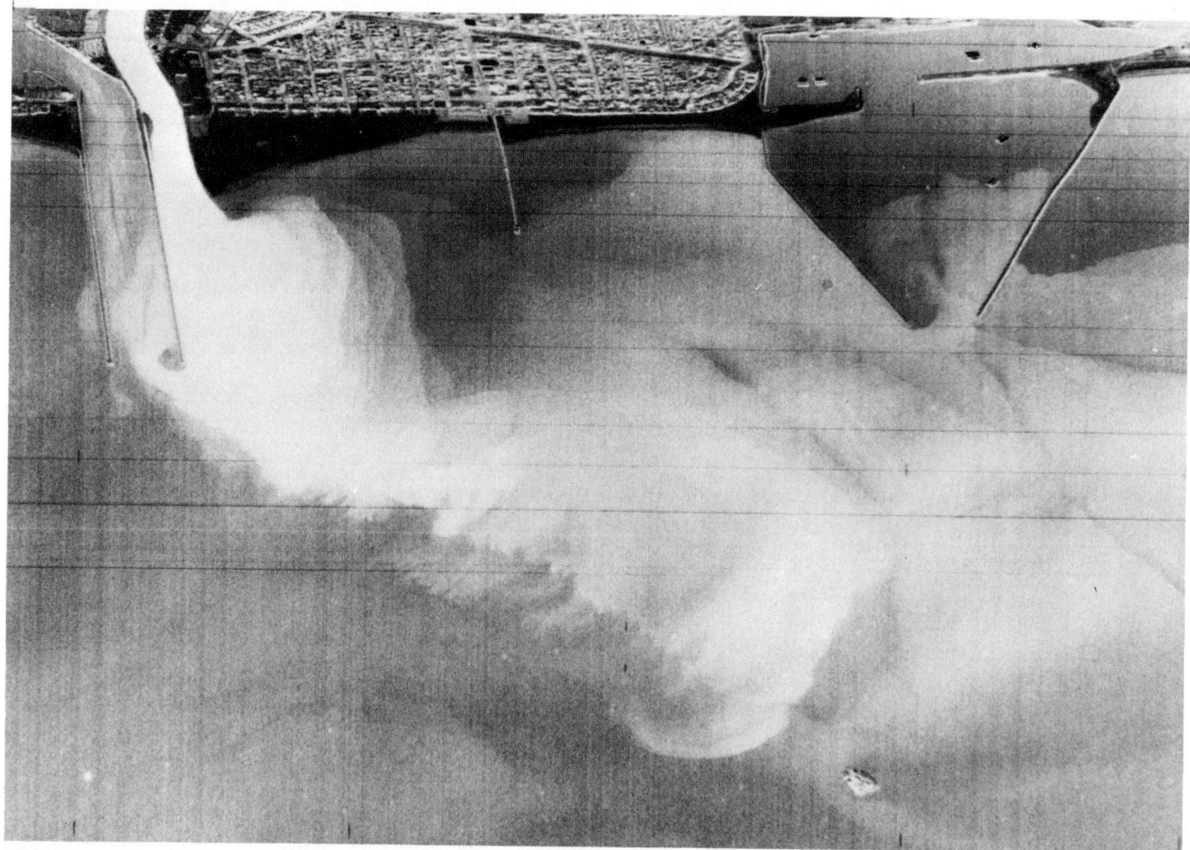

FIG. 15
Major industrial outfall, Gulf of Mexico. (Courtesy Texas Instruments.)

fact that almost all gases have characteristically strong and complex spectral signatures in this region. Instruments of the correlation interferometer type have also been constructed for use in the ultraviolet and visible portions of the spectrum. It must be emphasized that quantitative data obtained from any of these instruments cannot be relied on unless there is calibration capability provided for the system in whatever mode it is used.

Specific examples of instruments that have been built utilizing the above principles are given next.

Investigators at EPA's research centers in Las Vegas and Research Triangle Park have reported using LIDAR for detecting the height of an inversion layer in the St. Louis area under which air pollutants are trapped *(1)*. The lower the inversion layer, the more concentrated the pollutants are apt to be. This system utilizes a Q-switched ruby laser with a pulse energy of 1 J and a telescope with a 16-in. acrylic fresnel lens. This instrument was flown on a C-45 aircraft and positioned over an aerial camera port. The LIDAR was operated from an aircraft altitude of 10,000 ft.

A correlation spectrometer first developed by Barringer Research Ltd. *(2)* has been used for remote sensing of gaseous pollutants (e.g., SO_2). Ludwig et al *(3)* have modified the basic Barringer technique and have reported flying a gas filter correlation instrument to measure carbon monoxide; the instrument

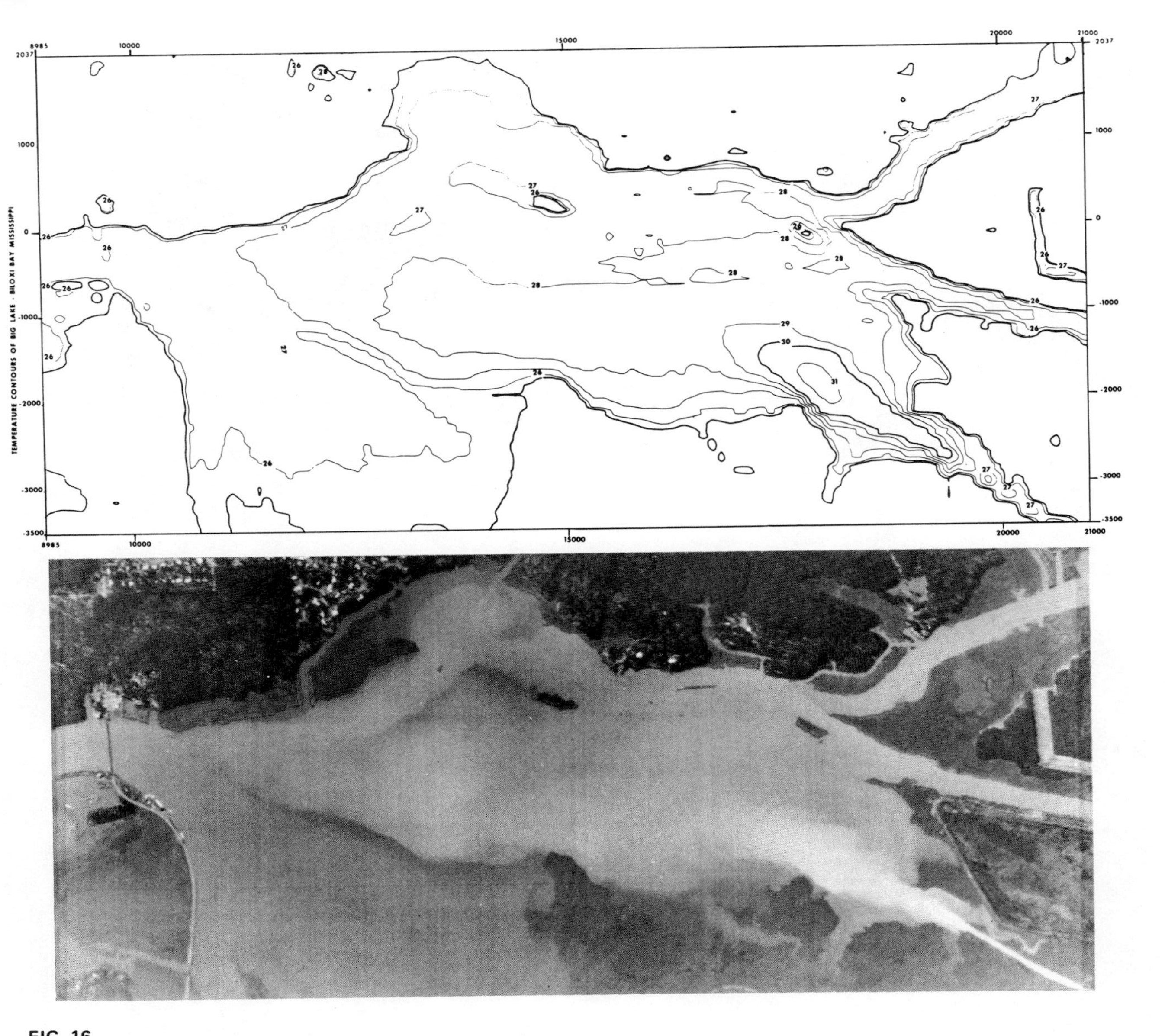

FIG. 16
Radiometric image and associated thermal contour map of Big Lake, Biloxi, Miss. (Courtesy NERC–Las Vegas/EPA.)

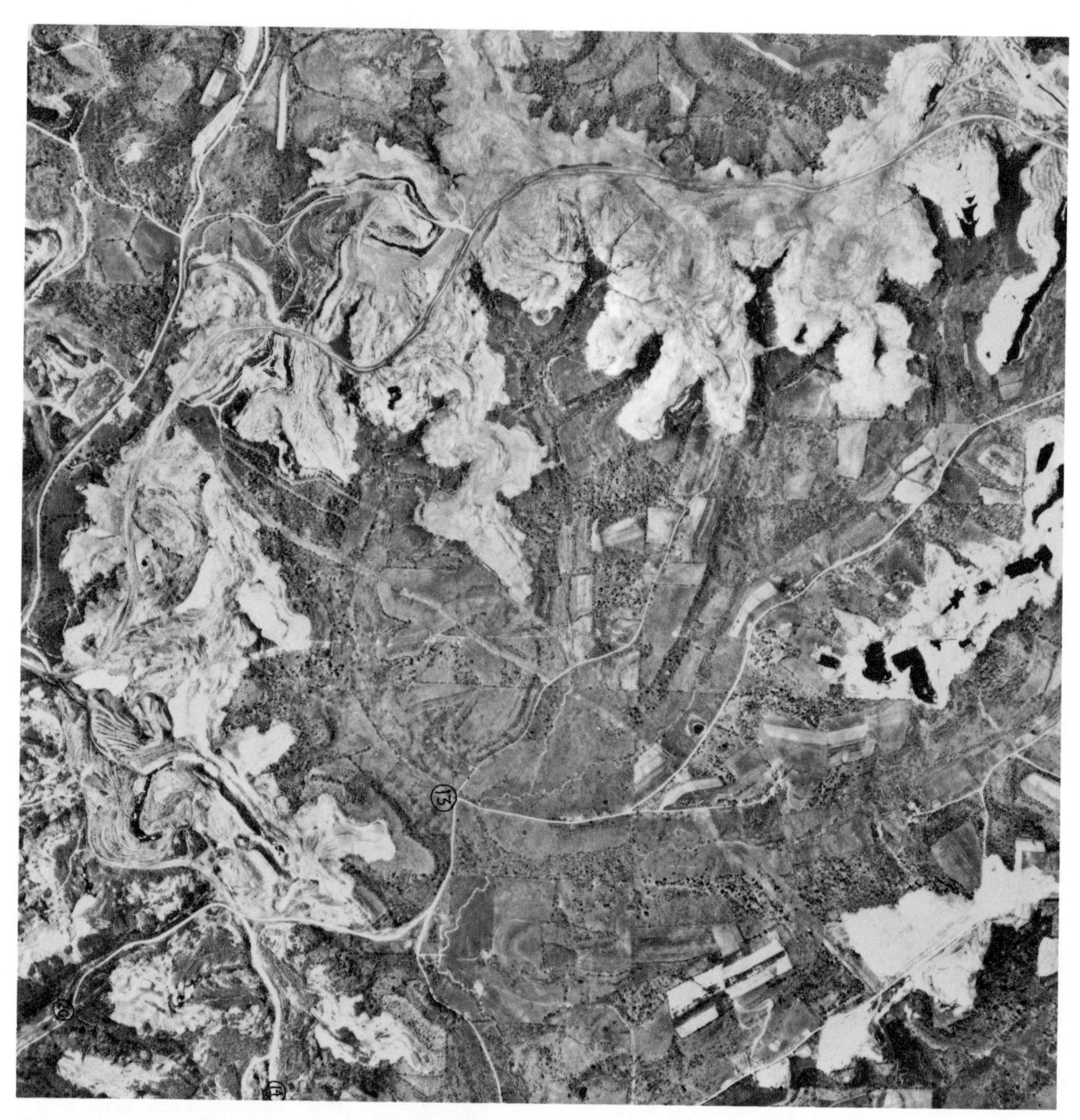

FIG. 18
Strip mine in Ohio. (Courtesy Wayne Pettyjohn, Department of Geology, Ohio State University.)

FIG. 19
Strip mine in Ohio. (Courtesy Wayne Pettyjohn, Department of Geology, Ohio State University.)

FIG. 20
Oblique aspect of strip-mine spoil pile. (Courtesy of Wayne Pettyjohn, Department of Geology, Ohio State University.)

operates in the 4.6-μm CO band, and CO determinations were made up to an altitude of 3000 m. Byer and Garbuny *(4)* have reported remote measurements of CO, NO_2, SO_2, C_6H_6, Na and Hg by absorption techniques using topographical reflectors or atmospheric Mie scattering as a distributed reflector. They have found that the use of topographical reflectors offers the advantage of a single-ended absorption measurement for ranges up to 10 km, with sensitivities less than 0.01 ppm for a 10-mJ, 100-ns transmitted laser pulse; the distributed Mie reflector permits absorption measurement over a depth of $c\tau/2$ (determined by the pulse length τ), and allows ranging by time-of-flight measurements. For 100-mJ, 100-ns pulses, sensitivities to 0.3 ppm at a 15-m depth are reported possible.

Remote Sensing of Land and Water Quality

Closely allied to photographic techniques in that the medium of record is film or heat-sensitized surface are multispectral scanners (MSS) similar to those previously mentioned and operating in thermal spectral bands from 3 to 15 μm. This class of instruments is useful in detecting heated effluents from power plants and industrial processes and in observing thermal anomalies of rivers, lakes, and streams. The MSS has also been used in detecting oil slicks in coastal waters and large freshwater systems. These instruments have the great advantage of operating day or night, since the thermal IR region (>3 μm) is an emissive region and is not dependent on the sun's reflected energy.

In contrast to cameras and line scanners, nonscanning radiometers are nonimaging sensors that measure emitted or reflected electromagnetic energy and display this information on a strip chart or magnetic tape. They operate in the same spectral domain as thermal line scanners and imagers, as well as at microwave frequencies, and are useful in recording surface temperatures. Radiometers operate by measuring the difference between the target radiation intercepted by the detector and a radiant energy reference level. Regardless of spectral region, all radiometers contain at least three basic components: (1) a radiation collection and focusing system that determines the receiving aperture and angular field of view, thus specifying the amount of radiation received by the detector; (2) a detector element that converts fluctuations in incident radiation into variations of an electrical signal; and (3) an amplifier and output element that transform the detector signal into a suitable form of presentation.

The target radiation intercepted by the receiving aperture is generally a broad-band random signal of thermal origin. The radiometer detector signal is generated from the "apparent temperature" of the antenna or receiver to which it is connected. This apparent temperature is related to the temperature of a pure resistance connected in place of the receiver, which would generate the same power spectral density as the receiver. This relationship can be expressed as

$$R_T = \frac{R_s}{k} \tag{12}$$

where R_T = receiver apparent temperature, °K
R_s = spectral density of radiation intercepted by the receiver, W per Hz
k = Boltzmann's constant

The radiation intercepted by the receiver is a function of the target characteristics, receiver and detector characteristics, and the absorption and scattering characteristics of the medium or atmosphere between the target and receiver. In making radiometric measurements, it is useful to refer to the apparent temperature of the target, which is the temperature the target would have if it were a blackbody radiating at the same rate as the target. It can be shown *(5)* through an approximation of Planck's radian law that the corresponding apparent receiver temperature is

$$T_T = \frac{\Omega_T}{\Omega_A} \tag{13}$$

where T_T = target apparent temperature, °K
Ω_T = solid angle subtended by the target at the receiver
Ω_A = solid angle of a pencil beam antenna or receiver with a specified power gain

The apparent temperature of targets is a function of their emissive or radiative and reflective properties. Metals, for example, are poor emitters of radiation but are good reflectors; when viewed from an airborne platform the received target radiation will be mostly reflected radiation incident on the metal object from the sky, which is at a very low temperature, typically 20 to 60°K. A metal object would thus have a very low apparent temperature when its actual physical temperature might be around 300°K. On the other hand, earth and vegetation are good absorbers and emitters of radiation and correspondingly poor reflectors. The apparent temperatures of these targets are therefore close to their own physical temperatures. The emissivity of water fluctuates widely with wavelength, being nearly that of a blackbody at optical wavelengths and falling off to about one third or one half in the X-band microwave region. Thus, at X-band a 300°K water body has an apparent temperature of about 100 to 150°K.

From the foregoing it is seen that although targets may have the same physical temperatures, differences in emissivity produce large differences in apparent temperatures and thus materials can be differentiated through radiometric techniques. In addition to surface temperature, microwave radiometers have also been used to detect oil slicks and water salinity.

The side-looking radar and radar altimeter–scatterometer, in addition to being day–night sensors, are capable of functioning in all types of weather as well, since they operate at microwave frequencies that are located in the atmospheric windows. Although these sensors were not specifically designed for pollution detection, they are nevertheless effective in monitoring strip- and pit-mining operations as well as other large-scale features of land and water surfaces. For example, side-looking radar can also be used to monitor extent and changes in large industrial waste ponds and detecting oil slicks.

The altimeter–scatterometer produces on-board magnetic tape records whose informational content contains both measurements of the distance of the instrument platform to the earth (altimetry) and the radar reflection properties of surface structures at various angles of elevation from the vertical or nadir direction (scatterometry). This instrument typically operates at radio wavelengths of about 3 cm. Measurements are achieved by the radar emitting a train of radiowave pulses, which alternate between short and long bursts separated by relatively long intervals of about 5000 μs. The emitted radiation spreads out to the ground below and also forward of the aircraft and is reflected back. Altimetry measurements to an accuracy of about 10 m are obtained by timing the arrival of the reflected short pulses.

The alternate longer pulses supply information on the radar reflection properties of ground surfaces; typically, the scatter cross section of the ground structure is illuminated out to an angle of 60° from the vertical, where scatter cross-sections are small and the distances large. To make these measurements, the long return pulse is typically divided into 10 samples corresponding to 10 arc-like patches of the surface areas being investigated. By comparing records of a succession of pulses as the spacecraft moves forward, a given patch of ground is viewed at different angles so that one can derive how a given ground element changes its scatter ability with various aspect angles of illumination. In addition, the altimeter–scatterometer is capable of transmitting and receiving horizontal and vertical polarizations. These are radio waves which generate electrical voltages that are maximum in a direction horizontal or vertical to the ground, respectively. Such capability aids in interpreting ground-slope directions and surface roughness such as encountered in strip-mining operations.

In contrast to other microwave systems (i.e., radiometric and radar techniques), the passive microwave imager constructs a picture of a viewed surface such that the light and dark intensities displayed on the image are related to the amounts of microwave energy radiated by the objects in the scene. This is similar to the imaging of thermal radiometric techniques discussed previously, but requires specialized apparatus to receive and record the longer, invisible wavelengths involved. To provide such picture taking, the imager uses a special radio receiver typically tuned to signals in the 8- to 10-GHz region (X-band). Signals intercepted by the antenna are transmitted to a detector and receiver. The receiver generates electrical signals for each of the radiations coming from the different portions of the ground scene being viewed. These signals are then recorded on film to re-create a photo-like image of the observed surface based on thermal (microwave) emission in much the same way as is done in IR imaging, and in contrast to light reflection utilized in ordinary photography. Equation (11) is equally valid here for determining the microwave power detected from an observed surface element, which extends over the entire receiving beam width.

The passive microwave imager is basically a microwave radiometer operating in a nonquantitative mode. It shares the advantages of other thermal radiation detection instruments in possessing all-weather and day–night operational capabilities and can produce brightness maps of surface features. Because it operates in the X-band region, this imager's resolution capabilities are four or five orders of magnitude less than for conventional photography or IR-produced images.

In the following, some of the instruments utilizing the principles of thermal radiation that have been constructed for detecting surface pollutants are discussed.

An investigation of airborne optical detection of oil on water has been reported by Milard and Arvesen *(6)*. They found for optical and near optical wavelengths (380 to 950 nm) that maximum contrast between oil and water occurred in the ultraviolet and red portions of the spectrum; minimum contrast is in the blue-green. Sky conditions greatly influence the contrast between oil and water and highest contrast was achieved under overcast sky conditions. The measurements were performed with a spectroradiometer, which measured spectral radiance, and a differential radiometer, which measured the difference in radiance between two wavelength bands or two polarization components.

Chandler *(7)* reports detecting oil on seawater with a passive microwave radiometer operating at 19.35 GHz. He also reports obtaining comparable results under laboratory conditions with a 3.0-GHz microwave radiometer. His work has shown that the value of the microwave radiometric temperature from an oil film on water is a function of film thickness.

Hickman and Moore *(8)* report laboratory investigations of laser-stimulated fluorescence of algae in solutions of varying concentrations. The presence of algae is a general indication of eutrophication. These investigations used a nitrogen laser lasing at 337 nm and stimulated the emission of algae at 680 nm. Using a 3-ns pulse, they extrapolated their measurements and determined that an airborne system should detect algae from an altitude of 100 m. Upgrading of certain parameters, such as peak power, pulse width, beam divergence, etc., would increase the operational altitude of this system.

CONCLUSIONS

The foregoing treatment of remote monitoring for environmentally significant contributions by industry has necessarily been brief. Because of this constraint, the major emphasis has been placed on operational monitoring systems, of which the camera and IR scanner are the principal examples. These systems are basically qualitative instruments and are useful in pinpointing areas exhibiting environmental anomalies. As such, they are commonly employed to look at terrain and water surfaces as well as haze, smog, and visible stack emission. In addition, the IR scanner when calibrated against a blackbody reference can be used to produce thermal maps (isotherms).

Semioperational monitoring systems, those that have been at least flight tested, include the many varieties of correlation spectrographs. These instruments can be quite sensitive, detecting gaseous pollutants in the ppm and in some instances even the ppb range. However, as a class they are subject to interference from closely lying spectral absorption lines of other atmospheric molecules so that their specificity is decreased. The advent of the laser and its use as an atmospheric illuminator has gone a long way toward eliminating the problem of interference.

Instruments such as the derivative spectrometer and correlation cell spectrometer have also been investigated in recent years. However, these, as well as other instrument developments, are generally not as far advanced as the class of correlation spectrographs, and thus could not be treated here.

Microwave systems (radiometers, radar) are useful in that they have all-weather and day–night monitoring capability. However, these systems have poor resolution compared to optical and IR systems; they are principally advantageous in monitoring relatively large areas such as occur in oil spills and possibly some strip-mine operations.

In summary, an operational remote-sensing system will be widely deployed and effective only when it is relatively inexpensive, reliable, sensitive, accurate, specific, and easily serviced. Thus, the degree of success of environmental remote-sensing instruments rests on the performance characteristics of the instruments themselves. Although many advances in sensor technology have occurred, the ideal sensor has yet to emerge. In the future, interest will focus on electro-optical instrumentation involving correlation or matched filter techniques and derivative spectrometry, where truly important developments are occurring in remote-sensing technology. For the near term, however, imaging systems will continue to be the principal methodology for remote environmental monitoring.

REFERENCES

1. Melfi, S. H., et al., Boundary Layer Investigations Using a Down Looking Airborne LIDAR System, Environmental Protection Agency—National Environmental Research Center,/Las Vegas, Nev.,.
2. Barringer, A. R., et al., Surveillance of Air Pollution from Airborne and Space Platforms, *Proceedings of Fifth International Symposium on Remote*

Sensing of Environment, Ann Arbor, Mich., 1968.

3. Ludwig, C. B., et al., Remote Measurement of Air Pollution by Nondispersive Optical Correlations, Joint Conference on Sensing of Environmental Pollutants, AIAA 71-1107, Nov. 1971.
4. Byer, R. L., and M. Garbuny, Pollutant Detection by Absorption Using Mie Scattering and Topographic Targets as Retroreflectors, *Appl. Opt.,* v. 12, no. 7, p. 1496–1501, 1973.
5. McGillen, C. D., and T. V. Seling, Influence of System Parameters on Airborne Microwave Radiometer Design, *IEEE Trans. Mil. Electron.,* Oct. 1964, p. 296–302.
6. Milard, J. P., and J. C. Arvesen, Airborne Optical Detection of Oil on Water, *Appl. Opt.,* v. 2, no. 1, p. 102–107, 1972.
7. Chandler, P. B., Remote Sensing of Oil Polluted Seawater, North American Rockwell Space Division, SD 70-377, 1970.
8. Hickman, G. D., and R. B. Moore, Laser Induced Fluorescence in Rhodamine B and Algae, Presented at the 13th Conference on Great Lakes Research, Buffalo, N.Y., Mar. 1970.

INDEX